"Read and heed *Unyielding*. It's born from the principles our nation required of Colonel Rempfer to ensure the integrity and readiness of America's armed forces. The author's meritorious victories during decades of professional dissent mirrored the unique heritage of our Air Force and constitutional founders. Buzz lived the code and donned the armor of integrity, which protected him from reprisal. May his call to erase the inequities and punishments over anthrax and COVID mandates unify the compassion of all Americans."

—Rod Bishop, Lieutenant General, USAF (retired)

"Defending our nation's armed forces was Colonel Buzz Rempfer's job as a fighter pilot. Every American should be grateful that he internalized his oath and used his fighter pilot skills to dutifully attack and help halt illegal mandates."

—Thomas G. McInerney, Lieutenant General, USAF (retired)

"*Unyielding* in one word describes Colonel Tom Rempfer's personal courage during his quest over twenty years to spotlight the physical dangers and injustices created by the mandatory military anthrax and covid vaccinations. Tom's deep sense of duty compelled him to challenge the mandates coming from higher authorities at great personal career risk. He came under intense pressure throughout this ordeal, but he never compromised his deep sense of duty to stand tall in the face of fire as his inspiring book chronicles. Tom's quest is to seek the facts and the truth and hold people accountable who intentionally ignore the 'truth.' Colonel Rempfer was not only a warrior in the air, but more importantly, a warrior in the moral fight for what is right versus wrong."

—Joe Arbuckle, Major General, US Army (retired)

"Buzz Rempfer shacks the target in his memoir, *Unyielding*! As the USAFs first pilot-physician to fly the F-22 Raptor, I know firsthand that fighter pilots are the first to admit their mistakes, and the debrief is where our lessons are learned and corrected regardless of rank. We have a tenacious commitment to excellence. Buzz once again hits the nail on the head and recognizes that doctors 'debrief' but often less transparently than fighter pilots. If there's one thing I learned as a dual qualified asset, it's that physicians and public policymakers should remember their oaths respectively—first, do no harm; and second, support and defend the Constitution of the United States. Buzz upheld his oath!"

—Jay T. Flottmann, MD, USAF (retired); Pilot-Physician (F-15C, F-22, T-38)

"Men of Honor. They were who I sought to emulate during my lifelong journey in uniform. The truest virtue of a warrior is revealed in his story. When you find those warriors like Col. Tom 'Buzz' Rempfer, you will know a man of honor stands."

—LTC Pete Chambers, DO retired), US Army (retired), Special Operations Flight Surgeon, Green Beret

"I have deep regrets for submitting to seven investigational anthrax vaccinations during my military career. Perhaps my admission will motivate others to read and scrutinize Colonel Tom Rempfer's (Ret) David versus Goliath manuscript, inculcated through a twenty-five-year saga. Colonel 'Buzz' Rempfer is a model leader; principled, self-sacrificial, professional, and unyielding in his pursuit of truth. His intestinal fortitude, and ability to assiduously sift through the myriad minefields of law, corruption, and intransigent leadership, produced a masterpiece. I firmly believe his analysis will be utilized to educate future generations of military leadership.

My twenty-year personal association with Colonel Rempfer includes listening to Buzz discuss the entire anthrax debacle with a reserve general during a memorable London flight. As is typically the case, Tom handled the discussion with exceptional grace, respect, expert communication, and empathy. He charitably tutored this 'very sharp' general on the institution's well-meaning, but woefully controversial, implementation of the mass vaccine mandate. In a nutshell, Buzz is in a league of his own. It is an honor to be a 'Pal' of this All-American Hero."

—Dr. Scott Keller (DSS); Lt Col USMC/USAF, ret.; F-16, F-18, F-5, TA-4, B-787, B-777, B-767, B-757, MD-80

UNYIELDING

MARATHONS AGAINST ILLEGAL MANDATES

COLONEL THOMAS L. REMPFER

Foreword by Dr. Philip G. Zimbardo

Skyhorse Publishing

Skyhorse Publishing books may be purchased in bulk at special discounts for sales promotion, corporate gifts, fund-raising, or educational purposes. Special editions can also be created to specifications. For details, contact the Special Sales Department, Skyhorse Publishing, 307 West 36th Street, 11th Floor, New York, NY 10018 or info@skyhorsepublishing.com

Visit our website at www.skyhorsepublishing.com.
Please follow our publisher Tony Lyons on Instagram @tonylyonsisuncertain

10 9 8 7 6 5 4 3 2 1

Library of Congress Cataloging-in-Publication Data is available on file.

Cover design by David Ter-Avanesyan
Cover illustration by Bob Englehart

Hardcover ISBN: 978-1-64821-045-7
eBook ISBN: 978-1-64821-046-4

Printed in the United States of America

DISCLAIMER

CLEARED
For Open Publication

May 15, 2023

Department of Defense
OFFICE OF PREPUBLICATION AND SECURITY REVIEW

The views expressed in this publication are those of the author and do not necessarily reflect the official policy or position of the Department of Defense or the US government. The public release clearance of this publication by the Department of Defense does not imply Department of Defense endorsement or factual accuracy of the material.

Unyielding was cultivated in the spirit of freedom of speech and honest academic inquiry. Just as his Naval Postgraduate School thesis was approved and published on the DTIC.mil website, Colonel Rempfer's objective in publishing this memoir includes the hope that the work might be thoughtfully considered in military and government realms, his target audience.

Prepublication security and policy review of this book ensured information damaging to the national security was not inadvertently disclosed. Military members have a responsibility to submit for pre-publication review any works intended for public disclosure. This author complied with those requirements since the book's fundamental premise is to follow the institution's rules, while requiring our government and military institutions to do the same.[1]

ENDORSEMENT BY ROBERT F. KENNEDY, JR., ESQ.

My uncle President John F. Kennedy observed that moral courage is a rarer commodity than physical courage in war. Tom Rempfer has both. Rempfer is the pilot you want flying your jet and the soldier you want beside you in the foxhole. Tom lives in the thrall of a relentless and noble idealism that has left him actually believing what he learned during four years at the Air Force Academy about honesty, hard work, and accountability. He has steadfastly refused to trade those values for expedience.

Tom was unyielding in his struggle to stop the Defense Department from coercively injecting an experimental, untested, unlicensed, worthless, and dangerous, zero-liability anthrax vaccine into a million military service members during the 1990 first Gulf War, and again starting in 1998.

The vaccine was a multimillion-dollar boondoggle to enrich crooked military contractors and their government cronies, and a monumental theft of millions of taxpayer dollars. The military brass, intelligence agencies, and public health regulators unleashed an arsenal of sinister tactics to force soldiers to take the illegal jab and to muzzle doubts, and marginalize, gaslight, silence, and destroy dissenters like Rempfer. American soldiers were the losers.

In retrospect, the anthrax vaccine program was a "trial run" for imposing experimental, unsafe, and untested vaccines on the entire population during the COVID era. The anthrax jab's striking parallels with the current COVID-19 vaccine fiasco deserve our attention. Understanding the tactics that the vaccine cartel used then to hide the vaccine's problems and bulldoze the opposition will help us protect ourselves in the future.

Tom's example as an ultimately successful advocate gives us hope that we may one day restore our constitutional rights and rule of law to America.

—Robert F. Kennedy, Jr., Esq., author, *The Real Anthony Fauci* and *The Wuhan Cover-Up*

ENDORSEMENT BY PETER A. MCCULLOUGH, MD, MPH

As documented in Colonel Rempfer's *Unyielding*, the genetic COVID-19 vaccine mandates on an unwilling fighting force has been a disaster for the US military for three reasons: 1) those forced to take the investigational injections every six months against their will have been mentally broken, 2) vaccine recipients face high rates of heart damage, blood clots, neurological disease, autoimmune illness, and malignancy, 3) the mentally strongest and fittest of our fighting forces have declined the vaccines and are facing dismissal or have left the service. The military should have been tracking the accumulating ranks of those who contracted the respiratory illness and featured natural immunity as the crisis progressed. By the colossal blunders of ignoring natural immunity and willful blindness to COVID-19 vaccine safety, our military leaders all the way up to the commander in chief will be held accountable and will face justice for the irreparable damage done to the men and women who have dedicated their lives to defending and serving our country. Colonel Rempfer's *Unyielding* commands this accountability of senior officials. May God bless them and guide them from further destruction of the mind, body, and soul.

—Peter A. McCullough, MD, MPH, coauthor,
The Courage to Face COVID-19: Preventing Hospitalization and Death While Battling the Biopharmaceutical Complex

DEDICATION

To

My wife who put up with me.

My family who supported me.

Our three children, niece, and nephew, who all enlisted and served.

Finally, to my country and military, for teaching the ideals that motivate this work.

CONTENTS

FOREWORD BY DR. PHILIP G. ZIMBARDO

My professional life involved studying and teaching about the social forces humankind endures, both good and bad. Ideally, some of your students internalize your best efforts and apply your teachings in their own observations of the world. Colonel Thomas "Buzz" Rempfer's book exemplifies that outcome, regardless of whether you agree with his methods or conclusions. The colonel's tale serves as a real-life Stanford Prison Experiment or another chapter in *The Lucifer Effect: Understanding How Good People Turn Evil.*[1]

Tom's "unyielding" journey serves as a testimony to the application of academic inquiry while also combating situational forces. His story unfolds continuously across an over thirty-year military career and through multiple professional dilemmas lasting over two decades. I am confident the sincere goal in telling his story is to help readers understand the factors that led to serious divides and mistrust in our armed forces during the anthrax vaccine controversy.

The commonalities in the ethical dilemmas posed by the anthrax vaccine controversy twenty-five years ago, and the subsequent equally controversial COVID-19 vaccine mandates today, illuminate the imperative to study these events diligently and dispassionately. By understanding and reversing the situational ethics gaps, and by not repeating the same mistakes, unity and trust might again be restored.

Col Rempfer's experience witnessed military leadership at the highest levels ignoring violations of the law, while turning a blind eye to illegal mandates of experimental vaccines on America's soldiers. After federal courts ruled the program illegal, the government failed to reverse punishments and invented a new Emergency Use

Authorization (EUA) law as a work-around to the court's injunction. The precedent EUA law application ensured "no penalty or loss of entitlement" for the troops who continued to choose not to be vaccinated. Almost twenty years later our government forgot that EUA legal precedent, jeopardizing COVID-19 vaccine mandates executed under the same law. As a direct result, this author had his livelihood placed in jeopardy again, twice in a twenty-five-year span over two different vaccine mandates.

Current and future public servants we entrust to ensure the government lives up to the high ideals expected by our citizens should read *Unyielding*. By following this "road less traveled" of courage exhibited by the bottom of the chain of command, we can hope to change the negative paradigm into a constructively positive one. Tom shares his saga since it was his final duty to explain this marathon-like journey of discovering the situational breakdowns, which fostered mistrust and division. The colonel's "unyielding" effort to help correct records for the previously punished troops further defied his bosses but defined his duty.

This book is a testimonial to that paradoxical challenge of righting wrongs, and offers a recipe on how to prevail through "unyielding" persistence. Though the author was able to fend for himself, overcome the obstacles, to survive and continue to serve, what about the troops and citizens who could not? Many were discharged and fired without justice before the final legal rulings. Could our nation be committing the same errors and fomenting the same injustices today?

Based on this lifetime account of ever-broadening controversial mandates spanning twenty years, we cannot say this has not happened before. By considering the lessons from *Unyielding: Marathons Against Illegal Mandates* future generations have a starting point to alter the trajectory of such controversial and divisive mandates before we allow them to be injected and infect our lives and history again.

Read this book and contemplate the content, pro or con. Reflect on how we as malleable human beings can be united, while overcoming the situational forces employed to divide us. Decide for yourself where the bad apples hang, and which actors fill the bad barrels. The parallels to current events deserve our attention. These behavioral patterns must be fully recognized to reverse current division and to restore trust in our public health establishment.

—Dr. Philip G. Zimbardo
Professor Emeritus
Stanford University
Yale University (MS, PhD)

ACKNOWLEDGMENTS

Role Model, Mentor, Flight Lead, and Dear Friend—Lieutenant Colonel Russell E. Dingle.

Mutual Support

Jen Bard, Esq.; Ted Doolittle, Esq.; Capt Kelli Donley; Col Phil Fargotstein, Esq.; Lt Col Jay Flottmann, MD; Lt Col Perry Forgione; Capt Jeff Frient; Col Juan Gaud; Lt Col Tom Gervais; Larry Halloran, Esq.; Lt Col Jack Heideman; Lt Col Scott Keller, PhD; Capt Wally Kurtz; Capt Enzo Marchese; Arnie Menchel, Esq.; Lt Col John Michels, Esq.; MSgt Rick Mischke; Sgt James Muhammad; Lt Col Dave Panzera; Lt Col Mark Perusse; Col Rick Poplin; Lt Col Dom Possemato; Col Lee Pritchard; Col John Richardson; Lt Col Larry Rizzo; Maj Gary Rovin, DC; Maj Dale Saran, Esq.; Lt Col Bruce Smith, Esq.; Col Sammie Young; Mark Zaid, Esq.; Col Jim Zietlow

Content

Bob Englehart, cover illustrator; Rebecca Schmid, author's proofreader; Skyhorse Publishing

PREFACE

Where do I begin? How about introducing myself from the eyes of my family? They teasingly liken me to Walter Mitty, I guess due to my sporadic lapses into absent-mindedness. However, I prefer to think it is about how I see the world: sometimes dangerous, but with an optimistic sense of courage and hope. My brother jokingly compared me to Forrest Gump. I'm not sure if it was the simple stuff or the jogging themes. I like to run, even marathons. Either way, he was right. My career was like a marathon. Like Gump, despite running into barriers, I adjusted or turned around, but kept going. My wife joked about the simple part, her favorite name for me being "Simple Tom." I do prefer the simple: no manipulation, no compromise. I see the world through the hue of idealism. This simple lens sustained me to endure all the bad and complicated things.

My ancestry possibly offers perspective on how I am wired. Like many Americans, my lineage springs from central and eastern Europe. Earliest family records reveal origins in the Black Forest region of present-day Germany, with later moves to Poland and Ukraine. My Prussian ancestors took advantage of the opportunities, settling in Bessarabia, west of Odessa. Tsar Alexander and Catherine the Great offered Prussian farmers opportunities for free land to farm, without taxation and conscription. When the good deals ended, my relatives took advantage of the Homestead Act in America, and found themselves relocated once again to the Dakotas. Historically they made a good decision, since Germans didn't fare too well in the upheaval that crisscrossed Ukraine before and after World War II. My family history tells a story about a clan that seized opportunity. It was in our blood to maneuver smartly, preserve life and liberty, all while balancing professional opportunities and experiences. I outmaneuver adversity many times in the pages ahead, just as my ancestors fatefully did from Europe to America.

In contemplating my professional experiences, I reminisce on the unhealthy, intersecting patterns we observed. I detail bad behaviors and illegalities underlying the anthrax vaccine and COVID-19 (coronavirus disease 2019) inoculation mandates. The bad behaviors and illegalities tethered the two ethical dilemmas, occurring over twenty years apart. Critical thinking required reflective contemplation. Particularly with recurring ethical themes, I found myself returning to the foundations of my education and training in order to navigate challenging circumstances. In telling you my story, I reflect across several decades to explain the professional disputes. I employ a marathon metaphor, and use "miles" as my chapters, in order to illustrate the lengthy effort. Throughout the marathons, I developed my rules of engagement (ROE). I summarize the ROE below and expand on them at the end of the book. The ROE helped me to endure, stay on course, never give in, and to be "unyielding," just as my parents, mentors, and leaders taught.

My story begins at a pivotal point in the middle of my military flying career, after surviving the Department of Defense (DoD) anthrax vaccine mandate. I attempted to get my military career back on track after the mid-career anthrax vaccine ethical dilemma, but suffered more turbulence. From there, I chronologically reflect back on my earliest education as a United States Air Force Academy cadet to add perspective on why I felt duty-bound to challenge higher authority. Simply put, they trained us to do so. Yet this directive inevitably cast a shadow on my subsequent progression and triggered the unexpected turmoil later in my career. The story ends almost forty years later in another professional upheaval, this time with COVID mandates targeting my civilian flying career. COVID-era mandates not only affected over two million military members, but also loomed distressingly over hundreds of millions of industrious citizens. Twice in my professional life my jobs were threatened, but I was never alone.

With COVID-shot mandates, Americans gained empathy over what military members endured those many years earlier with the anthrax vaccine predicament. This new, much grander quandary revisited the same themes, requiring the same tools and guidelines to challenge the legally questionable nationwide decrees. The lessons learned and tactics we used almost twenty-five years ago, dusted off for the latest conflict, may be timely and instructive for my fellow American citizens and our troops. But if not, put them in your quiver for the next battle.

What these experiences taught me is there will likely be a next encounter. The other side learns no lessons without accountability. If there is none, they will, almost fatefully, do it again. So be prepared. An iconic aviation novel, *Fate Is the Hunter* by Ernest Gann, chronicled the pilot's psyche and its struggles to control fate and fortune in overcoming the dangers of aircraft accidents. But in our *Unyielding* story, instead of accidents, mandates were the fateful hunter. We did not accept such a fate. We methodically challenged the inevitability of mandates, just as pilots do with accident avoidance. We piloted and controlled our fate like an Ernest Gann story.

As you read this "unyielding" journey, place yourself in our shoes. Run with us. How would you right the wrongs? This book encourages leaders to listen more, to reflect, to resurvey judgments, and to contemplate past abuses that fatefully resulted in their fellow citizens and soldiers to reject inequities. What should any human being do when cornered by a government we perceive violated the social contract, did not listen to their own people, and did not follow their own rules? Being "unyielding" was the answer for me. It is not about patriotism, heredity, or creed, because it is not nationalistic, genetic, or cultural. Unyielding reactions to injustices are a shared human quality, without borders and absent politics. Understanding such unyielding instincts may help readers and leaders to empathize with this approach and to avoid conflict.

I'm just a pilot. My operational mindset may be instructive in understanding this story. When I push up the throttles, my airplane accelerates. If I pull the throttles back, my craft decelerates. I point the nose up, and the plane ascends against gravity. I point the ship down, and it descends carefully. These are the simple mechanics that provide perspective on this unyielding mission. When the vector was wrong, I corrected it. This is a pilot's mentality—a continuum of reflection and correction. Pilots follow rules, instructions, laws, and expect our leaders to as well.

Based on these instincts, I learned and executed a simple formula to steer the ethical deviations back on course. This professional dilemma checklist protected me and my late role model, mentor, flight lead, and dear friend—Lieutenant Colonel Russell E. Dingle. It may prove valuable for current and future troops as they correct our nation's heading. Here's the ROE:

- Be professional at all times, and give no one ammo for anyone to use against you.
- Be reasonable and respectful at all times, even when your adversaries are not.
- Stick to the facts and rules, and avoid challenging purely discretionary matters.
- Be cognizant of your limits and skills, but if others do not follow the rules, prosecute.
- Use the oversight and reporting tools organizations provide to make the system work.
- Wisely, live to fight another day—maneuver smartly—outflank efforts to get rid of you.
- You do not have to lead every battle—work from the background to preserve your energy.
- Continually cross-check your own preconceptions to avoid the pitfalls of cognitive bias.
- Never give up, never quit, and avoid voluntary personnel actions that sacrifice redress.
- Most of all, despite the moral duty to be unyielding in the professional realm, always capitulate to win at home with your family—they are your highest duty and priority.

PROLOGUE

Unyielding is a tribute to Russell E. Dingle, Lieutenant Colonel, retired, United States Air Force (USAF). This prologue ends with a timeline for context on the book's thematic content. The book omits the antagonists' names where possible, instead focusing on lessons learned, broken processes, and bad behaviors versus the identities of low-level actors. As the lead protagonist, Russ wrote an introduction for this book's draft in 2003. We did not publish the work at the time, but the effort helped us to compartmentalize the struggle and document the lessons learned. The striking parallels to events several decades later revived its relevance. Russ's words below, and his stalwart leadership, served as an emotive force across the remainder of my career.

> Tom ran the Boston Marathon in 2003. During his run Tom realized that he and I surmounted a marathon of sorts in our battle against the disinformation perpetrated by our own government and military. Tom's book utilized his run to frame our multi-year journey that began in Connecticut looking for answers from our commander. Tom finished his run, but the outcome of our marathon continued. Our government and military may not have provided the answers, but they provided the tools to discover and win the truth.

Lt Col Russ Dingle passed away September 4th, 2005, after a valiant fight with cancer. As an Air Force officer and fighter pilot, Russ flew over two thousand hours in the A-10 Thunderbolt II, served as an instructor and commander, and earned multiple awards as "Top Gun." Lt Col Dingle's career was distinguished by noble advocacy for military members' health rights. He testified as an expert witness for the US Congress in 1999. Russ's exemplary career included over sixteen years of service as a pilot and captain for American Airlines. Russ will always be remembered by his fellow citizens, troops, and loving family as the intellectual heavyweight behind efforts for accountability, as well as for his courage, service, leadership, and honor. Russ's tireless example exuded idealism, juxtaposed with a healthy dose of cynicism due to the realities of human nature. He worried winning was a fantasy, but fought, nonetheless.

This prologue required a timeline of events, not only for Russ's and for my travails in challenging the anthrax vaccine program, but also to incorporate the comparable context of COVID inoculation mandates surrounding the pandemic from 2019 to 2023. A thorough understanding of the history, and the patterns of bad behavior, was required in order to fully appreciate the legal and ethical breakdowns during COVID and with the earlier anthrax vaccine immunization program (AVIP). The lessons not learned from anthrax gained renewed relevance when the entire nation faced COVID shot mandates. Predictably, the mandates were promoted through fear, despite unknown safety and efficacy. Embellishment of the threat overshadowed following the laws governing unapproved emergency use authorized (EUA) medical products. As Russ always said, the timeline proved crucial to understanding the depths of the wrongdoing:

1957: First anthrax epidemic in one hundred years occurred during a US Army anthrax vaccine clinical trial at wool mills in Manchester, New Hampshire (NH), where four workers died.

1970: US Government licensed a different anthrax vaccine without clinical trial efficacy data.

1972: FDA (Food and Drug Administration) assumed regulatory control of biologics (vaccines).

FDA had to re-license vaccines with proposed rules, final rules, and public comment.

1985: FDA never finalized anthrax vaccine, proposed rule that noted no required clinical trial.

US Army acknowledged limitations of the anthrax vaccine, asked industry for a new one.

1989: US Army testified about limitations of the existing anthrax vaccine to the US Senate.

1990: Illegal unapproved manufacturing changes to anthrax vaccine prior to the First Gulf War.

One hundred fifty thousand US troops inoculated with anthrax vaccine with inadequate recordkeeping.

1993: First active onsite FDA inspections of military anthrax vaccine manufacturer began.

1994: US Senate critiqued Army use of anthrax vaccine in First Gulf War as "investigational."

Anthrax vaccine considered, not studied, as a possible cause of Gulf War Illness (GWI).

1995: First warning letter issued by the FDA to the anthrax vaccine manufacturer for deviations.

1996: Manufacturer applied to the FDA for approval of "inhalation anthrax" vaccine indication.

1997: FDA issued a notice of intent to revoke anthrax vaccine license due to violations.

US Army acknowledged that the anthrax vaccine was not licensed for biological warfare.

1998: FDA inspected anthrax vaccine plant, finding the manufacturing process "not validated."

New law deemed investigational vaccine mandates illegal without a presidential waiver.

Defense Secretary illegally mandated anthrax vaccinations for all 2.4 million US troops.

1999: Congressional hearings began, ultimately finding anthrax vaccine "experimental."

The Department of Defense punished and discharged over one thousand refusers.

2000: The FDA continued inspections, finding additional violations of manufacturing practices.

The DoD continued inoculating over a half million troops, while thousands more fell ill.

2001: President George Bush directed a review of the anthrax vaccine and Gulf War Illness.

Initial recommendation to the Secretary of Defense—"minimize" use of anthrax vaccine.

Second anthrax epidemic in US history killed five from letters sent through US mail.

Initial government reports suspected anthrax spores originated from US Army stockpiles.

Dingle and Rempfer filed an FDA citizen petition to challenge the anthrax vaccine license.

2002: Letter attacks led to accelerated anthrax vaccine reapproval after a four-year closure.

More troops fell ill and were punished, imprisoned, fined, demoted, and discharged.

2003: US DC District Court imposed a preliminary injunction against military anthrax vaccine.

Basis included the citizen petition claim of no FDA license and the investigational use.

2004: DC District Court imposed a permanent injunction, ordered FDA to license the vaccine.

2005: The DoD used new Emergency Use Authorization (EUA) law to continue vaccinations.

The Federal Court granted an exception to the injunction only if vaccines were voluntary.

First ever EUA assured the court there would be "no penalty or loss of entitlement."

2010: FBI published report accusing a US Army anthrax scientist of letter attacks and lab leak.

Department of Justice findings affirmed motive was to save the failing anthrax vaccine.

Attempts began to correct military records after FBI revelations about anthrax origins.

2019: First military record corrected, including upgrade to a fully honorable discharge, restored rank, separation code allowed reenlistment, award of good conduct medal, and back pay.

The SARS-COV-2 virus from suspected lab leak in China resulted in a global pandemic.

2020: America implemented lockdowns and nationwide countermeasures to contain the virus.

2021: Rapidly engineered COVID EUA countermeasures allowed by public health emergency.

Nationwide EUA product mandates illegally imposed by presidential executive orders.

EUA COVID shot safety, efficacy, and mandate illegality harmed public health and trust.

Unconstitutional mandates struck down by Supreme Court and Federal Court injunctions.

2023: Congress halted DoD mandate—federal mandates rescinded—original shots deauthorized.

Government agencies assessed a lab leak tangentially related to US-funded research as the most likely origin of the pandemic.

MILE 1:
Part One
"Unyielding"
(2011)

My memoir about this marathon of a story spans four decades, from 1983 to 2023. Halfway through my professional marathons, I hit a pivotal juncture in my military career—in 2011. After serving for several years in a remotely piloted aircraft (RPA) or drone unit, I was invited to compete in a meritorious selection process to serve as our unit's next commander. Frankly, I was more than content serving as an instructor pilot and flying the midnight shifts with my fellow crewmembers. Command and rank were not my aspirations, but I followed the encouragement of my colleagues and threw my name in the hat. I knew I wasn't the top choice of the higher-ups in the chain of command, so I found it intriguing that I was even invited to apply for command.

Per my training, I prepared diligently. That's what they expected of us. Every young officer should aspire to lead their unit someday and to serve as a worthy steward for their troops. I had already hit some bumps in the road in my military career, so in light of all those past shenanigans, I was honored to be given the opportunity. In preparation for my interview, I reviewed the Air Force's Core Values, the Honor Code, the Oath of Office, our Principles of War, and my favorite pamphlets on military doctrine. I left no stone unturned.

Our operations were a twenty-four-seven, nonstop environment, so I had just completed a flying shift on the night of the interview. The interview went well and apparently lasted longer than all the others by almost an hour. I gave it my best shot, and it appeared the hard work had paid off. The inside word from the hiring board president was that all of the hiring board members selected me by an uncontestable high margin. Mission accomplished, or so I thought.

I certified as a qualified candidate, interviewed, and all the members of the hiring board evidently selected me as the top officer for the billet. I wasn't supposed to have the insider knowledge that they selected me, but the board president leaked it. Normally, the military tends to keep internal deliberative processes secret—maybe in case they change their minds. On this occasion, fortunately, the selection panel members couldn't keep it a secret that all the hiring board members selected me as the highest-ranking applicant by a significant margin out of a field of eight candidates. The hands of the leadership were tied—according to the rules. Perhaps that is why they leaked the results?

Several weeks later, I was ordered to report to a unit meeting in the early hours of the morning where the new squadron commander would be announced. Despite completing another all-nighter shift, I reported as ordered. When I entered the squadron conference room, I noticed they'd ordered a cake. It was odd, particularly in the midst of nonstop flying operations. Eating cake was simply not the priority. The meeting was promptly called to order and some unfamiliar faces were introduced. The newly minted squadron commander was announced as one of them. I was quiet and a bit shocked, having prior knowledge of the leaked results. Obligatory applause filled the room. Needless to say, it wasn't my name announced, and I didn't eat cake. I maintained my professionalism and went home to go back into crew rest for the next night's missions. The result was surprising to say the least. I put my best foot forward, and evidently prevailed in a meritorious selection process, only to be undone by politics or worse.

Being a process-oriented person, I evaluated the situation and determined my next steps. Something was amiss, but I wasn't aware of the magnitude at that point. I struggled with the decision to ask for some sort of higher-level review of the matter, since I understood the reality of the seniors in the chain of command wanting their chosen officer to be the next commander. Yet still, why did they conduct a meritorious hiring process? Why was a hiring board convened? Why put everyone through the process if, in the end, the bosses tossed the results and just did what they wanted? It was one of those ethical dilemmas they warned us about. What was I to do?

What I did was politely engage the Human Resources Office to figure out the rules pertaining to hiring, and that led me to the regulation called the Meritorious Placement Plan. Those were the rules governing the hiring process, and they were very clear-cut. The rules directed me to the inspector general's office, the "IG." I followed those rules and asked for an inquiry. Very promptly, the local IG got back to me and made it clear they wanted to "wrap it up quickly." The next day I received their swift dismissal email, which stated,

> This office has determined the allegations concerning the complaint are not substantiated. . . . A review was conducted in the selection process and it has been determined no misconduct was committed. . . . The Board's final selection for a candidate who scored within the 10% Rule is compliant with regulations.

I found the wording and rushed reply quite odd. Reasonably, no thorough investigation could have been conducted in less than a day. Also, inclusion of the "10%" figure piqued my curiosity. Sure enough, within the rules, the hiring official had the discretion to select someone other than the top scoring candidate if another candidate scored within a 10 percent margin, but only if certain circumstances applied. The unusual circumstances included that the highest ranked officer had pending disciplinary charges, financial troubles, or security clearance issues. Nothing like that applied. Therefore, I read the rules further and opted to elevate the complaint to the higher United States Air Force IG. That was my procedural right, again, according to the rules.

The Air Force IG was very responsive and promised to be in touch. In short order, though, I was summoned to our higher headquarters to have a one-on-one meeting with our general. I figured he might have caught wind that I elevated the complaint, and we would have some kind of professional discussion to resolve the situation. I have been guilty of being naive and idealistic in the past, and this turned out to be one of those times. Before I drove north for the meeting, my wife even told me, "You're going to get fired." Darn, I love that lady and all her no-nonsense common sense!

Sure enough, within about a minute of my arriving at the meeting, saluting in, and sitting down, I was promptly fired. The boss told me that challenging the chain of command would not be tolerated. The general gave me five weeks' notice before I would be involuntarily separated, but "not for cause." Instead, they called it "force management." That was a cypher for "we really don't like you and we're going to administratively remove you, hopefully blocking you from having any recourse or appeal." I saluted professionally. The general asked me if I had any questions. I responded, "No, sir," and departed for home.

Once I returned to my base of operations, I discovered I was grounded from my flight duties the day prior. I was also secretly removed from active duty status so that the general officer I had just met with would have the legal authority to fire me. The general's subordinate, a colonel, was the behind-the-scenes officer doing all the administrative coordination without my knowledge—setting me up for my exit. He then ordered me to finish my extra duty assignments. I was in the midst of working on writing a CONOP (concept of operations) for the standup of a new local drone detachment. I complied with his directives and put forth an extra-solid effort.

I updated the Air Force IG on what happened, which precipitated a letter from the IG informing our local headquarters that the USAF was officially launching an investigation. As it goes in the military, I knew it would not be a quick process—certainly not in time to save me from the five-week discharge deadline. As I did with the earlier anthrax vaccine issue, I placed my faith in the system, packed my bags, and got to work on finding another unit to serve—again.

Simultaneously, I appealed the "force management" firing to the general's boss, as the rules allowed. That, too, was of no avail. I expected it was going to be an uphill battle since it was unlikely anyone would accept responsibility for the potential process violations I suspected. The top general wrote me a response denying my appeal request. The message was consistent. He wrote, "Your *unyielding* demands that decisions must fully meet your satisfaction before you will accept them undermines leadership and shows a lack of respect." Wow, I had a feeling the top general did not know the full story. I loved his line about being "unyielding." I thought that's what they wanted us to be and how they trained us to conduct ourselves in times of adversity.

> 5. LTC Rempfer, your unyielding demands that decisions must fully meet your satisfaction before you will accept them undermines leadership and shows a lack of respect. I do not need to remind you that this is the military and your comments about dereliction of duties, negligence, and possible corruption are detrimental to good order and discipline.

Full letter at Appendix A.

The general added reference to incidents across several years related to other process problems I attempted to resolve for my unit members—all successfully. The general made clear to the IG that he harbored no "animus" against me for my unyielding efforts on behalf of my troops. Unfortunately, in those instances I was compelled to point out "dereliction of duty," "negligence," and "corruption" when they existed. Obviously, they had been taking notes.

The out-of-context history was all there to attempt to justify the firing, along with the classic "good order and discipline" catchall phrase. Anytime commanders employ that "Hail Mary" motto, it's a hidden and hopeful distraction to make everyone look the other way. The military tends to divert attention from the actual issues of contention. In this case, the issue was alleged command-selection violations or cheating. Instead of looking into serious allegations, they will often change the subject by implying misconduct, even if none occurred. Just the illusion of such allegations is normally sufficient to make observers avoid questioning senior leadership or the legitimacy of the underlying allegations. It's really poor form, but it's a time-tested tactic that works—and not just in the military. In my case, no one accused me of any actual misconduct, but none of that mattered at the time. It still meant that I had to find another flying job. Fortunately, I had a good reputation in our mission arena, so I put out the feelers.

For perspective, this wasn't the first bone of contention. Every time I stood firm on a personnel issue for the members of my unit, I prevailed. That rubbed them the wrong way. The members of my organization appreciated my persistence, and I never regretted holding a hard line against the incompetence or errors often directed at my unit in a prejudiced manner. Once again, however, I was reminded that memories were long, and often there was no reward for sticking to my guns, standing up for my troops, and doing what I believed was the right thing. The one right thing I always did, in the execution of resolving problems, was to be unquestionably professional and never give anyone ammo to say otherwise. It didn't mean they didn't try—but it wouldn't stick. In my case, the IG later confirmed any and all counter allegations were unsubstantiated.

Ultimately, the USAF IG, in findings signed off by the Department of Defense (DoD) IG, validated that the firing was an illegal reprisal following my complaint over the hiring process. Additionally, the USAF IG found six irregularities in the hiring process, to include the suspected alteration of the scoring by someone other than one of the hiring board members [letter at Appendix B]. Rarely were reprisal allegations substantiated. Fortunately, in my case, the IG took the extraordinary added step to thoroughly investigate the underlying hiring controversy.

The investigator could not provide me with a copy of the full report, but fortunately a JAG (judge advocate general) at our base legal office provided me copy of the document. The investigation revealed suspected lying, cheating, and forgery. The substantiated misconduct was so far removed from the expectations of military officers that the DoD IG could not ignore it.

> Dear Colonel Rempfer
>
> Our investigation of your allegation was conducted under the provisions of Title 10, United States Code, Section 1034, "*Protected communications; prohibition of retaliatory personnel actions.*" The investigation substantiated your allegation of reprisal. The IO also found that several "procedural violations" of DEMA Directive 25-6 were committed in the hiring of AGR Vacancy Announcement 2011-091A. The Inspector General of the Air Force has reviewed the report of investigation and approved its findings. Additionally, the Department of Defense Inspector General conducted a thorough review of the report, found that it adequately addressed your allegations, and concurred with its findings.

USAF IG summary of findings on substantiated reprisal and hiring procedural violations.

The DoD IG referred court-martial allegations for false official statements, dereliction of duty, and forgery against the officer suspected of changing the scores. But that officer's chain of command, at some undisclosed level, hurriedly dismissed the charges with a confusing and misleading response. The JAG gave me that whitewash document, too. The investigative process spanned over a year. Another full year transpired for the Air Force to grant corrections based on the substantiated reprisal. The remedy included continued military orders through retirement flying for the new unit that had taken me in, plus repayment of a withheld flight pay bonus.

I felt vindicated, but also held some disappointment with the fact that there was literally zero accountability. Nonetheless, I was thankful. I had substantiated reprisal report findings to help correct my professional records. Regardless of the runarounds, lack of accountability, and the reality that no one ever investigated lower-level investigators, I was grateful. I survived the attempts to end my military service and flew the remaining years of my career with professionals who provided safe harbor during those dark times.

About a dozen former unit members also transferred to the new squadron out of protest. I was eternally indebted to serve my final years performing our mission with such an outstanding group of aviators. We felt safe. That feeling should be an essential characteristic in any healthy military, government, or business setting. Thankfully, my colleagues and I outflanked the unhealthy territory and landed in a safe zone.

Ultimately, the Office of the Secretary of Defense (SecDef) reviewed the corrections case twice in the subsequent years, since reprisal findings were extremely rare. In 2014, the Defense Secretary made an unprecedented recommendation that the Air Force grant me a command credit in my professional records for the lost leadership opportunity. Command selection served as a key stepping stone for further advancement. The SecDef's office also recommended retroactive consideration for promotion to the next rank of full colonel. The Air Force's prompt compliance with the SecDef's recommendations was a reasonable expectation, but it did not happen.

Two years later, the Air Force's formal response stated that there were "intangible reasons" as to why I was not selected for command or promoted. I got that, too—there was history. Nevertheless, the excuse didn't go over well with the Defense Secretary's office as they tried to tactfully remedy the ethics violations and substantiated retaliation. Fortunately, the SecDef's office remained persistent and turned their recommendation into a 2017 directive. In the interim, I reached mandatory retirement, so the remedy processes became retroactive.

The SecDef's memo effectively ended the adjudication of the case and added perspective on the ignored hiring violations. The SecDef's office affirmed the "evidence is clear and compelling that [the] Applicant was improperly denied command," that "actions were arbitrary and capricious," that the conclusions were "not supported by the weight of the evidence," and that the "weight of the evidence clearly and convincingly supports the original IG conclusions" about the alleged violations. The ruling stated that there was "no discretion" to select another candidate and added that the "scores were manipulated to reduce the scoring gap to within 10 percent to enable the selecting supervisor to choose a candidate other than Lt Col Rempfer."

Admitting the "scores were manipulated" was code for good old-fashioned cheating. The SecDef's office commented on the "unreliable" memory of the officials involved in the hiring violations. In contrast, the IG found the testimony of those officers who helped expose the misconduct as "particularly compelling." The two unprecedented SecDef-level remands resulted in a directive that ordered the command credit update for my professional military records to allow me to compete for a retroactive promotion to the next rank. The SecDef's final oversight decision ordered a Special Selection Board for promotion consideration [letter at Appendix C].

The Special Selection Board took over another year, but it was worth the wait. I was selected for a meritorious promotion to colonel in retirement. The board bestowed a promotion date of 2012, as though none of the injustices ever occurred. That was actually the way the law worked to make people whole. The promotion was finalized by the Air Force on June 8th, 2021, over ten years after the hiring violations and firing occurred [letter at Appendix D].

Though I would have been honored to serve our unit members as their commander, I was grateful for the Secretary of Defense directive for the constructive command credit, which resulted in the retroactive promotion. Now that the full truth was known, I hoped my top general might be proud of me for being unyielding, sticking to my guns, and getting the truth exposed.

OCTOBER 2018
SPECIAL BOARDS AND SPECIAL SELECTION BOARDS
SELECT LIST

Svs Comp	Comp Cat	Name	SSN	Original Bd	Proj Grade	Projected DOR	PAS	Old PAS
V	A	REMPFER, THOMAS L.	7617	V0611A	Col	01-Jun-12	W60VF167	

Air Force message documenting the "Col" promotion and 2012 date of rank (DOR).

To this day, my now adult children often tease me, "Don't be 'unyielding,' Dad!" It's a fun family joke, and we all get a good laugh. Of course, they are right—one cannot be that way at home. However, in the professional realm, one can and should. That is what our troops require of us as their leaders. That is also what higher authority expects of us as officers. Senior leaders should defend mid-level officers when they professionally demonstrate this vital trait, even if it means exposing malfeasance or mistakes by others within the chain of command.

Perhaps the most important thing about being unyielding, or true to one's ideals, is that one can emerge on the other side of the professional dilemma without regrets. These were the traits drilled into us at the United States Air Force Academy. Since it is important to understand how it all started, I will share my earliest military memories of how my classmates and I were formally trained to be tough, unbending, and unyielding in matters of duty, honor, and ethics.

MILE 2:
The USAF Academy (1983 to 1987)

The decision to become a United States Air Force officer, and ultimately a fighter pilot, followed several fortuitous forks in the road. On a bleak New England day in 1983, during my senior year in high school, one of my classmates knew I was undecided about where to go to college. With the best of intentions, my parents really wanted me to go to the United States Air Force Academy. We compared Cornell University, with a military scholarship in engineering, to an Air Force Academy appointment. My dad helped me sketch out the pros and the cons of the schools. Cornell was my first choice, but the scholarship potentially precluded my serving as a pilot. Once the Air Force spent the money on me, they likely needed me to serve as an engineer.

Despite putting my acceptance in the mail to Cornell the night prior, a high school friend talked me into thinking it through for one more day. She drove me home during our lunchtime to get the two college admissions return envelopes out of the mail. The next day I switched the choices, dropped the envelopes back in the mail, and sealed my fate for the next thirty-plus years. I accepted the appointment to the Academy and began a journey dramatically different from the one offered in academia at Cornell. I did it for all the right reasons. I wanted to make my parents happy. The choice was smoother financially. I could be a pilot. I never regretted the decision and never looked back—well of course, except all those times I got homesick.

On my very first day in the summer of 1983, I found myself surrounded by hundreds of other eighteen-year-olds. A military bus picked us up and transported the new cadets to the grounds that would serve as our home for the next four years. The mountainous terrain and the plentiful pine trees were starkly different from the Northeast. The fact I didn't know anyone compounded the unfamiliarity. I did sense that the young men and women on that bus were no more at ease than I was. We got off the bus to the greetings of sharply dressed cadets who lined us up into formations and barked out our first marching instructions. A couple of years later, I was one of those greeters myself. In those subsequent years, I did my best to help the kids from shedding a tear as I lined them up—especially the ones like me whose parents were thousands of miles away and could not give us a hug goodbye on our last day as a civilians.

We haphazardly marched up a ramp entering the Academy grounds. The words at the top of the concrete wall in front of us read, "Bring Me Men." The inscription came from a Samuel Walter Foss poem. I later learned parts of the poem that were put to music when I sang with our academy singing group, the Cadet Chorale: "These are men to build a nation / Join the mountains to the sky / Men of faith and inspiration / Bring me men, bring me men, bring me men." Twenty years later, they removed those words, replacing them with the USAF Core Values: Integrity First; Service Before Self; Excellence in All We Do. I'm okay with that because those words meant more to me. But I still liked the poem, too.

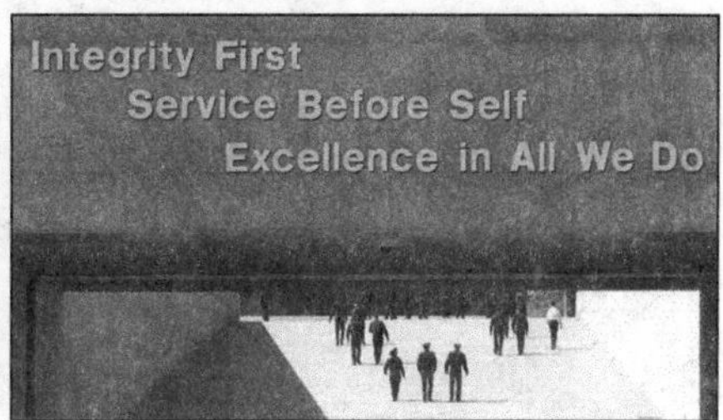

USAF Academy Entrance Ramp 1983 and 2003.

The words of the poem helped me to hold back an exasperating wistful feeling, no different from what I felt just five years earlier when I got homesick at summer camp so badly that my dad picked me up early. Something about those words and their inspirational message helped me to hold back the tears and made me realize I was about to become part of something that seemed special. No sooner did I lose sight of those words on that ramp than our little ragtag formation arrived in an area they called the "terrazzo" due to the marble strips that outlined its edges. The place was as impressive as it was beautiful, but still imposing at the same time.

A short march later, an ominous inscription on a wall just in front of us came into view. The cadet honor code read, "We will not lie, steal, or cheat, nor tolerate among us anyone who does." Over my career that followed, most would agree that I showed average intelligence and above average effort. My intellect was just low enough to internalize ideals like the institution's honor code, oath of office, and core values literally. I had the perfect IQ for believing in those principles, coupled with the stubborn tenacity to help the institution ensure they were honored.

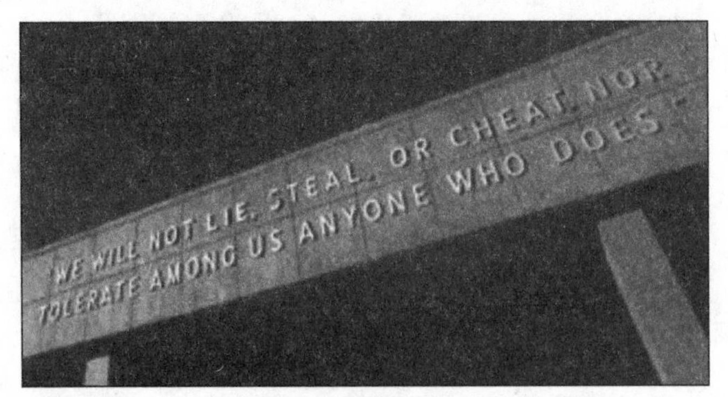

Honor code inscribed on wall at United States Air Force Academy.

The honor code was meant to be an ideal, and the Academy was the place that would break us cadets down and build us back up with such ideals as an integral part of our lives. I guess that's why one of the next stops was the barbershop. They marched me to the barber's chair and I took my seat. What had once been blonde feathered hair in high school was gone before I could say "so help me God." They cut it off, leaving only fuzz. As I marched to the next in-processing station, I glanced a frightened gaze into the mirrors we passed that people used to adjust their uniforms. I peered at myself, unrecognizable. I felt my head. I sported my childhood crewcut again. How embarrassing. That was their mission, to remold us as USAF cadets and future officers. The hairdo cut us down and built us back up uniformly—the Air Force way.

Before and after the Buzz Cuts.

They took us to uniform issue, another way of making us the same according to one standard. I kept thinking how proud my dad would be of me for getting all the free clothes. Little did I know my new friend, Uncle Sam, would get every last penny back over the next thirty-two years. It got better though when they marched us into the chow hall. The only price for the meal was sitting at attention and pouring drinks for the upper class students who trained us. The dining facility, called Mitchell Hall, was enormous. During the meal, the upper class cadets made us memorize lines from a book called *Contrails*, our pocket-sized bible of Air Force history.

We read about the officer the dining facility was named after. I memorized the quotes by Brigadier General William "Billy" Mitchell, what he said, what he did, and why the Air Force embraced him as our inspirational founder. US Army leadership had court-martialed Mitchell in the pre-World War II era for accusing his superiors of "incompetency, criminal negligence, and almost treasonable administration of the national defense." In the 1930s, Mitchell boldly challenged Army senior leadership over the misuse of airpower.

Ultimately, history vindicated the maverick once World War II proved Mitchell's arguments about the advantages of airpower. The

USAF honored Mitchell with this dining hall. The lesson learned—there was room in the military for people to stand up for what they believed in, for doing the right thing, and for professional dissent. At least that was what the officers, cadets, and the idealistic environment at the US Air Force Academy taught back then.

At the end of that first day, the seasoned cadets and officers in charge ushered us into a big meeting room called a lectinar. There, I repeated the oath of office with my classmates:

> I, Thomas Rempfer, having been appointed an Air Force Cadet in the United States Air Force, do solemnly swear that I will support and defend the Constitution of the United States against all enemies, foreign and domestic; that I will bear true faith and allegiance to the same; that I take this obligation freely, without any mental reservation or purpose of evasion; and that I will well and faithfully discharge the duties of the office on which I am about to enter. So help me God.

I wondered at the time, what if someone did not have a strong faith in God? Did it somehow diminish their oath? I wondered if the oath meant different things to different people. I figured, just as the Samuel Walter Foss poem at the entrance of the academy said "Bring Me Men," these were idealistic expressions, not necessarily to be taken literally. Instead, the oath was something you swore to and was grounded in everything you felt was sacred—your family, your higher power, and your country—take your pick or all of the above, like me.

After a grueling but rewarding summer of athletic and military training, my first academic year began with the study of the military profession. We learned that the fundamental characteristics of our profession were expertise, responsibility, and corporateness. We read the words of military philosophers such as Samuel P. Huntington from Harvard University. Huntington's classical study of military professionalism in *The Soldier and the State* identified these criteria for the profession of arms. I was often torn in subsequent years as I contemplated the academic ideals and how they compared to the reality of what I witnessed in the military. Despite the anthrax vaccine incidence of the institution veering off course, I met consummate professionals over those years, and knew we did our best to live up to the tenets as officers.

Much of the four-year academy experience was a surrealistic blur, but I do remember always being grateful for the unparalleled opportunities and experiences. I had key memories that endured, like joining the boxing team. I borrowed my own personal favorite quote from the Academy's boxing room wall. The quote shaped my unyielding spirit: "Tough times don't last, tough people do." My tough-as-nails coach, Ed Weichers, coined the saying. Such inspirations helped me to endure my experience at the Academy. This axiom also sustained me during my Air Force career and throughout the marathon against the ensuing anthrax vaccine mandate.

Boxing memories with best friend and corner man, Jeff Rhodes, class of 1987 top graduate.

The quotes and ideals I learned at the Academy gave me the fortitude to persevere when I knew something was wrong and needed to be fixed. This ingrained sense of duty was also a credit to one of my history professors, Captain John Poole. He taught us of the lessons of the Vietnam War, of false body counts and other lies that artificially shored up that conflict. The falsehoods rendered incalculable harm to the American military and our nation's credibility. Captain Poole insisted that we may someday face similar institutional lies, and must have the courage to stand up, to stop the damage those lies might cause.

This was the Academy experience, an idealistic one, born from the teachings and examples of wise officers such as Captain Poole and others who came before him. We read quotes from men like General George S. Patton: "If I do my full duty, the rest will take care of itself." We memorized quotes from Robert E. Lee: "Duty then is the sublimest word in the English language. You should do your duty in all things. You can never do more. You should never wish to do less." Quotes embossed on Academy statues by Austin "Dusty" Miller such as: "Man's flight through life is sustained by the power of his knowledge." I learned their quotes. They gave me direction, a vector. They molded my classmates and me to be future officers.

Concluding our four years at the academy, the Honorable Edward "Pete" Aldridge, secretary of the USAF, delivered our graduation speech in 1987.[1] His words were equally ideal:

> You will be in a position to identify and resolve irritants and establish positive programs in your organization . . . I encourage you to look for the factors that are irritating your people—the rules, regulations, and procedures that are getting in the way of performing their mission. . . . And, if a regulation or policy doesn't make sense, get involved; propose a revision that does contribute to your mission.

I didn't think too much about the implications of Secretary Aldridge's words at the time. On that day in 1987, we all just wanted to put the academy in our rearview mirrors. I didn't know the future challenge the words of the secretary of the Air Force would pose for many years to come. I trusted my leaders and believed in their motivational quotes. I hoped to live up to their words, the honor code, the oath of office, and the institution's core values. I trusted my leaders would back me up if I ever found myself in the "position to identify and resolve irritants."

Reflecting back on those early imprinting years, I realized the Air Force and the DoD failed uniformly as an institution to live up to the principles with regard to the anthrax vaccine program, as the subsequent chapters of this book will demonstrate. The program violated the oath of office and each of the Air Force profession's core characteristics and values. Most military leaders were not experts on the vaccine, and those that were deceived the higher leadership. The majority of military leaders failed the second criterion of responsibility by not protecting service members from a vaccine and a policy that was not based on a foundation of truth—and in fact, it was illegal. In doing so, it became clear the majority of our armed force's leadership violated the third requirement of corporateness as well.

As Huntington said, "The members of a profession share a sense of organic unity and consciousness of themselves as a group apart from laymen" and a "unique social responsibility." The foundations of my earliest military education stood in stark contrast to what I later witnessed with the violations of expertise, responsibility, and corporateness normally required of the profession of arms. The anthrax vaccine policy created a chasm in our professional military consciousness and violated everything military members held dear. Left uncorrected and undebriefed, the ethical abuses would have grave implications for many decades to come.

Rempfer from 1983 basic training to 1987 Academy Graduation with SecAF "Pete" Aldridge.

MILE 3: The Fighter Pilot (1987 to 1994)

"Mutual support" is a term that describes how fighter pilots watch out for each other. We often referred to it as checking each other's six o'clock. "Check six!" Whether while flying or in life, you knew who had your back. My late friend Russ Dingle was that guy, a quintessential fighter pilot. He was the one everyone trusted. He was one of our commanders, and held respect both up and down the chain of command. Russ was the aviator that everyone in the squadron wanted to have in their formation in combat. Russ seemed a natural born leader, scored tops in every gunnery competition, and outranked his peers in performance—in and out of the cockpit. Russ was also the guy who made the squadron beer everyone wanted to try. He built the squadron bar everyone wanted to hang out at after flying. Was it any wonder I followed Russ into the anthrax vaccine dilemma? No. Wingmen follow men of honor, men who exude the professional traits I was taught to emulate—expertise, responsibility, and corporateness.

About the same time I graduated from the Academy, Russ returned from an assignment in Europe after doing his part in winning the Cold War. His next duty was instructing pilots to fly the A-10 in Tucson, Arizona. The same principles I learned at the Academy in that idealistic setting were being lived operationally by men like Russ. They instilled the very principles in their students in the air that I learned in college. The same principles that Air Force Secretary Pete Aldridge professed to my graduating class in 1987 played out on a daily basis in the briefs and debriefs of the fighter, transport, tanker, and bomber missions around the world. Secretary Aldridge told us to challenge the nonsense. If you don't wave the flag when it needed to be waved, people could die. Pilots like Russ understood what Aldridge meant. Russ embodied the qualities that we were all trained to live up to, that our leaders demanded we strive to achieve.

Russ would be the first to acknowledge that we were about as different as two fighter pilots could be, but somehow that made for a great team. I followed his lead, and we found ourselves side by side in a quest for truth on a mission of honor for our fellow military members.

One day, several years before we were tasked to investigate the anthrax vaccine, we flew an A-10 training mission together. We got stuck near Fort Drum, New York, grounded by a nasty thunderstorm. We really wanted to get home to our families that night, but Russ was not going to break the local rules as our flight lead. Making the best of it, we spent the night getting to know each other better and sharing flying stories. That next day we flew a LATN (pronounced Latin, meaning Low Altitude Tactical Navigation) mission. Our A-10s hugged the terrain as we flew low-level back across the scenic green Adirondacks to our home base.

During the flight, Russ passed me "the lead," meaning he was following me. I took him on a slight diversion, to an area I used to live in as a young boy. Russ gave me the lead for several reasons. He knew it was important to nurture leadership qualities in young officers. He knew it was important to allow young pilots to "get out front" in a controlled environment. Passing the lead to a wingman in a formation of fighters gave the instructor pilot the opportunity to evaluate and mentor their "flock." He passed me the lead because of his understanding of leadership, mutual support, and trust. Russ knew that leadership and followership worked together, both up and down the chain of command. Russ and the great fighter pilots that trained me knew it was this trust that bred "unity of command"—the linchpin of any fighting force.

During that multiday mission, Russ shared with me the tale of how he became an Air Force officer. While growing up, Russ dreamed of flying. But, like many kids raised during the Vietnam War era, he decided against going the military route initially as a teenager. Instead, he went to the University of Maine at Orono to pursue his second "love," the outdoors. He earned a degree in forestry from the University of Maine in 1979. After graduating, Russ went to work part-time for the US Forest Service in Idaho. He also spent two and a half years doing what every kid from the east dreams of—skiing out west. Russ worked a wide range of jobs in Jackson Hole, Wyoming, to support his skiing habit.

He was having a blast, but felt that he was destined to do more. His thoughts returned to flying fighters and he began making plans. Russ had an opportunity to help his friend, Wayne, participate in the National Handicap Winter Olympics in Winter Park, Colorado. Russ would be Wayne's "coach." After the competition, they traveled through Salt Lake City to visit friends and ski Mount Snowbird and Alta. While in Salt Lake, Russ visited a local recruiter to see what chance there was to fly, and before he knew it, he joined the USAF. Russ took the Air Force Officer Qualification Test and "maxed" it. He passed the flight physical and entered the Air Force in 1981 as an officer trainee at Lackland Air Force Base (AFB), Texas. Ironically, Russ had never flown in or on an airplane in his entire life! No more bartending and skiing for this forestry major; he was off to fly jets for the USAF.

For Russ, the hardest part of joining the Air Force was not the commitment or the hazards. Instead, it was giving up a life that was comfortable, carefree, and familiar. This was something I never had, having entered the Air Force directly out of high school. Russ said he lost many friends who could not understand or accept what he wanted to do. He set out on his journey alone. That's how I began at the Academy too, initially alone. Yet the many friends we made in the military changed all of that—particularly those who continued to support us over those disappointingly dark and sometimes lonely years of the anthrax vaccine dilemma.

Russ did very well in USAF pilot training. He finally learned how to study. We were both near the top of our pilot training classes, and we both earned fighters for our first assignments. Overall, the military life was a good fit for Russ and me alike. We were very linear, structured, and comfortable within a defined hierarchy. On the other hand, Russ maintained that he still had a definite "wild" side where structure was fluid and boundaries were limitless. Perhaps this was why Russ took to being a fighter pilot so well and felt at home in the "wild blue yonder."

After a year and a half at his initial duty station, Russ became the first-ever lieutenant instructor pilot in his European fighter wing. He seemed to have been born with the fighter pilot mentality. In Russ, this manifested as confidence, not arrogance. He did not have the "top gun" swagger that one would imagine. He did not run around with the proverbial big watch, bragging to the world. Instead, he was confident in his abilities, never awed into a state of reverence by other's exploits, but always confidently, privately thinking that he too could attain or do what others had done if given the chance. Maybe this was another difference between us because I seemed to just strive to survive. I had to work hard in fighters, more so than Russ. The difference was not so much confidence, but maybe skill. The final product though was the same—success.

Russ thought of himself as a jokester, which may have been true in his younger years. Nonetheless, what I saw by the time I met him was a humorous manifestation of sarcasm. Russ, like any good fighter pilot, and like any good mid-level leader, liked to complain. But Russ had solutions, too. Russ was definitely an insightful man who could differentiate common sense from nonsense. Russ admitted he was never too skilled at being tactful. He suffered a disease many fighter pilots and leaders share—straightforwardness. He called the nonsense for what it was.

Perhaps that is why younger fighter pilots are drawn to this kind of intellectual honesty that lacks the veneer of tact. Perhaps young officers that are raised in idealist institutions, like the service academies, gravitate to leaders that exude frank, unblemished, and simple honesty in all they do and say. Maybe this was why I liked Russ. He was a straight shooter, the ideal flight lead, mentor, and friend. He was the definition of the military officer they trained us to admire.

These qualities served Russ well as a fighter pilot. He and those he flew with knew these were the requisite traits of good officers and fighter pilots. The world that Russ and I shared, airborne in Air Force fighters, was truly life and death on a daily basis, whether in peace or war. The "game" we played was always in close proximity to the "edge" based on the complexity and speed of the equipment, and the complexity and seriousness of the endeavor itself. The fluidity was incredible, the variables infinite. Yet the structure was inviolate, so we had to follow rules.

In flying, the rules of engagement protected us. If we failed to honor them, we might screw up and die, or potentially kill someone else. Russ hammered these truths as he briefed me before flights and debriefed me afterwards. We had both lost good friends over the years based on the dynamic environment we operated in and its inherent risks. Russ knew what the consequences of mistakes were, and they demanded that the entire operation be undertaken in a straightforward manner—with no nonsense allowed. It did not always look like it, but that is the way he conducted his life, his flying, and even his time at the bar telling his flying stories.

Lieutenant Colonel Russell E. Dingle and the A-10 "Warthog" Thunderbolt II.

In the fighter world there is the flight lead and the wingmen. It matters not the rank of either, the flight lead is in charge. Russ illustrated this philosophy through a "war story" about the first pilot he "busted," or failed, when he served as an instructor pilot. He was a first lieutenant, and his student was a full colonel. The colonel screwed up, and Russ busted him on a training flight. The colonel was a little miffed, but that was the contract—Russ was in charge. The colonel understood that he had screwed up, ate a little crow during the debrief, and they pressed on from there. Russ respected the rank, and the rank respected Russ as the flight lead.

Feedback generally occurred in the flight briefing or debriefing, which could sometimes be a raucous affair. The flight debriefing was where you checked your ego at the door, along with your rank. Facts were laid out regarding personal performance and achievement, or the lack thereof. It was there in the briefing and debriefing room that honor, integrity, and ethics were mandatory. In the debrief it was fairly impossible, and absolutely impermissible, to quibble one's way out of the mistakes made during the mission. That wasn't condoned, approved, or tolerated—just like at the Academy. Quibbling didn't cut it because it dodged the truth.

According to Russ, when a pilot discovered that something was "wrong," that pilot was eminently qualified to identify and deal with it because that was something they did on a daily basis. The problem was identified in a straightforward, truthful manner. Then, a solution was formed and implemented for the betterment of the mission. A fighter pilot operated this way in the air and on the ground, for any task or duty. In fighters, we often referred to it as "debriefing to win." One of our Top Gun Fighter Weapon School jocks authored an article on the mechanics of "debriefing to win." Used effectively, debriefing to win allowed pilots to identify root causes of problems. Often, we knew it was not simply the wingman's fault. We sought root causes. Was it our own misperceptions about the mission, or how we mistakenly conducted critical aspects of the flight? That was what thorough, honest briefs and debriefs were all about, without sugarcoating, whitewashing, or quibbling about anything.

We trustingly employed these very debriefing tools and thought processes in the fall of 1998 when our commander tasked his wingmen, Russ and Buzz, to identify problems, or perceptions of problems, related to a newly announced mandatory medical program for the entire armed forces. The DoD's anthrax vaccine immunization program (AVIP) sought to inoculate our troops with a seemingly tried and tested vaccine. The goal was to protect service members against a traditionally scary threat, and our boss wanted to lead the charge. Our mission of discovery rapidly revealed the old vaccine was known to be not adequate, not approved properly, and the manufacturing process was not validated by our government's vaccine regulators.

Russ did not want to be thrust into the dogfight of the anthrax vaccine dilemma any more than I did, but that was the last mission we flew together—a mission without end. Russ later commented that our formation flying in combating the anthrax vaccine program was the last act that resembled what we knew about how the military was intended to work. The mission lasted the rest of Russ's life and for the entirety of my thirty-two-year career, and beyond. At the time Russ joked, "I care, but not enough to get involved." In the case of the anthrax vaccine, something really struck a chord because he and I realized that we could never walk away. We soon discovered that the rank and the nonsense had entered the briefing room as honesty and candid risk assessment were shown the door. The world we knew of integrity and ethics was turned upside down, replaced by the politics of force protection.

With our review of the anthrax vaccine program, we entered a parallel universe where the common sense of our training went out the window as quibbling and rank took over. Russ said the anthrax vaccine tragedy rose to the level of a constitutional crisis. He did not like the idea of having to drop his plow and take up arms against his government or the DoD, so we both picked up our pens instead. We would use the tools and the training we were given by our leaders to fight an unconventional battle. Unfortunately, the battle we pitched was against those forces.

With the exception of a few friends and our families, Russ and I again found ourselves alone just like when we both first joined the Air Force. Regrettably, the anthrax vaccine ended both of our A-10 flying careers. Take the shot or be grounded and get out. That was the ultimatum, despite the facts. Russ said he missed the people, the esprit de corps, the camaraderie, but that we made the right decision. He said he would miss his integrity more if he blindly followed a lie. He discovered most everything about the anthrax vaccine turned out to be a lie.

I agreed, and followed my flight lead. Russ lamented, without regret, that waving the warning flag on the anthrax vaccine was hard because of the personal losses. But it was also the easiest thing he ever did, since it was a flag waved with the truth behind it. Shortly thereafter, we lost Russ to cancer, but the fight waged on in his honor for the many decades that followed.

MILE 4:
Anthrax Vaccine 101 (1995 to 1998)

Losing Russ many years ago summoned another sad memory from one of my A-10 flights over Bosnia. Russ and I deployed with our squadron and served in Operation Deny Flight to assist the United Nations in operations throughout the former Yugoslavia. Flying out of Italy, my flight was tasked by an airborne command post to divert to the middle of the Adriatic Sea to search for an F-16 that went missing. On that dreary day, I provided my best mutual support for another flight lead as we coordinated the search and rescue for the F-16 pilot, call sign Bane 01. The pilot possibly pulled too many g-forces and may have blacked out during his maneuvers.

We searched and searched, but there was no sign of life, just an oil slick in the water. My focus that day shifted from providing mutual support for the downed F-16 pilot, or Viper driver, to checking my lead's six and twelve o'clock. There was a lot to keep an eye out for since we were so low, with a murky horizon, rough seas, and other search and rescue assets above, below, and all around. We did not find the pilot. When I returned to our base, I discovered the missing fighter pilot was one of the first men that instructed me at the Academy, Mark McCarthy.

I had not seen Mark for years. He was one of those cadets and officers that I really didn't know personally because he was a couple of years my senior. I just remembered him: what an impressive professional he was and how we all looked up to him. Years earlier, during my basic training at the Academy, he had challenged me and another cadet one night to see if he could beat us at push-ups. He picked on the wrong two cadets, but he gave it a good try. He was proud of us and told us so. We looked up to cadets and officers like Mark who were humble role models. The Air Force lost a great man. Mark McCarthy's tragic mishap over the Adriatic Sea reinforced my mission to always provide solid wingman mutual support in all my duties.

Fighter pilots sometimes push the outer edge of the envelope. Yet one thing I learned about fighter pilots over the years was that, despite their aggressiveness, they pushed the edge in single-seat airplanes without putting others in harm's way. I reflected on the day that Major McCarthy died, and I always hoped he'd be proud of us—as proud of us as the day we beat him doing push-ups. Mark had something go wrong. It was likely something physiological. It wasn't because he took unnecessary risks or imposed unnecessary risks on others. It just wasn't his day. Mark, and those who trained him, did not take broken jets airborne. Nor would they allow such conduct from others. I pondered this and how it applied to Russ, me, and our fight for truth about the anthrax vaccine. With the anthrax vaccine, unnecessary risks were thrust upon us.

Before I met Russ in our duties as forward air controllers in the A-10, I spent several years after pilot training flying F-16s like Mark. During those years, principled officers taught me to know my limitations and always honor the rules of engagement (ROE). I met great pilots with call signs like Killer, Mukes, Shooter, Tav, Sarge, Speedy, Cheese, Boomer, Gumby, Fifi, T.O., Rover, Turk, Swamp Thing, Wedge, Moose, Blacky, Budd, T-Bag, Mace, Trip, Smiling Al, Spreader, Mongo, Wally, Kav, Gafi, Buster, Bones, Bone, and Calvin. There are many more! My call sign was "Buzz," from my days on the Academy boxing team. Calvin coined the name, saying I attacked my opponents like a "buzz saw." But that's not how I remembered it. I led with my face most of the time—and took a lot of headshots. My face gave my opponent's gloves such a workout during nationals in 1987 that I ended up in the hospital. Luckily, I have a really hard head. I took a lot of punches, but never got knocked down or out—ever.

When I found myself pitted against the dilemma of the anthrax vaccine, the choice was similarly hard-headed. I reacted in accordance with my training. I faced the threat and counterattacked the policy as I would any enemy or bandit that lacked honor, not worrying about the consequences. Many of those men we respected from our past flying fighters expressed their pride in us as we methodically attacked the anthrax vaccine policy in the years that followed.

Rempfer and the F-16 Fighting Falcon.

Between my A-10 assignments, I flew the F-117 Stealth fighter—another lucky opportunity. I knew I would be back to fly with Russ and the old crew again in the A-10. I launched off with my fighter pilot techniques, stories, lessons, philosophies, and bag of tricks that Russ had helped to fill. Undoubtedly, the F-117 ranked as the coolest jet I flew for my nation. The highlight of my F-117 tour came in the fall of 1997 when the United Nations weapons inspectors were ejected from Iraq. They were there to search for weapons of mass destruction (WMD). Without inspectors, how were we as a nation to know what capabilities Saddam Hussein possessed? That was why our DoD sent in F-117s to draw a line in the sand and to get our inspectors back in country. In a November 25, 1997, news release, the Air Force News Service documented the mission.

> KUWAIT—Six F-117A Nighthawk fighter aircraft from the 49th Fighter Wing, Holloman Air Force Base, N.M., landed Nov. 21 in support of Operation Southern Watch. . . . "The flight was a real thrill. It was the first time I've ever seen the sun go down and come back up in one mission," said Capt. Tom Rempfer. "We traveled about 500 knots ground speed which is high sub-sonic and had a great tanker crew from McGuire, AFB, N.J. It was a good mission."

[Note: Air Force News Service article excerpts were later scrubbed from all military websites.]

I remembered our deployment flight so vividly, lasting both day and night. I gobbled all my "Go" pills over the almost fifteen-hour mission, so when I finally landed, my mind was far from tired. When an Air Force News Service reporter stuck a microphone in my face, I couldn't help but reminisce about the breathtaking flight. Of course, the next four months living in the sands of Kuwait weren't quite as awe-inspiring. We did not successfully get the inspectors back into Iraq at that time. Our line in the sand was merely a brief moment of military muscle flexing. Since our saber rattling did not work, our military and civilian leaders were asked how they would ensure the troops were being protected from the very threats we opposed—like anthrax.

This question and its answer must be placed in context by remembering the tragedies that encompassed those times. Our troops suffered a myriad of attacks on foreign soil. Accountability and force protection became buzz words in the Pentagon. Genuinely, everyone wanted to protect the troops. Understandably, no one wanted their career to become a casualty for failing to do so. Secretary of Defense William Cohen had fired an Air Force general over such a controversy—the terrorist attack on Khobar Towers in Saudi Arabia in 1996. At the time, Cohen said:

> Personal accountability is not simply a question of assigning blame. It involves understanding the obligations of leadership, defining command responsibility, and clarifying the high standards of performance that we expect from commanders who are entrusted with the safety of our troops . . . force protection is first and foremost the responsibility of the commander on the scene.[1]

Cohen made it clear: force protection was first and foremost! Brigadier General Schwalier, the Khobar Towers commander, got the axe. It was a controversial move and one protested by our own extremely popular Air Force Chief of Staff, General Ron Fogleman. It was widely recognized that General Fogleman then resigned to protest the dismissal of Schwalier.[2]

No doubt, the defense secretary took his responsibilities seriously in anticipation of the questions about how he would protect us from biological and chemical threats now that the United Nation's inspectors could no longer do so. That was the background we understood when Cohen deployed our flying unit and then announced the anthrax vaccine program in December 1997. Within weeks after we landed in Kuwait to draw the line in the sand, the vaccine drew a line in the sand for medical deterrence and biodefense. Over the years that followed, it became clear from lieutenant to commander that it was imprudent to mandate an antiquated and notoriously inadequate vaccine. They pulled it off the shelf to outflank the questions. It worked, initially. The problem was they made it a mandate, contrary to the laws governing vaccines.

United States Air Force Captain Tom Rempfer speaks to reporters on the flight line of Langley AFB November 19 after he piloted one of ten F-117's from Holloman AFB, New Mexico to Langley. The stealth fighter bombers will continue on to Kuwait 11/20. IRAQ USA STEALTH

Rempfer and F-117, as a pilot in Kuwait. (Reuters)

Very few questions were raised initially, and the military public relations machine insisted it was the right way to protect our troops. No one of any rank or significance would stand in the way of the defense secretary's vaccine mandate. Most initially ignored possible problems. The poorly thought out, doctrinally unprecedented, force-wide mass inoculation program with an old vaccine shielded all but one question from a curious Pentagon reporter early on:

> **Q:** My question was just, is this another step up from a biological arms race where you've got moves, countermoves on each side?
>
> **A:** I don't know. It could be, but I'm not prepared to answer that.[3]

There were no follow-up questions for this anonymous briefer. There was no outrage by people that didn't understand the implications of this answer. What did he mean, "It could be?" How could he be "not prepared to answer that"? It was poignantly clear to those of us who spent our holidays deployed that we were there to do a job. We didn't need quick fixes that only inoculated our leaders from force protection accountability questions. It seemed the anthrax vaccine was too important and unprecedented of a program for a DoD briefer to not be prepared to answer straightforward military doctrine, "biological arms race," and policy questions.

At the time, even our F-117 commander did not support the plan, privately guaranteeing us we would not have to worry about the shots. Perhaps he was aware the program was being implemented in phases, and he could just put it off for the time being. If the threat was so grave, and we were the ones on the front lines in Kuwait, why procrastinate? From my boss's tone, I sensed he, too, felt it was a force protection façade. I respected, followed, and flew with this man that winter. That boss had the insight to call the panacea for what it was—a quick fix.

Defense Secretary William Cohen first broadcast the cure-all in 1997. Cohen hoisted a five-pound bag of sugar, simulating the threat of anthrax for the cameras.[4] If it were anthrax, he claimed, it would kill half of Washington, DC. At the time, experts contended that Cohen's bag of anthrax fear bomb rhetoric overestimated the lethality of anthrax by one hundred times. Years later, an intentional anthrax lab leak, meant to create fear to save the failing anthrax vaccine program, killed only two people in Washington DC due to a lack of early treatment with simple antibiotics. But then, no one questioned that the vaccine left a great deal to be desired. It was still considered a possible cause of Gulf War Illness (GWI). The military lacked medical explanations for GWI, and lacked scientific and legal foundations to justify mass inoculations. So, the military media machine cranked up a fear and hype campaign, taking advantage of a poorly understood weapon and our innate desire to believe our leaders.

The essential element of trust, vital to any military operation, was in question. Based on the DoD's track record of failures with their medical programs over the decades, this was a particularly important asset to not compromise further. The military experts were not prepared to answer the simple and obvious questions about a biological arms race escalation. Some resorted to inventing falsities about past widespread veterinary use of the vaccine, or intimidated the reluctant with threats of military discipline. Others resorted to scare tactics by embellishing the threat. This pattern repeated itself throughout the military, building on the misinformation.

No one seemed to have the intestinal fortitude to call the ruse for what it was—dangerous. Those were my first thoughts on the issue. Policy-wise, it was doctrinally imprudent to imply to senior military leaders that their troops were protected when the vaccine was known to be ineffective against some strains of the disease. It also seemed medically unwise to use a vaccine known by scientists and doctors to cause high levels of adverse reactions. And it was ethically dangerous to turn a vaccine program into a sales campaign for perceived political cover.

The military provided us early on with extremely unprofessional "education" materials. The handouts looked like something my ten-year-old could have crafted. Our doctors did not seem to know specific details about the vaccine, and the shoddy brochures didn't help. The Americans I served with justifiably questioned the program's announcement. That was our gut reaction, and the haphazard presentation and sloppy sales pitch strengthened this impression.

Lacking the old Cold War to fight, Cohen and many other politicians seemed absorbed with broadcasting the new threats of "bioterrorism" and "biological warfare." The anthrax vaccination program appeared to be their first attempt to stake out their new "war"—against an odorless, tasteless, scary, and intangible enemy. It did seem scary, but this threat was not new. Our leader's application of the informational instruments of power smacked of propaganda to the troops.[5] The vaccination policy announcement was accompanied by a mandated review of the quality of the vaccine. In order to sell the program, Cohen presented a four-point review designed to win over any skeptics. The required four points included: (1) supplemental testing, (2) tracking vaccinations, (3) communications, and finally, (4) an independent expert review.

Yet, if the vaccine was "safe, effective, and FDA approved," as their elementary-school-quality brochures maintained, why was this review even necessary? The whole thing seemed very "Hollywood." The "four-point review" later turned out to be window dressing. The sales pitch came with all the gimmicks—coffee mugs, pens, mouse pads, and flashlights, all designed to sell the policy. All their gimmicks advertised the message: "Anthrax Kills–Vaccination Protects." To the troops on the line, the entire dog and pony show seemed completely absurd.

Something was wrong. We didn't need pens, paraphernalia, and slogans to convince us to take a vaccine. We just needed a doctor to tell us he knew what he was talking about, but none seemed to have any experience or expertise with this vaccine. Even the "independent expert" appointed by Cohen to review and approve his mandatory force wide inoculation program later turned out to be an OB/GYN from Yale University who admitted he "had no expertise in anthrax." The DoD apparently intentionally chose a doctor who "accepted [the assignment] out of patriotism." The DoD made it "very clear that they were looking for a general oversight of the vaccination program."[6] We didn't question the doctor's patriotism, but the DoD appeared to take advantage of his lack of expertise in anthrax to purposely ensure limited scrutiny of the program.

There were many threats to worry about. Anthrax was not a new one. The vaccine was not a new protection. Why now? Who would ask this question? The answer was probably "no one" after the General Schwalier and Fogleman departures. Cohen had made an example of Brigadier General Schwalier, and was surely embarrassed by four-star General Fogleman's professional dissent. Fogleman's resignation meant only one thing—both were gone. Everyone got the message and kept quiet. I thought a lot about General Fogleman. He was a good man. He tried to do what he felt was honorable. He tried to stand up and defend a man he knew should not be the scapegoat of a tragedy. Fogleman belonged to a service that owed its roots to General Billy Mitchell, a military maverick who had the courage to speak up when things were wrong.

The first time I met General Fogleman was several years before he was the Air Force chief of staff, and before I was much of a fighter pilot. One night in 1990, we knew he'd be eating dinner with our commanders at the Kunsan Air Base Officer's Club in Korea. At the time, he was in charge of our higher headquarters, 7th Air Force, located at Osan Air Base. I organized a group of our lieutenants to execute a preset plan of action. Our group, referred to as the "LPA," or Lieutenant's Protection Association, all met at my "hooch." We had friends in the security police (SP) squadron who returned a favor for giving their troops incentive rides in the backseats of our F-16s. They lent us an armored personnel carrier (APC). They even loaned us one of their SP's to drive, a sharp-looking airmen with an M-16. We briefed the SP to go into the club and wait for General Fogleman to come out to the lobby. We previously coordinated with one of the club's servers to tell General Fogleman that he had a phone call in the lobby.

As the plan came together, all the lieutenants waited in the APC with a cooler of beverages, wearing our fighter pilot party suits. Those outfits alone seriously challenged our maturity in the eyes of any observer. The general showed up in the lobby as planned. The razor-sharp SP snapped a salute to the general as he approached, and gave the pre-briefed greeting: "Sir, Airman Jones reporting to take you on a spirit mission." We could see the general swing his arm in what appeared to be a "hot damn" sort of motion followed by a counter salute. He came out to our APC. We handed him a drink, and he returned the kindness by giving us the biggest smile we'd seen on a general's face. I don't think our bosses were happy with us, but then again, they were lieutenants once, too. Our spirit mission joy ride with the general only lasted an hour, during which we painted Panton Paws (our squadron symbol) on the taxiways at the end of the runway. The general had a good time, but we could tell he knew he could not act like a lieutenant all night. I guess that's why he was the general. We delivered him back to the officer's club.

A subsequent Air Force chief of staff briefed our F-117 squadron in Kuwait in anticipation of striking suspect WMD targets in Iraq. Our bosses woke us up to report to the chapel in the early hours of the morning. The new chief of staff gave a pep talk about our upcoming missions, but it just didn't match up to the pre-Gulf War motivational speeches delivered by the Air Force and Army generals that we'd heard about from Desert Storm. Ultimately, our missions were scrapped. Someone evidently decided it wasn't the right time.

We ended up packing our bags. We never got the anthrax shot. We never prosecuted our targets. Both were fine with me. We all realized that drawing the line in the sand would have to wait for another day. That day would actually happen less than a year later on the eve of Congress's impeachment trial during an operation called Desert Fox. But, for us, it was over.

8th Fighter Squadron, the "Blacksheep," and F-117.

We flew home, and all I wanted to do was see my family. I wanted my wife to see my pathetic attempt at growing a mustache. I wanted to hug my kids. I wanted to forget the empty saber rattling and retreat. I wanted to get as far away as possible from the politics of medicine, and the façade of force protection vaccination program that I sensed was not based on sound doctrine or a solid foundation of truth. Unfortunately, there were some things I could not outrun.

That's where it started for me. I tried to look the other way for a while. I returned stateside and within months, sailors on Navy ships such as the USS *Stennis* were refusing the vaccine. Parents were vocally expressing their concerns about the program. I wished there had been a military official somewhere who would stand up for those kids. It had begun. I hoped I could just avoid it and go back to my old A-10 unit, to Russ, and to my old fighter pilot friends.

Rempfer with the A-10 "Warthog" Thunderbolt II.

MILE 5: Tiger Team Alpha (1998 to 1999)

I zipped up my helmet bag, gathered my flight gear, and prepared to leave Kuwait in the spring of 1998. During and after the deployment, the officers from my former A-10 unit in Connecticut contacted me. They wrote asking what was up with the anthrax shot. I replied, "No worries gents, if a deployed Stealth Fighter Squadron commander in a potential combat zone protected us from receiving the vaccine, you have nothing to worry about." How wrong I was.

After finishing my F-117 Stealth Fighter tour and getting re-qualified to fly the A-10, I returned home to my former base. Unluckily, I was instantly injected into the front lines of the anthrax vaccine dilemma. Earlier that spring, our unit's flight surgeon briefed the anthrax vaccine program to the unit. He said that anthrax vaccine included a six-shot series, was expensive, and that the Air National Guard was an extremely low priority. We wouldn't be seeing it at our base for a long time. But before I'd set foot back on the base, members of the pilot cadre collected information from government documents that disputed the safety, efficacy, and legality of the vaccine mandate. I discovered that the officers looking into the policy were skeptical of reports and stories that did not have official citations or references. The efforts were very professional and were spearheaded by my flight commander, Major Russ Dingle.

Russ served as the intellectual heavyweight and flight lead in our struggle against the illegal anthrax vaccine program. He did everything by the books. Russ obtained a copy of Senate Staff Report 103–97, often referred to as the *Rockefeller Report* [excerpts at Appendix E].[1, 2, 3, 4] The report's subtitle read, "Is Military Research Hazardous to Veterans' Health? Lessons Spanning Half a Century." The Senate Veterans' Affairs Committee determined the anthrax vaccine should be considered "investigational" or "experimental," and could not be ruled out as a causal factor in Gulf War Illness maladies. The report shaped our understanding of the DoD culture that we faced from the outset of our fact finding with observations such as these:

> "While I have not yet determined the reason for this apparent aversion to full disclosure by DoD, the staff working on this issue from our committee has been constantly challenged by the Department's evasiveness, inconsistency, and reluctance to work toward a common goal here."
>
> "I can only conclude, Mr. President [of the Senate], that when dealing with the Department of Defense on this issue, you have to ask the right question to receive the right answer. I do not believe they understand that we are only seeking the truth in a way to help our veterans."
>
> "For at least 50 years, DoD has intentionally exposed military personnel to potentially dangerous substances, often in secret."
>
> "DoD has repeatedly failed to comply with required ethical standards when using human subjects in military research during war or threat of war."

When I returned as an A-10 pilot, this was the background to explain Russ's cautionary state of mind based on what he discovered about the DoD's disturbing historic health policies. Simultaneously, the unit compiled a roster of pilot volunteers to deploy for a preplanned rotation to the Persian Gulf, relieving the active-duty forces of which I had just been a member. Of course, I eagerly volunteered. I didn't mind going back. My previous F-117 fighter squadron friends gave me a going away party once we returned stateside, and even embossed my departing gift plaque with the words, "Buzz, we need a volunteer." Sure, I was a volunteer for almost anything, just not for a vaccine suffering from so many unresolved questions.

Initially, the DoD guidelines affirmed that the vaccine was not required unless you were spending more than thirty days in the theater.

Most pilots would be going for less than three weeks, so we would not be getting the shots. It became apparent that several officers, based on their research of the vaccine, would not be taking the shot under any circumstances. The commander found out, and behind the scenes started to modify the policy in order to force vaccination. Why would someone manipulate a mandate, in conflict with the DoD's official message guidance, in order to push his own people into a precarious career-ending corner?

Amazingly, events came to a head the first month I rejoined my old unit in September of 1998. The wing commander announced his new policy regarding the anthrax vaccination program: All officers, regardless of mobility status, would begin the anthrax shot series in October whether they were deploying to the Persian Gulf or not. To be fair, perhaps he did not understand the depth of the controversy and simply wanted his officers to lead by example. However, the announcement was not received like that.

Considerable resistance surfaced in part due to premature implementation of the policy compared to the DoD's message traffic that required a much later timeline for compliance. It became apparent there was another agenda. The commander chose to challenge his officers, rather than understand their concerns and look out for their best interests. No one understood why we were doing this to ourselves when one of the mottos of the Air National Guard had always been "procrastination is the key to air power!" Sometimes impulsive and poorly thought-out policy needed an independent party to non-concur or kick the can. The National Guard, technically assigned to the states, provided the federal active-duty force this healthy and judicious "common sense check."

I spent my time that month moving into a new home. I also ended up, sadly, paying my respects to an old friend by attending the funeral of Major Tom "Boomer" or "Gumby" Carr. He perished in an F-16 accident in Florida. Boomer was one of those free-spirited fighter pilots and was truly the best of the best. He just graduated from Fighter Weapons School, which is the Air Force's Top Gun training program. Boomer was a role model and one of the fighter pilots that looked after me in

Korea. He wasn't in my squadron, the "Pantons," but was one of the few "Juvats" with whom I'd developed a lasting friendship. Following our assignment at Kunsan, we both returned to the states and served in F-16s at Homestead Air Force Base in Florida until Hurricane Andrew rolled through destroying our base, a whirlwind to our lives.

The hurricane forced us out of Florida, a state we loved. Soon thereafter we both "pulled the handles," a pilot term for ejecting, and got out of the active-duty Air Force. To add perspective, over the years we were disheartened as we witnessed many of the most respected officers and most competent fighter pilots being "passed over" for promotion, while less admired and less proficient officers were promoted. We decided that there was another life for us outside the full-time Air Force, but we both committed to continue to serve as part-time citizen soldiers.

Boomer accepted a pilot slot in the federal Air Force Reserves, and I joined the Air National Guard. Boomer had become a close friend to my family. He was an honorary godfather to my second son. Tragically, his wife contacted us in September of 1998 when Boomer's fighter pilot career and life ended. Boomer bravely maneuvered his jet aggressively to avoid hitting another F-16 during a practice low-altitude bombing run. Catastrophically, his jet ended up in an unrecoverable attitude, with a vector into the ground. Boomer tried to eject once he had steered clear of the other pilot, but it was too late. We buried Boomer that month, and then I returned to Connecticut and my own duties once again as an A-10 fighter pilot.

Coming from a friend's funeral, I wasn't up for a bunch of nonsense. But that's precisely what I ran into. Our wing commander held a meeting on September 27th, 1998, interrupting a unit picnic to convince us the vaccine was okay. We knew it wasn't, and he knew we knew. He realized the pilots had been doing their homework. He knew many of his pilots had no intention of getting the anthrax vaccine on principle, based on what their preliminary research uncovered. He assured us that those who chose not to get the shot would be treated equally. In other words, a pilot would receive the same punishment as a supply officer, and flying status would not be used as a punishment tool for pilots. We were supplied with the same sloppy, unprofessional brochures that I received the previous winter in Kuwait. Everyone was dismayed. We were there to enjoy the annual family picnic, not to be served a bunch of baloney.

I wrote my commander a formal letter a few days later. I respectfully requested a bottom-up review of the anthrax vaccine policy. The commander appeared to take my letter seriously, or so I thought. He formed "Tiger Teams," and for a short while the shots again became optional, in accordance with the actual DoD policy, since we were not scheduled to be in the Gulf for more than thirty days. The military routinely did this, creating short notice task forces—such as our Tiger Team—to solve problems. However, the aviator culture, and our unique experience in problem solving, posed a troubling situation. If a commander disingenuously tasked officers to look into something serious but then he couldn't accept the answers or questions that resulted, it presented a difficult ethical dilemma for both the boss and the investigative team.

Major Russ Dingle and I were the two pilot participants for Tiger Team Alpha. Russ realized what the DoD was up to before anyone else. To avoid disobeying an order, he announced his intention to leave the unit—but only after completing his performance report duties and serving on Tiger Team Alpha. He committed to sharing his research on the anthrax vaccine and developing a list of questions for the commander to send to higher HQs. Tiger Team Alpha's research revelations stood diametrically opposed to the DoD's rhetoric.

Later we established that our position actually mirrored the historic DoD stance on the anthrax vaccine. The DoD knew there was a high systemic adverse reaction rate, questionable efficacy, and most notably that the vaccine was experimental. These were official DoD positions our leaders either did not have the expertise to know or didn't expect anyone to uncover. Most importantly, our position, based on Russ's research, mirrored the DoD's previous official assessments according to their own documents. Now, they were just making it up to sound good and support a new policy.

According to our professional military education, we learned that ethical dilemmas fell into several conflicting categories: short-term versus long-term, integrity versus loyalty, and individual versus organizational.[5] We charged that the anthrax vaccine problem challenged all three of these ethical dilemma classifications. It was arguably a shortsighted, reactive policy, lacking honest foundations, which threatened the long-term integrity of the military institution. Individual rights were violated in order to accomplish organizational goals. Thus, the ethical predicament began. The commander had a choice to support his officers in exposing the truth, or to blindly enforce an evidently illegal policy in order to loyally support institutional objectives.

I still don't know why I was chosen to assist Russ with Tiger Team Alpha. Maybe it was because our commander knew I was recently off active-duty. Maybe it was simply because I had written him the letter. It mattered not; I respected him for giving us a chance to get our concerns addressed. This particular commander was an active-duty officer on exchange from the Pentagon and USAF to the Air National Guard. He was also a USAF Academy graduate. I trusted that he would honorably evaluate our analysis, being born from the same ethical fabrics and codes.

Russ, Tiger Team Alpha's lead researcher and our most experienced flight commander, had already made up his mind. Based on his research, he was not going to order his men to submit to a mandate that he felt was illegal and immoral. Russ performed the bulk of Tiger Team Alpha's research and worked very hard to ensure the information presented was factual. He only cited material that included government documents or peer-reviewed publications. The team member's initial list of questions ultimately evolved into a document that was to be no more than two pages, so that the commander could present it to our Pentagon leaders.

We developed fifteen questions with supporting information. Our supporting information included an FDA inspection report. This report identified microbial contamination in the batches used for the vaccine sitting over at our medical clinic (FAV 030). The very first line of the FDA inspection report confirmed that the anthrax vaccine manufacturer had lost its FDA manufacturing process validation just six months earlier for failure to comply with FDA standards or "current good manufacturing practices." When we saw this report and the dozens of serious quality control deviations, we just scratched our heads and asked, "Why were we doing this to our troops?" We initially believed the commander valued our inputs, so surely he would do the right thing.

This initial inquiry uncovered the fact that the US Senate had concluded the vaccine's use was considered experimental and was a possible cause of Gulf War Illness. The discoveries demonstrated the vaccine was improperly licensed, its effectiveness was unknown, and its adverse reaction rates were unusually high. What the people who were protecting the anthrax vaccine told our DoD and civilian leaders conflicted with the facts. The snappy briefing bullets for commanders to tell their subordinates didn't match our research. We used common sense and our training and recommended our commander call a timeout. Or, as a fighter pilot would say if he was in the midst of an unsafe practice aerial engagement: "Knock it off." Instead of following our investigative team's advice, instead of protecting our troops from a questionable policy that we would later prove was illegal on multiple levels, our commander grounded us from our flying duties. He ordered our resignations and further shattered our trust.

The government documents Russ later uncovered proved the FDA never "finalized" or completed the licensing of the vaccine. The vaccine also had never been tested according to the requirements of the law, because it had "not been employed in a controlled field trial." And, according to vaccine records there was "no meaningful assessment of its value against inhalation [or inhaled] anthrax"—its intended use [1985 Federal Register entry at Appendix F]. Instead, these early documents established only effectiveness "for individuals in industrial settings who come in contact" with the organism (cutaneous anthrax). In March 1990, Colonel Takafuji of the Army Surgeon General Office and Colonel Russell of Fort Detrick described the anthrax vaccine as a "limited use vaccine . . . unlicensed experimental vaccine."[6]

The facts Russ uncovered contradicted the DoD's rhetoric that maintained the vaccine was fully safe, effective, and FDA approved. The contradictions meant the DoD was not being fully truthful to its commanders and it was illegally experimenting on its troops. This reality possessed me that day, and every day after, as we tenaciously researched the program. We were not alone on this journey. Many came before us, from suffering Gulf War veterans, to concerned lawmakers, and even intellectually honest forces within the DoD. Despite all the history, we were abandoned by our chain of command and now faced the anthrax vaccine dilemma anew.

In early October, our commander hand-delivered Tiger Team Alpha's questions to the director of the Air National Guard, or so we thought. According to our commander's assurances, the shots would be delayed until the answers came back. The boss later admitted he forwarded a different letter summarizing our inputs up the chain of command. One of our subordinate commanders snuck us a copy of that letter. We discovered the commander reduced our fifteen questions to four and told the general above him he thought of us as "hard liners." He said the unit would be better off when we were gone. We were suddenly not too confident about getting answers to our questions. Indeed, the commander did not wait for the answers and abandoned our team and his own people, even after ordering the inquiry. Everything challenged Air Force and Academy training.

By the November Unit Training Assembly, also referred to as a "drill," it was apparent that the answers to the Tiger Team inquiries would never arrive. We were told that the anthrax debate was over, that our questions could not be answered, and that the shots would begin. The policy changed again. The vaccine would not be mandatory for all officers. Instead, only the pilots deploying to the Gulf required the shots. The previous thirty-day, in-country requirement was changed to one day. Feeling cornered by our top leadership, sixteen vacancies appeared on the volunteer deployment list. Unfortunately, the only option was to stop volunteering, which was tragic since that was our entire purpose as volunteer citizen-soldiers.

In early December, the squadron leadership called in the "experts." They arranged for Dr. David Huxsoll, dean of Veterinary Medicine at LSU and a former commander of the US Army's biological warfare laboratory, to appear at the unit to dispel our concerns. As this meeting approached, all unit members were provided with a guidance sheet of what they could and could not ask. Russ attended the event, as did my wife, since I was at work for my civilian airline job. Russ grilled Dr. Huxsoll, who basically agreed the vaccine had a lot of problems.

By our December drill, and as a direct result of the sudden vacancies on the deployment roster, the commander announced yet another policy change. All pilots, whether deploying to the Gulf region or not, would either take the shots or be grounded from their flying duties. We would have to find other jobs. They forced us into a Hobbesian choice, an untenable option of equally objectionable alternatives. They told us to quit, or our fate would be out of the commander's hands. They relayed a message from our state's commanding general officer—anyone refusing the vaccine, or trying to transfer, would never work in the military again in any capacity.

Our commander even held a meeting where he implied that any pilot who refused the vaccine, and therefore could not deploy, was a "traitor."[7] Despite inferring vaccine refusal equated to treason, the commander admitted he owed his troops answers.[8] In that briefing, the commander admitted that the efficacy data "does not exist"—despite being required by the FDA. Years after our commander dismissed his aviators for being distrustful of the government, while slinging traitorous accusations and suggesting that we should find other jobs, a federal court ruled that the mandatory use of the anthrax vaccine was illegal for the precise improper licensing reasons that our Tiger Team asserted to our commander from the outset.

Within a month, a policy letter came out designating a deadline of the January drill for vaccination. All pilots not in compliance would be grounded and discharged. What happened to answering the questions? Since when did refraining from vaccination until legitimate safety, efficacy, and legality questions were answered mean one was a "traitor"? Under the Constitution, such allegations of treason applied to "levying war," or giving enemies "aid and comfort."[9] The framers intended to prohibit the "numerous and dangerous excrescences," which "disfigured the

English law of treason," in order that "ordinary partisan divisions within political society were not to be escalated by the stronger into capital charges of treason." Unfortunately, our leaders missed those constitutional classes just as they ignored our warnings about the illegal order.

Despite earlier assurances that flying status would not be used as a punishment for refusal, our grounding became the coercive tool to leverage compliance. A group of subordinate commanders also met with us individually to poll our intentions. They ordered us not to talk to the media about the controversy within our squadron. What was going on? What did this program have to hide for grown men and officers to act in such a manner? As a member of the military for almost sixteen years at that point, I had never been subjected to veiled threats or coercion. The officers that I expected to follow into combat reached an all-time low. I naively expected our leaders to honor the training they provided and review their policies and actions.

Russ and I contacted our elected representatives. The initial letters we received back either maintained they would contact the DoD, or just repeated misinformation provided by the DoD congressional liaisons to the civilian congressional staff. By our deadline on January 10th, 1999, nine pilots elected to not take the vaccine. One of our pilots decided to transfer to another non-flying position, so he was not included in the "official" numbers. The squadron commander issued a letter confirming eight losses. Later, the DoD denied this happened, minimizing the official losses to two. Once we realized the DoD was misrepresenting our principled groundings and losses as combat assets, we drafted another letter for our elected representatives. All eight pilots signed the letter attesting that it was the policy that forced us out of the cockpit and the Air National Guard. My squadron mate, Enzo Marchese, coordinated the effort. He was our youngest captain and a former enlisted airman. Like Russ, he could see the baloney for what it was.

The National Guard also later incorrectly reported the losses to a congressional interviewer, and a DoD spokesperson reported it in a Pentagon news briefing. We wondered why. In response to my naive frustrations, Russ would cynically say, "No one cares, Buzz." Russ repeated this over and over again. But of course, I knew he cared. That caring made for an "element"—a two-ship—a formation. We stood firm and did not leave our flying unit right away. With no due process, our boss called us at home and ordered us to report in, resign, and out-process immediately. We were allowed a maximum of four more hours on the base. We also refused that order.

One of my old friends, an Academy classmate, came up to me on one of those tortuous days in the winter spanning 1998 and 1999. He asked me what I was going to do. My answer was simple: I was going to do what we had been trained to do. I was not a martyr. I was just an operationally oriented pilot who did not fly broken jets. I had no aspirations to be a leader against the anthrax vaccine mandate, but could see a bad order for what it was. I also witnessed the conduct of our leaders and knew something was terribly wrong. Russ was the leader I respected in this ordeal. I knew where my duty would take me and who I would follow. I told my Academy classmate and squadron mate that I was going in a different direction than our chain of command. My classmate joined our formation, too.

Something disquieting happened that winter, something we were warned about. Our leaders dangled a continued career flying fighters in front of us. They tried to get us to ignore what our guts and fighter pilot instincts told us was a bad order. A bunch of knucklehead fighter pilots, not trained in law or medicine, called the baloney for what it was. Many others did too, both before and after. We looked at the facts, the law, and the documents no differently than we looked at our checklists when we preflight our fighters. We knew when something did not pass the "common sense check." We do not take broken jets airborne.

The lawyers had missed it. The doctors had missed it. Perhaps they were afraid to point out the faults and malfunctions during the design and implementation of the program. What did it say about our military culture when the doctors and lawyers either missed or would not speak up about something so compelling and so obvious? How were those doctors, lawyers, and officers, all who swore various oaths, seeing this so differently? Or were they overtly and intentionally breaking their oaths of office in order to support policy and advance their own careers?

We memorized the oath. It was the basis of everything we were supposed to believe. The oath is about supporting and defending the Constitution. The Constitution is about the law. It follows then that those who defend the law should not violate it. It is logical and academic. The violations of the oath and the law were obvious. The law specifically said the vaccine must be approved "for its applied use" and must be licensed by the FDA, which it was not. In order for the DoD to mandate it, it would need proper approval or a presidential waiver. The DoD knew the vaccine wasn't approved for "inhalation anthrax." The DoD didn't allow an optional program with informed or prior consent as required by law. The DoD didn't acquire the presidential waiver required to mandate the vaccine. The DoD blatantly broke the law and violated oaths of office at the highest levels. Russ always said this was a constitutional issue. He was right, again.

As we got deeper and deeper into the anthrax vaccine dilemma, we discovered that the unprofessional threats and attempts at coercion we received from some in the military chain of command were most likely tied directly to those officials' unease over their own willfully blind disobedience of the law. The leadership's reaction to our legitimate questions made the dilemma much bigger than the anthrax vaccine. Our fellow refuser, Captain Enzo Marchese, described the disappointing reaction we received from our military leaders as "mafioso" type tactics. Some of our commanders told our fellow guardsmen that we dug our materials off "Nazi websites." Others told our enlisted troops we were "cowards." We were left in a state of disbelief.

Enzo was right: This was not the leadership by example we were accustomed to, not the tactics of professional military officers. No answers. Just unsubstantiated, behind the scenes, thug-like unprofessionalism. Our maintenance commander even wrote a derogatory opinion editorial in a local newspaper titled "Pilots' decision to quit Guard does not uphold oath."[10] The support officer wondered "what these eight pilots would do in the face of real combat," exclaiming he "certainly would not want them as my wingmen." That was a weird thing to say as a non-flying maintenance officer. Ironically, we were given verbal orders for our resignations. In contrast, this officer earned a promotion. Indeed, the tactics of disparagement and ridicule, not to mention supporting the mandate, were good for career progression. This was not the military I served loyally. Either I misunderstood my training and oath, or something was terribly wrong.

A confirmation of these tactics came from a congressional staffer who had contacted us. Suddenly, he wanted to check on us and ensure we were okay. He sent a cryptic email one night asking, "What is your status? I've heard you were summoned, told to resign, and separated. If this is so, please describe those events to me in detail, including times, names, and ranks of who spoke to you. Did you get any indication what precipitated the change from transfer to non-mobility to out-processing? Thanks, and I'll try to call you Monday." The next day this staffer, who had already interviewed our state's National Guard generals, clarified his earlier note.

In doing so, he reinforced Enzo's theory about the mafioso tactics, and how our refusals spurred their vindictive ire. He confirmed, "Yes, it was from me. I agree the timing looks odd, and I plan to have another conversation with the general who said 'only two' of those refusing were doing so over the anthrax alone." According to the staffer, the general added, "things would be OK 'as long as they keep quiet' and he would make sure the pilots didn't just transfer to another unit and cause the same problem there." The staffer helped by calming us with, "Please assure your colleagues I mentioned no names, made no accusations, and betrayed no confidences in my meeting Friday. So, I was completely surprised, and fully prepared to be completely pissed, to learn your resignations were requested the same morning. Like you, I suspect it was due to the coverage. Keep me posted. Thanks."

Unsurprisingly, the commander later admitted to Enzo and me in his office that he was getting pressure from "outside Connecticut" to get us out of the unit. From my first recitation of the oath in 1983, to those dark days in early 1999, I had never seen such conduct. Fortunately, a congressional committee led by one of our state's US representatives, Christopher Shays, seemed intent on finding the underlying cause of what was going on, as well as protecting us from this bullying behavior by our military leaders. As fate would have it, his staffer was our first contact. As the weeks dragged on in those early days of the fight, we turned our trust to their turf, to our elected representatives in the United States Congress.

We began a series of trips to Washington, DC. That was where our commander and the anthrax vaccine policy came from. On our first visit, Enzo and I traveled in our military uniforms to walk Capitol Hill for the very first time. We believed our concerns, though disregarded by our own military bosses, should be professionally voiced to our leaders. Before I left, my mother-in-law performed my tarot card reading. I pulled the card of vengeance. She cautioned me that it meant that we should not seek revenge. That sage advice guided me over the years. Vengeance was not our mission, and accountability was someone else's duty. I kept hearing my wife tell me to make this principled stand worthwhile and not just go away. "If you're gonna give up your flying job over this, you better follow through and fight it, so others don't get hurt!" I reassured her, "See, I do follow orders. Yes, Ma'am!" Optimistically, professionally, we flew to DC.

On one trip, Enzo and I made a visit to Connecticut Senator Christopher Dodd's office for a meeting with his military legislative assistant. He kindly met with us in the senator's oak paneled conference room, at a huge table, with plaques and pictures everywhere. It was rather imposing. We were simply happy to have someone listening to us. Enzo and I walked the staffer through the events in our unit, always trying to concentrate on the larger issues of our safety, efficacy, legality, and doctrinal or policy concerns. We avoided discussing the petty events that transpired in our unit. The staffer said he followed the controversy on the news, but the DoD made it clear that the pilots left the unit because they "didn't want to deploy." He specifically cited the Pentagon news briefing where the spokesperson launched this insinuation. To set the record straight, we provided Senator Dodd's office with the letter, signed by all eight of the unit's pilots, stating that we were leaving the unit due to the anthrax vaccine. We spent a lot of time outflanking the DoD's misinformation.

As Enzo and I were wrapping up, the staffer gave us his contact information so he could keep Senator Dodd informed. The gracious staffer frankly stated that he "didn't realize people did things on principle anymore." Enzo and I left the office and looked at each other, amazed. Why else would we have refused the shots, allowed ourselves to be grounded from our A-10 jobs, and accepted the humiliation of our commanders calling us traitors and cowards? Why would we go to Washington, DC, on our own time at our own expense? We kept the senator's office informed as the years went by, staffer after staffer. At least they always listened. That day had a profound impact on us. We traveled to Washington, DC, to do what we believed was our citizen's duty, to "lobby" against the anthrax vaccine program, against an illegal DoD policy.

On a subsequent trip, we encountered a gracious doctor working on the staff for our state's other US senator, Joseph Lieberman. The doctor unequivocally concurred with our conclusions. We also talked with several plain-clothed USAF officers assigned to assist the senator's office as DoD liaisons working on Capitol Hill. We figured these camouflaged officers would likely get word back to the chain of command that we were trying to undermine their policy. Although we were concerned about the DoD's reaction, we believed the policy was illegal and knew we had a duty to inform our elected officials. We never heard from anyone. The DoD seemed to be "hands off," probably to not highlight the severity of the alleged violations.

Captain Enzo Marchese as an E-2 Airman in 1985.

Russ and I kept making trips to DC. We did not want to disappoint our colleagues, officers like Enzo who had also taken a stand. Enzo was young and wanted to start a family. Based on the lack of long-term proof of safety and the unanswered reproductive questions, he decided he was not going to risk taking the vaccine. We committed to learn more, not only about the unanswered questions but also about whether we were doing the right thing. What began with our favorite question, "based on what?," evolved into answering the questions for ourselves—based on this—based on the DoD's own documents that proved the program was illegal.

We wanted the pilots who refused the vaccine with us to know we would not give up. The quest to look for the larger meanings of the anthrax vaccine dilemma became the most fulfilling part. We studied, listened, and learned. We remained committed and continued our Tiger Team Alpha research as citizens and communicated our findings to leaders at all levels.

Russ and I expressed well-documented concerns to our lawmakers, their staffers, and on occasion, even to DoD officials. On each visit to Washington, DC, we shared increasingly compelling documents—the DoD's and FDA's documents—which showed our position was the DoD's previous position until the mandate announcement. Their documents did it for me. Russ would dig them up, we would analyze them together and distribute them, and then we would ask the questions that never received any answers. Instead, military leaders would just offer spin and quibbling excuses designed by those who were supposed to be in charge of the program.

Some of these papers were documents showing problems with the vaccine as early as 1985, when I was just an Air Force cadet and the USAF Academy. The documents were minted on US Army letterhead. They could not explain the disparity between what the documents said versus the new snappy sound bites and sales pitch. This was not the military I had joined a decade and a half earlier. At that very time, in the middle of my studies at the Academy, the DoD acknowledged the highly reactive, potentially unsafe, and ineffective nature of the anthrax vaccine. A serious disconnect existed between the dated documents and what our leaders told us now. Since these were their documents, only one of two conclusions could be drawn: either they were not telling the truth, or they simply did not know the truth.

A soldier always hopes that if their life and career in the military are to mean something, it would be because they exemplify courage in open combat. The last thing anyone seeks to become is a pariah to his or her own military leaders or—heaven forbid—a whistleblower. The extraordinary decision to challenge military leaders and their policies pretty much wrecked any hope of recognition or promotion. Then again, we had not asked for the job or the imposition.

Our Founding Fathers did not ask for their jobs either. They were pushed into a corner, through taxation without representation, by bullies who abused their power and discretion. The bullies had to be held accountable for their actions. Based on the Founders' courageous examples, they spelled out the duty for all who would follow. We knew what our job was, and we chose to utilize the tools given to us by our Founding Fathers, our commanders, and by our military institutions. Our risk was considerably less, but hopefully they'd be proud of us, nonetheless.

Though grounded from our flight duties, we enjoyed the mutual support of former military colleagues. Together, we planned and executed a multi-azimuth, multi-angle, or multi-prong attack using the same diverse tactics we would have employed against any other enemy threat array in order to maximize our probability of success. Our new target—the illegal anthrax vaccine immunization program—required this multidirectional, unconventional, and collaborative approach against our own institutions by employing classic military strategy, principles, and tactics. Mutual support boosted each of our vectors as a vital force multiplier.

MILE 6:
Mutual Support
(1999)

One of our colleagues was also a mentor throughout the anthrax vaccine dilemma, Colonel John "JR" Richardson. He attended Harvard University's National Securities Studies Fellowship coincident with our duties for Tiger Team Alpha. JR was a USAF Reserve F-16 pilot and Gulf War veteran. I first met JR while I was serving on a temporary duty assignment at the Pentagon. We stayed connected over the years. He was always there to help younger fliers find assignments and provided us with advice. So, when I knew things were not going well in our unit, I called him up to seek his counsel. His first reaction to me was, "Well, Buzz, it's an order, and you need to follow it, or else you know what you need to do." JR was a very successful officer and obviously, based on his selection to go to Harvard as a reservist, he was on a "fast track." We referred to those kinds of folks that were being groomed to be the service's senior leaders and eventually generals as "fast burners." JR didn't know me well, but he extended the professional courtesy to review what we had discovered.

I provided some of Russ's initial research findings to JR. I included the Senate report calling the vaccine "investigational" and a possible cause of Gulf War Illness. We shared the Army scientist's critical safety and efficacy reviews of the vaccine, and the FDA inspection report confirming the manufacturer lost its validation due to quality control problems. JR was notably disturbed. He relayed our concerns to some of his own Air Force Reserve mentors, but no one seemed interested in getting involved. JR rapidly became troubled that the vaccine policy could backfire on the armed forces and hurt the integrity of the military institution.

From the outset, Russ and I were convinced that the dilemma was really not about the vaccine, but instead about the larger ethical ideals involved. When the ideals—intended to be the foundations of the institution—were jeopardized to promote policy, we knew what we had to do. JR joined our formation. He was professionally opposed to embarrassing his military leaders, so he worked behind the scenes. JR was a litmus test for Russ and me. If he had rejected our work, or refuted it, it would have altered our own decision-making processes. But he didn't. Instead, he validated our fundamental premise that the military leaders were blindly repeating a mantra that was easily refutable with readily available congressional, medical, and military documents.

JR entered the fight and remained involved out of principle for many years to come. He acted out of a sense of duty and gave "mutual support" to Russ and me. According to our earliest training, we knew this was fundamental in the theory of flying fighters—where two pilots could "cover each other's six," allowing the other to prosecute an attack with the confidence that their vulnerable zones were covered by their wingman.[1] We used our comrades and other resources to provide mutual support, ensuring we stayed on target and that our vector was true. JR provided the intellectual mutual support we needed to stay on course. The words of Captain Edward V. "Eddie" Rickenbacker, America's ace of aces with twenty-six kills in WWI and the first flyer to earn the Medal of Honor, defined JR's role: "Whenever you're over the lines you have to keep twisting your neck in all directions, or you're sure to be surprised." JR kept his head on a swivel, helping us in the years to come as we journeyed across the line into uncharted enemy territory. Together, our team owned the element of surprise due in large part to JR as our erudite wingman.

JR attended a civilian college and participated in the Reserve Officer Training Corps (ROTC) program. This was just one more avenue for officer commissioning, different from my four years of Academy education or from Russ's sixteen-week post-college commissioning program through Officer Training School (OTS). The three of us followed different paths into the military but all became Air Force officers. We were shaped according to the same standards and now were tested by the anthrax vaccine controversy. JR used his well-read education and "policy wonk" background, plus his Pentagon experiences, to paint the picture for us of the five-sided "puzzle palace." That term came from the officers who worked in the Pentagon due to its puzzling and perplexing internal processes. JR helped us to find non-flying jobs and thanks to him, we outflanked efforts to get rid of us by leaders who were unable to answer our questions.

That was how we continued to serve after the dismissals and groundings from our previous flying jobs. My new role was at a desk job with good people at Hanscom Air Force Base in Massachusetts. Russ and I, along with several others in our crew of tossed misfits, also became Air Force recruiters for the Air Force Academy. JR exhibited more of the same "mutual support" by putting us in touch with various Air Force leaders so that we could present our case to them concerning the vaccine's problems. Despite our efforts, many senior leaders looked the other way. One of JR's general officer mentors wrote back to me early-on about the controversy:

> Buzz: Thanks for the note. I get plenty of info on a regular basis on this subject, from both sides of the issue. My belief is that each and every one of us needs to make a personal decision about taking the series. As long as the company policy is to take the shot, the only recourse, in my view, is to SIE and work from the outside.

Bewilderment was the only word to describe my reaction. JR and I both pointed out the military institution's apparent illegal conduct, yet the rote answer was to follow "company policy"—without comment on the questions. To top it off, the viewpoint that an officer's alternative to company policy, which we charged as patently "illegal," was the advice to "SIE" (Self-Initiated Elimination); in other words, to quit. Needless to say, that didn't happen. It sure would have been convenient to strip us of our uniforms and get us on the "outside," out of the USAF altogether.

In my dogged pursuit of mutual support, and to verify my own analysis, I reached out to another retired senior military official. I engaged him respectfully and professionally, and was grateful to receive a reply note from this good man who I considered a role model. I didn't know if he would remember me, but I hoped to share what we learned about the vaccine. We needed advice. We needed help, so we hoped he might get involved and help us too. He replied:

> Buzz: I apologize for taking so long to respond to your letter of 15 February on the Anthrax inoculation. I am afraid I am on the opposite side of the issue from you. Three years ago, before a mandatory program, when the Joint Chiefs first discussed this issue in the Tank [secure meeting room], I went to the Surgeon General and asked for all the available information on the existing vaccine, which has been in use for many years. . . . Bottom line: All the evidence points to a safe and effective serum. Given the threat, US troops should have the protection.

I couldn't blame the general for not wanting to get involved in this fight, but it was worth checking in to see if we could get him to join our formation. He had obviously read the talking points, repeated their sound bites, and seemed wedded to the boilerplate script. He supported the program before it was a program. I was disappointed but appreciated the dialogue. It was evident from his note that he didn't read the information I provided. Our documents, the DoD's documents, didn't support his comment that "all the evidence points to a safe and effective serum." His conclusions conflicted with GAO findings and the DoD's prior opinions, pre-1998. His response validated my larger concerns about a dearth of in-depth review by senior officials prior to launching of the program, or after. At least he didn't encourage me to quit as the only viable alternative if I didn't agree with "company policy."

Unfortunately for the DoD and some of its general officers that looked the other way, or didn't review our research, we didn't look the other way or quit. We became the embodiment of our nation's expectations of military officers in our fight against "company policy," the anthrax vaccine mandate. Instead of directly fighting our DoD, and instead of "SIE," our "only recourse" was to carefully engage the anthrax vaccine program indirectly from the inside and the "outside."

Our fight extended beyond the Air Force. A family in Pennsylvania, the O'Neils, asked for my mutual support. They wanted to protect their son from further anthrax inoculations and to extricate him from the US Marine Corps. The family sued the DoD and the Department of the Navy, which oversees the Marine Corps. Their goal was to help their son, but also to halt the emotional distress the anthrax vaccine controversy created for the Marine's mother. After speaking to the O'Neils, I felt an obligation to help communicate their concerns to the DoD.

I received a federal subpoena to appear in their case in order to testify before the chief judge for the Western District of Pennsylvania on how the anthrax vaccine policy was driving good military members out of the service. I sought to explain to the judge why the policy was driving parents to sue the DoD in order to get their children out of the military. In that case, *Private First Class (PFC) O'Neil v. Secretary of the Navy*, O'Neil's attorney, Mr. Louis Font, informed the Honorable Judge Ziegler that my testimony to the court "of course, would be identified as not being official policy of the DoD." That covered me as far as any ethics charges the DoD might attempt to levy against me after the fact. If I implied that I was representing the DoD's position, and my testimony was contrary to DoD policy, I could be guilty of an ethics charge. I did not misrepresent anything, especially related to the ethical problems associated with the anthrax vaccine mandate. Everyone clearly realized I was not representing the DoD position.

I reported to court on December 2nd, 1999. I wore my uniform so that my testimony was as professional as possible. Attorney Louis Font unequivocally verbalized the disclaimer to Judge Ziegler that I represented my own opinions. DoD's counsel even called me to the back of the court and gave me a letter from the General Litigation Division of the Air Force approving my testimony. They required that I "clearly state on the record" that my views "do not represent the views of the Air Force or the Department of Defense in any way." I complied.

My wife's stepfather, Bill, drove me to the court. Bill was a retired US

Army Colonel. In fact, I used to call him "Colonel." He liked that. He was definitely old school Army and was a Vietnam veteran. I don't think he ever completely understood why I was challenging the DoD, however, Bill said he was proud of me before we entered the court. When I went inside, I met the O'Neil family. PFC O'Neil sat by his mother and father, looking sharp. I realized at that moment that my mission had shifted from flying fighter aircraft on combat missions; my new mission was to protect our troops like the young O'Neil. Together, we met the opposing "element," fighter pilot speak for a two-ship, the team whose job it was to protect the opposition's policy.

A US Naval officer in charge of the Navy and Marine Corps' anthrax program was tasked with defending the vaccination in the Pennsylvania court. On the stand sat the Navy lieutenant commander, the equivalent of my Air Force rank of major. I was extremely surprised to see her in court because I recognized her from one of the early congressional events that I had attended. She was not wearing a uniform on that day sitting in the audience in the US Congress. Maybe she donned the "incognito mode" then like the DoD staffers who wore civilian clothing while working in our elected officials' congressional offices. Perhaps it was her day off or she had run out of clean pressed uniforms. I wasn't sure why she wasn't in uniform in the halls of Congress. But, like me, she did wear her uniform in the Pennsylvania court. She sat to the judge's left, took the oath, and her duty to protect the anthrax vaccine program began.

The Navy lieutenant commander was part of a small cadre of officers whose mission was to defend the anthrax vaccine. Her charge was to implement the program for both the US Navy and the US Marine Corps. It was an enormous responsibility. And she was sharp, too, ready for the task. She seemed well prepared by her fellow anthrax vaccine program tutors. The equally sharp US Navy defense attorney asked if the vaccine was being used for an approved purpose. The witness responded by literally quoting the precise words from a memo written by FDA Commissioner Dr. Michael Friedman in March 1997. The DoD obtained the Friedman memo years earlier to secure consent from the FDA on the legal status of the vaccine. The Navy officer responded to the defense attorney's questions by stating, "The anthrax vaccine is **not** being used in a manner that is **inconsistent** with its approval by the Food & Drug Administration." That sounded like a double negative. My mother always warned me about those, so I took note. I had heard of the memo, but was not aware of its pivotal significance until that hearing.

My colleague, JR, would later put his Harvard education to work by thoroughly analyzing and documenting the timeline of events, to clarify what transpired for all involved.

> September 1995: The US Army developed a plan to obtain FDA approval for the anthrax vaccine for use against aerosolized or "inhalation anthrax." The plan's text included: "This vaccine is not licensed for aerosol exposure expected in a biological warfare environment."[2]
>
> October 1995: The US Army meeting minutes concerning the "inhalation anthrax" approval included comments by a Brigadier General: "BG Busby addressed a need to make the case that anthrax is currently the principal biological warfare threat."
>
> September 20, 1996: DoD applies through the manufacturer to FDA to get "inhalation anthrax" on the vaccine's product label and license, via an IND [investigational new drug] application (#6847)—the IND was not approved.
>
> March 4, 1997: Pentagon official writes: "DoD has long interpreted the scope of the license to include inhalation exposure, including that which would occur in a biological warfare context." This was undoubtedly false as evidenced by the DoD's "inhalation anthrax" investigational new drug application.
>
> March 13, 1997: the Friedman memo agrees: " . . . while there is a paucity of data regarding the effectiveness of Anthrax Vaccine for prevention of inhalation anthrax, the current package insert does not preclude this use . . . Results from animal challenge studies have also indicated that pre-exposure administration of Anthrax Vaccine protects against inhalation anthrax. Therefore, I believe your interpretation is **not inconsistent** with the current label."

> December 1997: the same month Secretary Cohen announced the anthrax vaccine program, internally the DoD concluded that: "Anthrax and Smallpox are the only licensed vaccines that are useful for the biological defense program, but they are **not licensed for a biological defense indication.**"[3] [DoD capabilities report at Appendix G]

Russ and I had found the Friedman memo in documents received through the Freedom of Information Act (FOIA). I was amazed at the Navy officer's ability to recite it almost verbatim, especially the double negative part. The DoD memo contradicted the documentary trail with this leading remark, "DoD has long interpreted the scope of the license to include inhalation exposure, including that which would occur in a biological warfare context. Please advise whether the FDA has any objection to our interpretation of the scope of the licensure for anthrax vaccine." Friedman, who had only been on the job for four days, and was ultimately never confirmed, gave the DoD the bureaucratic license and go-ahead they needed to launch their program. More likely than not, the DoD needed the memo to counter the historic documents that clearly demonstrated the military previously understood the vaccine was not licensed properly.

Friedman's memo also included, "while there is a paucity of data regarding the effectiveness of Anthrax Vaccine for prevention of inhalation anthrax, the current package insert does not preclude this use." There, he admitted the scientific information was lacking. He was also implying that the DoD could use the vaccine for inhalation anthrax since the product's label "does not preclude this use." This was tantamount to giving a green light to use aspirin to cure cancer simply because the product's label, or package insert, failed to preclude that indication.

The memo exchange was absurd, and the documentary trail was clear—the DoD knew the vaccine was not properly licensed both before and after the Friedman memo exchange. It was evident there was a time when the DoD and the manufacturer tried to do things right by applying to the FDA to include inhalation anthrax on the product label. However, the regulatory compliant approach was apparently abandoned, and the quick fix was employed—the Friedman memo.

Under continued direct examination, the defense counsel asked the Navy officer to explain the government exhibits once again. Yet again, the officer was right on cue explaining the memo from Friedman: "Correct, sir. The next letter . . . indicates that basically the DoD's use of the vaccine to protect forces against a potential threat of a biological warfare attack is **not inconsistent** with its license." She repeated the double negative script a second time. It was distinctive because the wording "not inconsistent" was so awkward. She must have mistakenly left out the part where Friedman wrote, "There is a paucity of data regarding the effectiveness of Anthrax Vaccine for prevention of inhalation anthrax." This was why the DoD resorted to using animal data to attempt to prove efficacy. The key question was: how could the vaccine be approved for use in a biological attack, or for inhalation anthrax, if there was a "paucity" of data?

According to Friedman's memo, the proof was based on animal challenge studies. Nonetheless, animal data did not meet licensing requirements under Title 21 of the US Code—the Food, Drug, and Cosmetic Act—until 2002.[4] The circularity of the answer was remarkable. Maybe these officers were unaware that the DoD was party to later changing the law to allow animal data, rather than human data, to garner full licensure of vaccines. Regardless, they keenly outflanked any concerns by the judge. He seemed unaware of the "animal rule" or drug licensing requirements. They implied it would be unethical to gather data from human testing. But wasn't it also "unethical" to mandate a vaccine coming from a non-validated, shut-down manufacturer?

The Navy officer's memory appeared to fade altogether under cross-examination. Mr. Louis Font, a US Military Academy at West Point graduate, asked the Navy officer a series of questions about other quick fixes that gave the anthrax vaccine program an appearance of propriety. He asked about the DoD's designated independent expert, the Yale University OB/GYN, who reviewed the program. When asked if she was aware of the "expert's" letter telling Congress he had "no expertise" in anthrax, the officer's memory faded, answering, "No, sir, I am not." Louis Font also asked, "All right. Now have you attended congressional hearings concerning the anthrax vaccination program, ma'am?" The officer admitted, "Yes, sir, I have."

Louis Font followed up, "Were you present May 25th of 1999 or thereabouts at a meeting of some hundred physicians at Detrick? Detrick, Maryland. Were you there?" Again, the officer admitted, "Yes, sir, I was in attendance." Attorney Font asked her if she was aware of GAO (Government Accountability Office) reports. She replied, "Sir, I have not seen the GAO reports." Several more exchanges ensued that either elicited other indefinite responses or were "outside my realm of knowledge." Mr. Font was clearly frustrated in his attempts to get straightforward and candid answers. Font and the Navy officer jousted back and forth, and her bottom line was, "Sir, this is a commander program." Font clarified, "Alright. This is a commander's program as opposed to a medical program. Right." The naval officer affirmed, "Absolutely, sir."

The message was clear. It was a commander's program. This added a crucial perspective in understanding why the DoD sent an unqualified military officer to Pennsylvania instead of a medical doctor guided by the Hippocratic oath. This seemed to be a key aspect to the ethical dilemma over the anthrax vaccine—DoD doctors and lawyers failed to ensure a satisfactory vaccine was used in a legal way. As a result, the DoD program was nothing more than a military operation run by commanders—not a force-wide medical program run by doctors. Removing the doctors and the lawyers from the operation attempted to skirt the medical and legal obstacles.

Frustrated, the O'Neils' attorney dropped his medical and legal questioning, and concluded by reading a quote from a Department of Defense publication called the *Armed Forces Officer*. Mr. Font explained, "It was a document that was given to me within my first week at West Point in 1964." Font asked the Navy officer if she was familiar with concepts in the document, "born of the American Revolution," as he read:

> Within our school of military thought, higher authority does not consider itself infallible. Either in combat or out, in any situation where a majority of military trained Americans become undutiful, that is sufficient reason for higher authority to resurvey its own judgments, disciplines, and line of action.[5]

Even the Navy officers wanted to read the quote for themselves. The judge quickly intervened, "It is not relevant." But it was relevant to us. I realized that Mr. Font, like me, actually took his academy training seriously. He still remembered the quotes they ordered him to memorize so many years ago. He was trying to apply that training because he sincerely believed there was "sufficient reason for higher authority to resurvey its own judgments, disciplines, and line of action" when it came to the anthrax vaccine irregularities. Mr. Font offered the warning, despite the judge's determination that it was not legally "relevant." Font was trying to explain why military members were "undutiful" when it came to the anthrax vaccine. It also applied to the conduct in the court that day. Two different military agendas were being served. Mr. Font and I were on one side, and the Navy officers were across the "line of action."

I wondered if I was being too hard on the Navy officers. My empathy soon vanished as I recalled another transcript from a meeting that the Navy anthrax representative admitted she had attended at Fort Detrick with the doctors on May 25th, 1999. During this meeting, one of the top DoD doctors on anthrax vaccine, Dr. Renata Engler, a colonel from the US Army's Walter Reed Army Medical Center, opened her remarks with, "the actual efficacy for the prevention of inhalational anthrax [is] not known but presumed based on the existing data." Interestingly, that was how the transcript read, but the PowerPoint slide was more complete: "Actual efficacy for the prevention of inhalation anthrax NOT known but presumed, based on existing data for prevention of disease (e.g., primate data)." The slides also specified "primate data" for monkeys and confirmed it was "not known" to be effective in humans. Rather, effectiveness was "presumed." That was the point, and that was why the vaccine was investigational and could not be mandated without direction from the president of the United States. That was the law.

Colonel Engler had a reputation for being honest about the limitations of the anthrax vaccine and the harm it was doing. Perhaps that is why a doctor like Colonel Engler was not sitting on the witness stand that day in Pennsylvania. Dr. Engler's comments added perspective on the enormity vaccines were to play in the future, helping to answer the question of why the DoD was digging in so deep in order to defend the anthrax vaccine. Anthrax vaccine was just the start, with as many as thirty to fifty new vaccines in the next five years.[6] Dr. Engler advised her fellow doctors and clinicians to not be quick to deny the anthrax vaccine could be the cause of illnesses. Dr. Engler admitted, "One of the docs I talked to said it couldn't be anthrax." Instead of encouraging and condoning such anxious denials, she honestly mentored the group of doctors:

> There are things that as I get older, as an immunologist, I am humbled ever more about the things we don't understand. I think we cannot make a presumption; we should just report and then the cards fall where they may.
>
> When I hear doctors say, well, if I report, the system might make me lose my job. That really worries me . . . I think our leaders need to send a message that honest inquiry, honest concern about a patient is never a bad thing, and if you report an adverse reaction to anthrax, you are not hurting the anthrax program; you are helping the anthrax program.
>
> We have to be honest about what we don't know. I don't think there is anything wrong with saying I don't have perfect answers.
>
> We have to recognize that these are young men and women who are calling us from pay telephones during the daytime because they don't trust that their telephone message might be recorded, and they are telling us that they are experiencing these events, that nobody is believing them. When they go to the medics . . . they are discarded . . .

> I think the honest answer here is . . . In biology, immunology, human beings are not widgets, and we have to deal with that reality or we are not worth our salt, frankly.

The transcript of the doctor was honest and candid. Too bad Dr. Engler was not testifying in the Pennsylvania courtroom in December of 1999. Dr. Engler was not there to be "honest about what we don't know." No, the DoD sent the Navy program manager instead of a doctor. And, despite the judge's initial assurance that the hearing was only about a discretionary discharge, that he wouldn't be reviewing the merits of the anthrax vaccine, he did just that.

On the morning of December 3rd, 1999, the District Court Chief Judge Ziegler held that: "(1) decision of Commandant of the Marine Corps to deny hardship discharge to serviceman was not arbitrary and capricious, and (2) serviceman was not entitled to a preliminary injunction."[7] The young Marine's discharge had been denied.

The judge went further. He found "the vaccine is not an experimental drug," and that the "evidence establishes that the anthrax vaccination program is lawful." This was the amazing part to me. The judge specifically said he would not make such a determination. Accordingly, he did not hear evidence or expert arguments from which to make such a ruling. The DoD also did not reveal or admit their participation in the investigational new drug application process.

The judge dismissed the court late in the day and showed up early the next morning with the extensively prepared pages upon pages of "findings of fact and law." The judge relied heavily on the US Navy lawyer's assurances that the young Marine would go to a base where he would not have to receive any more anthrax vaccine. The judge wrote, "The Marine Corps has attempted to accommodate PFC O'Neil by assigning him to . . . Quantico, Virginia, where he may never receive further inoculations, and where he will be closer to his mother." I felt the judge had been misled on this account, and evidently, news reports soon agreed:

> *Marine Corps Times*, December 20, 1999, page 8
> . . . A Navy lawyer is accused of misrepresenting the government's side in a federal lawsuit. . . . The Marine Corps has attempted to accommodate Pfc. O'Neil by assigning him to . . . Quantico, Virginia, where he may never receive further inoculations and where he will be closer to his mother. . . . However, Marines who begin the anthrax series at another installation and then are transferred to Quantico continue to receive the inoculations on schedule until the six-shot series is completed.

Within a month the Marines decided to cut their losses in the O'Neil case and approved the young man's discharge. The DoD's win in Pennsylvania was lost. The commandant of the Marine Corps granted PFC O'Neil a special exemption for humanitarian reasons. I was happy with the outcome for the O'Neil family, yet saddened that the military lost another good Marine. I wasn't sure if the Navy officers were intentionally evasive, but all they'd done was successfully protect the program temporarily and keep O'Neil in the service for a few more weeks.

Once the folly of the court became public knowledge, it appeared the four-star general commanding the Marine Corps authorized the hardship discharge himself. The tongue-tied testimonies of double negatives merely helped to further expose the modus operandi of legal quick fixes upon which the anthrax vaccine program was based. No different than the Navy representatives believing the vaccine's use for inhalation anthrax was "not inconsistent with its license," I was less than unimpressed with the performance of the Navy tag team. Their taxpayer-funded trip to Pennsylvania only served to successfully manipulate a court and judge.

Bill took me back to the airport and again said he was proud of me. He expressed his equal pride for the former Army lawyer who represented the O'Neil family. He watched the hearing with military precision. He was unable to understand the testimonial vagaries over the GAO reports, the licensing for "inhalation anthrax," the adverse reaction rates, or the "independent expert's" recantation letter to Congress. Bill commented that the testimonies were either misleading or the officers were incompetent. As Russ would say, "they can't have it both ways." Even Bill, the good soldier and colonel, knew something was very wrong.

After testifying for the court in Pennsylvania, I got a call from my Hanscom Air Force Base commander while still at the airport. He said, "Buzz, I need you to come in and talk with me." I asked him what it entailed. He responded that it was my "activities" dealing with the anthrax vaccine in Pennsylvania. I asked him where this was coming from and he told me, "the command structure," but didn't elaborate. So, I traveled home and soon thereafter reported in.

I arrived at the base and reported to my commander in my full military uniform. The colonel informed me he was ordered to counsel me and document my personnel file for wearing a uniform in the Pennsylvania federal court. According to my commander, Pentagon officials told him to reprimand me for wearing my uniform because "Air Force sanction of the cause . . . may be inferred." The letter of counseling he asked me to sign said wearing the uniform in the Pennsylvania court "would appear to be in violation of Air Force Instruction 36–2903."

I was surprised. For the Pennsylvania hearing, I had received a letter specifically approving my testimony from the General Litigation Division of the Air Force at the Pentagon. I complied with their disclaimers by confirming my testimony did not represent that of the DoD or the US Air Force. As a result, my wearing of the uniform and conduct could not be wrongly inferred. My commander did not know any of this. I explained that the judge in Pennsylvania subpoenaed me, and that I had received the Air Force's approval. Essentially, the Air Force was attempting to sanction or intimidate me for appearing in uniform in a federal court in Pennsylvania under subpoena from a federal judge, despite my prior approval by the DoD.

An Air Force attorney friend in the Pentagon helped me to understand the internal DoD dynamics at play. The attorney warned me that a group of Air Force Pentagon officers was trying to figure out how to punish me for wearing my uniform in federal court. Overall, I wasn't concerned, and figured it only appropriate and professional to wear my uniform in front of the federal judicial branch hearing—especially since it seemed that I was the only uniformed person that was telling the full truth. The sanction was just an intimidation tactic designed to try to stomp out the professional dissent I was expressing. I didn't sign the letter. My boss asked for a memo for record explaining why. I complied by preparing a formal response, rebutting the reprimand letter with a special request for a top-to-bottom, "tooth-to-tail," review of the program.

In the letter, I affirmed that I tried to not involve my leadership at Hanscom Air Force Base in the dilemma. Hanscom was not forcing the anthrax vaccine on anyone. They also graciously let me join their ranks, and I did not want to cause any problems for my new bosses. But, based on this attempted sanction, I respectfully requested meetings with top Air Force leadership to begin a process of critically analyzing the anthrax vaccine program based on our discoveries of the illegal conduct. I never heard from anyone on any of it again. At the end of the day, I was proud to offer mutual support to the O'Neil family and the Pennsylvania federal court.

I remembered how scared my parents were early on in the anthrax vaccine dilemma. My Mom was afraid the DoD would punish me, too, but they never did, successfully. Still, she worried because she knew what military leaders were capable of, as she saw so many others get punished. She witnessed so many lives and careers ruined. She told me she knew something was wrong when she worried more about me now than she did when I flew fighters like F-16s, A-10s, or F-117s. She confided that she only trusted Russ now, and that other friend, JR.

Baron Manfred von Richthofen, the infamous "Red Baron" and the leading ace of WWI, with eighty kills to his credit, once said, "One has to know one's flying partner." It was true. Though we didn't all initially know each other personally; we knew each other professionally. The same role models and heroes raised us. The same principles and ideals were our foundations. Richthofen also wrote, "The aggressive spirit is everything." It was this we also shared, Russ, JR, and me. The natural aggressive fighter pilot spirit carried us through it all as we fought the anthrax policy. We did it because we knew its foundations were dishonest. We knew that "no answers" to legitimate questions were unacceptable. JR provided mutual support in the form of many mentoring sessions, insisting that we shape and temper our attack on the anthrax vaccine policy. One such example came in the form of a book he sent that influenced our efforts.

JR gave us a copy of *The Hunters* by James Salter.[8] Salter based the book on his own experiences flying F-86 fighters during the Korean War. He penned the story of Captain Cleve Connell, a fighter pilot. Salter depicted Cleve's goal to chalk up five kills and become an "ace." But on one fateful mission, Cleve discovered the depths of his own courage and honor after a fight over the Yalu River. His goal had been to shoot down the top North Korean pilot, "Casey Jones," dubbed as such because he always led the "train" of enemy fighters into the sky. Cleve "only wanted his chance, nothing more." As a fighter pilot he had journeyed to the Korean peninsula for "a climax of victory, but in a way . . . He wanted more, to be above wanting it, to be independent of having to have it." Cleve's words lamented his desire for personal success as a fighter pilot versus wanting his professional aspirations to be about something more. He opined the same loneliness that Russ and I had felt at times, but also why being fighter pilots suited us:

> You lived and died alone, especially in fighters. Fighters. Somehow, despite everything, that word had not become sterile. You slipped into the hollow cockpit and strapped and plugged yourself into the machine. The canopy ground shut and sealed you off. . . . You were alone. At the end, there was no one you could touch. You could call out to them, as he had heard someone call out one day going down, a pitiful, pleading, "Oh, Jesus!" but they could not touch you.

Cleve's words reminded me of our fate, and the mentoring mutual support offered by JR. I hoped I had learned the lessons offered to me as I tried to internalize Cleve's story. In the book, Casey Jones, the legendary elusive enemy MIG fighter pilot, was later shot down in flames after a harrowing dogfight with Cleve. Cleve successfully prevailed and gunned down Casey Jones. He sent Casey Jones into the ground north of the Yalu from where he had come. Cleve and his wingman, Lieutenant Billy

Hunter, ran out of fuel that day, prosecuting the fight until it was done. They attempted a high altitude, fuel-conserving return to their airbase, their gas gauges on empty. Cleve flamed out, but made it to the field. He glided in to claim his rightful place, amidst an eager crowd in the middle of the runway, having shot down Casey Jones. Cleve's wingman glided to a crash, dying on impact one mile short and right of the field. When Cleve's gun film was reviewed to confirm the Casey Jones kill, it had not run. The camera was broken. No one could confirm the kill. Except on that day, Cleve thought fast, acting instinctively. Cleve credited the kill to Billy. The base personnel crowded Cleve's jet and questioned his victory, since Hunter had died and could not confirm the kill. Cleve decisively retorted, "I can confirm it. Hunter got him." Salter wrote: "Billy Hunter would have his day as a hero. . . . He had kept his pledge."

I got it. I understood JR's point. The fight was just about the fight and not about anything else, least of all our own personal victory. I tried to put this lesson to use and focus on the tangible objectives ahead that were achievable. I realized our professional goal of getting the hundreds of punished military members' records corrected had not yet been achieved. They had not done anything to us except take our jets away. In the years of the anthrax dilemma, they sentenced others to jail with felony convictions. They punished, fined, and dishonorably discharged our fellow troops. Our DoD leaders made those troops pay with their careers for not wanting a vaccine that came from a shuttered and cited manufacturer. They did so despite the vaccine being highly reactive, described by experts as unsatisfactory, and experimental—all in violation of the law. Our leaders were protecting the institution above the personal rights and freedoms guaranteed by law for their troops. Our job would be to prove and fix the injustices.

JR would later say that "Blind adherence to a falsity meant that the defining characteristic of the militaries we've fought over the past one hundred years had become a prerequisite for service in the US military." To us this was no longer about the anthrax vaccine at all. Instead, it was about what kind of a military we would be as we emerged on the other side of this fight. JR was every bit the fighter pilot behind the scenes that we were. He enjoyed the camaraderie of our intellectual formation flying as much as any fighter he flew or squadron where he had the honor to serve. He left a testimony to the importance of our work for me on the inside cover of *The Hunters*. JR's passage included an inscription from the Royal Air Force Memorial in Canterbury Cathedral. It captured our camaraderie, "We few, we happy few, we band of brothers." The phrase was borrowed from Shakespeare's early fifteenth-century tale *Henry V*. King Henry raised the spirits of his Englishmen before they battled the French against great odds.

Just as King Henry prevailed that day against a confident enemy, in time we too would prevail in the anthrax vaccine dilemma—in part due to our camaraderie, idealism, and virtue of cause that our adversaries lacked. We would use every ounce of the training we received, every tool in our "bag of tricks." As one of the Royal Air Force aviators that had come before us, Air Vice Marshall J. E. "Johnnie" Johnson, said, "The only proper defense is offense." We would go offensive against the vaccine and program because they violated the ideals we were trained to uphold. And as we implemented our plan we would, "See, decide, attack, reverse" as the "Black Devil of the Ukraine" and history's top ace, Major Erich Hartmann, had directed. We would do so methodically, executing a multi-azimuth attack using a variety of weapons. We would fight a war of attrition on the battlefields of our choosing. We would conduct informational guerrilla warfare. As the Red Baron instructed, "Fighter pilots have to rove the area allotted to them," and "when they spot an enemy they attack and shoot him down, anything else is rubbish." We had not chosen the fight, but we would fight, we would "rove the area allotted" to us. Anything else would be "rubbish." Our allotted area would be a non-militant intellectual realm because no other would be tolerated. We roved cautiously, but confidently, in our allotted cerebral lane.

I recalled my mother's cautions about our DoD leaders. She did not trust them anymore. I also recalled the words of my mother-in-law who warned me not to be vengeful. Both were right in cautioning me to fight an honorable fight. Both were with me as I made my choices. My wife, Louise, was my most ardent supporter, as was Russ's wife, Jane. Both had seen our commanders abandon us. She made me promise that if I was to "flameout" my own career and be grounded from flying fighters, that I must fight to stop others from facing the same fate. More importantly, she reminded me that our kids might serve someday, and that she would not allow it unless we made sure the DoD leadership was held to the same high standards they expected of all military members. Our mutual support had to fly far beyond our generation to help future troops, too.

The mutual support of my family and friends emboldened me and gave us strength. They were all we needed to sustain us in the quest behind us and with that which we knew laid ahead. Even JR's initial dutiful words about our requirement to follow orders quickly evolved as he too saw the deceit and the test. JR advised us to "go public" with our concerns, something of which the military leadership seemed fearful. JR knew we had exercised every option within our chain of command, all to no avail.

JR counseled again, having exhausted the remedies within the chain of command, "You know what you need to do, Buzz." Our mutual support morphed into an orchestrated multimedia campaign to counter the DoD's misinformation. We did something we were trained never to do by engaging in unconventional information warfare against our own DoD. We went to the press and executed a multi-azimuth attack against the military's illegal and dishonest anthrax vaccine mandate. I just hoped that we would not run out of fuel to sustain our marathonic tactics, like the ill-fated end to Cleve Connell and Billy Hunter's final mission.

MILE 7:
Information Warfare (1999 to 2000)

The best weapon in our information ammo box was the fact that our arguments came from the DoD's old records and documents. For example, congressional staffers provided access to subpoenaed DoD and FDA archives documenting "significant deviations" from FDA standards that were noted during inspections of the anthrax vaccine manufacturing facility.

The documents exposed information from as early as 1990: "Anthrax prod. fac. [production facility] was observed to be in a state of general disrepair in that there was: (A) Paint peeling from the walls (B) Exposed light fixtures (C) Cracked ceiling (D) Exposed raceways (E) Dirt & filth & dust on overhead pipes (F) Cluttered work space."[1]

From 1993: "There are insufficient personnel to assure compliance with current GMP [good manufacturing practices] regulations, e.g., failure to report changes in manufacturing, failure to maintain calibration records adequately, failure to adequately validate equipment used in the formulation or testing of product."[2]

From 1994: "There is no annual review of production batch records [for anthrax] . . . Raw [anthrax vaccine] material stored in an unapproved warehouse."[3]

From 1995: "The company did not inform FDA of the procedural and equipment change . . . facilities and equipment were not adequate . . . SOPs [standard operating procedures] did not exist for many procedures . . . SOPs were incomplete or incorrect . . . SOPs were not adhered to . . . Frequent contamination during vaccine manufacturing was documented but not investigated."[4]

From 1996: "The firm had not completed cleaning validation studies for routine cleaning procedures on multi-use equipment . . . Validation studies to demonstrate microbial retention and compatibility have not been conducted for sterilizing filters."[5]

From 1997: FDA admits the anthrax production facility not inspected because "it comes under military inspection."[6] The FDA issued a "Notice of Intent to Revoke" (NOIR) citation to Michigan Biologic Products Institute on March 11th, 1997[7] [FDA NOIR at Appendix H].

In the midst of these disturbing findings, we were grounded from flying duty in 1998. This was the very same time the FDA affirmed, "There are no written procedures, including specifications, for the examination, rejection, and disposition of Anthrax," and that "prior to August 1997, the [redacted] filters used for harvest of Anthrax vaccine were neither validated nor integrity tested." FDA found, "This filter is the only sterile filtration step in the Anthrax manufacturing process," and that "There is no written justification for redating lots of Anthrax vaccine that have expired."[8]

Despite the revocation notice, the DoD approved and mandated the anthrax vaccine for all troops in 1998, with the headline, "DoD Anthrax Vaccination Program Proceeds Well."[9] The DoD did so despite the FDA declaring, "The manufacturing process for anthrax vaccine is not validated" [FDA inspection report invalidating the manufacturing process at Appendix I].

DURING AN INSPECTION OF YOUR FIRM (I) (WE) OBSERVED:

1. The manufacturing process for the production of Anthrax Vaccine Adsorbed is not validated

FDA inspection report finding the anthrax vaccine manufacturing process "not validated."

"Calculations" was the first chapter in a timeless military strategy text, *The Art of War* by Sun Tzu. Written over two-thousand years ago, the manual's simple military lessons on strategy are still studied today. Sun Tzu's principles pragmatically apply in situations where victory is the goal, whether that is in war, in business, or in life. More importantly, a leader might also be able to avert disaster by applying the principles.

Sun Tzu revealed his vision on calculations: "A general who listens to my calculations, and uses them, will surely be victorious, keep him; a general who does not listen to my calculations, and does not use them, will surely be defeated, remove him."[10] Our goal was to encourage our leaders to listen to our calculations.

Sun Tzu's warnings reminded me of flying an air combat maneuvering sortie in an F-16. Those were my favorite training missions, whether over the Yellow Sea, the Gulf of Mexico, the Adriatic, or the Persian Gulf. But the key to living, flying, and fighting another day was to do it safely. We all had too many friends who lost their lives in hypoxia mishaps, midair collisions, controlled flight into the terrain, or gravitational-force (g-force) induced loss of consciousness. This was not an all-inclusive list of what could go wrong. No one wanted to become a statistic. I knew safety was the foundation in all things, both flying a safe jet and keeping good situational awareness while engaged in air-to-air combat. "Safety first" was more than just a slogan.

If safety was compromised, or you just had that horrible feeling when your hair stands up on the back of your neck, the ROE (rules of engagement) obligated you to call a "knock it off" over your radio to terminate the engagement. You commanded it three times, "knock it off, knock it off, knock it off!" Once you resolved the problem, or other adverse safety conditions, and regained sight of all flight members, the engagement could proceed, or a new engagement could be set up. The engagement always started anew with the carefully calculated radio command to "carry on"—or "Fight's on!"

That is the way we were trained, so when Russ and I discovered there were fundamental problems with part of our military operation when we were researching the anthrax vaccine, we instinctively did what we were trained to do. We called a "knock it off." Unfortunately, the anthrax vaccine and the mandatory program were unlike any military operation the troops in the trenches, or any soldier in uniform, had ever experienced. We had a simple safety issue, an operational program with legitimate safety and legality problems, which evolved ultimately into a catastrophic malfunction—an ethical dilemma. Instead of following the sage philosophies of Sun Tzu, or following our time-tested military training which dictated the requisite "knock it off," our military leaders tried to convince everyone it was okay to "carry on." It wasn't.

During those years, my friends and family always gave me grief about how I could take any obscure topic, like calculations and math, and turn it into a discussion of the larger lessons to be learned from the anthrax vaccine dilemma. It was comical at times. From Sun Tzu's calculations to safety paradigms, I could relate everything to the anthrax vaccine. It seemed so simple to me. A fighter pilot instructor's job is a constant calculation of the risk-benefit-reward equation. While flying at five-hundred plus knots in a perilous environment, bad decisions can be fatal. An instructor had to have complete situational awareness, or "SA." You might say to your student who had lost sight in a dogfight to "carry on," because you knew the situation was safe. No "knock it off" was required if the instructor had sufficient SA. The instructor was likely patiently getting in position for a gun kill to give his student a necessary lesson in aerial combat.

But, when it came to the anthrax vaccine, our leaders didn't have "SA." They did not have the answers to the questions raised by our research. All they had was boilerplate rhetoric from the Pentagon that conflicted with their own historic documents. They could not explain the stark dichotomy. At some level, the decision was made to arrogantly bully their way through. The leadership of the military was violating the very rule of engagement they had mandated for their troops, their officers, and their fighter pilots. They were taking a broken jet airborne. They were letting an unsafe situation develop. Military leaders ignored the required "knock it off." They failed to command it. Our training and instincts told Russ and me what to do: "Fight's on!"

Ultimately, JR was right. We had done everything we could at the unit and service level. Even going over the heads of our bosses, or "jumping the chain of command," by engaging lawmakers and judges proved ineffective. So, we used our training. Over the years, through our military studies, it was academic that you prosecute an attack where you perceive a weakness. Sun Tzu, Clausewitz, and other military philosophers and tacticians throughout history employed this strategy. The quotes from the World Wars' top aces—Richthofen, Johnson, Hartmann—reiterated and added aggressiveness to this tactic, "The aggressive spirit is everything . . . The only proper defense is offense . . . See, decide, attack, reverse." More recently, the US military studied the concept of attacking weaknesses by confronting our enemies through "asymmetric" strategies. The US Army War College defined such asymmetric attacks as, "acting, organizing, and thinking differently than opponents in order to maximize one's own advantages, exploit an opponent's weaknesses, attain the initiative, or gain greater freedom of action."[11]

We acted aggressively and instinctively in 1999. We engaged in an "asymmetric" attack against the anthrax vaccine. Our commanders specifically ordered us not to contact the press. They called pilots one by one into their offices to be interrogated by a group of lieutenant colonels. They tallied whether we were taking the shot and emphasized we were not to talk to the media about the events in our squadron. Going to the media seemed to be an Achilles heel ripe to exploit. What did we have to lose after being grounded from the duty they trained us to perform?

Asymmetrically—aggressively—we went offensive. We decided to counterattack by reversing on them and exploiting their fears by talking to the press. We contacted Russ's friend, Tom Donnelly. He was the father of an ailing Gulf War veteran, Michael, and had media contacts. Michael later passed away from amyotrophic lateral sclerosis (ALS), or Lou Gehrig's disease. His disease was directly attributable to his first Gulf War service exposures according to presumptive disability rulings. Mr. Donnelly gave us the name of a *Hartford Courant* reporter, Thomas "Dennie" Williams.

We later came to know "Dennie" all too well. Dennie and the *Hartford Courant* became our equivalent of the DoD's Armed Forces Press media machine. The *Courant* served as our conduit to the public to dispute the anthrax vaccine misinformation coming from the Pentagon. Talking to the press was a decision we did not take lightly. Honestly, we were anxious, nervous, and uncomfortable. We knew it could result in our immediate expulsion from the military. But we knew what we had to do. So did Dennie and the *Hartford Courant* in launching a genre of articles questioning the DoD's use of the anthrax vaccine during the first Gulf War and after.

One of the other pilots in our squadron, the 118th Flying Yankees, agreed to rendezvous with Dennie, Russ, and myself. Major Dom "Poss" Possemato was a great man who also knew something was not as it should be, or he never would have agreed to come. We all met at a trailer style diner in downtown East Hartford, Connecticut. Everyone was wearing overcoats. The image of four secret agents handing over state secrets was truer than we would ever know. We chose to talk publicly because we realized the chain of command, which was supposed to embody all the principles we learned over the years, had something to hide. This decision was not easy, but in hindsight it may have been the best protective measure and cover possible. By going offensive in the public realm, we asymmetrically put the DoD leadership on the defensive.

Dennie broke his story about our expulsions on January 15th, 1999.[12] The story hit the newswire and was picked up nationwide. Dennie wrote, "Eight veteran combat pilots from the Connecticut Air National Guard—almost a quarter of the 103rd Fighter Wing—are resigning to protest a requirement that they be inoculated with a controversial anthrax vaccine. . . . They stepped forward to publicize the plight of others nationwide because they are sworn as military officers to be responsible for the health and safety of those they supervise."

Dennie also interviewed our commander. Our boss candidly confirmed "eight pilots have resigned in response to the controversy. He said the ninth pilot's resignation was only partially due to his opposition to the vaccine." Dennie captured the commander's version of events: "All the pilots were shown what he considers convincing scientific evidence that the vaccine is safe and effective." The commander uncandidly added that he "urged anyone concerned about safety to show him any evidence that the anthrax vaccine was unsafe or ineffective, and no one could."

To our collective disappointment, the commander carefully acknowledged our principled intentions, which he was painfully aware of, yet he denied knowledge of the quantity, quality, and veracity of the research presented by Tiger Team Alpha. We knew he read our work, because he painstakingly cut and pasted our analysis, while omitting the scientific discrepancies, in his letter to our top Air National Guard general. He also threw in a little nugget that the pilots would "rather spend more time with their families, and at their more lucrative full-time jobs." This was the first indication of the DoD's public relations tactics to subtly disparage any who questioned their program or were reluctant to take part in their biological loyalty test. The false and malicious service-dodging dig helped motivate us and refill our fuel tanks early in the fight.

The *Hartford Courant* published a cartoon caricature of two pilots defensively ejecting from an aircraft due to their distrust of anything emanating from the Pentagon. Russ and I did not have this preconceived notion, but it was a witty cartoon illustrating the DoD's track record on trust. Years later, the same cartoonist illustrated the offensive versus defensive *Unyielding* cover.

Hartford Courant editorial cartoon by Bob Englehart—same illustrator as the *Unyielding* cover.

The cartoon put a smile on my face, but it wasn't the way I wanted my service memorialized. Instead, I reflected on the principles of war and Sun Tzu's wisdom. "Therefore, one who is skilled in warfare principles subdues the enemy without doing battle. . . . Do not attack an enemy that has the high ground; do not attack an enemy that has his back to a hill. . . . Calculate the situation, and then move." We did not attempt to "subdue" through battle, for our numbers and resources were infinitesimally small in comparison. Instead, we held the "high ground." The DoD did not attack, except with words of repute. I found it fascinating that apparently there was too much truth within our accusations, questions, and concerns for the military leadership to dare to highlight them, or us, with a counterattack. They let us continue to serve, while dishonorably jailing, fining, demoting, and discharging others. The intolerable inequity was only explainable by the reality that at the core of the issue the military institution knew it was breaking the law.

Instead, the charlatans responsible for the anthrax vaccine program battled on the turf we chose, while attempting to misrepresent our intentions and resolute position. Army Lieutenant General Ronald Blanck, by now the Army Surgeon General, wrote an opinion editorial for the entire military to read on every post, base, and ship across the world. He titled it, "Ignore the Paranoiacs, the Vaccine Is Safe." Blanck claimed, "The truth has to compete with suspicion, fear, and paranoia." Blanck was right, but who was telling the "truth," and who was exaggerating the threat to create "fear" and "paranoia"? Blanck's ridiculous rhetoric continued, "This is not unlike stories circulating that minorities' right to vote is about to be taken away, or that travelers are in danger of being drugged and having their kidneys stolen for transplants. Some people prefer to believe that their government is deliberately making service members sick."[13]

Wow. "Kidneys stolen for transplants"? Even the *Army Times* editors saw something was amiss with the inane rhetoric. The extreme analogy about the fear of organ theft was too much for any intelligent person or editor to endure. I called them up and asked why they would print something so disparaging without asking for evidence. The editors understood and listened over the years that followed. In time, we would earn their trust and deference. They eventually accepted our opinion editorials as well. They gave us a voice to counter the DoD's drivel.

This was information warfare. The DoD's only weapon was disparagement. It was the DoD that was guilty of distributing the misinformation. The DoD's public relations spokesperson responded to questions about losses from our fighter squadron. In a Pentagon press briefing in January 1999, he insisted, "It's safe and reliable . . . It works and has no side effects." Reporters queried about our fighter unit's losses. The DoD presenter claimed, "I think eight or nine people have resigned rather than take the shots . . . during exit interviews, six of the eight Connecticut pilots said anthrax was only one of many factors that entered into their decision to resign."

That was a misrepresentation, but perhaps the PR representative did not have the facts. All the pilots were grounded due to the anthrax vaccine mandate. We got that in writing. The DoD envoy continued, "Some may have found that the pressures of staying in the Air Guard and training were hard to balance with their family or business lives . . . Some may not have wanted to deploy to the gulf for personal reasons."[14] There it was again—the subtle disparagement—don't address the problems, but instead question your troop's commitment to serve their country. And General Blanck was not the lone detractor. General Charles Krulak, then the Marine Corps commandant, chided, "People are petrified that their penis is going to fall off, yet it is the safest vaccine ever given to American citizens."[15] Uninformed, less-than-professional comments such as these demonstrated the overt patterns of disparagement versus a professional discourse.

Looking back across the horizon of years adds perspective to what we witnessed. The crafty comments by senior DoD officials were most likely not innocent mistakes, nor were those of our commander. They were carefully orchestrated by experts in misinformation—by our own Pentagon public relations team. The innuendo that pilots could not handle the "pressures," "may have not wanted to deploy," were concentrating on their "business lives" or "their more lucrative full-time jobs" was a classic attempt to divert attention from the seriousness of what we had discovered—the DoD wasn't telling the full truth about the anthrax vaccine. The DoD's and our commander's spin was insulting, but at least at times it carried an element of professionally cloaked class. A less savvy attempt to shoot the bearers of bad news came from the manager of the DoD's Anthrax Vaccine Agency. Its June 1999 newsletter, titled "Right to the Point," stated:

> Much of the hand-wringing and bizarre allegations about the vaccine is coming from a vocal minority of people who think the "field" is where a farmer works and "Gortex" is one of the Power Rangers. Most of these folks have never spent a single moment in harm's way and have no appreciation of what that sacrifice means—and they openly resent the limited budget currently used to finance our nation's defense. . . .
>
> Unfortunately, those of us who actually have to fight our nation's wars cannot afford such childlike optimism about the world we live in. Other groups believe that we are spreading a virus through vaccinations that will weaken our military and allow the uprisal of the New World Order. I don't make this stuff up ladies and gents . . . it's too rich even for Hollywood. . . . See you on the high ground. . . . For those who have had to fight for it, freedom has a special flavor the protected will never know.[16]

Something was undoubtedly out of control when support personnel pontificated from the Pentagon with war rhetoric against front line operational troops when we are all supposed to be on the same team. Some of the materials I read from our Air Force's Air University addressed the tactics of deception. In fact, Russ attended the Air Force's Tactical Deception Officer Course during his active-duty years in order to incorporate tactical deception into his old unit's tactics. Therefore, we had an idea of what was going on. We always believed we were in trouble as an institution when such venomous tactics were employed internally against our own people.

My Air Force readings quoted a West Point professor who queried, "Is it unethical for a military officer to mislead the enemy?" He rhetorically replied, "We can confidently say no. The requirements of the practice of truth-telling extend only to fellow participants in the practice."[17] His point was unambiguous—you can deceive the enemy, but we do not deceive our own people. For some reason, our institution lost sight of this fundamental distinction, at least with respect to the anthrax vaccine program. Years later during COVID, such deceptive tactics would become commonplace, not only by the DoD, but also by the federal government against its own citizens.

Sun-Tzu wrote: "Warfare is the way of deception." But we do not conduct war against our own troops and certainly not against citizens. Sun Tzu saying "a military operation involves deception"[18] didn't mean that information warfare operations should be directed at our own military forces by our DoD leaders. Someone had undoubtedly missed the point of their ethics classes—they weren't supposed to deceive their own commanders, Congress, and media. But, as Sun Tzu said, we held the high ground. Mostly, the DoD tried to ignore us, without doing battle.

That summer I deployed for military reserve duty to an Air Force base in Indian Springs, Nevada. I told my kids I was going to "Area 51," and that I couldn't tell them what I was doing because it was "Top Secret." They loved that, and the fact that Dad was still in the Air Force, still trying to serve in any capacity. After being forced out of my flying job, I continued to work on special projects and deployed to Nellis Air Force Base in Las Vegas. In the summer of 2000 though, I found myself working, ironically, with military and government scientists on protective biological warfare detection systems at the Nellis satellite base in Indian Springs.

During my time at Indian Springs, I picked the brains of the worker-bee scientists from Army labs like Fort Detrick and Dugway Proving Ground about the anthrax vaccine dilemma. They humored me by listening, graciously telling me they agreed with me that the assumptions behind the vaccine program were baloney. I told my Indian Springs colleagues what we'd been up to, about our multi-azimuth information warfare attack against the anthrax vaccine program. Talking to military and civilian colleagues was one of the ways we continued our strategic targeting campaign, while at the same time quality control testing our tactics and arguments.

As Sun Tzu instructed, we sought to subdue "the enemy without doing battle." We therefore engaged the DoD with our minds, utilizing the tools the institution furnished. We also used the mechanisms our government instituted to prevent abuses of power. We did not think of our own DoD as an enemy in the classic military sense, but instead as an adversary we hoped to convert. This was our "strategy" or goal. According to Clausewitz, strategy is the "combination of individual engagements to attain the goal of the campaign or war."[19] Clausewitz taught that warfare had three main objectives: First, "to conquer." We accomplished this objective by waging an information war without getting punished. Second, "to take possession of his material." We accomplished this by acquiring the DoD's documents, emails, and utilizing the tools the DoD provided. Finally, "to gain public opinion."[20] Public opinion was an arena the DoD unquestionably cared about, so our goal was to expose its institutional arrogance and policy blunders in the public realm through information warfare.

An example of the arrogance in defending the policy despite obvious problems occurred during one of my earliest intellectual engagements with my ex-commander. One night at the club, after flying, he insisted, "Buzz, you'll never win." He said we would never stop the anthrax vaccine program because it was too far along and part of something larger. He came to our unit from the Pentagon and had the insight that this was indeed part of a larger campaign. This was also the most likely reason the DoD would not capitulate. But he also must not have read or been inspired by Clausewitz because, "even when the likelihood of success is against us, we must not think of our undertaking as unreasonable or impossible; for it is always reasonable, if we do not know of anything better to do, and if we make the best use of the few means at our disposal."[21]

Sun Tzu also advocated, "When doing battle, seek a quick victory. . . . No nation has ever benefited from protracted warfare." This applied to the DoD because we knew the only way to overcome them was to outlast them, to illuminate the DoD's documents and bad behaviors over the long-term. Our determined objective was to expose their modus operandi, their M.O., and modify their behavior. This was the promise we made to our families, as well as to ourselves and our colleagues, that would make our fight, our lost jobs, worthwhile.

We hoped the effort would negate future punishments and illnesses, as well as stop institutional deceits and ethical breakdowns by our DoD. Our campaign against the DoD would have to be unconventional. There was no familiar smell of aviation fuel, no smell of oxygen or tasting the rubber from our masks. The DoD stripped us of those familiarities, but they could not strip our knowledge of the USAF's tactics and training or stop our innovative academic methods.

We knew that there were many ways to attack an adversary. Fighter pilots in air-to-air combat try to destroy their opponent before they merge by using standoff weapons, such as radar-guided missiles. If this fails, they use heat-seeking missiles at closer ranges, or worst case, if entangled in close combat with another fighter, they resort to gunnery tactics. When attacking targets in air-to-ground combat, fighter pilots use standoff tactics through high altitude ingress routes that allow their aircraft to fly above both anti-aircraft artillery and surface-to-air missile threats. If lower altitude operations become inevitable due to weather, aircraft terrain-mask and use onboard countermeasures such as chaff to foil an enemy's radar or flares to confuse heat-seeking infrared missiles. If these layers of defense fail, a pilot can jink, or erratically maneuver, to create an acceptable miss distance from an enemy warhead or artillery gun projectile.

In an aerial clash, a pilot might try to offset so that the sun is at his back in order to surprise the adversary. In a ground attack, multiple aircraft may try to overwhelm an enemy air defense array of radars and their operators in order to create confusion. These were the tactics we were raised on, taught by those that came before us. In this fight, just as our tools were different, so were our tactics. There was no reason to use "standoff" tactics. Instead, we wanted the DoD to know what we were doing. We wanted answers to our questions. Therefore, we attacked head-on, went "offensive," using the DoD's own principles of war: Objective, Offensive, Mass, Economy of Force, Maneuver, Unity of Command, Security, Surprise, and Simplicity.[22]

The army scientists at Indian Springs seemed fascinated. I drew out our "multi-azimuth attack," which effectively incorporated the principles of war. Normally such actions by military officers against their own chain of command's policies would have been viewed as seditious or even mutinous. With the exception of disparaging slurs and accusations of treason, the DoD tried to ignore us, feigned ignorance, misleading the media and Congress about our efforts and intentions. This only refueled us, reaffirming that our attack was on target. For the Army scientists at Indian Springs, I drew the anthrax vaccine on the inside of a target triangle. The triangle was how we labeled a target on our attack maps in preparing for an F-117 strike against a high valued asset, an F-16 attack on a radar site, or how Russ and I might plan a two-ship close air support A-10 sortie against enemy antiaircraft artillery at the forward edge of the battle area.

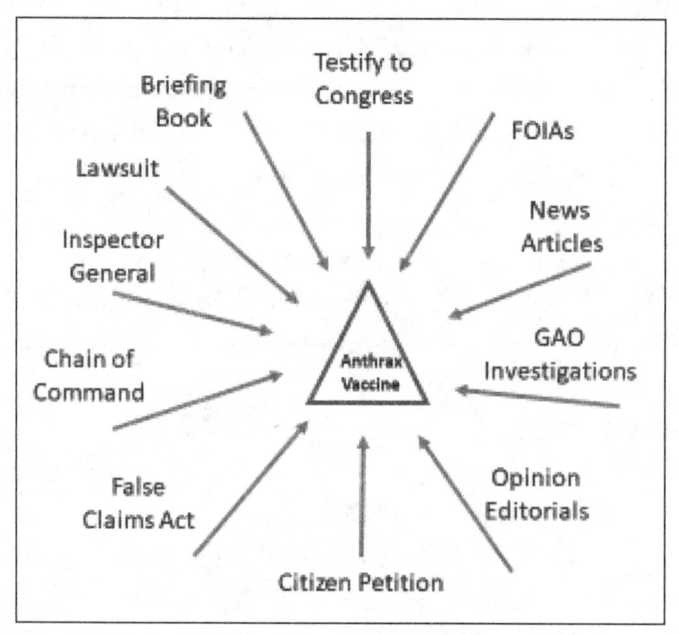

Depiction of multi-azimuth attack against the anthrax vaccine target triangle.

I explained in military terms how our plan of attack involved overwhelming the DoD across the long-term through multifaceted angles or azimuths of attack. Our "angles" and azimuths included writing opinion editorials, facilitating news articles by the media, assisting the GAO with some of their many critical investigations, testifying about our concerns to Congress, and filing complaints with the DoD inspector general. We relentlessly targeted the anthrax vaccine program from a multitude of directions and with a variety of intellectual munitions.

We used any and all governmental mechanisms available to have the program deemed illegal, with the ultimate goal of forcing the DoD to correct the records of the punished troops and care for the ill. I thought about the scientists' reactions to our multi-azimuth attack and our idealistic goals. They seemed astonished at the DoD's lack of reaction. I explained our tactics to any who would listen, from commanders to colleagues, legislative staffers to media contacts. Many thought it was fascinating, but no one could put their fingers on why the DoD ignored us. It didn't add up. Was it because the DoD didn't want to highlight Tiger Team Alpha's findings and their critical documents, which we used as our weapons? It mattered not. What mattered was that we executed the principles of war in this unconventional conflict right out of their textbooks.

We used the DoD's principles of war—the principle of "Offensive." Our offensive attack in the media expectantly should have resulted in a counterattack of prompt punishment by the DoD if not 100 percent true. A *USA Today* opinion editor published an opinion letter based primarily on his fascination with the DoD's inaction and my "persistence." The letter was titled "Anthrax for troops amounts to 'quick fix.'" It was published in *USA Today* on May 9th, 2000.[23] It ended by agreeing, "Protecting the troops is paramount, but having to take a 'quick fix' vaccine of questionable safety, effectiveness and doctrinal necessity is beyond the call of duty for any trooper." The media seemed intrigued that we directly engaged, but the DoD did not counter.

We used the DoD's principles of war—the principle of "Economy of Force." We were more economical in our expenditures than the DoD, allocating approximately one dollar for every $50,000 the DoD spent. JR masterminded a website for a few hundred dollars. In contrast, the DoD officials testified, "The Department has programmed $74 million between fiscal year 1999 and fiscal year 2005"[24] for anthrax vaccine education. Another colleague, Sergeant William Mangieri, a biological warfare specialist, decorated for bravery in the Gulf War with a Bronze Star, coordinated a magazine article. With Mangieri's help, we were allowed to show our analysis of FDA data on adverse reactions. We countered the DoD's misrepresentations:

> *The New American*, "Vexing Vaccine," by William F. Jasper, November 20th, 2000:[25]
> The Clinton administration is forcing an unproven, experimental anthrax vaccine on the military, risking our soldiers' health, destroying their morale, and damaging military preparedness. . . . Further refuting the Defense Department's propaganda and rhetoric, Major Russell Dingle maintains that the mounting evidence shows adverse reactions to be far more common than has been reported. . . . He points to the March 24, 1999 and the October 3, 2000 testimony by FDA officials . . . FDA reported a dramatic increase from 42 adverse reactions by March 1999 for 634,000 shots, to 1,561 reactions for nearly 2,000,000 shots given by October 2000 to about 500,000 service members receiving the partial vaccination series. "Put in context, this tripling of vaccinations caused an almost forty-fold increase in reported adverse reactions. These adverse reaction rates are almost fifty times the annual rate when compared to all vaccines administered to the population at large," he says. As a reference, Major Dingle points to the FDA's testimony on July 21, 1999 where FDA officials admitted that only 11,000 to 12,000 [adverse reaction] forms are filed annually nationwide for a population of 250 million Americans. . . . Major Rempfer adds: "The only thing that ever should have been forced out of the military in this tragedy is this dishonorable policy and questionable vaccine."

The statistical significance of the adverse reactions we analyzed for *The New American* was further validated beginning in 2003 with approximately seven hundred adverse reactions reported in six months. They were higher than any vaccine on record, at least up to that point.

One of the army scientists at Indian Springs asked me if I'd be willing to talk with another scientist they knew who was involved with the anthrax vaccine at Fort Detrick. I gladly accepted, as I was always anxious to meet or talk to someone who could answer our questions. I called the man, a lieutenant colonel in the Air Force, and he explained to me that he regularly briefed the issue on Capitol Hill. I asked if he was familiar with any of the regulatory problems, the known experimental use for inhalation anthrax, or some of the other areas Russ and I were researching concerning the never-finalized anthrax vaccine license or unapproved and illegal manufacturing alterations from the Gulf War period. He was not aware of the irregularities.

I debriefed the scientists I was working with at Indian Springs. They were not surprised. They explained that the people in top bureaucratic positions often only know what they are told to brief. They explained the differences between the worker-bee scientists in the trenches and the more politically oriented scientists in higher positions. We saw these problems with the DoD's anthrax vaccine promoters to be illustrative of the same cultural problems at the FDA, and in all probability within the entire US government bureaucracy—the briefers often aren't the experts.

The GAO specifically documented a root cause of the expertise deficit in one of their investigations. GAO stated, "Until 1993, FDA inspectors did not inspect the MDPH [Michigan Department of Public Health] facility where the anthrax vaccine was made. According to FDA, access was not granted because its inspectors had not been vaccinated against anthrax." Though FDA would later try to show proof of inspections based on peering from behind locked doors, through observation ports, it was evident that a lack of proper inspecting occurred for years. This was in direct conflict with FDA's charter to guarantee that "vaccine production is very highly regulated to ensure that the products are of consistent quality and safe and effective for the purpose(s) for which regulatory approval was granted."[26] Unsurprisingly, the DoD took full advantage of the regulatory breakdown that occurred at the FDA with respect to oversight of the anthrax vaccine. These failures and abuses of discretion most likely contributed to the FDA tacitly waving the magic wand of authorization without the proper investigational new drug approvals or waivers in place, and despite the fact the manufacturer remained non-validated.

Despite the DoD's best attempt to steer us off target, we attacked only the anthrax vaccine—no other vaccines. This posed a problem for the DoD and helped to explain their unwillingness to engage us. A retired US Army officer and Boston University professor wrote about the DoD's dilemma in an article called "Bad Medicine for Biological Terror"[27] for *Orbis* in the spring of 2000. Excerpts from US Army, retired, Colonel Andrew Bacevich's article:

> Pentagon officials blame the Internet and a malicious campaign of disinformation for misleading 'our youngsters.'[28] . . . Yet the most vocal and impassioned critics of the vaccination policy are anything but kids. Indeed, a disproportionate number of refusniks are pilots. Many are combat veterans, field grade officers in their 30s and 40s, currently serving in the Air Force Reserve or Air Guard. From the Pentagon's perspective, the specific character of opposition complicates matters considerably. To the extent that those questioning the vaccination regime are mature, well-educated professionals who when not flying military jets are engaged in responsible civilian careers, it becomes difficult to dismiss opponents as naïve, misinformed, and easily manipulated.

JR knew Colonel Bacevich and arranged a meeting. More than anything, I wanted to thank the professor for his thoughtful article, which I shared with my new scientist friends that summer. I was grateful for the continued service in "Area 51," and the opportunity to check my biases with these new biodefense insider friends. The experience affirmed we were on target.

I remembered running around the airfield in Indian Springs, watching the F-16s and other fighter aircraft do practice approaches. I was disappointed that I was not flying anymore, but I harbored no regrets over why. I was most disappointed over the confidence-shattering reasons we were forced out. They were not based on the principles of those who raised us as military officers and fighter pilots. We were forced out over what we believe to be a patently illegal order, buttressed only by misinformation from our leaders. Paradoxically, we had not been punished. At least thoughtful minds like Bacevich's made people like those equally thoughtful scientists think.

We continued to engage the misinformation with a multi-azimuth, information-warfare campaign through dozens of TV interviews and opinion editorials in order to expose the illegalities. We believed it was less than honorable to intimidate our troops into taking off the uniform. It was gratifying to know the scientists and Bacevich agreed. When intimidation didn't work, the DoD jailed troops in lieu of the requirement to "resurvey its own judgments."[29]

It was unacceptable, so we continued our information-warfare marathon. Though the adversary did not engage us, they could not ignore questions from academic authorities and media outlets. Nor could they ignore inquiries by civilian leaders or duly elected authorities. Our tenacious legislative leaders and their staff members would have to continue the information war for the truth in yet another prong of our coordinated multi-azimuth attack—Congress.

MILE 8:
Unproven Force Protection (1999 to 2000)

The Founding Fathers' sacrifices gave us the right to speak our mind. That right is protected under the law, specifically by the First Amendment to the Constitution:

> Congress shall make no law respecting an establishment of religion, or prohibiting the free exercise thereof; or abridging the freedom of speech, or of the press; or the right of the people peaceably to assemble, and to petition the Government for a redress of grievances.[1]

Russ and I spoke publicly and told the truth. We backed it up with facts. The fact was that the DoD wasn't telling the truth about the vaccine. The DoD systematically rewrote the anthrax vaccine's history and their own critiques of its unsatisfactory nature. They conveniently forgot what they'd written in their own documents. They were not telling the truth to our commanders, troops, or Congress. This was unacceptable to us and hopefully to all Americans.

It was also disturbing that our military was trying to create the impression our military members were protected from the threat of anthrax with this vaccine when their internal assessments of the vaccine's efficacy and indemnifications showed otherwise [Indemnification at Appendix J]. It was this dichotomy that we objected to—between what the DoD was saying versus their previous objective analysis of the vaccine's experimental and inadequate nature. For us, the debate became one of principle. As we learned more about the constitutional processes that were intended to prevent such abuses of power, we became more encouraged.

In line with our form of government, the Founding Fathers empowered three separate branches of government to prevent such abuses. One of the Founders, John Adams, said, "A Legislative, an Executive and a Judicial Power, comprehend the whole of what is meant and understood by Government. It is by balancing each of these Powers against the other two, that the effort in human nature towards Tyranny, can alone be checked and restrained and any degree of Freedom preserved in the Constitution."[2] Thinking about checks and balances brought to mind the day that Russ and I both received a letter from the Congress of the United States, inviting us to testify about our anthrax vaccine mandate experiences.

The House Government Reform Committee's National Security, Veterans Affairs, and International Relations Subcommittee convened the first in a series of hearings on the anthrax vaccine program. This was the same committee whose staffers were trying to look out for us. They specifically wrote and requested, "In your testimony, please discuss your experiences with the AVIP, including a description of any information on the vaccine provided to you by the Connecticut Air National Guard (CTANG) or any other military source, and a chronology of any discussions or meetings within your unit with your superiors regarding the AVIP."

The premise of this hearing, and those that would follow, was simple. The anthrax vaccine was still being studied as a potential causative or contributing factor in Gulf War veterans' illnesses. Congress prudently measured the program against this standard and held that "any expanded use of the same vaccine should be undertaken only with the greatest care and only to the extent necessary."[3] This was actually established in public law during the previous congressional session, the 105th Congress. Of course, we also knew this from the senate staff report titled *Is Military Research Hazardous to Veterans' Health?* The report maintained that anthrax vaccine was a possible cause of Gulf War Illness according to the Army Surgeon General.[4] This was the same general officer that thought anthrax vaccine concerns had something to do with organ theft. Lieutenant General Ron Blanck also testified that day, March 24th, 1999.

In our testimonies, we followed the committee's tasking by submitting a chronology. We were careful to not focus on our own military unit's bad behaviors and their irresponsible accusations of treason against us. Instead, we explained the differences between what we discovered in our research for Tiger Team Alpha versus what the DoD was telling their commanders, troops, the media, and our Congress. The contrasts exemplified the program's inherent lack of integrity. Another difference included the doctrinal departure that caused me to question the mandate in the first place. I called it a dangerous "façade of force protection."

We exercised our First Amendment rights to educate the public, the media, and Congress. Every good military operation requires multiple avenues of attack. So, in the spirit of keeping the public-awareness aspect of our multi-azimuth efforts alive, we submitted a simultaneous article for the opinion section of the *Baltimore Sun* newspaper. It was published on the Sunday preceding the hearing. The article was a precursor to our hearing testimony. The *Baltimore Sun* titled it, "Anthrax vaccine offers no sure cure in warfare" with the byline "A fighter pilot questions the effectiveness of the military's immunization program and calls for an independent review."[5] We intended the essay to spur a policy debate, one we were gravely concerned was missing in the launch of the vaccine program. The article, like my congressional testimony, focused on doctrine. It asked readers to contemplate the "dangerous façade of protection by using a vaccine that might not protect against all strains of anthrax," and how "it might be wise to step back and determine if a new doctrine that mixes biology and bullets is prudent."

The hearing began with questions from the chair of the committee resembling our own. Representative Christopher Shays opened with: "Why now? Why this vaccine? Why a mandatory program? And why would active-duty, Reserve, and National Guard personnel jeopardize their military careers, and even their liberty, rather than take the vaccine?" He went on: "the missing element of the mandatory anthrax vaccine program is trust." As Russ and I approached the testimony table, we passed the Air Force surgeon general, Lieutenant General Charles Roadman. Russ politely greeted him, only to receive a glare and scowl in return. As we took our seats Russ leaned over to me and asked, "Did you see that look?" I leaned back and said, "Roger that—he must have gotten a sneak peek at our testimonies."

Russ and I listened to this gentleman's own testimony and to the questions and answers that followed from committee members. General Roadman actually said the anthrax vaccine was required "body armor." It was amazing how far the rhetoric about the vaccine evolved since 1985 when the DoD sought unsuccessfully to get the unsatisfactory vaccine replaced. Then, it would not "safely and effectively protect military personnel against exposure to this hazardous bacterial agent." It was "highly reactogenic, requires multiple boosters to maintain immunity and may not be protective against all strains of the anthrax." By 1996, it was "experimental" and even the DoD didn't consider it fully or properly approved. In 1998, the anthrax vaccine was so hazardous it required indemnification. Yet by March 24th, 1999, at the hearing, the vaccine was elevated to "body armor." It was crystal clear to us that we were there for all the right reasons.

General Blanck gave his testimony and firmly held belief that "the threat is real. I believe the threat is greater today than it was two years ago, five years ago, ten years . . . we have a way to counter the threat and to offer protection to the men and women in uniform and it is the fully FDA approved anthrax vaccine."[6] Blanck offered similar testimony to a House Armed Services Subcommittee by stating, "The bottom line is very, very clear: If we are attacked with this agent, and we have a force that is vaccinated and protected, our soldiers, sailors, airmen, and marines will largely survive. If they are not vaccinated, they will inevitably die."[7] Blanck wasn't alone. Another official testified, "The Defense Department received unequivocal evidence in 1997 that Iraq had weaponized anthrax . . . and the anthrax vaccine is as necessary for force protection as a flak jacket or a helmet. . . . If you don't get inoculated, you're going to die."[8]

The problem was General Blanck's statements conflicted with General Accounting Office assessments. Later hearings included the independent

GAO analysis that appeared to agree with our trepidation of accepting the threat hype at face value. The GAO reported, "The nature and magnitude of the military threat of biological warfare (BW) has not changed since 1990, both in terms of the number of countries suspected of developing BW capability, the types of BW agents they possess, and their ability to weaponize and deliver those BW agents." [9] The GAO consistently reiterated the conclusion that "the nature and magnitude of the anthrax threat has been stable since 1990 and has not changed materially in terms of the number of countries suspected of developing a BW capability, the types of biological agents they possess, or their ability to weaponize and deliver such agents." [10]

Blanck's claim that the unvaccinated would surely die was also proved false in the years ahead. Post-exposure treatments with antibiotics were fully effective in the aftermath of the 2001 anthrax letter attacks for those who received prompt treatment. Blanck's testimony conflicted with his own concerns about safety from the 1994 Senate Veterans' Affairs Committee Staff Report, which acknowledged a possible link between the anthrax vaccine and Gulf War Illness:[11]

> Although anthrax vaccine had been considered approved prior to the Persian Gulf War, it was rarely used. Therefore, its safety, particularly when given to thousands of soldiers in conjunction with other vaccines, is not well established. Anthrax vaccine should continue to be considered as a potential cause for undiagnosed illnesses in Persian Gulf military personnel. . . . Records of anthrax vaccinations are not suitable to evaluate safety.

Most importantly, the senate report effectiveness data exposed the mandate's illegality:[12]

> The vaccine's effectiveness against inhaled anthrax is unknown. Unfortunately, when anthrax is used as a biological weapon, it is likely to be aerosolized and thus inhaled. Therefore, the efficacy of the vaccine against biological warfare is unknown. . . . The vaccine should

> therefore be considered investigational when used as a protection against biological warfare.

After listening to the generals, I gave the committee our chronology in the form of written testimony for the Congressional Record.[13] We didn't want to get into the disappointing details of how our leaders in the Air National Guard reacted to our team's research, but we did feel it was important to put this chronology on the public record. We voiced our larger concerns through our verbal testimony to our representatives, which was aired by C-SPAN.

I testified, "Out of respect for the military and my chain of command, I am not here today in uniform. My professional dissent on this policy brings me to Congress only after attempting to resolve my concerns through my chain of command. I believe it is my duty to continue to speak out against the dangerous doctrinal precedents and questionable effectiveness presented by the anthrax policy." I asked, "What suddenly mandates the use of this outdated vaccine? . . . Why force us to take a vaccine that was not intended to combat inhalation exposure to anthrax? . . . And finally, could it be dangerous to erroneously imply to our top military and civilian leaders that we can withstand a biological weapons attack through defensive posturing? Why have we prudently avoided this path for the preceding three decades?"[14, 15]

Russ and I later obtained a copy of the DoD's National Guard Bureau Office of Policy and Liaison summary of the hearing. The DoD liaison's synopsis referred to the day as, "A fairly contentious hearing where the DoD was clearly put on the defensive. The Air National Guard representatives, dressed in civilian clothing, stated up front that they were there as civilians, loved the ANG, but could not support the anthrax vaccination program. . . . Although they took great exception with what DoD had testified to, they represented themselves well and were the star attraction."[16] The DoD liaisons transcribed several of the same snippets we wrote down in our notes. Perhaps they were also wary of the DoD testimonials emphasizing the inevitability of an anthrax attack? The DoD liaison's notes excerpted quotes from Lieutenant General Roadman testifying, "Not a matter of if we will be exposed to anthrax. It's a matter of when."

The internal DoD notes continued by quoting the committee chair's question: "Why does DoD talk so glowingly about the manufacturer?" Dr. Sue Bailey, the Assistant Secretary of Defense for Health Affairs, responded, "Yes, there were a number of deficiencies, but not significant enough to close it down. Improvements and facility renovations have been made." But in fact, the FDA had "not validated" the manufacturer, causing them to "voluntarily" close the plant. Over the years, we witnessed the DoD attempts to quibble about this fact repeatedly. In a later congressional hearing, FDA officials admitted to Representative Shays that the manufacturer would not be allowed to continue to operate due to the documented deficiencies. The DoD liaison's notes continued by identifying the real core question: "Is DoD paying for the renovations?" The response by Dr. Bailey included, "Not in my purview. I'll get back to you." In fact, the DoD paid sixteen million dollars for the initial anthrax vaccine plant renovations, with many millions more added in the years that followed. Surprisingly, shortly thereafter, Dr. Bailey retired from the Pentagon and joined the anthrax vaccine manufacturer's board of directors.[17]

The hearings served as an opportunity for multiple DoD officials to read the boilerplate script in their testimonials to Congress.[18] One testimony by Dr. Bailey effectively admitted that human efficacy data for the anthrax vaccine didn't exist, but left out the fact that the law required such data. Instead, she repeated the rhetoric, "Conducting lethal challenge studies in humans is considered unethical and . . . directly determining the efficacy of the vaccine in humans against aerosol exposure to anthrax spores is not possible." She specifically mentioned that animal tests "demonstrated the efficacy of the FDA-licensed anthrax vaccine against inhalation anthrax," but later FDA-Army workshops confirmed correlates of immunity between humans and animals were only theoretical. Either way, animal efficacy was not permitted under the law at that time.[19]

Russ and I thought long and hard about a final excerpt in the DoD notes from the testimony about giving up our "rights" when we joined the military. To the contrary, we absolutely agreed that we gave up certain personal freedoms, but we gave up no rights. We were in the military to defend those rights for all Americans. And military leaders, such as generals Blanck and Roadman, were the men charged with the responsibility to defend and protect one area of those rights for us—our military medical rights under the law. We trusted these leaders with the stewardship of those rights. These senior officers, as well as Dr. Bailey, were responsible for ensuring we were given only satisfactory, non-experimental, fully approved vaccines and drugs that came from "validated" manufacturers. Instead, both the USAF and US Army surgeons general hyped the inevitability of the threat under the false imperative of good order and discipline to justify the program. That was the gimmick—we saw through the smoke.

Worse yet, they came to Congress dressed in all the military trappings we'd left at home. They came to indoctrinate our elected officials in the belief that troops must take their shots; that they must follow questionable orders or get out of the military. But what Russ and I, and millions of other military members, really needed in order to take their vaccines was the truth—something missing from their congressional education effort. Our experience with the anthrax vaccine program was that the truth was the missing element, as Representative Shays expressed in his opening comments. General Blanck testified that vaccines were a term of employment. That made us think about the dozens of shots the military ordered us to take over the years. So we agreed, but insisted that vaccine programs must be run competently and legally. Our troops deserved no less.

I shared a story with Russ about my return from Korea in the early 1990s. The military lost my shot records. I was ordered to resubmit to all of the required immunizations. Years later I discovered the military recovered the original dates for my electronic immunization record, and therefore only updated the yellow fever vaccine on that date [career shot record at Appendix K]. My wife witnessed the aftermath of my retaking all those shots back in 1991 before we took a long weekend drive. I couldn't move or turn my neck. She had to drive. At the time,

I was unaware of the risks and my rights. I ignorantly and obediently submitted to all the shots the military mandated, again, which was silly since the medical officers should have known I already received them as a combat-ready F-16 pilot. You could not be designated combat ready without them. Fortunately, my wife helped me through the adverse reaction—from doing my duty.

Suffice it to say, I absolutely satisfied General Blanck's terms of employment and proved I was not anti-vaccination. We were not even anti-mandate, as our decades of following orders and military mandates proved. We were simply anti-illegal-mandate. This was a standard everyone in or out of uniform should have applauded. My immunization record documented my adherence to vaccination historically and for many years to come. Nevertheless, accepting poor recordkeeping was not a term of employment—nor was blindly following illegal orders. Instead, we required medical competency. Yet, General Blanck's alternative prescription was: "Those that refuse should be separated where they can go and be free to make choices." That was too convenient of a means to rid the military of critical thinkers who contested illegal mandates.

At the end of the day, Russ and I were honored to do our duty in response to the congressional committee's request. Upon returning home, I had a message from a USAF staff judge advocate who tracked our situation and helped me prepare for the hearing:

> Buzz: First, let me say how proud I am to wear the same uniform you do. Congratulations on your effort. Second, take a moment and reflect on what you and the others just did. You did "participatory democracy"—that is, the right of the American people to seek redress from their elected officials. You reminded us all that rank is not a birthright and that "it ain't true just because the brass says so." For all the spin, sham, and drudgery, America and her government are still pretty wonderful things. That's because citizens, like you, feel they have the God-given right to stand up and yell "bullshit" honestly and openly. It is the protection of that very right that motivated me to put on the

> blue—and it's what keeps me in uniform to this day. Congratulations, Tom. I am honored you let me participate in a very small way.

Even one of our Connecticut Air National Guard commanders wrote to us: "Buzz, Russ, Great job in DC yesterday. Thanks for all your hard work and dedication. It is appreciated and respected by more people than you might sometimes think. Hang in there guys!" Encouraged by the notes from our legal colleagues and former commanders, we hoped we were making a difference. That was our first experience with "checks and balances," but it was not our last.

Dingle and Rempfer testifying to the House Government Reform Committee.

My academic imprinting on military doctrine at the Air Force Academy inspired my testimony to Congress based on the previous decades and past leaders that understood the biological weapons "taboo." The logical extension is that messing around with petri dishes for defensive biological research would at a minimum be unwise, and should be pursued with the greatest of caution. I first learned about these concepts while reading one of my old academy texts and the editorial work of a professor emeritus from the Air Force Academy, Brigadier General Malham Wakin. General Wakin articulated the doctrinal concept referred to as the Chemical and Biological Warfare Taboo ("CB Taboo").[20] The military appeared to have forgotten the taboo's cautions.

As we researched and realized the anthrax vaccine program was so poorly thought out, I hypothesized that the doctrinal shift of implying we could vaccinate against biological warfare was equally flawed and likely not well thought out either. Even if we could temporarily protect against a certain biological threat, it was academic to envision our troops being potentially more vulnerable if US enemies targeted them with diseases other than those against which they were vaccinated. The case could certainly be made that the anthrax vaccine could be seen as escalatory in nature, even if well-intentioned. I tried to engage our DoD officials early on to discuss these doctrine, strategy, and policy concerns, but received the classic "stay in your lane" blow off.

Governments around the world, the United States included, codified the CB taboo. Leaders of the international community signed on to the Hague Convention as early as 1899, with the express purpose of establishing laws of war to prohibit the use of projectiles to disperse asphyxiating gases or chemical weapons.[21] Following World War I, the Geneva Protocol of 1925 specifically added bacteriological materials to the list of weapons outlawed in warfare.[22] The US complied with the spirit of these protocols during World War II, though the Cold-War era that followed resulted in significant escalation of biological research, both offensive and defensive. This period culminated with a presidential directive in 1969 by Richard Nixon to stem biological warfare and research escalation. The President's National Security Decision Memorandum 35 (NSDM-35) specifically directed the destruction of all offensive biological weapon stockpiles.[23]

By 1970, with NSDM-44, President Nixon reaffirmed commitments prohibiting offensive use of biological weapons and committed to exclusively defense-oriented research, such as development of immunizations as a prophylaxis against biological toxins.[24] By 1972, the United States formally signed the Convention on the Prohibition of Development, Production, and Stockpiling of Biological Weapons. The convention memorialized the US presidential national security memoranda into international agreements. The treaty, ratified by 1975, prohibited stockpiling of biological agents intended for any purpose other than peaceful prophylactic development. By the turn of the twenty-first century, the United States in particular expressed concerns about compliance and verification elements of the Biological and Toxin Weapons Convention

(BTWC). The United States later withdrew from the protocol pending future agreements related to verification and monitoring in subsequent BTWCs.[25] Sadly, we would have to relearn the same lessons that resulted in conventions and protocols in the years to come.

Around the time of our congressional testimonies, a *PBS NewsHour* camera operator commented to me that the anthrax vaccine policy seemed ludicrous at its core. In an off-air conversation he asked me, "Why are they even going there? What's changed?" The DoD's answer was, "It's the threat." The "threat is real" rhetoric justified it all. Was that the threat that the GAO confirmed had not changed in twenty years? The one fact that was indisputable was the DoD was the owner of the intelligence that validated whether or not the threat was real and therefore the issue was difficult to debate.

The previously accepted US policy and doctrinal stance did not legitimize biological warfare by implying we could defend against its dynamic nature. It was critically important to discuss and—at a minimum—should have been open to debate. It became obvious they didn't want to listen to, or answer, the questions on policy or doctrine. Failures to listen, a lack of a willingness to learn and reflect, not following the rules or instructions, and nonadherence to the law, all became repetitive themes. That's why citizens and service members objected. That's what people do when their legitimate concerns are ignored by dismissive people in power.

We needed discussion about the doctrinal departure and the shift away from the previous CB Taboo. Brainstorming and dialogue might have encouraged senior leaders to envision the dangers of defensively posturing through vaccines versus a more comprehensive, non-escalatory use of external protective garments, detection systems, and non-invasive medical therapies—the essence of my 1999 congressional testimony. It was professionally and painfully evident to us in 1998 that the DoD would entertain no debate, perhaps because their policy was a poorly thought-out quick fix. This reoccurring theme defined the DoD's modus operandi when debating anthrax vaccine. It offered a perspective on why the unanswered questions were ignored or squashed.

Beyond Congress, Russ and I continued to debate the DoD on doctrine through the media in both opinion articles and public commentary. Exercising our constitutional First Amendment right of freedom of the press led to an essay titled *Biology and Bullets, a Dangerous Doctrinal Mix*. I published it later as an opinion article with the *Hartford Courant* with the title "Anthrax Vaccine is a Force Protection Façade."[26] The dubious policy shift to vaccinate against biological weapons signaled a troubling tacit acceptance of weapons outlawed by international treaties. Coupled with an apparent lack of a coherent deterrence strategy, this trend undermined decades of international restraint in the biological warfare arena. It also begged the question if inevitable neglect was possible regarding comprehensive traditional protections using detection devices, external garments, and post-exposure antibiotics. Our First Amendment efforts hoped that some think tank in the bowels of the Pentagon, government, or civilian realm might actually stop and say, "Yeah we need to be careful with this stuff so we don't end up causing a pandemic someday." Unfortunately, no one apparently read our essays or heeded the warnings.

These were serious issues that we never stopped discussing. In the late 1990s, I cornered the Air Force's top doctrine expert with these same questions. The expert major general was the commander at the United States Air Force Doctrine Center. His reaction: the anthrax vaccine was not his issue. As the years went by, willful blindness seemed to be an all-too-common and convenient response. When asked for his views on Brigadier General Wakin's book and the concept of the CB Taboo, he said he "hadn't thought about it." His confidence-shaking responses were reminiscent of the DoD briefer in 1997 responding to a reporter's question about "a biological arms race where you've got moves, countermoves on each side." Surprisingly, the Pentagon briefer responded, "I don't know. It could be, but I'm not prepared to answer that."[27]

I politely reminded the general that his office published pamphlets on doctrine, and respectfully affirmed that biodefense strategy and the anthrax vaccine's inclusion in those policies were his issues. After all, "Doctrine" was part of his office's name. He asked me to share my concerns with an aide in his office. I did, but I never heard from him or the aide again—another all-too-common trend. Just as the DoD doctors weren't practicing medicine, and the DoD lawyers weren't practicing law in order to protect our troops' health rights, the DoD doctrinal experts appeared to be equally unprepared to answer serious policy questions. People were looking the other way, both from their professional responsibilities to stop bad and illegal medicine, and later with the ongoing punishments and illnesses resulting from the breakdown.

The doctrinal departure and willful blindness we witnessed motivated me to accept another invitation to testify to the full Committee on Government Reform in Washington, DC. Russ and JR helped me prepare my testimony for the hearing on October 12th, 1999, titled *Defense Vaccines: Force Protection or False Security?* It was a team effort no matter who was on the pointy end of the formation's spear. When I arrived in DC, I proceeded by subway over to Capitol Hill. I was wearing my uniform this time, but didn't feel the least bit out of place since there's every imaginable uniform walking around the capitol. Two weeks earlier, I'd witnessed our fellow servicemen and women in civilian clothing, again trying to tell the truth about what they'd discovered regarding the safety and readiness risks associated with the anthrax vaccine. Shedding the uniform was meant by them as a sign of respect toward the military institution that raised each of us. But after observing the DoD officials in uniform, rewriting the history of the events and the vaccine, this time I decided to go head-to-head with them in full uniform.

Too often the uniform had a subtle, intimidating effect on congressional staffers, most without military experience, in the post-Gulf War, pro-military environment. Regardless of the extent of these "effects," I donned my uniform and stripped this advantage away from our adversaries in line with the latest parlance of military planners—for "effects-based targeting." Such targeting processes endeavor to change the enemy's behaviors and compel them to comply with our will. Our military doctrine gurus would add that, "Effects-based targeting is distinguished by the ability to generate the type and extent of effects necessary to create outcomes that facilitate the realization of the commander's objectives."[28] Rather than capitulate by not wearing my military garb, this time I shed my civilian clothes instead, and proudly put on the uniform that found me in this ordeal in the first place. Being insubordinate was not the objective. By testifying truthfully in uniform to the legislative branch of government, my goal was to present myself as professionally as possible, while targeting the anthrax vaccine program's dishonor with both the precision of truthful testimony and the armor of military attire.

Once at the Rayburn Office Building, which housed many of the US congressional representatives' offices and staffs, I reported to the congressional cafeteria where I met JR and one of the other service members invited to testify. His name was Sonnie Bates, a major in the USAF. Sonnie was also in uniform. Sonnie served as a C-5 pilot at Dover Air Force Base in Delaware. Sonnie did not submit to the vaccine after witnessing the ailments of the troops that surrounded him in his squadron. My spunky wife and kids even attended an earlier rally at his base, carrying a sign that said, "My husband will take a bullet for America but not the Anthrax vaccine." A reporter quoted her saying, "I am here in support of my husband and to show support for Maj. Bates."[29] Our supportive families made the otherwise lonely duty survivable.

Before the hearing, Sonnie and I engaged in a quick "skull session" to review the themes of each other's testimonies. We jokingly referred to our meeting as a "skull session" because we knew that was part of the problematic nature of the DoD's anthrax vaccine program. The skull session term was slang in the Pentagon to describe a briefing or quiz session of staff officers and senior officials when they needed to get ready for a hearing on Capitol Hill, or for other high-level meetings. The staff would fill the "skull" of the senior official with all the sound bites and briefing bullets that his or her brain could hold. We looked at this process as a part of the problem that led to the anthrax vaccine mess. Basically, senior officials never really knew the facts, and instead they simply repeated the boilerplate script provided to them in these "skull sessions." As a result, the briefing bullets that ignored the truth about the seldom-used, known-to-be inadequate anthrax vaccine became a mantra within the DoD and on Capitol Hill. Tiger Team Alpha's mission was to fill our own skulls with the facts about the illegalities—the full truth.

We left the cafeteria and went upstairs. Approaching the committee's hearing room, an officer came up to me wearing an army uniform and "full bird" colonel rank. The colonel handed me a styrofoam coozie that you'd use to keep your favorite beverage cold. The colonel said, "Look what the Pentagon is spending the taxpayers' money on, Buzz!" We only had a moment to meet, but it was nice to see that this colonel wasn't afraid to call out the wasteful propaganda.

The coozie was part of the "education campaign." It said "Anthrax Kills—Vaccination Protects." There was even a skull and bones. The coozie included website addresses and a toll-free number. I was face to face with the DoD's tools of anthrax vaccine propaganda—paid for by the US taxpayer. The DoD purchased little cardboard suitcases full of these embarrassing trinkets, such as pens, mouse pads, pocket knives, flashlights, posters, and even credit-card-size wallet inserts just in case you suddenly needed to call the DoD's anthrax vaccine hotline.

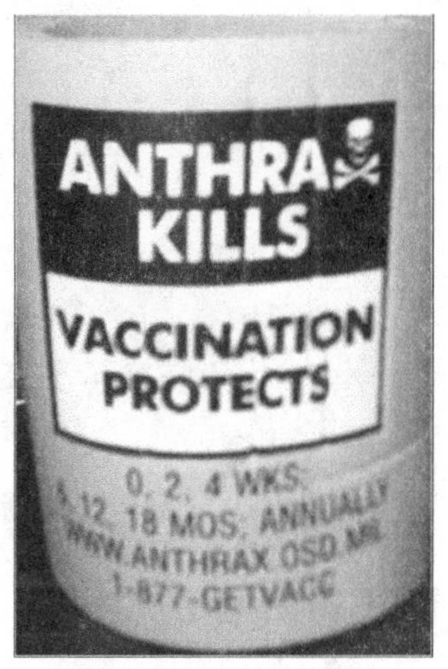

The DoD coozie promoting the anthrax vaccine.

I wondered how much all of this cost. I wondered why, in my decade and a half of military service, I'd never received a single knick-knack to promote vaccines for influenza, measles, mumps, rubella, or any other public health inoculation I freely submitted to. Was there a difference in those programs? Were these ornaments really necessary for the anthrax vaccine? Yep, the answer was pretty obvious for folks like me, Russ, Sonnie, and JR. And that's why we were all there within the halls of the US Congress that day—to uncover the sham. The mandate replaced the truth with giveaways, snappy slogans, and public relation tricks. The campaign was as unprecedented as our counterattack on Capitol Hill. We settled into our assigned seats—the hearing began. The Honorable Mr. Burton, a representative from Indiana and chair of the House Government Reform Committee, began the hearing.

> The Committee on Government Reform will be called to order. . . . We're here this afternoon to discuss the development of the US defense vaccine policy . . . the anthrax vaccine program . . . the importance of informed consent, the concerns about vaccine ingredients, purity, and the long-term safety concerns. . . . Is it viable and appropriate to use vaccines as a defense mechanism? Will it be possible and practical to develop vaccines to protect against all known and potential biological threats?

Representative Burton asked the same questions we asked, in vain:

> The Defense Department would have us believe that the concerns raised about the anthrax vaccine are minor and by a small and vocal group. . . . Either the Defense Department is being less than forthcoming about the objections being raised or they have their heads buried in the sand. . . . There are currently another eighteen vaccines in development under the Joint Vaccine Acquisition Program. . . . Are we going to vaccinate our military to death? Maybe we need to look at other approaches to dealing with the biological threat. For instance, with good detection equipment and protective gear . . . I hope we can . . . get the full story on issues raised and, by doing so, take action to begin to restore trust in the ranks and restore and preserve the careers that have been destroyed.

Representative Burton, in accordance with the protocols of the Congress, deferred to another member, Representative Henry Waxman of California: "I now recognize Mr. Waxman, the ranking minority member." Mr. Waxman stated, "In the case of vaccines against biological weapons, the threat is also severe. . . . If you are infected with anthrax, you die. There is no treatment." I was surprised to hear this from Congressman Waxman. Antibiotics were proven effective in animal tests—then verified after the anthrax letter attacks. To incorrectly claim "There is no treatment" was right off the DoD's script to justify the vaccination panacea.

Congressman Waxman continued by claiming, "We did not learn the identity of one witness, Major Rempfer, until this morning." JR just smirked, having helped me with the testimony, and said, "They're scared of you, Buzz." My Harvard-trained historian, JR, reminded me that it was Waxman who sponsored the National Childhood Vaccine Injury Act of 1986 (NCVIA).[30] It paved the way for routine vaccine manufacturer liability shields. I remembered thinking that they had no reason to be scared of us. After all, we were all on the same team, and we were simply reading from the DoD's previous script and their forgotten documents. We all swore the same oath, defended the same constitution, so they had no liability in hearing me out.

As the DoD panel finished their testimony, Congressman Shays felt compelled to ask, "Who is regulating the DoD?" A non-answer promptly came from the FDA's Dr. Kathryn C. Zoon, director of the Center for Biologics, Evaluation, and Research (CBER). Essentially her answer was: "You are, sir." Dr. Zoon insisted that the FDA only exercised, "control over the manufacturer, which is BioPort. We don't have control over the users." In other words, she said that the FDA had no power to tell the DoD what to do as the user of the drug, even if they were using it "off-label." If the DoD used the wrong dosage, or used it for a mandatory off-label or unapproved indication for inhalation anthrax, then this was experimental and violated the law. Shays retorted, "Who is responsible then? Who is going to make sure that DoD abides by the protocol, if you don't do it?" Dr. Zoon replied, "We don't have the authority." Mr. Shays was perplexed at the abdication. He questioned, "That is your testimony?" Dr. Zoon repeated, "We don't have the authority."

Representative Shays continued by demanding, "Well, who is going to protect our men and women if you aren't going to do it? Who? Who has the authority?" Representative Shays did not buy Dr. Zoon's answers. He explained: "I honestly don't believe that. . . . And, Dr. Zoon, for you to say that you have no authority is the most amazing thing I have ever heard at a hearing, because the FDA has the obligation, whenever it licenses a drug, to make sure it's used the way the protocol requires, and you don't allow the military or anyone else to deviate from that. That is your requirement."

Representative Burton concurred: "We've had hearings on . . . pharmaceutical companies who have had the wrath of the FDA come down upon them because things weren't being used in conjunction with what the FDA specifies as the way it should be done; and that's why I concur with Mr. Shays, because I have heard it before that you do come down on them, you close down companies. You pound them on the head with a meat cleaver, for crying out loud, and yet you say you have no authority over the military."

In September 1999, Dr. Zoon had actually previously admonished Dr. Bailey for the DoD's use of the vaccine outside of the prescribed approved labeling based on information received from "congressional sources." The FDA reiterated their previous position to the DoD, from the December 1997 meetings, writing that they "strongly recommend the Anthrax Vaccine Immunization Program follow the FDA approved schedule." These exchanges were a window on why troops were caught in the dilemma in the first place, and why we reported to Congress.

Russ had always said, "DoD was playing loose with the law," and now we were seeing the FDA and the DoD trying to explain the inconsistencies. It was unambiguous—people were conflicted and weren't doing their jobs. Dr. Zoon, like Dr. Bailey, retired and joined the anthrax vaccine manufacturer's board of directors, the same company they promoted while working as government employees. It appeared to those of us caught up in the mandate dispute that they were rewarded for pushing the manufacturer's deviation-cited product on unwilling troops.[31]

Representative Burton recognized Major Sonnie Bates when the second panel testified. Sonnie wasted no time getting to the issues, detailing for the committee about a dozen troops from his own base who were suffering serious anthrax vaccine ailments. Sonnie explained, "Our leadership seems to be desensitized, and that is not an attack on my chain of command. I believe there are people so close to this issue, they are so deep in the woods, they can't see the forest. I'm a new guy. I've got a fresh set of eyes, and I can see the forest. It is as if it were snowing in the summer, and nobody wants to acknowledge it."

I was proud to follow Major Bates with my testimony. Representative Burton said, "Thank you, Major Bates. I know that you and Major Rempfer had to take some risk to come here today. We appreciate that, and we'll do our dead-level best to make sure you're treated fairly. We appreciate your bravery in coming forth." I'd been at this quest for almost a year, and the DoD already let me transfer from my former flying job in the guard unit to the desk job in the reserves without punishment or adverse administrative documentation, so my efforts were simply bolder, not brave. Sonnie did have reason to worry. He was still active duty as a full-time officer. The DoD did control his fate on a similarly full-time basis. Bates was the brave one, because the DoD could mess with his life and career more. In time, sadly, the vindictive hammer of DoD injustice was inflicted against Major Bates. The chair recognized me, "Major Rempfer":[32]

> Thank you, Chairman Burton, members of the committee. I open my testimony with the core values of the United States Air Force: "Integrity first, service before self, and excellence in all we do." I'm not here to speak about the safety of the vaccine or the efficacy. Instead, I'm here to discuss another reason for the growing retention problem generated by the anthrax vaccination policy: integrity, its relationship to this policy, and how it extends to doctrine.
>
> After exhausting all avenues within my chain of command and communicating with hundreds of service members for the past year, I've concluded that the root cause of the negative reaction to the anthrax vaccination policy is a sense that the professional standards demanded of military personnel have been consistently violated by those implementing this program. It is not, as DoD officials assert, simply a failure to educate the troops. Instead, it is a failure to communicate the truth, the whole truth, and nothing but the truth.

I included the institution's core values in my testimony because they were one of the inspirational standards for our service in the profession of arms. Our military literature reminded us that the core values "serve as beacons vectoring us back to the path of professional conduct," and that they "allow us to transform a climate of corrosion into a climate of ethical commitment."[33] My written testimony for the Congressional Record also included ten specific detailed and documented ethical breaches of the anthrax vaccine program based on statements made by DoD officials. I only verbalized three in my oral testimony that day for the Congress.

My first example countered the DoD's mantra from the original 1997 anonymous press briefing launching the anthrax vaccine program: "It's been licensed since 1970 and has a proven safety record. It's been documented." In contrast, I specifically quoted a letter written by Dr. Zoon, who'd dodged responsibility for overseeing the DoD in this very hearing. A letter she wrote for the FDA in April 1998 stated, "Clinical studies conducted on the long-term health effects of taking the anthrax vaccine have not been submitted to the FDA." I reiterated that the GAO confirmed this lack of long-term safety proof in their April 1999 testimony. The US Army, the very month of this hearing, in October 1999, admitted they would finally conduct the required long-term study.

Next, I highlighted the farcical contention made by Dr. Bailey in the first anthrax vaccine hearing in March of 1999: "The safety of our AVIP was also confirmed by an independent review of the program." She referred to the DoD's independent expert, the Yale OB/GYN who had "no expertise in anthrax." I added additional examples of contradictory DoD statements, and contrasted them to "the whole truth" in order to illustrate the growing ethical concerns of service members. Examples of those accountable for the misinformation included both the assistant secretaries of Defense for Public Affairs and for Reserve Affairs, the secretary of the Army, the director of the Air National Guard, and one example from the Secretary of Defense himself.

I also attempted to explain how the misrepresentations placed "commanders at all levels in an untenable position, either implementing a questionable policy or sacrificing their careers. Consequently, the anthrax vaccination policy has turned into a biological loyalty test." I explained that each of the examples I cited demonstrated a "breakdown of intellectual honesty, which is the linchpin of integrity and doctrine between commanders and their troops. Without honesty, doctrine is merely dogma." I noted that Congressman Shays referred to the policy dilemma as a "medical Maginot Line." My examples illustrated the "acute dichotomy between what defense officials are telling Congress and the information readily available in government documents, congressional testimony, medical research, and news reports."

Finally, I explained that the contrast created "an ethical dilemma for service members whose core values required the questioning of immoral orders." Russ insisted I close with a quote from *Soldier and the State* by a famed Harvard scholar, Samuel Huntington. He rhetorically asked, "What does the military officer do when he is ordered by a statesman to take a measure which is militarily absurd when judged by professional standards?" Huntington answered, "The existence of professional standards justifies military disobedience." That was our dutiful counterpoint to their tiresome soundbite of "good order and discipline." Our professional standards were truly clear to us: "Integrity first, service before self, and excellence in all we do." I was happy to have it done. I said my piece once again, and was grateful to participate.

As I got up from the table where I delivered my testimony, the hearing room cleared out and three uniformed officers surrounded me. I recognized one officer from the last hearing, an army colonel who served as the director of health care operations for the Office of the Army Surgeon General. He called me "Buzz" and said, "Don't you think we have people who get paid full-time to work doctrine issues in the Pentagon?" I was surrounded, but at least they protected me from talking to any reporters. I didn't want to do so while in uniform. Maybe that was their goal in closing ranks on me in the first place. I remember being fascinated that this gentleman knew "Buzz" was my call sign. The only way he could have known this was if he was monitoring the anthrax vaccine internet traffic where I signed my postings with my full name, plus my fighter pilot nickname. I politely responded, "Yes, sir, I hope those doctrine experts are looking at this issue."

The hearings culminated in Congress's sole report on the program. The Government Reform Committee unanimously adopted House

Report 106–556 on April 3, 2000.[34] Titled *Unproven Force Protection*, the report recommended suspension of the anthrax vaccine program due to its "experimental" status. The committee concluded the anthrax vaccine program was "a well-intentioned but over-broad response to the anthrax threat. It represents a doctrinal departure over-emphasizing the role of medical intervention in force protection." They recommended "development of a second-generation, recombinant vaccine." The elected officials cautioned the "safety of the vaccine is not being monitored adequately. The program is predisposed to ignore or understate potential safety problems due to reliance on a passive adverse event surveillance system and DoD institutional resistance to associating health effects with the vaccine."

The committee also affirmed the DoD's forgotten position that the "efficacy of the vaccine against biological warfare is uncertain. The vaccine was approved for protection against cutaneous (under the skin) infection in an occupational setting, not for use as mass protection against weaponized, aerosolized anthrax." The committee recommended that "the force-wide, mandatory AVIP should be suspended until DoD obtains approval for use of an improved vaccine," and to "accelerate research and testing on a second-generation, recombinant anthrax vaccine." Most significantly, Congress warned, "while an improved vaccine is being developed, use of the current anthrax vaccine for force protection against biological warfare should be considered experimental and undertaken only pursuant to FDA regulations governing investigational testing for a new indication." Translation—they found the program illegal.

The law, Title 10, Section 1107 of the US Code, was passed by Congress the same year the anthrax vaccine manufacturer was shut down for quality control deviations, and the year the anthrax vaccine program launched. This law directly applied to the DoD's use of the anthrax immunization. The law read: "In the case of the administration of an investigational new drug, or a drug unapproved for its applied use . . . the requirement that the member provide prior consent to receive the drug in accordance with the prior consent requirement imposed under the Federal Food, Drug, and Cosmetic Act may be waived only by the President."

The congressional committee's compelling and unprecedented report, which reiterated the conclusions of the 1994 Senate report, similarly deemed the DoD's use of the vaccine "investigational." This meant it violated the law. It was black and white. Congressman Shays, along with his fellow representatives and their staffs, analyzed over a hundred thousand pages of documentary and electronic records and held five subcommittee hearings, one in the full Committee on Government Reform, encompassing twenty hours of testimony from forty-six witnesses. One of the premises of the report was that public law, written prior to the initiation of mandatory vaccinations, held that the anthrax vaccine was a possible cause of Gulf War Illness (GWI).[35] The magnitude of those inoculated—over one hundred fifty thousand, more than any other Gulf War exposure—required that the vaccine be studied. Instead, the DoD misleadingly implied that past studies ruled out any connection. Those study misdirects didn't sway the Congress's findings. Their report declared the anthrax vaccine "investigational," which meant the program was illegal.

I was grateful the congressional officials also took pause over the manufacturing and policy issues, which were Russ's and my original concerns. The committee concluded in their executive summary findings that the vaccine policy was a "doctrinal departure over-emphasizing the role of medical intervention in force protection." They went on to document my testimony by writing, "Others question the necessity of the program, asking whether it betrays a lack of confidence in deterrence and other force protection elements, and suggesting a vaccine program makes anthrax attack more, not less, likely."[36] By implying we could protect against the threat, this façade of force protection and potential legitimization of abhorrent weapons of mass destruction was the greatest folly of all. Even if we could protect against these threats in the near term, it was inconceivable that we would be able to over the long-term. With the almost inevitable "moves, countermoves on each side,"[37] a biological arms race might ensue.

Representative Shays referenced my testimony in his press release announcing the report, stating that "service members see an important difference between the physical body armor worn in battle, which can be removed, and medical prophylaxis, which cannot. The body armor that our Department of Defense refers to is perceived by many service members as 'tin foil armor'." The chairperson added, "We would not ask our armed forces to fight a battle using rifles and tanks designed in the 1950s. We should not ask them to risk their lives relying on 1950s-vintage medical technology . . . when modern science has the capacity to produce a safer, more effective vaccine." Representative Shays added that "Pejoratively labeling any challenge to AVIP orthodoxy mere supposition, the response indulges in the same absolutist arguments and hyperbole that mark the program's flawed education and communication efforts."

Representative Shays concluded, providing Russ and me and others some high cover, with these reassuring words: "These people were not troublemakers or conspiracy theorists. They were not malingerers looking to duck tough duty. They are among the most levelheaded, dedicated, and patriotic Americans it has ever been my privilege to meet. They deserve a better vaccine, and a better vaccination program." He closed with, "To many men and women in the armed forces, the Department's simplistic, one-sided approach to this complex issue only reinforces the legacy of suspicion and mistrust inherited from past military medical misadventures: atomic testing, Agent Orange, Gulf War drugs and vaccines." [38]

Federal court rulings later vindicated the *Unproven Force Protection* report by the US Congress that documented the illegality of the anthrax vaccine program. Encouraged by our dauntless representatives, we did not let up. At the time we had no idea that our marathons against bad orders and disciplines had just begun. Delirious and motivated by the victories, we continued our multi-azimuth attack with the help of our elected officials and the press.

MILE 9: The Military Times (2000)

On July 28th, 1988, ten years before Defense Secretary Cohen announced the anthrax vaccine program, he criticized the DoD for emphasizing public relations (PR) over "safety." The accusations occurred during Senate hearings on the DoD's "Safety Programs for Chemical and Biological Warfare Research." Cohen criticized the DoD's "public relations program." He added, "This is not a matter of PR as far as we are concerned. It is a matter of safety . . . questions of adequate safety guidelines, enforcement, oversight, and monitoring."

I wondered how we got away with waging the information war against our defense secretary's vaccine program, unless it truly was a PR stunt based on a "foundation of sand." That was a congressional staffer's portrayal. We both saw the unprecedented vaccine mandate as a dangerous façade of protection. If it wasn't "body armor," as the Air Force surgeon general said, we believed our national leaders were being misled. Everything Russ had uncovered in the medical libraries across the land defied their script and made the DoD's vaccine program appear to be an illusory PR campaign, versus a leading-edge biological weapons protection program.

We knew the vaccine was of questionable safety, and that the plant continued to be cited by the FDA. In the fall of 2000, the FDA once again inspected and cited the manufacturer.[1] Deviations included, "The design and construction . . . do not assure sterility of products filled . . . lots failed initial sterility testing for release or for stability testing. . . . Investigations into these initial sterility failures are incomplete. . . . Investigations are incomplete, inaccurate, or not conducted. . . . There is no assurance equipment is operating as designed." The FDA specifically cited the manufacturer for not investigating deaths and adverse reactions allegedly caused by the vaccine. This was a breach of the manufacturer's obligatory responsibilities under FDA rules.

Based on the ongoing deviations from FDA good manufacturing practice requirements, Russ formulated a complaint to our state attorney general's office, specifically to its fraud division through the state's auditor of public accounts. Russ's complaint focused on the unfunded mandate and liability burdens imposed on our state if National Guard members fell ill from the vaccine. I submitted an additional complaint detailing the expulsions of 25 percent of our pilots from our guard unit without due process, without discharge boards as required by state statutes, and without the proper vetting of the unanswered questions formulated by Tiger Team Alpha. Connecticut's attorney general, Richard Blumenthal, began an inquiry.

Interestingly, in response, the military cited the Perpich Supreme Court precedent to our attorney general, implying our state had no right to challenge the DoD policy. Our attorney general shared this response with us and we, in turn, respectfully related the "Constitution 101" lessons we learned from a great general at the National Guard Association of the United States (NGAUS). NGAUS is essentially a lobbying group for all the state guard units. They work in a stunning stone building near Union Station in Washington, DC. Their senior defense fellow, policy advisor, and Capitol Hill liaison, Brigadier General David McGinnis, US Army, retired, asked us to visit him in DC when the anthrax vaccine controversy first broke.

Russ and I didn't waste much time. We traveled to DC and met with General McGinnis. He was an impressive man. He was what we expected from our leaders. But what impressed us most of all were his bookshelves with copies of *The Federalist Papers* and assorted books on the Constitution. He expertly explained that the framers of the Constitution, the Founding Fathers, created two military forces for our country: one federal and one state. These two militaries were the federal full-time and state part-time forces, otherwise known as active-duty and the militia. The total force included them and us—the part-time citizen soldiers. The two distinct forces augmented each other, but also acted as a check and balance. Every school kid learns that "checks and balances" are a basic ingredient in our democratic republic's form of government. General McGinnis told us that the Founding Fathers actually anticipated the need for checks and balances in the military too. General McGinnis said it was "Constitution 101."

Since we were lifetime members of the National Guard Association, and every guardsman was encouraged by their National Guard leaders to join this lobbying effort, General McGinnis may have felt obligated to walk us through the constitutional foundations of our nation and our military. But there seemed to be more to this man's lessons. He was genuinely concerned about what he witnessed. General McGinnis also shared the thoughts of one of the Founding Fathers, Alexander Hamilton. In 1788, Hamilton authored an article for the *Daily Advertiser* to explain and encourage the passage of the Constitution. Hamilton reminded early Americans, "If the representatives of the people betray their constituents, there is then no resource left but in the exertion of that original right of self-defense. . . . The citizens must rush tumultuously to arms, without concert, without system, without resource; except in their courage and despair. The usurpers, clothed with the forms of legal authority, can too often crush the opposition in embryo."[2] Of course there was betrayal, but we also had many supporters and resources, so it was not a constitutional crisis, yet. We used our brains as our "arms" in intellectual self-defense.

General McGinnis's point was that if the federal government abuses its power, then these "usurpers" needed to be held accountable, and the task falls to the "citizen." The general seemed genuinely interested in our work, enjoyed our questions, and kept pulling his books off the shelves to show us quotes and passages from the Founding Fathers. It seemed we found another who cared. He pointed out the *Perpich v. Department of Defense* Supreme Court case.[3] The 1990 Supreme Court case upheld the constitutional check and balance served by the National Guard. The justices wrote, "Members of the State Guard unit continue to satisfy this description of a militia. In a sense, all of them now must keep three hats in their closets—a civilian hat, a state militia hat, and an army hat—only one of which is worn at any particular time."

General McGinnis gave us the confidence to know that we were doing the right thing. He explained that we were wearing the "state militia hat" in providing our professional dissent over the anthrax vaccine. Our commander protected the Pentagon's policy while wearing the "hat" of the federal military, most likely because he was a full-time, active-duty officer on exchange from the Pentagon to the National Guard. In contrast, we defended our rights as citizen soldiers by first wearing our "state militia hat" and then the "civilian hat" after being grounded from flying.

General McGinnis didn't give us much hope that the National Guard Association would help us. He let it slip that he'd been told not to question the anthrax vaccine on company time. Still, we took consolation in General McGinnis's counsel that day. It was only after our "Constitution 101" session, and our subsequent work with the attorney general, that we realized the importance of our visit and our work ahead. The general provided reassuring words in his follow-up note after our visit, reminding us, "Would you believe the Founding Fathers intended you to fly high cover on the Federal Military establishment? Hang in there. Dave McGinnis."

General McGinnis not only encouraged us, but he also put an end to some of the misinformation by the DoD political commissars. In their association magazine, NGAUS obtained DoD admissions from the Anthrax Vaccine Agency about our "well-documented" Tiger Team Alpha research. *National Guard Magazine* quoted one of the DoD's PR representatives with this admission:

> Until finding out about the pilots' research in libraries, Defense Department officials assumed most of the misinformation about the vaccine was coming from the Internet. They had not considered outdated but well-documented information as a problem.[4]

In other words, our research was spun as "outdated," because we discovered the DoD's previous official position from before they started to modify the available medical research with new conclusions. It was the DoD's anthrax vaccine proponents who changed the information in order to support the policy. The DoD was the source of the updated misinformation, not us. They misinformed our commanders, our troops, our Congress, and our media. The Founding Fathers' idealism did not condone such conduct. We continued our well-documented research, knowing that the wrong people were being punished for simply reminding the DoD what they'd forgotten.

General McGinnis also gave us the tools to help our attorney general respond to the DoD's dismissals of our concerns by citing the Perpich precedent. If the federal government abuses its power, then these "usurpers" needed to be held accountable, and the task falls to the "citizen."[5] Our attorney general realized something was wrong by the attempted dismissal of his legitimate concerns, and wrote a myriad of inquiry letters to the DoD beginning as early as December 27th, 1999. Attorney General Richard Blumenthal wrote:

> I have been contacted by members of the Air National Guard who raise highly significant issues regarding the Anthrax Vaccine Immunization Program. . . . Based on the uncertain state of knowledge regarding the Anthrax vaccine, and potential legal and policy consideration, I would request that you suspend the AVIP until there is a scientific consensus on its safety.

Blumenthal invited us into his office to meet with him and two of his assistant attorneys general running his fraud division. Once seated, he explained that he served as an enlisted Marine, and that he was not satisfied with the answers his office received from the DoD. It felt good to have elected representatives like Congressman Shays and our own attorney general feel equally concerned about the unanswered questions and unsatisfied with the PR campaign. In response, US Army officers from the anthrax vaccine agency wore their uniforms and hit the sales trail with their PR bling. In addition to engaging in damage control by visiting Connecticut and personally briefing our attorney general's office, they also came bearing gifts—a suitcase of trinkets paid for with taxpayers' money—more beer coozies, mouse pads, flashlights, pens, pocket knives, and posters displaying their PR motto: "Anthrax Kills, Vaccine Protects."

During this same period, we maintained our frontal assault through a variety of news outlets. We scheduled "editorial boards" in order to offer major newspaper editorial staffs a review of the information that the DoD conveniently forgot about when the anthrax vaccine program was launched. We articulated the previous DoD position concerning the vaccine's unsatisfactory reviews and its investigational legal status, as well as the questionable safety and effectiveness. We exclusively used government documents, most printed on US Army letterhead.

One particularly important target for us was the publishing company that put weekly newspapers on the newsstands of every ship, base, post, and fort in the US Armed Forces worldwide: the Army Times Publishing Company. The company later changed its name to the Military Times. My earliest involvement with their editors was in reaction to General Blanck's "Ignore the Paranoiacs, the Vaccine Is Safe" opinion piece.[6] From my earliest calls to these editors, they saw the disparagement tactics for what they were and took what the DoD liaisons briefed at editorial boards with "a grain of salt." Yet by the winter of 2000, the anthrax vaccine dilemma had been dragging on for over two years, so the editorial staff finally gave us an opportunity to tell our side of the story—which was also the DoD's story before their rewrite.

The Military Times entertained the DoD's pitch on at least three prior occasions in their boardroom. Now, we were given a chance to counter the misinformation of the Pentagon staff responsible for the anthrax vaccine policy. Just prior to the December 2000 gathering, one of the editors sent me a message asking, "Tom, can you tell me who's coming Tuesday?" I responded with our lineup: "Thomas L. Rempfer . . . Russell E. Dingle . . . John J. Michels (USAFR JAG) . . . Dale Saran (Captain, USMC, JAG) . . . Redmond H. Handy (Colonel, USAFR) . . . John Richardson (Col, USAFR, Harvard)." I added, "Frankly, Kent, a gathering like this is historically unprecedented in civil-military affairs . . . this should be admonished by the military leadership. I predict it will not be though as they will want to do anything possible not to highlight what we're telling you." The editor queried, "How do you mean by 'admonished'?" I replied, "I've always wondered how we cannot be punished or admonished—the reason is our message is based on facts, DoD's facts." The editor got it, responding, "Ah—I see. I've often wondered the same . . . I find the JAGs' participation in this intriguing . . . quite a group—I really appreciate you pulling this all together, Buzz . . . Again, thanks—I'm looking forward to finally meeting you."

We descended upon DC. JR and I came in the night prior to finalize the "briefing book" that he and Russ and I had been diligently compiling. Dr. Meryl Nass, an internal medicine doctor and anthrax vaccine expert from Maine who'd testified for the committees, assisted with the medical section. I hammered out the timeline section. Russ and Sammie Young, a retired USAF Reserve colonel and former FDA regulator, put their expertise to work on the history and regulatory sections. They outlined why the anthrax vaccine met the specific requirements of the Federal Food, Drug, and Cosmetic Act as an "adulterated" drug. Colonel Redmond Handy, who testified for young troops at federal disability hearings, did his part to update the editorial board on various ill patients at the Army's Walter Reed Army Medical Center. Our JAGs, Captain Dale Saran and Lieutenant Colonel John "Lou" Michels, provided our legal analysis. We consolidated the briefing book, with a mountain of government documents for attachments. We spent the night at the Fort Myers VIP quarters in Arlington, Virginia—just up the Potomac from the Pentagon. We figured if we got booted out of the military after this editorial board, at least we could enjoy some fine general-officer-style furnishings during our last days in uniform.

The billeting at Fort Myers was located just across a small lane from the DoD's joint chiefs of staff homes. We peered out the windows as we typed, wondering if they had any idea what the documents said or what they meant. Did the four-star generals even know the truth, or did they only know what was put in front of them in the daily briefs, on the slides, the soundbites from the skull sessions, and the boilerplate script they all mimicked so well? I hoped they didn't know. It was less disheartening to believe that the truth-blocking agents who protected the anthrax vaccine program had misled them. JR dubbed these agents as "gate guards." The term described what we faced from JR's Machiavellian worldview. That's why we came to DC again that winter and were willing to put it on the line again publicly. We were trying to reeducate our military leaders and warn them they were being misled. We worked hard, well into the night.

On the morning of December 19th, we woke to reveille at Fort Myers and departed for the Military Times headquarters after a diversion to a print shop to copy our briefing book. We expected a one-hour audience with the managing editors for all their publications: *Army Times*, *Air Force Times*, *Navy Times*, and *Marine Corps Times*. The editors were appalled at the disparity between what DoD officials briefed previously in their editorial boards compared to the story we told. Delivering just the documents and facts lasted four hours. The time granted was as unparalleled as five military officers from multiple services speaking out against a DoD policy.

I opened the meeting with the timeline of events, a thorough understanding of which allowed everyone to understand the tragedy. Russ expertly briefed the regulatory regime as though he'd been working for the FDA for twenty years. I was proud of Lt Col Lou Michels, too. He rode in slightly late in his long leather overcoat, right out of an old western movie. He briefed the black-and-white legal issues related to the illegal experimental use of the vaccine. As one of our assistant attorneys general termed the illegal mandatory use: "it was academic." Captain Dale Saran corroborated Lou's legal analysis, wearing his dapper Marine Corps uniform.[7] JR offered an eloquent policy analysis and a military culture dissertation from a civil-military affairs perspective, expressing the dangers of allowing the military to not only violate the laws, but also mislead our Congress and media. I also closed the meeting, following Colonel Handy's update on the ill patients he was in contact with, and provided our briefing book to the editors. Once again, we all said our piece and hoped it might make a difference. It did.

Within two weeks, the Military Times published an editorial designed to send a message to the incoming administration of President George Bush, specifically addressed to his newly appointed defense secretary, Donald Rumsfeld. Rumsfeld previously served as the Pentagon's top civilian in the Ford administration. He was also an ex-fighter pilot, a former congressional representative, and even headed a pharmaceutical company, Searle Corporation. We felt confident we had a leader about to run the Pentagon that would assert the constitutional prerequisite of civilian control of the military. As a former fighter pilot and legislator, not to mention with his expertise in the biologics industry, we hoped for an opportunity to resolve the anthrax vaccine dilemma once and for all. The Military Times evidently felt the same way based on the words and recommendations they published in the *Army / Navy / Air Force /* and *Marine Times* in their editorial column on January 8th, 2001:

> Congratulations, Donald Rumsfeld . . . we thought we'd offer a little advice. . . . Scrap mandatory anthrax inoculations. The anthrax vaccine inoculation program has sewn dissent in the ranks and added to the mistrust some troops harbor for their commanders. The vaccine is already in short supply, its manufacturer continues to struggle to win federal certification; and questions persist about the quality, efficacy, safety and legality of the shots. Suspending the program now, pending further review and until such time as a ready supply of safe vaccine is available, is the only prudent course. If the new vaccine production line can be certified and other problems worked out, the program can be revived—if it is really needed. But if problems persist, the whole mess can be laid to rest as a failure of the prior administration. Love your troops. . . . Fathers and mothers, husbands and wives, sons and daughters all, you will hold their futures in your hands. May you use that power wisely.

The Marine Corps officer and JAG, Dale Saran, who helped us with the Military Times briefing, was also a former helicopter pilot. Initially, when assigned to defend Navy, Marines Corps, and Seabee anthrax vaccine refusers, he believed the DoD's misinformation. He figured his clients must be malcontents and that their concerns were due to misinformation. After defending the refusers, Captain Saran joined us that day in DC to speak out publicly. In defending his client, he came to realize the root source of the problem. Dale wrote us all with heartfelt words after the editorial board:

> All, I wanted to echo . . . I never did think I would feel so privileged to know and have worked with so many Air Force officers. . . . Okay, intramural fights aside, you all are warriors of the highest order and I am proud to be a part of our group of rabble-rousers. I thought each of you contributed something beyond your knowledge and intelligence and hard work . . . something the DoD (unfortunately) cannot—a credibility and sincerity of purpose that can come only from standing up for something you know is right, though all else in the world tell you you're wrong. I believe the quote is that "one man with conviction is a majority." I was proud to sit next to each and every one of you—no matter the outcome. It was also nice to put a face to all of the email addresses. I hope that even if we do not make the footnotes in history that perhaps someday I get a chance to explain to your kids how truly courageous their fathers were . . . Dale

Colonel Redmond Handy also expressed his gratitude for being invited to participate and sent a "reminder for an electronic version of the total document we used for the *Air Force Times* meeting. Thanks for the info. I'll put it to good use. Redmond." Colonel Handy would spread the word of the briefing book at press conferences and to others, such as a retired lieutenant colonel F-16 squadron commander and friend, Thomas Heemstra. That colleague, in turn, published a novel using the briefing book's contents as his primary reference. The word was getting out.

Soon thereafter, another veteran colleague, Frank Schmuck, who I'd trained at the Air Force Academy, ensured our briefing book made it to veteran philanthropist H. Ross Perot, a Naval Academy graduate. Mr. Perot was one of the few high-profile outspoken advocates for those suffering from Gulf War Illnesses. In time, we discovered that Perot sent the briefing book to the highest offices in the land at the White House. Russ and I also provided the briefing book to our Connecticut attorney general's office. They informed us they intended on corroborating the research and writing a follow-up letter to the DoD and FDA. Knowledge is indeed especially precarious if the people you serve and command catch you promoting falsities. That was our tactic with the briefing book. We wrote it in an effort to turn even tighter inside the DoD's decision-making processes. Fighter pilots "lead turn" their adversaries to stay in a position of advantage in aerial combat. That's why we "lead turned" their efforts to falsely frame our findings as internet misinformation.

Within months, our Connecticut attorney general, in what was obviously a politically controversial maneuver as a high-profile Democrat, took a risk by writing to the new Republican national administration. He recommended they terminate the anthrax vaccine program and blame it on their predecessors. In his press release on March 22nd, 2001, he recommended the DoD and FDA "eliminate the military's controversial Anthrax vaccination program."[8] He was essentially blowing the whistle on the previous executive branch by saying, "Unfortunately, replies to my repeated requests from the DoD and from the FDA, under the prior Administration, were cursory and unresponsive." He wrote to both President Bush's new defense secretary, Rumsfeld, and the acting FDA commissioner. Blumenthal explained, "In response to my request, Principal Deputy Assistant Secretary Charles L. Cragin wrote that 'the threat of biological warfare using anthrax is real.'" The Connecticut attorney general compared our brief to the DoD script used to sell the program. He chose not to be steamrolled by the sales pitch and scary soundbites. He continued:

> Despite the federal government's past stonewalling and its apparent efforts to avoid and evade the truth and the law—prior to the current Administration—I am sending this letter to you to highlight important information regarding the AVA's [anthrax vaccine absorbed]

> questionable safety and efficacy and the lack of legal authority surrounding its use in the AVIP.

The attorney general was doing his part to "fly high cover," though politically risky. We were proud of him, and figured the Founding Fathers would be too. His press release elaborated:

> In effect, the military is forcing its personnel to serve as human guinea pigs for an unlicensed drug that has not been proven to be safe or effective.
>
> Suddenly in 1997, DoD and the FDA, with no change in the facts or the law, reversed themselves and with the stroke of a pen wiped out the protections afforded our members of the Armed Services by clearing the way for DoD's mandatory mass inoculations.
>
> The plain fact is that the AVA is still an investigational drug and should not be used without appropriate informed consent. I call upon the DoD and the FDA to cease and desist from their illegal conduct and to abandon plans for Anthrax Vaccine inoculation of the Armed Forces.

Attorney General Blumenthal was gracious in this acknowledgment of Russ's and my concerns, which were shared by thousands—perhaps millions—of dutiful troops. He wrote that he "was prompted to investigate the program by current and former Connecticut Air National Guard pilots, including Major Russell Dingle and Major Thomas Rempfer." In his letter acknowledging our concerns, he reiterated the unanswered questions in a twenty-page document saying, "Their concerns are justified. The safety and efficacy of the Anthrax Vaccine Absorbed ('AVA') in current use has, over the past two years, been subject to close scrutiny by the military, medical, and scientific community."[9] He critiqued the "quick fix" Friedman memo:

> Dr. Friedman wiped out ten years of DoD analysis and twenty-five years of FDA law designed to protect the safety and well-being of the

> citizens of the United States. There was no justification, legal, scientific or otherwise, for this action.

Our attorney general asserted that, "Mandatory vaccination of troops with a biologic product not licensed for its current use violates the Federal Food, Drug, and Cosmetic Act and 10 USC § 1107." In addition to this valiant effort, he also later advocated for Russ's posthumous promotion.[10] Attorney General Blumenthal did his part in flying high cover by concluding: "The Department of Defense should make inoculation voluntary as the United Kingdom has done, or properly invoke the President's powers as required by statute."[11]

The attorney general's efforts provided Russ and me with some high cover as well. Since our entrance into the dilemma originated at the state level, and our whistle blower complaints remained active until the ordeal was resolved, the last thing the DoD wanted to do was highlight how we originally became educated—our commander's official Tiger Team Alpha tasking. Further, the attorney general's gutsy move provided the Military Times the credibility they sought to move forward with their exposé. Within a week, articles were printed in all of their papers and were shipped to every DoD installation on the planet. The cover story splashed:

Titled "Shots in the Dark: What the Government doesn't want you to know about Anthrax Vaccine," the cover stories pictured a hand covering up the facts about the vaccine.[12] The facts were laid out on the frontpage so that senior leaders could understand how they'd been misled. The article focused on the Blumenthal letters and a new refuser, USAF doctor John Buck.

The Military Times explained that "The Defense Department [was] on the verge of having to further curtail or altogether suspend its anthrax vaccination program because of dwindling supply." The paper also informed readers about "prosecutions of service members who refused the shots." They zeroed in on a court-martial at Keesler Air Force Base in Mississippi and "the case of Air Force Capt. John Buck, the first military doctor to refuse the shots." The reporters introduced Russ, as well as our

MarineTimes

What the Government doesn't want you to know about Anthrax Vaccine

PLUS Limited vaccine supplies are running out 14

ArmyTimes

What the Government doesn't want you to know about Anthrax Vaccine

PLUS Limited vaccine supplies are running out 14

NavyTimes

What the Government doesn't want you to know about Anthrax Vaccine

PLUS Limited vaccine supplies are running out 14

AirForceTimes

What the Government doesn't want you to know about Anthrax Vaccine

PLUS Limited vaccine supplies are running out 14

Military Times papers' cover stories, April 9th, 2001.

attorney general, Richard Blumenthal, regarding his appeal to Rumsfeld and the FDA to "abandon the anthrax vaccination program."

The publisher framed the cover story as a cover-up with excerpts from the DoD's own documents. They included snippets such as, "There is an operational requirement to develop a safe and effective product which will protect US troops." It cited admissions from the DoD's official request for proposal that acknowledged, "There is no vaccine in current use which will safely and effectively protect military personnel against exposure to this hazardous bacterial agent" [RFP excerpts at Appendix L]. The DoD's documents confessed "The vaccine's effectiveness against inhaled anthrax is unknown," that the "efficacy of the vaccine against biological warfare is unknown," and that the "vaccine should be considered investigational when used as a protection against biological warfare." That meant the vaccine mandate was illegal.

The paper recognized that the "anthrax threat is real, but overstated." They recounted then-Defense Secretary William Cohen's embellishment of the threat and how "government experts later wrote in *Archives of Internal Medicine*, a medical journal, that Cohen had overstated the effect by 100 times." The paper also clarified for readers that "The vaccine has never been clinically tested on humans," and referenced our original Tiger Team Alpha tasking by our chain of command to research the program. They wrote, "Air Force Reserve Maj. Thomas Rempfer was in the Air National Guard in 1998 when he and fellow pilot Maj. Russell Dingle were tasked by their commander to research the vaccine and develop questions and answers for Air Guard leaders. Their research led them to conclude the vaccine wasn't safe."

The journalists referenced our team's "December meeting with Army Times reporters and editors," and noted that "Rempfer pointed out that the only controlled clinical tests of the vaccine on humans actually involved a different vaccine." They corroborated the assertion by citing an April 1999 report by the General Accounting Office, which concluded, "the current vaccine differs from the vaccine used in the Brachman study in three ways: The manufacturing process changed. The strain of anthrax . . . used to grow the original vaccine was not the same. [and] The ingredients in the vaccine were changed." Despite the regulatory anomalies, the reporting revealed that "Still, a license was granted for the military's vaccine in 1970 without data on its effectiveness." The article exposed the fact that "The manufacturer did not notify the FDA of key changes," and the FDA "documented numerous violations in the manufacturing of the vaccine."

The editors cited Russ's explanations of the regulatory deviations. They wrote, "Dingle, the Air Force reservist, explained the issue." They quoted Russ's explanation that "Before the Gulf War, 7,500 doses was the largest batch they had made." Russ described that the Pentagon "saw the need for more vaccine because the United States had sold [anthrax] spores, equipment and technology to the Iraqis." The paper explained

Russ's and Blumenthal's point that "while the changes increased production capacity, the FDA was never notified, as required by law." This more likely than not led to the subsequent findings by the FDA, immediately prior to the anthrax vaccine mandate, that "the manufacturing process for anthrax vaccine is not validated."

The exposé also uncovered the DoD's attempted workarounds related to "modifying the FDA license so the anthrax vaccine could be administered in fewer doses and so it could be approved for inhalational anthrax." The license modification went forward and "automatically put the vaccine in the experimental category—an 'investigational new drug,' in the parlance of the FDA." The news article explained that these problematic items led to Connecticut Attorney General Blumenthal's letter to Rumsfeld and the FDA alleging "four reasons why the anthrax immunization program is illegal." The four items were summarized as follows: "The anthrax vaccine has not been proved safe or effective for its intended use in that [it] has never been licensed for protection against inhalational anthrax." Next, "the vaccine is not being manufactured in accordance with either its site license or product license." Also that "the vaccine is not being administered according to the license." And finally, "since the vaccine has not been tested on humans, there is no basis for concluding that it is safe and effective."

Within days of the cover story, US Marine Corps Major General Randy West, Special Assistant to the Secretary of Defense for Anthrax and Biological Defense, offered a terse and bullying response to the Military Times. Civilian appointees seemed unwilling to defend the policy in print, so West got the last word. We weren't concerned. This was their M.O., and we expected a baseless and disparaging attack. General West's reply included the following barb: "This article's innuendo, misinformation, and misleading headlines contradict this opinion." Understandably, the DoD didn't like the well-researched article or the headline, but all they had to counter it were empty accusations of misinformation, but without any substantive explanation.

Just as with the extraordinary undue deference the US judicial system provided the DoD traditionally, we also saw the media—knowing who helped "butter their bread" in the defense arena by providing regular stories—tended to give the last word to the DoD. That was okay. We put more "chinks in their armor," as JR would say, and we moved on to the next spear in the multi-axis attack. We had journeyed deep inside the DoD's area of operations. We turned tighter, faster, and with a truer vector than the DoD. The editorial staff at the Military Times let us in their doors because they knew something was very wrong. There was a blatant breach, a chasm between the DoD's soundbites and the full truth. That's why they published "Shots in the Dark."

The Military Times was also trying to get the word to commanders, troops, and policymakers. Few likely noticed outside of military circles, and we heard that sales ultimately suffered that week. Many of the half-million troops who had already accepted the vaccine were prone to not read more disturbing details about what they already knew deep down inside. We understood that element of human nature. General West and the other gate guards were fully aware that the majority of our troops didn't trust the program, so they most likely were happy that the cover story wasn't recognized for what it was outside their own DoD community.

Soon we would see Generals Blanck and West retire and move on, transporting their DoD bio-warrior expertise into outside "consulting" adventures and working indirectly for the anthrax vaccine manufacturer, BioPort. BioPort's press release captured their newly found civilian opportunities by publicizing the biological threat of anthrax from the "outside." *PRNewswire* titled their press release, "New Report Calls for Protecting At-Risk Civilians from Anthrax, Raises Possibility That Bioengineered Strains May Be Resistant to Antibiotics." It introduced the authors, "Lieutenant General Ronald R. Blanck, D.O., US Army (Ret.), President of the University of North Texas Health Science Center and former US Army Surgeon General . . . and Major General Randy West, USMC (Ret.), former Special Assistant to the Secretary of Defense for Anthrax and Biological Defense."[13] The men appeared to capitalize on their new private sector jobs to continue the same mission they performed for the Pentagon.

The government established ethics rules regarding such revolving-door activities, but that was someone else's job to scrutinize. The general's actions were no different than one of the other partners at BioPort—the former chair of the joint chiefs of staff (CJCS), Admiral William Crowe. The chairman had endorsed President Clinton upon retirement, and was awarded the ambassadorship to England. Admiral Crowe was later granted a no-investment co-ownership of BioPort at the time of the lucrative DoD anthrax vaccine contract. Our mission was not to cast judgment on men such as Crowe, Blanck, and West, or those that followed them out the revolving door and cashed in on the paradigm tolerated by the top of the chain of command.

We had nothing to offer but the truth, and the Military Times had nothing to gain by exposing the DoD's cover-up. As one of the world's earliest warrior leaders, Alexander the Great, said, "History shows the military commander who best analyzes, decides, and controls the speed of the engagement prevails in nearly every conflict."[14] I knew we would prevail in the end. We entered the fight "jinking," as fighter pilots say, in and out of the DoD's territory. We rested briefly within the walls of their fort. We prepared our briefing book on their post. We went offensive, avoiding the "fur ball" of close combat, then "bugging out." Our "hit and run" tactics controlled the fight for a time. Our wins were based on the truth and convincing others of the same. At least our victories would never be on the backs of the harmed, punished, discharged, imprisoned, or ill troops. That was exclusively the DoD gate guard's territory.

MILE 10:
The OODA Loop (2000 to 2001)

The media is sometimes referred to as the "fourth estate." The idea being that when politics gets in the way of the other three branches—the executive, legislative, and judicial checks and balances—the press would hopefully step in and expose injustices. But what if the branches of government or its agencies attempted to co-opt the press? That's what the DoD's PR people did.

An example of the blur included comments concerning the anthrax vaccine by Mr. Kenneth H. Bacon, assistant secretary of defense for public affairs: "It's proven itself safe and reliable. It works, and it does not have side effects. . . . We have given now I think shots to nearly 170,000 people in the military. . . . All these people are fine . . . I'm not aware that there are problems with this vaccine. . . . I've had three shots. My hair is growing more robust than ever. (Laughter) I sleep better. I eat better, run farther. It's been nothing but a great experience. (Laughter)."[1, 2]

Prior to his public affairs position in the DoD, Bacon was an assistant news editor for the *Wall Street Journal*. His career with the journal spanned twenty-five years. Bacon also served in the US Army Reserve from 1968 to 1974. Our early dealings with the media made us aware of the crafty nature of the DoD's carefully picked people to defend the vaccine policy. By utilizing a former media man as the Pentagon's point man for public relations, the DoD shrewdly incorporated insiders who knew how to handle their own colleagues on the other side of the microphone. We understood why there were no follow-on questions, only laughter. I hoped it wasn't really funny to those reporters. Maybe their laughter, captured in the DoD transcript, was instead a nervous embarrassment over Bacon's comments. Real people were sick at Dover Air Force Base, in Michigan, and elsewhere. One of Dover's own officers, Major Sonnie Bates, tried to stand up for ill troops. Sonnie personally witnessed that people could not "run farther," as Bacon joked. Sadly, the DoD forced Sonnie out of the USAF over the anthrax vaccine. His discharge represented one of a thousand troops whose honorable stand earned only injustice.

Dr. Renata Engler knew that Dover wasn't having a "great experience." Colonel Engler, chief of immunology for a US Army Medical Center, spoke at the conference on the anthrax vaccine policy at Fort Detrick, Maryland in May 1999. During her address she described, "Potentially more than twenty-five individuals from [the] same location, having received anthrax vaccinations around the same time and from [the] same lot, [and the] growing 'belief' that anthrax has caused potentially long term, indefinite, untreatable disease." She reported a "fear of [the] military medical establishment" where "affected service members fail to report [reactions]."

Dr. Engler described a major breakdown in military medical culture where troops were afraid for their jobs and scared of their own doctors' receptiveness to their illnesses. Publicly, DoD officials all across the board mocked the situation. Perhaps the reporters were intimidated to ask the tough follow-on questions of a former colleague. All we knew was it always seemed to work. The follow-on questions were not asked, and DoD spokespersons continually concocted unchallenged or diversionary excuses about the mounting anthrax vaccine controversy.

The lack of questions and non-answer answers were everywhere. Once, an anonymous DoD spokesperson was asked about the closing of the anthrax vaccine manufacturing facility due to quality control problems and about the FDA revoking their validation. The DoD's anonymous response was, "That's another one of those urban legends or something that just keeps cropping up. We planned to shut the plant down to modernize it."[3] The fact was Kathryn Zoon, the FDA's director of the Center for Biologics Evaluation and Research, confirmed if "corrective actions proved to be inadequate, they would run the risk of having their license revoked."[4]

Just as the FDA has a revolving door with pharmaceutical companies, the DoD is known for the same. The DoD is known for their powerful press division, but to embed the media apparatus and current or former media professionals within the DoD was more visionary indeed. The DoD prided itself on its ability to control all aspects of the battlefield. Embedding media provided a powerful means of doing so, just as with the disguised DoD liaisons in Congress wearing civilian clothes that don't seem to identify themselves as military members. A reporter from the *Baltimore Sun* called it "access." Whereas truth and facts were our only goal, the DoD's embedded access with the media and Congress was more likely about information control.

All this made me think about a concept a military general had explained to me once called the "OODA loop." The acronym stood for: Observation, Orientation, Decision, Action. I later studied the concept more thoroughly, discovering that an Air Force fighter pilot and officer coined it. His name was Colonel John Boyd. Our learned friend, JR, talked of Boyd as well, so I read some of his quotes off our military online university website. "Machines don't fight wars. Terrain doesn't fight wars. Humans fight wars. You must get into the mind of humans. That's where the battles are won," wrote Colonel Boyd.[5] Boyd used the OODA loop to articulate the act of outthinking and outmaneuvering your adversary, just as a fighter pilot might do in aerial combat. The OODA loop was Boyd's fighter pilot description of decision-making cycles, but it obviously applied beyond the dynamics of dogfighting in aerial combat.

Colonel Boyd maintained that every individual and organization, every military and command structure, operates within an OODA loop. The "cycle speed" of the decision loop or cycle is based on the individual's or organization's mental capacity or capability to deal with information and a changing environment. John Boyd asserted that one could paralyze an enemy by operating inside their OODA loop, meaning that the individual is operating at a faster cycle speed than their adversary's.[6] Boyd's goal was to operate inside an opponent's decision cycle, thereby generating confusion and disorder based on the ability to anticipate the other's actions. Was our DoD implementing Boyd's OODA loop in the anthrax vaccine dilemma? To us it seemed obvious; the OODA loop was an integral part of the DoD's misinformation operations.

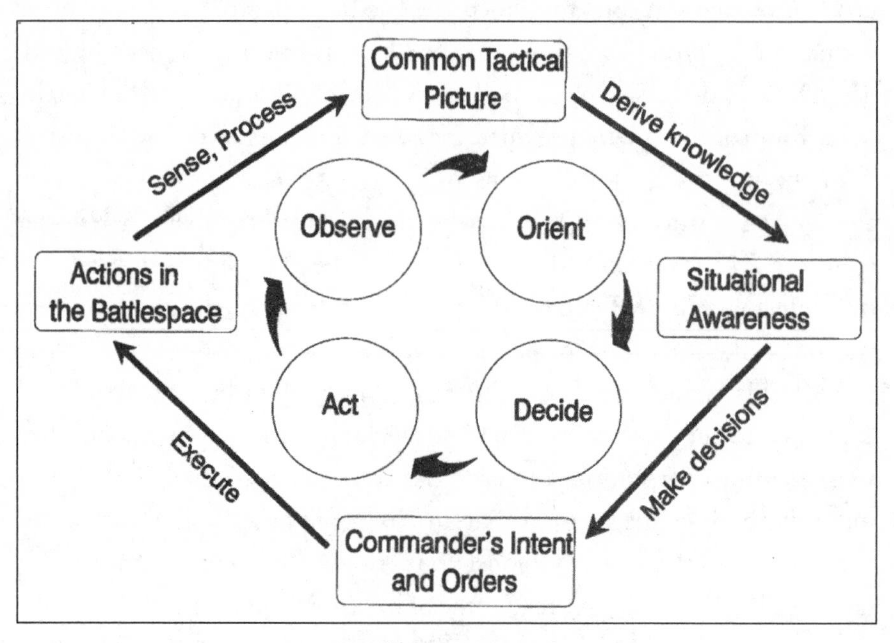

Boyd's OODA Loop.[7]

The way Boyd's inspirational ideas played out within the DoD public affairs and the Pentagon press corps was intriguing to me. It was twofold in its application. One, the DoD could anticipate the tough questions, knowing the inherent vulnerabilities of their policies and programs, and concoct snappy responses or diversionary, disparaging quips in order to throw reporters off track. Two, the DoD could embed the minds and experience of former reporters within their own staffs. The DoD co-opted insiders from news organizations in order to fashion a collegial "we're all on the same team" environment. Such an atmosphere served to paralyze the oversight potential of the fourth estate. Couple this with a reporter's desire for access, plus an editor's necessity for easily digestible soundbites, and it was conceivable to see how the DoD exploited the media's weaknesses and entered their OODA loops. The press as an ally, if manipulated, might hesitate to question the DoD's misguided efforts to protect the troops.

Over and over, we saw the DoD manipulate the press, anticipate their questions, and confuse them as Boyd had taught. Circular non-answer answers seemed to be the favorite tool, with diversionary answers being a close second. Humor never hurt, and derision worked as well to throw the media off target. Russ, JR, and I attempted to enter the media's OODA loop by informing them intelligently and honestly, and taking the time to answer their questions. Our tactics were different than the DoD's. From the outset, Russ and I were getting "into the minds" of DoD leaders and the media, because "that's where the battles are won"—whether attributed to Boyd's OODA loop or not. We observed the DoD's modus operandi. We oriented our efforts based on our constitutionally based citizen-soldier duties. And finally, we decided to act.

Our OODA loop was pretty simple. We met the DoD at the "merge" with the media. Sometimes we even "lead turned" them and beat them to the "merge." Over the years of challenging the DoD's leaders, we not only realized that the DoD couldn't answer our questions, but they also responded poorly to our entry into their OODA loop and that of the printed and internet press. The DoD wasn't accustomed to their own troops mixing it up with them in the OODA loop of the media. Normally the DoD would crush such activity unless they couldn't stand the scrutiny of their documents and didn't want to highlight the factual dichotomies.

We employed the concepts of Boyd's vision in other realms as well in response to the manipulative efforts by the DoD's misinformation warriors. We stayed ahead of their OODA loop and exposed their "dupe-loop," a twist of phrase coined by my squadron mate, Enzo Marchese. While we channeled Boyd's vision with truth, law, and the DoD's own documents, the DoD duped their own commanders and troops, as well as the lawmakers and the media. We went one step further within the DoD's OODA loop by publishing papers based on "Operational Risk Management" (ORM) tools as defined by Air Force regulations.[8] We also worked with the Air Force's Air University, exercising academic freedom, to respond to the DoD articles that hyped the threat and the moral imperative of the vaccine—we embedded within the DoD OODA loops.

The Air University placed an extensive essay of ours on their military website. The paper documented the compounding policy process breakdowns inherent in the anthrax vaccine program. The paper successfully underwent a security and policy content review by the secretary of the Air Force and was approved for public release and unlimited distribution.[9] The paper remained online for years, but censorship attempts temporarily prevailed—the paper vanished. A polite letter to the school's accreditation board fortunately fixed the suppression—for now. A Connecticut newspaper also cooperated, setting up an "anthrax project" section on their website. Ultimately, over seventy articles resided on the site, documenting the extraordinary work of T. Dennie Williams, the *Courant*'s Gulf War Illness and anthrax vaccine investigative reporter.[10] Dennie also crafted an article in the *Connecticut Law Tribune* in tribute to Russ after his passing, titled "Anthrax Avengers."[11] Courageous journalists like Dennie helped to expose the dupe-loop.

We learned the DoD's dupe-loop existed in many other circles as well. The DoD provided "free help" and expertise to congressional offices. These officers, liaisons, and "defense fellows" were placed within the halls of Congress wearing civilian clothing, blending into the staff, and addressing military concerns for constituents. Other DoD officials worked within the nation's scientific bodies, such as the National Academy of Sciences and the Centers for Disease Control (CDC). The DoD's tentacles were fascinatingly far-reaching. But we found we could operate uncontested within the DoD's dupe-loop as well—most likely because they didn't want to highlight our presence or what we were saying. We were only doing what they were doing, and we were using the tools they'd given us. We were telling the truth with their documents and articulating the DoD's official position prior to the genesis of the anthrax vaccine program.

The DoD underlings rewrote the science and the rhetoric to conveniently conform to Defense Secretary Cohen's program. It was stovepiped from the top down and never properly formulated from the bottom up. This idiosyncratic reality was captured in news articles before the launch of the program. One revealed, "A review of events leading to the

Clinton vaccine decision reveals that the proposal was pushed by a small group of scientists, businessmen and policy makers who largely shared the same views as they struggled to do something, anything, about a threat whose dimensions were potentially terrifying but frustratingly unclear."[12]

Another article documented how policy decisions surrounding the anthrax vaccine program "departed from normal departmental practice this spring." It specifically cited that the "meetings were unusual in that we were starting at the top instead of trying to staff an issue from the bottom up."[13] It was unmistakable that anthrax vaccine was a program wrought with upside-down decision-making, doctrinal departures, and quick fixes. They used "independent experts" who weren't experts. Supplemental testing was terminated due to "inconsistencies." Memos were crafted to obscure the vaccine's known experimental status from a never-confirmed FDA commissioner. We identified these fundamental dysfunctions in the DoD's distorted dupe-loop.

Based on this Boyd-inspired observation and situational awareness, we oriented and decided to act. We challenged one of the Washington-area papers to publish a point-counterpoint essay between the DoD officials and us. The *Baltimore Sun* agreed after conceding that it was the only way to end my persistent calls to their toll-free numbers. The man the DoD sent to answer our essay was the same gate guard messenger who replied to Senator Blumenthal—Charles Cragin, who served as Deputy Under Secretary of Defense for Personnel and Readiness. With our common knowledge of Cragin's tired rhetoric, we executed our OODA loop on the perimeter of DoD's Washington, DC, informational battlespace.

The *Baltimore Sun* published Russ's and my essay on January 30th, 2000.[14] Our essay listed detailed excerpts from the DoD's documents. The Pentagon's *Early Bird* publication, published by our friends at the *Military Times*, was a summation of military stories from around the world. The *Early Bird* carried the essays, titled "Military split over anthrax shots." It included excerpts such as Army Surgeon General Ronald Blanck's past testimony that "anthrax vaccine should continue to be considered as a potential cause for undiagnosed illnesses in Persian

Gulf military personnel." It quoted a US Army colonel and biological researcher for the textbook *Vaccines*, concluding, "The current vaccine against anthrax is unsatisfactory."

The counterpoint essay by Mr. Cragin didn't respond to any of the facts we listed, despite the paper giving him an advance copy. Even the *Baltimore Sun* editor was surprised to see the same old boilerplate rhetoric from the DoD, despite Cragin's opportunity to respond to our points. Cragin's soundbites included, "Anthrax vaccine is as safe as most common vaccines. It has had an excellent safety record . . . we have an obligation to give our personnel the best protection available from all anticipated threats—anthrax is one of those threats; and the vaccine offers safe and effective protection." I hoped for new insights, but we'd heard it all before.

After the opinion editorial exchange, I was relieved to know that we weren't missing something, and they really had nothing but empty, easily refutable rhetoric. It appeared the DoD was merely bluffing their way through the dilemma by hyping the threat and offering false assurances about the vaccine's easily disproved protective qualities and safety profile. Not too many years later, the anthrax letter attacks proved to the entire nation that simple antibiotics fully protected against inhaled anthrax, even after symptoms of infection. As usual, the DoD's Cragin had nothing but fear to convey. Once again, no one else from the DoD's PR machine reacted.

January of 2000 was a good month in exposing the DoD's dupe-loop and beating them at their own OODA loop games. After we completed work on the *Baltimore Sun* essay, but before it was published, the *Washington Post* contacted me. I was grateful that they too published a front-page "Outlook" section opinion editorial on the exact same day that the *Baltimore Sun* essay hit the newsstands. The *Baltimore Sun* editor complimented me on getting published in two Sunday opinion sections in Washington, DC–area papers on the same day. The coincidence of unique essays being published on the same day was just that—double-OODA-loop coincidence.

Like the *Sun*'s point-counterpoint format, the *Post* coincidentally employed the same. Excerpts from my essay and a counterpoint article by a retired Army general included:[15]

I have been a loyal member of the US military. Yet today, along with many other service members and members of Congress, I find myself challenging the Pentagon's anthrax vaccine program. Some will see my act as one of disloyalty. It is not. My loyalty is to a military institution of integrity, not to policies that lack integrity. As the Pentagon digs in to protect a policy under fire, debating this issue outside military channels . . . it is also my duty to let my chain of command know when something doesn't pass the common-sense test. As military leaders employ court-martials and even imprisonment to enforce this questionable policy, a contagious groupthink at the highest levels has equated critical thinking with a lack of good order and discipline. . . . While service members must sacrifice personal freedoms, they entrust certain rights, such as their safety, to their commanders.

Sticking Point

In Defending Its Troops Against Anthrax, The Pentagon Has Injected Distrust Instead

By James Terry Scott

I've always believed that the military tries to do the right thing. That's why I never really questioned its decisions to protect me against various diseases—even if I sometimes wondered what the odds of contracting those illnesses might be. Thirty-five years ago, as I prepared for duty in Vietnam, I accepted the painful gamma globulin shots to reduce the risk of hepatitis, even though at the time my fellow soldiers and I had serious questions about their efficacy. I accepted them just as I accepted the forced separation from my family and the rigors of combat. I certainly never considered inoculations to be the most dangerous part of the contract between soldier and state.

So you might think I'd have no understanding for the 300 or so men and women who have refused to take the anthrax vaccine that the Department of Defense has ordered for all 2.4 million active-duty and reservist personnel. But after reading and hearing the confusing and contradictory claims about the vaccination program, and observing the positions of both the Defense Department and the resisters harden into intractability, I've come to believe there's an important message that both sides need to hear:

When you see you are digging yourselves into a hole, put down the shovel.

See ANTHRAX, B4, Col. 1

Terry Scott, director of the national security programs at Harvard's Kennedy School of Government, retired as a lieutenant general in 1996, having led the U.S. Army Special Operations Command. During the Persian Gulf War, he was deputy commander of the Army's 24th Mechanized Infantry Division, which led the "left hook" attack in Iraq.

Why Am I Resisting the Vaccine? The Military Trained Me To

By Thomas L. Rempfer

For almost 17 years—as an Air Force Academy cadet, an officer and fighter pilot, and now a reservist—I have been a loyal member of the U.S. military. Yet today, along with many other service members and members of Congress, I find myself

military channels is my only recourse.

A year ago, as part of the program to inoculate all service members against anthrax, I was asked to roll up my sleeve and take the first of six shots. And like hundreds of other service members, I said "no"—in my case, because of unanswered questions about the vac-

Rempfer's *Washington Post* opinion editorial,
"Sticking Point"—with David Suter illustration.

Lieutenant General Terry Scott, US Army, retired, director of the national security program at Harvard's Kennedy School of Government, wrote the counterpoint to my *Washington Post* Outlook piece and illustrated the internal divides within the DoD. During the Persian Gulf War, General Scott was the deputy commander of the Army's 24th Mechanized Infantry Division, which led the "left hook" attack in Iraq. Though written as a counterpoint, General Scott's essay bolstered our viewpoints: "The Defense Department has failed to establish a credible argument for administering the anthrax vaccine. . . . What's to be done? Starting now, the Defense Department must rebuild the credibility of military medical policy and military medicine. . . . So, here's my message to the Department of Defense: Put down the shovel."[16]

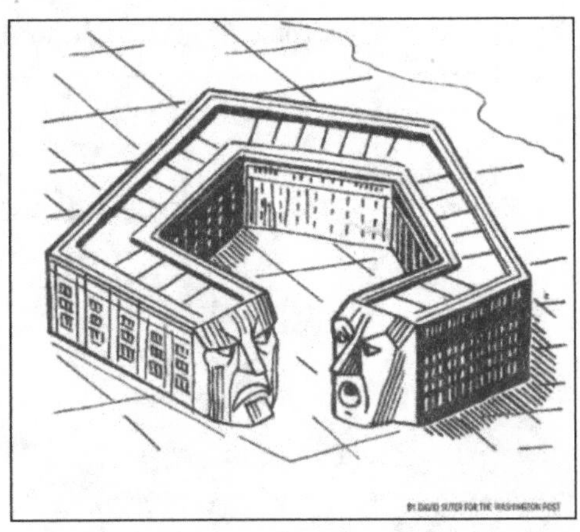

Washington Post's David Suter illustration accompanying General Scott's counterpoint.

After the *Baltimore Sun* and *Washingtons Post* articles were published, two key new players entered our fight by contacting me through the *Post*'s editorial staff. Both were retired Air Force colonels. One was the ex-USAF medical corps officer and ex-FDA regulator who joined our formation, Sammie Young. The other was a retired USAF fighter pilot and DC insider, Craig "Raven" Duehring. Sammie continuously proved to be vital to the scientific and regulatory validation of our concerns based on his experiences as a medical corps officer and former FDA regulator. He assisted our continued research on manufacturing practice violations.

Mr. Duehring later earned a post on the Bush administration transition team, becoming Mr. Charles Cragin's replacement in the Pentagon. Our entrance into the DoD's dupe-loop paid off. Officials on the inside of the DoD's and the FDA's OODA loops were contacting us and allowing us to get further inside their own decision cycles. These officials helped us to understand the most important aspect of the DoD's and the FDA's dubious dupe-loop—awareness of their tactics. This added to our SA (situational awareness) of their M.O. as we continued to fly high cover over the federal military establishment.

We agreed with General Scott's advice to "Put down the shovel and go back to making your units the best they can be." Though sage advice, if we dropped our shovels along the way, we never would have met good people like Sammie or Raven. Not to mention, we were kicked out of our squadrons, along with hundreds of other aircrews. If we "put down the shovel," we would never be able to fly and serve again. History showed that the DoD would not "put down the shovel," admit their errors, or right the wrongs until ordered to do so. This was something our top DoD leaders seemed culturally incapable of doing. Dropping our shovel was not an option when we considered that the military had a long track record of not properly or promptly caring for the people they harmed by medical mistakes. Our mission became something larger than the anthrax vaccine. By not dropping our shovels or allowing ourselves to be buried, we hoped we might, borrowing the DoD's own euphemisms, "change the paradigm" and "transform" the institution. The DoD needed to protect their people before protecting policy in the future.

The *Washington Post*, *Baltimore Sun*, *Hartford Courant*, *Military Times*, and many other publications had earned our gratitude for exercising their constitutional duty (freedom of the press), and for facilitating our rights (freedom of speech). With their mutual support, we had entered the DoD's deceptive dupe-loops and invited them to enter ours. What the DoD didn't realize was we intended on fixing their mistakes. Our goals were to expose and modify the unacceptable dupe-loop tactics. Mostly, they tried to ignore us, an exemplification of their M.O. on how they responded to veterans' health issues for decades. We had every intention of fixing that dupe-loop too. We were pitted across two sides of a house divided, in a multi-bogie, multi-dimensional engagement. While they tried to bury us through diversion and disparagement, we kept on shoveling too, using the institution's tools to keep its course true.

I think Colonel Boyd would have been proud of our Tiger Team. He mentored, "Tiger, one day you will come to a fork in the road and you're going to have to make a decision about which direction you want to go. . . . In life there is often a roll call. That's when you will have to make a decision. To be or to do? Which way will you go?"[17] Our vector was clear—we'd keep doing what we were trained to do—lobbying all who would listen about the DoD's bad behavior.

MILE 11:
Lobbying (1999 to 2001)

Our country's birth owes its origins to outrage over absurdities. One of the original patriots, Thomas Paine, wrote the *Common Sense* pamphlet, which paved the way for the Declaration of Independence. Paine saliently called for self-determination on our continent from the unrepresentative taxation by an island-based empire. Paine followed the pamphlet with a series of sixteen papers titled *The American Crisis*. The first of these papers opened with the distressing words, "These are the times that try men's souls."[1] The absurdity, the "common sense," and the trying times applied to our dilemma. *The American Crisis* included the astute premise that "A bad cause will ever be supported by bad means and bad men."[2]

Though we knew that there was an element of well-intentioned desire to protect our troops rooted deep in the foundations of the anthrax vaccine policy, the righteousness of that intent had long since been outweighed by the willfully blind ignorance of the illnesses, the law, and the subsequent wrongful punishments. Paine's prophetic words applied in all measures to include the "bad" nature and intentions of the gate guards as they confused top entities of the DoD, various branches of the government, and the media in order to defend the illegal conduct.

One of the gate guards, a US Army colonel and researcher, provided a potential example when answering questions by Congressman Burton about different strains and animal studies:

> **Burton:** You authored the only peer-reviewed efficacy study on anthrax in the 1999 edition of the medical textbook *Vaccines*. You wrote that the

> current anthrax vaccine is unsatisfactory for several reasons, including that there is evidence in rodents that the efficacy of the vaccine may be lower against some strains of anthrax than others. Did you write that?
> **Colonel [US Army researcher]:** Those statements were made in reference to an idealized vaccine, a goal that we are all approaching . . .
> **Burton:** But you said the current anthrax vaccine, this was in 1999, is unsatisfactory for several reasons . . .

It really didn't matter because the important thing was that the US Army researcher admitted that proof of effectiveness of the vaccine was based on "animal models." This drove to the core of one of the legally questionable aspects of the mandatory vaccine policy. The DoD sponsored legislation to get vaccines fully approved based on animal models, their so-called "animal efficacy rule." The emphasis related to strains was disturbing, but the confusion this created only helped to solidify the DoD "expert's" admission that the effectiveness of the vaccine had not been studied for inhalation anthrax—which was the purpose of DoD-mandated inoculations. It was more than a technicality. The law required a vaccine to be approved for specific purposes. For inhalation anthrax, the vaccine was experimental. The vaccine needed this specific approval and a finalized license by the FDA. The anthrax vaccine didn't have either one.

Russ and I appreciated that the elected official waved the flag on the confusing testimony about an "idealized" vaccine. We added this testimony to what we provided to our DoD inspector general's office as further examples of ambiguous statements to protect the program instead of our troops' rights. We hoped that our Defense Department would hold itself or its officials accountable instead of hammering lower-ranking military members. To top it off, this US Army researcher was apparently the DoD official who briefed the investigational new drug application for inhalation anthrax on multiple occasions to get an "idealized" indication for inhalation anthrax approved, since animal data was legally insufficient for full licensure.

This drove to the core of the illegality of the order. The US Army researcher maybe hoped he was well inside the representative's

OODA loop, but fortunately they knew something was wrong. The testimony seemed inconsistent with the researcher's *Vaccines* textbook article, but it wasn't the first time the colonel committed testimonial peculiarities. This was the same gate guard who also provided curious testimony in a Canadian court-martial over anthrax vaccine.

The US Army colonel and researcher was previously dispatched to testify in a Canadian court-martial of an anthrax vaccine refuser, Sergeant Michael Richard Kipling. He reportedly wore civilian clothes and testified about of the anthrax vaccine in front of Canada's chief military judge, Colonel G. L. Brais. He gave vague and unsure answers. He appeared to deny knowledge of the primary purpose of the 1996 investigational new drug application prepared by the US Army for the anthrax vaccine manufacturer to obtain a new licensed indication for inhalation anthrax. The US Army colonel and researcher responded to the defense counsel's interrogation:

> Assistant Defence Counsel Duncan: Q. the drug was licenced for cutaneous anthrax only and that there has been a subsequent amendment for coverage for inhalation anthrax, would you agree with me or disagree with me?
>
> Colonel [US Army researcher]: A. I'm not aware of that.

The documents, Russ discovered and JR analyzed, revealed on several occasions that the US Army colonel and researcher appeared "aware" of the inhalation anthrax application. One was on October 20th, 1995. The same US Army researcher and colonel presented a briefing at a meeting held by the Joint Program Office for Biological Defense (JPOBD). The purpose was to develop a game plan for "Changing the Food and Drug Administration License for the Michigan Department of Public Health (MDPH) Anthrax Vaccine to Meet Military Requirements."

According to the meeting minutes, the US Army researcher "presented a briefing covering the three topics: (1) evidence for a reduction in the number of doses of anthrax vaccine, (2) evidence for vaccine efficacy against an aerosol challenge [inhalation anthrax], and (3) progress towards an in vitro correlate of immunity." The minutes documented that the US Army researcher "agreed that the surrogate animal model needed to be established." He acknowledged that "there was insufficient data to demonstrate protection against inhalation disease." One of the meeting's briefing slides, titled "Immediate Objectives for Anthrax Vaccine Licensure," stated the purpose of the application was to "obtain a [FDA] Product License Application Supplement approval for a specific immunization schedule change . . . and for a labeled indication change."

At a follow-up meeting on February 9th, 1996, the same researcher presented another briefing titled "Research Plan to Support Reduction in Dosage of Licensed Anthrax Vaccine (AVA) and Indication for Aerosol Exposure." The researcher appeared integrally "aware," and the meeting minutes showed that he discussed the need for the study to show a correlation between animal and human immune response to the vaccine—a recognition that the vaccine never demonstrated efficacy for inhalation anthrax in humans, a requirement for licensure.

After the application was actually filed with the FDA in September 1996, another briefing titled "Supplement to AVA [anthrax vaccine adsorbed] License" was presented by the same US Army researcher and others on November 10th, 1997. The briefing slides appeared to detail three changes sought for the license (including an indication for inhalation anthrax). The researcher may have been "aware" of studies intended to obtain FDA approval for these changes. A final document included an update to the application specifying only "inhalation anthrax," dated January 29th, 1999. The update included an attendance roster from a December 15th, 1998 "meeting between BioPort and the Center for Biologics Evaluation and Research." The US Army researcher from USAMRIID was listed on the roster next to Robert C. Myers, the principal head of the manufacturer, BioPort Corporation. Given the documents, it was odd to witness that he was "not aware" of what the assistant defense counsel was questioning him about.

According to court transcripts, the US Army researcher's memory of the issues declined. He stated, "I know that there have been studies dealing with trying to reduce the number of doses and to look at the route of administration." His knowledge of the inhalation anthrax part faded:

> **Q.** So are you saying, sir, that you're not familiar . . . ?
> **A.** No, no. I don't know that—I'd have to look back at the documents.
> **Q.** Okay. So you're not saying the drug is not in an IND status . . . ?
> **A.** You know, I'm not clear what you're saying in terms of . . .
> **Q.** So you're saying you're not sure?
> **A.** That's right.

The US Army colonel and researcher told the court he'd "have to look back at the documents." That's what Russ, JR, Lou, Sammie, and I had done. We discovered those documents created questions compared to the testimonials. What was not clear was if the testimonials to the US Congress and the Canadian court potentially obstructed justice, so we documented our discoveries in the form of another DoD inspector general inquiry, and also provided the documents to the DoD IG in a meeting they requested in DC. The US Army researcher's 1994 *Vaccines* article about the "unsatisfactory," versus "idealized," vaccine was updated at the same time we were grounded and discharged in a new edition with the same critical verbiage.[3] Concurrently, another study was published in a British medical journal, *The Lancet*, suggesting a link between Gulf War vaccinations and Gulf War Illness.[4]

Our allies knew something was amiss and the Canadians knew something was wrong as well. The DoD had not been successful in getting inside their OODA loop. Judge Brais dismissed the case against Michael Kipling. He deemed the anthrax vaccine: "unsafe."[5] As well, a Canadian newspaper reported that some of the vaccine administered to Canadian troops came from lots that had been "placed under quarantine" or had "since failed testing."[6] Canadian troops received suspect doses that may not have been "potent enough to protect them from biological warfare" or were from "a batch that was highly contaminated with gasket particles."

A Vancouver news article also documented that "extra testing was initiated after the manufacturer . . . was cited for poor quality control and record keeping and for substandard manufacturing practices" and that "in 1997, the FDA gave notice that it would revoke the manufacturer's licence if it did not comply with regulations." We filed complaints concerning the US Army researcher's testimony, but the DoD IG ignored, forwarded, or dismissed our best efforts. Fortunately, the Canadian military saw the anthrax vaccine for what it was—"unsafe."

Though the Canadian military appealed Kipling's case, in what appeared to be an effort to "save face" for themselves and the DoD in the United States, within a year the case against Kipling was dropped entirely. That seemed to be a pattern as well: lose, appeal, and then withdraw—all for appearances. Our Canadian allies dropped the issue and never punished a single soldier for not accepting the US anthrax vaccine. It reaffirmed that Canada's actions were certainly not a vote of confidence of the clouded testimony of the US Army's researcher.

I was embarrassed at the possibility that Kipling could have been found guilty, like Dr. Buck. What would Canadians think if they knew that the US military officer, who came to their country in civvies to get their soldier punished over the anthrax vaccine, was a US Army colonel and chief anthrax vaccine researcher, but testified that he was "not aware" of its problems? What would the American people think if they knew their military officer provided forgetful testimony to a Canadian military court-martial to get a foreign soldier punished over our anthrax vaccine?

Years later, Canada's military appeared to learn the lessons and reversed the same errors with COVID mandates. Judicial review decided it was "wrong to accept that requiring someone to choose between vaccination and severe consequences like termination did not violate their charter rights." A military judge ruled for the troops that were "left with no meaningful choice between complying with the law and exercising their liberty."[7]

JR postulated the entire ordeal exemplified "a bipartisan mess, requiring a bipartisan solution." The political paralysis created through the bipartisan mistakes involved both political parties and marred five presidential administrations. The paralysis of the process was further poisoned by the DoD's consistent involvement and interference in the OODA loops of the many government agencies and branches whose job it was to fix the problems. These military lobbyist gate guards in civilian clothing paralyzed oversight and stained the integrity of the DoD.

Over time such a paralysis, in line with human nature, became more expensive and embarrassing to fix. As a result, the problems were ignored and compounded with one quick fix after another. This was the anthrax vaccine dilemma. We believed the blame rested squarely on the entities within the DoD that were responsible for the deceit and abuse of power—the gate guards. Even worse, they subsequently took advantage of the political paralysis to perpetuate the illegalities. The "bad cause" and "bad means" persisted due to abusive manipulations of the system by "bad men," despite efforts by many good lawmakers and scores of diligent reporters.

In our own attempts to lobby legislators and their staffs on Capitol Hill, one of our fellow reservists, who served as a "defense fellow" liaison to a congressperson's office, explained how the DoD anthrax vaccine program gate guards would come to Congress and lobby members and their staffers in a secure area known as "the vault." It was there where congressional members received what they later termed as the "rotting brain" brief. The DoD scared the staffers and elected representatives with pictures of an autopsied brain with a meningitis infection, implying it was due to a lethal anthrax exposure. The "threat" and the infected brain, though having nothing to do with anthrax infection, was convincing enough to make the majority of the congressional leaders look the other way from the obvious "technical" glitches in the implementation of the DoD's anthrax vaccine program. The DoD's lobbying pushed the "protect the troops" rhetoric and of course added, "soldiers can't refuse to wear their helmets."

But we kept lobbying too. During an interview with Paula Zahn on Fox News' *The Edge* program, another Air Force colleague, Technical

Sergeant William Mangieri, and I tried to break through the soundbites. Sergeant Mangieri was responsible for nuclear, biological, and chemical warfare defense training at Stewart Air National Guard Base in Newburgh, New York. Mangieri initially supported the DoD's anthrax vaccine program, but as base personnel asked him about the anthrax vaccine, he conducted his own research and "found a pattern of inconsistencies, constant misrepresentations by the DoD and poor program administration and management," according to his later congressional testimony. The DoD drove Mangieri to join our formation.

Mangieri courageously voiced similar concerns to Paula Zahn, and she asked me to wrap it up with final thoughts. I expressed the same sentiment as the DoD. Our duty was also to "protect the troops." That was my counterpoint to their soundbites. Like the DoD documents that supported our side of the story, we borrowed their soundbites too. We had the law and the DoD's own documents on our side, so we didn't give up. I made similar points in a CNN interview, explaining "we discovered that the previous government position on the vaccine was that it was experimental," and "we also discovered that the vaccine came from a plant that had lost its FDA validation."[8] The DoD should have been protecting the troops from those realities, but did not.

During our many lobbying trips, staffers and lawmakers reviewed our DoD documents. They became similarly outraged at the deceptions. Over time, the obstacles, as well as the support, were truly bipartisan—no different from the root causes of the problems. The policy, initially implemented by the first Republican Bush administration during the 1991 Gulf War, was renewed by the Democratic Clinton administration through Republican Defense Secretary Cohen. There existed a bipartisan reluctance to admit these bipartisan mistakes. Perhaps it became politically expedient to feign ignorance versus exposing bipartisan transgressions. An understanding of human nature and fear of governmental liability added perspective on this reluctance. Officials rationalized that it could hurt the DoD if the errors were exposed.

Even the elected representatives who also served in the armed forces admitted the approval problems. Representative Steve Buyer from

Indiana stated, "It was never FDA approved against inhalation. And I'm gonna tell you you're right, it's not."[9] The US Army officer admitted the limitations of the vaccine as did the US Army's own documents. Those pages confirmed the vaccine's "higher than desirable rate of reactogenicity, and, in some cases, lack of strong enough efficacy against infection by the aerosol route of exposure."[10]

Another continual ally on the Senate side of Capitol Hill was our Democratic senator, Christopher Dodd. In our five years of lobbying against the anthrax vaccine, Senator Dodd's civilian staff members graciously met with us in his offices in Washington and Connecticut during almost a dozen meetings. We never observed a DoD liaison in the senator's chambers, and the senator wrote several letters to ask the Department of Defense inspector general (DoD IG) about the status of our inquiries. One, dated January 22nd, 2001, inquired about the DoD IG not investigating one of our complaints. Instead, we learned the DoD IG referred it to the Office of the Secretary of Defense, specifically to the senior advisor to the deputy secretary of defense for chemical and biological protection who was responsible for the vaccine decree.

Senator Dodd's letter added, "Majors Dingle and Rempfer are concerned, therefore, that the allegations they are making have not been, and may not be, adequately addressed. These allegations are serious, and they need to be addressed." Our complaint remained unresolved over two years later. We augmented the complaint with additional information over the years as our research turned up more irregularities, such as the suspected adulteration of the vaccine. In time, the DoD IG acknowledged that how the first complaint was handled was "inappropriate."[11] Senator Dodd and his staff's steady efforts epitomized that of a lawmaker versus a "politician."

Senator Dodd did not turn a willfully blind eye or allow the DoD to mislead his office concerning our predicament. Senator Dodd's office remained diligently dutiful, committed to a process of patient inquiry. As fighter pilots, we often referred to someone hawking the fight, ready to help, as a fellow fighter pilot offering "detached mutual support." Senator

Dodd supported us as a reliable wingman in our anthrax vaccine marathon. We remained nonpartisan, educating legislators and their staffs equally on both sides of the aisle.

Our own legislators continued to write letters mirroring our concerns. Representative Shays joined the call for the correction of military records for any troops punished over the anthrax vaccine. In a press release, he wrote that the disciplinary actions should be reversed "against those whose legitimate health concerns prompted a decision to opt out of the program." According to a *Hartford Courant* article, he "urged Cohen to review all of the several hundred disciplinary actions against those resisting anthrax inoculations," and added, "Without an assured supply of vaccine, continuing to order any more soldiers, sailors, airmen, and marines to start the shots they may never finish constitutes military malfeasance and medical malpractice."[12]

My representative, Nancy Johnson, joined the oversight early on with a letter to Defense Secretary Cohen. She expressed "vehement opposition to dishonorably discharging service members who leave because they fear the health consequences of the mandatory anthrax vaccine." She added, "Every soldier is well aware of the dangers that come with serving in the military, but I doubt they ever expected their government would force them to take a vaccine that has unknown long-term side effects" [Rep. Johnson press release at Appendix M].

Representative Johnson also submitted written testimony to a House hearing reiterating her call to "suspend the mandatory aspect of the program and to reverse the dishonorable discharges of servicemen who have refused to take the vaccine." She cosponsored legislation to suspend the mandatory program. Most of our state's delegation in Congress at least tried to help by calling for the review of the program and the correction of the adverse personnel actions. Nonetheless, without a mandate from Congress, despite members and their staffs privately acknowledging the lack of full approval, there was little that individual lawmakers could do.[13]

After a congressional map redistricting, my new representative, Robert Simmons, chair for the House Veterans' Affairs Subcommittee on Health, helped us too. He provided our Operational Risk Management

(ORM) analysis to the DoD senior leadership for comment. Congressman Simmons told me in his office that four-star DoD generals tended to only listen to other four-star generals. For protocol purposes, the retired Army Reserve colonel explained he carried the four-star-rank equivalent as a Congressional Representative. Coming from him, we hoped that the operational risk management paper would jolt the DoD to reflect on its conduct, hold itself accountable, and do the right thing. Such vision and courage would hopefully elicit trust in our leaders in the future and ensure that we never experimented on our troops again.

Absent such personal four-star general-officer-level accountability, we felt confident that someday a federal judge would complete the judicial review and correct the wrongs if the DoD didn't have to courage to right it themselves. In time, the DoD responded to Congressman Simmons's letter. Though Representative Simmons wrote the memo to a senior civilian assistant secretary of defense, the reply came from a one-star army general. I'm not sure what was more insulting to my congressman. Was it the fact that accountability was non-existent? Or was it the runaround from an Army officer serving in the office responsible for the anthrax vaccine program? Or was it the mere one-page, evasive response to our forty-plus pages of detailed analysis? The DoD disregarded a congressional committee chair's inquiry. This was not the way the system of government was supposed to work. Instead of firm accountability, we witnessed dodging, derision, denial, delay, discharge, and disregarding ill troops as the order of the day.

Russ and I weren't lawyers, but we knew how to use the tools the military gave us. We followed their codes, rules, policies, and regulations, no differently than lawyers do. It was based on this training that Russ and I wrote the operational risk management analysis of the DoD's anthrax vaccine program using the Air Force's policy guidance.[14] The paper analyzed the anthrax vaccine program based on the long-term risks to DoD credibility, as well as the liabilities the Department of Veterans Affairs would inherit through disability claims.[15] We applied our operational logic to the dilemma and published the analysis to protect the institution from such recurring illegal conduct and future liabilities. Our point was simply that if, long-term, our DoD was found guilty of the illegal conduct we identified and analyzed with the anthrax vaccine program, what impact would this have on our reputation as a military? Especially if everyone realized that our leaders were aware of, but decided to ignore, the illegalities from the outset.

In addition to Representative Simmons, we provided the operational risk management paper to our military chain of command based on our own Air Force chief of staff, General John P. Jumper's memorandum directing the Air Force's senior leaders and commanders to ensure the complete integration of ORM in their areas of responsibility.[16] There was a perceived potential conflict of interest because General Jumper was also involved in key decisions related to the anthrax vaccine prior to Desert Storm. The chief of staff may have been aware of the suspected illegal and adulterating changes made to the vaccine's manufacturing process. General Jumper's involvement with the vaccine coincided with the increase in production capacity in order to accommodate the DoD's needs. The unapproved changes included altering the filtration system, using different fermentation equipment, different sterilization procedures, and new chill tanks.

The FDA was eventually notified of some of these changes after the fact. They were unaware of others until Russ's research was brought to the attention of the Department of Justice, Congress, and GAO officials, spurring inquires in 2000. DoD involvement to some degree was apparent from a review of declassified documents.[17] We turned these documents over to our DoD inspector general's office for review, trusting in the system, in hopes they could get to the bottom of all the unanswered questions. In good faith, we continued to faithfully use the ORM tools our military provided and sent our inquiries to the top of the chain of command.

The tools of ORM supported the logic Russ and I applied to our earliest evaluations of the anthrax vaccine program. One was the "Logic Tool." Another, the "What If Tool," examined possible consequences.[18] We combined the logic and what if tools, offering two propositions and a categorical syllogism.[19] For example, what if the government understood that investigational or experimental products "cannot be mandated"[20, 21]—what if the DoD previously acknowledged the investigational and experimental nature of the anthrax vaccine—then the mandate's illegality was the logical categorical conclusion.[22] The common sense intellectual exercise was healthy for us, but unfortunately the DoD would not listen to logic and uniformly ignored our lobbying.

We concluded that violating the law was unacceptable conduct by the DoD's senior leadership, and it violated Thomas Paine's "common sense." Other viable and legal risk-control options existed. Arguably, this perceived good cause turned bad through the use of bad means by otherwise good people. Paine would hopefully be proud that we lobbied against the bad behaviors that "try men's souls." We just needed more good people like our Connecticut senators and representatives. We needed more good means and good people for our good cause.

MILE 12: The Perot Factor (1999 to 2002)

Some of our lobbying efforts were more effective thanks to a good man and his good means. Mr. H. Ross Perot joined our formation one fateful day. Mr. Perot was a philanthropist, one who provided goodwill to his fellow veterans and service members. My academy friend, Frank Schmuck, provided our briefing book to Mr. Perot. Frank had his own bout with Gulf War Illness after his 1991 service in the Persian Gulf. He advocated for veterans with Mr. Perot's help. Mr. Perot inquired of Frank in the spring of 2001 about the anthrax vaccine controversy and Frank referred him to Russ and me. My phone rang, followed by a distinct Texan's voice.

"This is Ross Perot. Is this 'Buzz'?" I replied affirmatively. He came back, "Is it okay if I call you Buzz?" "Yes, sir." I affirmed again. Mr. Perot asked me to tell him about the anthrax vaccine. I started succinctly from the beginning. I'd lost a few listeners over the years, so I didn't overdo it with details—just the facts—improper original licensure, known to be unsatisfactory based on the DoD's pre-Gulf war documents, alleged adulterating unapproved manufacturing changes in 1990, never properly studied as a possible cause of Gulf War Illness due to a lack of record keeping, and illegally mandated contrary to law for a known unapproved purpose. It seemed overwhelming, but I followed my own rule to "keep it simple"—and Mr. Perot listened.

Following a brief conversation, Mr. Perot asked me if it was okay if he called back after he discussed the issue with a contact in the Pentagon. I responded affirmatively again, and we hung up. About an hour later, Mr. Perot called back. My wife ordered me to let her answer the phone to see if I was really talking to "RP," as Frank referred to him. She answered the phone and he humorously teased her about marrying an Air Force pilot versus a Navy pilot. She gave him one of her smart-alecky responses and then handed me the phone with a loving, proud smile.

After all the years on this flight path, it made me feel good to have my wife grin with pride as I started answering Mr. Perot's questions again. He hit me with a couple of the classic inquiries, probably because someone who was reading the script at the Pentagon had offered them up in defense of the program. Someone tried to rein him back in by throwing out some of the "soldiers can't refuse to wear their helmets" rhetoric. But Mr. Perot, who could taste a lemon when one was shoved down his throat, could sense something was off. He asked me to send the new information we compiled, so we emailed his secretary our updated briefing book.

Within a day, Mr. Perot called back, asked a few more questions, and complimented us on the briefing book. I concluded our conversation with the bottom line: the drug was "adulterated" under the law, but no one is doing their job and properly deeming it as such. I added that adulterated drugs can't be transported in interstate commerce, not to mention be mandated for use on human beings. He snapped back, "Well, Schmuck said you'd be the man with the answers on this subject. This is just plain stupid." He was right. Mr. Perot was a straightforward guy. This dilemma needed a gigantic dose of that quality. Perot listened to the DoD's pitch and realized quickly that things weren't adding up. Mr. Perot was exceptionally good at business. Mr. Perot knew that America's troops and taxpayers were being swindled.

The easy thing for Mr. Perot to do would have been to not get involved. An even simpler path would have been to not ask for our information in the first place. That may have been the simplest road, but it was not the logical one for a veteran, a Naval Academy graduate, and a philanthropist—who by definition actively promotes human welfare and provides goodwill to his citizens. Mr. Perot found himself inextricably journeying down the same road as Russ and me, and asked if he could provide our briefing book to his contacts in the White House.

Mr. Perot emailed our extensive briefing book on the anthrax vaccine dilemma, plus our supporting documents, to his contact. He included his own concerns over the investigation of Gulf War Illness in a package to his fellow Texans who had successfully lassoed the White House office and executive branch. Mr. Perot's letter, dated March 22nd, 2001, was addressed to Mr. Karl Rove, Senior Advisor to the President of the United States, The White House:

> **Attached (Exhibit A) is the information you requested about the people in the Pentagon whose mission has been to ignore the Gulf War illnesses and label them as stress.**

> **In addition, I have included (Exhibit B), a booklet prepared by two Air Force Academy graduates on the anthrax vaccine problem, and a packet of news stories (Exhibit C) that reveals a great deal about the company in Michigan, Bioport.**

Excerpts from letter by H. Ross Perot to Karl Rove at Appendix N.

Mr. Perot detailed in his exhibit on Gulf War Illness how "Clinton officials established a policy that the soldiers were suffering from the effects of stress and have attempted to make the science conform to this conclusion." Russ and I realized that the parallel efforts of Mr. Perot, which predated our own on the larger problems of Gulf War Illness, shared our conclusions. He called out the postfix nature of creating cover-up science in order to protect policy. In our case, it was the DoD that created pseudoscience after 1997 to make the anthrax vaccine appear "safe, effective, and FDA approved." Our research proved these assertions were fundamentally false.

Similarly, Mr. Perot and his group of Gulf War Illness researchers determined that the DoD and the administration disingenuously created new science to discount the maladies Persian Gulf War veterans suffered. Mr. Perot witnessed the same M.O. we observed with respect to the anthrax vaccine, but over a larger scope of issues and time. Mr. Perot informed Mr. Rove that the previous leadership expressed the view that, "There is no Gulf War syndrome, and we will never know what caused the illness." He wrote, "All officials who came into positions of responsibility over this issue during the Clinton administration had to subscribe to the view as a loyalty test."

Mr. Perot elaborated: "Most of the negative research leaders are still in place. In the final months of the Clinton administration, many of them were moved from appointed positions to permanent civil service jobs, where they can continue to obstruct the investigation." Perot exposed that "others have moved out to consulting positions in universities and military contractors, such as RAND, where they continue to influence policy and orchestrate the cover-up under lucrative consulting agreements or grants from their former colleagues."

Mr. Perot listed a series of names remarkably familiar to Russ and me in order to forewarn the new administration officials about the same decision-making, dupe-loop tactics we discovered. Mr. Perot endeavored to assist the new White House leadership from becoming embroiled in the cover-up of Gulf War Illness by the residual "negative leadership" who advocated the "stress theory." Mr. Rove received the data, and rapidly tasked appointees within the DoD to investigate the information. In a letter dated April 25th, 2001, stamped "W00554 01," and received by the office of the secretary of defense's White House Section on April 27th, Mr. Rove passed Mr. Perot's concerns on to Deputy Defense Secretary Paul Wolfowitz:

Here is material which has been sent to me by Ross Perot regarding the Gulf War Syndrome, as well as some material on the Anthrax vaccine problem.

I do think we need to examine the issues of both Gulf War Syndrome and the Anthrax vaccine and how they can be dealt with. They are political problems for us.

White House Advisor Karl Rove memo to DepSecDef Paul Wolfowitz at Appendix O.

I pondered the presidential advisor's words and his thought about "problems." I thought about flying fighters. I had problems with my jets before. Once I had an engine momentarily "cough," or stall, while I was flying an F-16 in central Florida. Stalling an F-16 engine is particularly problematic because you only have one engine strapped between your legs. After almost a thousand hours of F-16 flying, it only happened to me once. I kept it from being a problem by dealing with it. I did the safe maneuver, performed a precautionary landing at a satellite field, called in maintenance, and fixed the problem. I knew the only way the problem could be dealt with was to do it right. I had to land and get it fixed. I took no chances by limping a broken jet home. When flying, you deal with problems and safety issues in the way in which you were trained: using the checklist and following the procedures. If you did otherwise you could kill yourself or, worse yet, hurt others on the ground. I thought about how problems only get bigger if you try to fly hardware that is known to be malfunctioning. As General Scott warned, you need to know when to "put down the shovel," and stop digging the hole deeper.

We hoped that the president, Mr. Rove, Mr. Wolfowitz, and their appointees would ensure that the anthrax vaccine "problem" and Gulf War Illness were "dealt" with properly—without consideration as to the "political" implications. We knew that the previous leaders had not. Our commander forced eight of his pilots to go away rather than dealing with the root problems. General Blanck and West and others all had a chance to deal with the known safety, efficacy, and legality problems of the anthrax vaccine, but they chose to arrogantly protect the program instead of the legally prescribed health rights of their troops.

The checklist these people chose grounded hundreds of pilots and found hundreds more discharged from the armed forces, some with imprisonment and felony convictions. They left thousands more suffering adverse reactions. If they had chosen the safer path, none of these losses or illnesses would have occurred. Russ and I flew each sortie looking forward out our windscreens, "checking twelve o'clock." We also kept an eye out behind, learning from the past, "checking six o'clock," and not repeating past mistakes or patterns. The important thing was Mr. Rove knew they had "problems" that needed to be "dealt" with, so we put our trust in him.

The patterns were no different than the denial following the eras of World War II nuclear testing and the use of Agent Orange in Vietnam. Like those controversies, similar insinuations surfaced that America's ill troops' problems were all in their heads or were undeterminable.

The difference this time was that people like Mr. Perot were identifying the cover-up for the attention of the highest offices in the land near real-time. There were most likely gate guards, negative leaders, and stress advocates during earlier dilemmas, but what they didn't have was Mr. Perot.

Mr. Perot also identified a seemingly inconsequential actor on the anthrax vaccine stage, a contracting officer. Mr. Perot astutely surmised that, "though only a functionary," the contracting officer "probably knows how the corruption works and, if placed under oath with some pressure, would probably tell all." Internal emails from the contracting officer documented communications from BioPort as early as May of 1999. The emails showed supplemental testing results of the anthrax vaccine produced data that "continue to be all over the board." The contracting officer recommended the DoD "suspend any further potency testing" or else the results "must be reported to the FDA" [contracting officer memo at Appendix P].

This was the same testing that the DoD's General Eddie Cain later testified about to Congress. Email communications revealed, "Mitretek Systems Inc. identified an inconsistency" and "based on this inconsistency," the DoD "suspended supplemental testing and sent a 'Tiger Team' of subject matter experts to help resolve the problem." Cain's subsequent testimony to Congress didn't include the regulatory problems documented in the email communications. Mr. Perot was likely right that under subpoena, the contracting officer "would probably tell all."

Mr. Perot's note to Mr. Rove also relayed concerns about many of the studies put in place to support the negative leadership's goal to push the stress theory. Among the studies, Mr. Perot specifically mentioned the Presidential Special Oversight Board (PSOB) for DoD Investigations of Gulf War Chemical and Biological Incidents. Mr. Perot maintained that the board's leader, former Senator Warren Rudman, "intimidated the scientist staffers, altered their reports, ignoring their objections, and concluded . . . that stress is the main cause of Gulf War Illness." In the very first public hearing, Rudman opened with the comment, "Our board knows and recognizes our veterans are sick, but we don't know why."[1]

This statement, along with Mr. Perot's evidence about the diversionary "we'll never know what this is" pretense, matched our observations.

For Russ and me though, we always tried to find independent corroboration for any theory. Ours came from interviewing one of the members of the president's Special Oversight Board, Dr. Vinh Cam, of Greenwich, Connecticut. Because Dr. Cam lived in our area, I arranged a meeting with her. We met for a cup of coffee, and she explained the disturbing process and inner workings of the Presidential Oversight Board. Dr. Cam told me, "It's a cover-up." Mr. Perot obviously concurred, and Russ and I couldn't help but agree. It didn't take a rocket scientist to see the corrupt nature of the processes, the product, and the patterns.

Dr. Cam's experience as an immunologist obviously exceeded Senator Rudman's expertise. Her status as the only independent, non-DoD affiliated member of the panel was an important element of the equation to add transparency.[2] By December of 2000, the official report came out. Dr. Cam publicly objected to the PSOB's conclusions that "stress is likely a primary cause of illness" as a "blatant misrepresentation."[3] Rudman provided an official reaction to Dr. Cam's dissenting comments, but upon a detailed analysis, his reaction merely echoed the already well-established M.O. where the DoD and their surrogates attack and disparage the messenger.

Reuters captured the controversy in an article titled "Panel Finds Stress a Main Cause of Gulf War Syndrome."[4] The article and report relied upon the Institute of Medicine's (IOM) Committee on Health Effects Associated with Exposures during the Gulf War that was taken out of context earlier that year in September of 2000. This earlier report determined that there was no scientific evidence that long-term health problems suffered by Gulf War veterans could be linked to exposures such as anthrax vaccine. Upon further analysis, Dr. Harold Sox,[5] chair of that IOM committee, also said, "evidence was inadequate to determine whether an association exists."[6] But, I remembered that Dr. Sox told me personally at a Gulf War Illness conference in that same year that the DoD wouldn't release the data required to formulate conclusions.

Transparency concerns corresponded with documented internal DoD emails. After Congressman Shays's second congressional hearing in May 1999, Brigadier General Eddie Cain emailed colleagues in the Pentagon about GAO testimony given during that hearing. He worried:

> . . . two key areas in which we came up flat were the GAO's assertion that #1, the anthrax vaccine licensed was NOT the one tested and #2, how can DoD say that reported desert storm illnesses were not caused by the anthrax vaccine when we have no record of who received the shots. If we cannot answer these questions we (DoD & the Administration) are in big time trouble.[7]

In his internal office memos to the other Army officers involved, General Cain fretted about the "media picking-up GAO's as opposed to the DoD's soundbites." General Cain was talking candidly in his emails about the facility making the vaccine the troops were being forced to take. To say the least, it wasn't a normal vaccine production operation. He admitted, "And if you think Congressman Shays was critical of the current relationship between FDA & DoD, wait until he finds out that DoD is calling all the shots on site"[8] [email at Appendix Q]. Wow, "DoD is calling the shots on site." The cozy relationship was unsettling, but even Mr. Perot's access and influence failed to break up the decades-old, DoD-supported cottage industry.

Other than Mr. Perot, few were willing to take on the formidable, seemingly insurmountable inertia of the DoD's biomedical machine. Undeterred, he asked us to assist the USAF captain and physician, who took his Hippocratic oath to heart, Dr. John Buck. I respected the courage of this lone military DoD doctor that stood up to the anthrax vaccine policy. Dr. Buck did not want to be injected into the dilemma, even asking not to be ordered to receive or administer the vaccine. Yet Dr. John Buck's chain of command chose to test him instead. While serving as a physician at Keesler Air Force Base in Mississippi, Dr. Buck contacted Mr. Perot once he realized he was being forced into a court-martial over the vaccine. The military was unable to plea bargain a discharge, and Dr.

Buck's case went to trial in May of 2001. Dr. Buck wanted to stay in the military, but also believed the issues and facts deserved their day in court.

We assisted Dr. Buck and his civilian attorney, a retired officer and Air Force Academy graduate by the name of Frank Spinner. We provided the documents and the proof that the DoD previously acknowledged the vaccination to be experimental, and therefore the mandate was illegal. Ultimately, the military judge did not allow any of the evidence collected to be viewed by the jury after four days of pre-trial hearings. The judge instead relied only upon the expert testimony of one of the DoD's gate guards, the deputy director for clinical operations for the anthrax vaccine immunization program agency. This gate guard was a pharmacist, not a medical doctor. He told the judge about his fascination with the program while still a graduate student. The eager pharmacist said he gathered "professional articles on the product, and, since I've come into the agency, I have turned that inspect file into several three-ring binders."

What was amazing to Russ, JR, and me was that over the same period of time, we'd each collected three or more file cabinet drawers of DoD documents and articles on the limitations of the vaccine. Perhaps this was why the pharmacist only had several three-ring binders. He must have ignored the pre-1997 negative government assessments of the vaccine. Fresh from graduate school, but not medical school, he apparently monitored the defense secretary-level program. Within two years he became the top gate-guard defender of the program at the first and only court-martial of a military doctor. More amazing still, in the two or more years that followed, this pharmacist gate guard would become virtually the lone anthrax vaccine public defender. He became the man that the "agency" and the Army relied upon to protect their program, while all the others quietly vanished after high-speed passes—especially doctors. The pharmacist testified about the nature of his role in the Pentagon: "We are sort of the center of the bullseye that keeps track of the implementation of the program." Translated—he managed the DoD's dupe-loop.

During the pharmacist's testimony against Dr. Buck, he employed the boilerplate script very well in response to the judge's questions about the legally irrelevant Freidman memo.[9] Just like the Navy officers in the

Pennsylvania courtroom, the pharmacist had the double-negative verbiage down pat. He confidently explained that the use of anthrax vaccine for inhalation anthrax was: "not inconsistent with the package label." The pharmacist must have been reading our Connecticut attorney general's and our thoughts as well on the Perpich Supreme Court precedent. That ruling delineated the "hats" military members wore, but his testimonial twisted the analogy to effectively confuse the judge. The pharmacist embellished, "I would use the analogy—we talk in the military about wearing several hats and I think the vaccine wears a few hats in this regard. The licensed hat is for prevention of infection—the general indication."

Confusing the judge on the specific question about whether the vaccine was approved for "inhalation anthrax," the pharmacist referenced the "specific indication for inhalation anthrax." He stated, "It was very clear, I think, in their mind that this was a second hat." It was too bad he got the analogy wrong. If he remembered the Supreme Court precedent as well as he did the "not inconsistent" part of the legal fig leaf Freidman memo, he would have recalled that the high court said of the "three hats in their closets" that "only one . . . is worn at any particular time." In trying to analogize the anthrax vaccine's license as including many "hats," he was wrong. Only one "hat" is "worn at any particular time," and in the case of the anthrax vaccine, the approval hat didn't fit. The indication or use for inhalation anthrax had not been approved.[10]

Buck's attorney, Spinner, cross-examined the pharmacist by asking, "Are you familiar with 21 C.F.R. [Code of Federal Regulations], section 10.85?"[11] The pharmacist replied, "I have heard the numbers." Spinner followed up, "Are you familiar with the requirement established under that rule regarding advisory opinions?" He answered, "I'm not fluent in that portion of the C.F.R." It wasn't surprising that he was not "fluent," but he should have been familiar. He chose his words carefully. That regulation did not allow for the Friedman memo to approve a vaccine for specific use by the DoD. Friedman's memo was a personal opinion that did not undergo the federal rule-making process. It did not serve as an official opinion of the FDA, and court precedents previously upheld this fact as well. We knew the Friedman memo was the legally irrelevant quick fix, but the DoD's "expert" confused the judge about the licensing "hats."

We were frustrated over the pharmacist's testimony, because a doctor's career was at stake based on uniformed and seemingly misleading testimony. He was sent to represent the DoD on a medical and legal issue, relevant to both the oath of office and the Hippocratic oath. Where were the doctors to testify against this doctor? They were absent, except for the one brave doctor sitting as the defendant in the kangaroo court at Keesler Air Force Base in Mississippi. Dr. Buck requested I write him a character statement to be used only if convicted—for the sanctioning and sentencing phase of the trial. I proudly complied. I emphasized to the jury that I understood the letter would be utilized only if they convicted Dr. Buck. I explained his conduct was based "on his beliefs about the requirements of the aforementioned oaths and values." I cited the oath of office and the Hippocratic oath, and explained the evidence proving the anthrax vaccine mandate was patently illegal. But the judge never allowed the jury to see this evidence. The judge decided the legality of the order could not be questioned and therefore evidence about the illegality of the mandate was not admissible in the trial. The judge declared:

> I find, as a matter of law, that the order given the accused in this case was a lawful order. Accordingly, defense motion is denied. The government request that the defense be precluded from presenting to the members, on findings, evidence concerning the safety and efficacy of the vaccine is granted.

Following those pre-trial hearings, Dr. Buck was convicted, fined, and confined to quarters on the Air Force Base. His punishments were carefully constructed in order to preclude a civilian federal legal review through an appeal process. Something was terribly wrong when a pharmacist could be the sole expert and last standing gate guard defending a failing program. The DoD abandoned a doctor attempting to live up to his oath and the "practice of medicine."

Russ and I were dismayed by the proceedings and the inability of the military judicial system to provide a check and a balance to force the DoD to "resurvey its own judgments, disciplines, and line of action."[12] We wrote an essay for a Biloxi, Mississippi, paper in response to their editorial attempting to rationalize for the community about the injustice meted out to Dr. Buck.[13] Our article for the *Sun Herald* was published on June 13th, 2001, titled "Intellectual honesty in the ranks sets America's military apart from others." Their editorial, titled "In the end, discipline had to win," couldn't know that the vaccine program would be shuttered a few years later for the very reasons Dr. Buck unsuccessfully attempted to argue in court. We informed readers that the "jury never heard this evidence—The USAF judge would not allow the documentary evidence to be presented in Dr. Buck's court-martial to corroborate his contention that the anthrax vaccine is experimental, making the mandatory program illegal."

The relevance of that evidence, scribed on military letterhead, went to the core of the debate raised in the editorial—the oath of office. We discussed the compounding dilemma for Dr. Buck since he was also bound by the Hippocratic oath in his medical profession. These oaths were intended to be complementary. The paper's editorial proclaimed, "Woe to the branch of the armed forces that does not hold its officers and enlisted personnel every bit as accountable for their actions." We retorted that we "agree wholeheartedly, for if it is the Defense Department that is breaking the law, the responsible leaders should be held accountable for their actions and their abuse of power and discretion." Russ and I never heard from anyone about our call for accountability over the abuses of power and discretion by the DoD. We came to realize that to do so would only cast light on the abandonment of various oaths which we hoped to illuminate.

Russ and I gained more perspective shortly thereafter and realized the entire trial seemed rigged. We met in DC with the military assistant to Dr. Buck's senator, Mississippi's Trent Lott. Mr. Perot arranged the meeting, coinciding with the day Dr. Buck was sentenced. Two military officers in civilian clothes accompanied Lott's staffer. The officers told us that a military officer's duty was to resign from the armed forces if he or she felt an order was illegal. They insisted the only option available to an officer was to fight such a policy from the outside. Additionally, we discovered that Air Force generals and representatives from the surgeon general's office had been in Senator Lott's office the day prior confirming that the senator would not interfere with the sanctioning of Dr. Buck. Upon receiving the assurance of the senate majority leader's office, we better understood why the Air Force leadership proceeded with the conviction sentencing of Dr. Buck despite the absence of evidence being allowed in his trial.

We debriefed Mr. Perot that the staffer agreed he would not take the vaccine if he were still in the military, and that he applauded our efforts in attempting to stop the vaccination program. Both active-duty military officers agreed that they were similarly troubled by the anthrax vaccine policy. Our debrief added that the staffer maintained the only means of challenging the vaccine program is to apply "political leverage" through the media and Congress to "compel the DoD to obey the law" of informed consent.

Senator Lott's staffer called the *Washington Times* while we were in Senator Lott's office and asked Mr. Rowan Scarborough to work on a story for us after hearing facts on the vaccine's alleged adulteration and unprecedented adverse reaction rates based on FDA testimonials to Congress. He informed the reporter that these facts were "new" and "worthy of the front page, above the fold." The staffer maintained that the only way to stop the DoD policy was to get the controversy on the front page of the *Washington Times* or *Washington Post.*

We told Mr. Perot that the civilian staffer complimented our efforts, but referred to the entire exercise as a "poker game where the man who played the better hand would win." The senator's staff disclosed to us that high-level DoD officials had briefed their office in the preceding days concerning the Dr. Buck trial and the anthrax vaccine. Senator Lott's staff also related the fact that the prosecution in Dr. Buck's case "would not be pursuing confinement." Contrary to the previously imprisoned enlisted Marines, the DoD instead only wanted a guilty verdict and forfeiture of pay. This was prior to the ruling, and the pact promised no prison time.

As we departed Senator Lott's office, we were assured the staffers would assist us in "orchestrating a sustained public relations campaign" in order to highlight the controversies surrounding the anthrax program. But unsurprisingly, we never heard from Lott's liaison or the military officers again. And of course, Mr. Scarborough never responded nor inked anything on the front page, above the fold. I remember how Russ and I took a flight home that night, sitting in quiet disbelief after the comical meeting with Lott's three military men.

We pondered the "parallel universe of military ethics" that embodied the anthrax vaccine dilemma, where military officers contended we must obey illegal orders and that our only alternative was to resign. The DoD powers in our nation's capital coordinated carefully to allow a lone heretic doctor to be tried and punished. People on both sides of the aisle, and both sides of the Potomac, had forsaken Dr. Buck with backroom deals. They didn't want the truth of their willfully blind negligence to be exposed. The leaders, stewards of our oaths, deserted a doctor.

Reflecting on attending Dr. Buck's pre-trial hearings in 2001, I remembered sitting in the judge's courtroom in the early morning prior to his ruling, denying the admissibility of the evidence. I listened through the paper-thin walls of that Keesler Air Force Base, Mississippi courtroom on that humid May morning. I heard what the judge said in the one-way conversation during an early morning phone call. He was loud, so loud it surprised me. He specifically said, "This could hurt the Air Force." I realized incredulously that the entire dilemma boiled down to competing loyalties and oaths. The judge chose to hurt the doctor in defense of the Air Force.

Dr. Buck tried to live up to his oath of office to "support and defend the Constitution of the United States" and his Hippocratic oath to "give no drug, perform no operation, for a criminal purpose." But the good doctor was abandoned by his leaders that were living up to an oath of loyalty to the chain of command and the military institution instead. The oath to the chain of command was an unstated and unwritten oath of loyalty, yet was surely subordinate in hierarchy to the two scribed and sworn oaths. A blinding loyalty oath defined the conundrum of the anthrax vaccine dilemma, as the sworn oaths were ignored, dismissed, and violated.

I remembered what JR had said: "Blind adherence to a falsity meant that the defining characteristic of the militaries we've fought over the past one-hundred years had become a prerequisite for service in the US military." In an unnoticed warning sign for our democracy, Dr. Buck was sacrificed for not blindly adhering to the falsity. Though punished, denied promotion, and fined $21,000 for not following the anthrax vaccine party line, the good doctor later supported operations in Afghanistan. He complied willingly. Dr. Buck was awarded a medal for his distinguished service and received an honorable discharge from the USAF. We lost a good man, the only man, who adhered to his oath of office, Hippocratic oath, and our core values.

The court-martial of Dr. Buck was a travesty of justice. An Army pharmacist took the stand against an Air Force physician. The government's witness used carefully chosen and ostensibly misleading words, while all the doctors in the entire military looked the other way. How did a non-doctor come to sit in judgment of a practicing Air Force medical doctor? Where were the rest of the DoD doctors in the ethical impasse of the anthrax vaccine dilemma of conflicting oaths? Where were our military doctors while an alleged adulterated, improperly approved drug was illegally mandated on our troops for a known experimental purpose in violation of both oaths? It's inexplicable that Mr. Perot was the lone advocate for this courageous doctor while Dr. Buck's own military and elected officials collaborated to feign ignorance.

Somehow the anthrax vaccine mandate, and the loyalty oath to the chain of command, won out as the leaders of the DoD surgically removed the Hippocratic oath and the doctors from this military operation. Where were the military lawyers and the lawmakers as the interrelated oath of office legal issues were judged inadmissible? Where was the constitutionally guaranteed right of due process while the evidence was barred so that it would not threaten or challenge the program? Military leaders ensured our nation's lawmakers were cut out of the equation as well, securing preemptive handshakes of noninterference. Doctors, lawyers, and lawmakers all became irrelevant—conveniently cut out of the loop. The quandary of competing oaths, the upside-down nature of the unstated loyalty oath taking precedence over the sworn oath of office and the Hippocratic oath, defined the anthrax vaccine dilemma and predicted future injustices as well.

This distorted reality would only be made clearer over two years later at another congressional hearing where both Mr. Perot and GAO testified.[14] I listened to the hearing broadcast and its congressional testimonial exchange on January 24th, 2002. After Mr. Perot testified about the problems with the anthrax vaccine, a representative specifically noted his statement about the "lack of credibility with regard to what we've done with our anthrax investigation, and that the anthrax program is still a problem."[15] GAO investigators also revealed at this hearing that the Department of Veteran Affairs had data linking the anthrax vaccine to Gulf War Illness, but had not released this data to the public. Trying to clarify the unanswered Gulf War Illness-vaccine linkage issue, Representative Christopher Shays asked:

> In your testimony, you said according to studies in both the UK and the US, veterans of the Gulf war who reported receiving biological warfare inoculations for anthrax or other threats were more likely to report a number of symptoms than non-Gulf war veterans who did not report receiving such inoculations. This pattern was observed in . . . unpublished data collected by the US Department of Veterans Affairs. Why do you think the VA has not published its finding regarding the link between advanced symptoms and the anthrax vaccination?

The GAO's managing director, Nancy Kingsbury answered, clarifying the bottleneck:[16]

> I don't know why they didn't publish it. We are aware of it. We have asked them. They said to us what they said to you this morning, things about the analysis not being completed and that sort of thing. I'm

> not in a position to second-guess it. We consider it to be valid, useful information that ought to be in the public domain.

Russ attended the hearing. While speaking with Mr. Perot during one of the breaks, Mr. Perot motioned to the assistant secretary of defense for health affairs, Dr. Winkenwerder, to join the conversation. He introduced Russ and told Dr. Winkenwerder that Russ and I were the experts he should be working with on this issue. If Winkenwerder actually collaborated with us, it would conflict with the institutionally favored "stress" theory and the variety of indeterminate diversionary studies. Each of those studies was built on the obscurity of earlier results, possibly designed so that in the end no answers could be drawn. Russ and I became experts on seeing through this medical flak and doubted Winkenwerder would follow Mr. Perot's counsel.

Through this haze, it became plain to see that conclusions actually could be drawn if the data points were released. The circularity was astonishing—the DoD said they lacked data, and therefore no conclusions could be drawn. Taking it out of context, the DoD would say there was no evidence showing links between the vaccine and illnesses. And all the time data did exist, but wasn't released. This was why so many felt frustrated with what they believed was an attempt to create convoluted conclusions of competing science. In the end, no conclusions could be drawn.

Little did we know, at the very same time that our Tiger Team in Connecticut raised concerns to the chain of command about the unresolved issues swirling around the renewed use of the anthrax vaccine and Gulf War Illness, the DoD formed expert panels to ensure that the problems were relegated into obscurity. The DoD officials feigned ignorance by insisting, "We don't know why," or "stress is likely a primary cause." We knew, and they knew, this was the zero-liability position and the politically favored outcome to ensure no accountability. This is most likely why the cover-up transpired, and it lent an element of logic to everything that made no sense. To admit that we as a military made our own troops sick would prove too costly.

But as military men, Russ and I both knew that it didn't matter what the liability involved: You don't take broken jets airborne. Our troops and their leaders have an obligation to challenge illegal orders and stand up to institutional wrongdoing when ill-conceived and illegal programs attempt to take flight. These are the lessons that were born from World War II. American troops knew that blindly following immoral and illegal conduct was unacceptable. Lessons from prior military conflicts applied as well. We knew that the cliché, "I was just following orders," didn't cut it.[17] So, we pressed on for ourselves, and for all Americans like Dr. Cam, Dr. Buck, and Mr. Perot, in an attempt to live up to our oath. We connected the dots, albeit slowly, and in doing so, it helped us to understand the institutional loyalty justifications behind why so many behaved so antithetical to our training. Understanding it, though, didn't justify it.

That's why Mr. Perot continued to ask if Russ and I would go to Washington, DC, and meet with people he believed could have an influence on our issue. Though highly unorthodox for currently serving military members to directly challenge the DoD policy on Capitol Hill, especially directly through the White House, we believed the depth of the deceit required our continued dissent. Mr. Perot set up an appointment for us with Mr. Charles Abell, a staff member for Senator John Warner, the chair of the Senate Armed Services Committee. Mr. Perot also informed us that Mr. Abell was being appointed to report to the Pentagon by President Bush as an assistant secretary of defense for force management policy for the new administration.

We knew we were taking a risk to meet with a future DoD assistant secretary in order to challenge official policy. Our colleague JR warned us that Abell was a career Army officer, but he too agreed to go meet with him as well, as we all put our faith in the system once again. We created an updated version of our briefing book and headed to Capitol Hill one more time. We were also able to secure the sage presence of our former FDA regulator mentor, Sammie Young.

Upon arriving at the Senate Armed Services Committee staff chambers, we were all seated at a conference table. I opened the meeting, introducing our team and our larger concerns which transcended the anthrax vaccine. JR gave a precautionary talk on the dangers of the erosion to civilian control of the military, since we knew Abell was taking a trip through the revolving door and on a vector to be a civilian leader over his former Army comrades.[18] Sammie laid out the regulatory debacle encompassed by the anthrax vaccine with credibility as a retired full colonel in the medical corps, as well as with his experience as an ex-FDA regulator. The meeting was generous in time, but Abell cut it off as scheduled, leaving us with the parting comment, "Mr. Perot said I'd be impressed, and I am."

We saw ourselves out, and I turned to JR and Sammie and offered them one of Russ's catchphrases, "Thanks for participating in government, gentleman!" After Mr. Abell was confirmed into his new DoD post, an article came out in the *Armed Forces Press* titled "Why Civilian Control of the Military?"[19] The article stated that the "framers" of the Constitution "separated the responsibilities for the military, placing the responsibilities firmly in civilian hands." The exposé explained that "Article I, Section 8 of the Constitution states that Congress shall have the power 'to raise and support Armies' . . . and 'to provide and maintain a Navy.'"

The editorial elaborated, "Article II, Section 2 states, 'The President shall be the Commander in Chief of the Army and Navy of the United States, and of the Militia of the several States when called into the actual Service of the United States.'" The DoD's media arm affirmed, "Military members swear 'to support and defend the Constitution of the United States,'" and that "one of the more successful aspects of that document is civilian control of the military."

Maybe Mr. Abell took our thoughts seriously and heeded JR's larger concerns that the anthrax vaccine dilemma was a poignant example of the dangerous trend away from civilian control. Maybe Mr. Abell coordinated the article with his new civilian comrades, knowing the obstacles that needed to be surmounted by the new team providing civilian control over the DoD. Regardless, the important thing was to do our job, exercising the "competing loyalties" required by Articles I and II of the Constitution. Our job was to defend the ideals of that "document."

Mr. Abell was in touch once more, asking me for some more information, but soon fell into the abyss of the Pentagon's email labyrinth. Perhaps Abell knew he and Defense Secretary Rumsfeld and others could reassert civilian control of the military without ever publicly acknowledging our concerns. Maybe there were many factors of which I was not aware. The assistant secretary of defense didn't owe me a follow-up. Our goal was merely to get the information into the hands of the people that could protect the new administration from also becoming co-opted into the dilemma. We only hoped that the patterns that eroded civilian control of the DoD would or could be stopped. I never heard from Abell again after he wrote:

> Date: 5/4/01 6:59:58 AM Eastern Daylight Time
> From: Charlie_Abell@armed-services.senate.gov (Charlie Abell)
> Tom. thanks for this input and for your call yesterday . . . my new e-mail address is charlie.abell@osd.mil. I begin my new job today.

Those early days working with Mr. Perot were a welcome breath of rejuvenation for Russ and me. We needed to get our motivation fuel tanks refilled. Mr. Perot provided that fuel. Mr. Perot's faith in our work did the trick. We found someone who didn't owe anything to anyone—someone who was willing to tread the harder path. Perot wasn't a fighter pilot, but he acted like one. Russ and I knew that being a "fighter pilot" was more a state of mind, and Mr. Perot fit the profile. Though a Navy man, he was welcome in our formation and turned out to be an invaluable wingman, support officer, door opener, and common sense check for our efforts.

Mr. Perot made our arduous path easier to tread. He forced people to listen to us through his reputable influence. People who previously might not have taken the time to receive or return our phone calls actually met with us. Mr. Perot, and his allies in the fight, had been on the path to expose and treat Gulf War Illnesses long before Russ or I knew how to spell anthrax vaccine. While on the path together, we were able to sniff further down the trail of the DoD's attempts to obscure the actual causes of Gulf War Illness. Mr. Perot made a difference. He changed the dynamics of the classic DoD dupe-loop when it came to veterans' health ordeals. No longer could they ignore us. Together, our formation committed to continue to fly high cover.

In addition to our parallel efforts on Capitol Hill, Mr. Ross Perot continued to open the doors of top legislators outside our own state. Mr. Perot arranged a meeting for Russ and me with aides for Senator Daschle, the Senate Democratic Party leader, and Representative Richard Gephardt, the House Democratic leader. Russ and I gave their staffs copies of the *Military Times* cover story, and a distilled briefing book with highlights of the dilemma. Our goals concentrated on the correction of military records and proper care for the inoculated ill. We didn't know if Gephardt took a stance on the issues of Persian Gulf War Illnesses. But we had been tracking Daschle's involvements. We noted his testimonials as early as November 16th, 1993, during a hearing titled "Are we treating Veterans right?"[20] In Daschle's opening statement he testified:

> I find a remarkable similarity between much of what is now being discussed with regard to this issue and the history we've experienced on Agent Orange now for more than twenty years. There was a denial on part of DoD and this Government about its responsibility related to veterans' health for more than twenty-five years. The burden of proof was put on the veterans themselves. They were the ones who had to draw the connection. They were the ones that had to find the science. They were the ones who had to continue to press the Government into action. . . . The bottom line, Mr. Chairman, is that these people deserve care. We are providing that care, I think, to the extent that we can at this point, but so much more work needs to be done. More must be done to determine what is causing these problems and to compensate veterans disabled by this. So much more priority needs to be given to these veterans' health. . . . Thank you, Mr. Chairman.

Senator Daschle was correct about the "remarkable similarity" of the "denial on part of the DoD." It applied with the anthrax dilemma as well. As the senator said, "more work . . . must be done to determine what is causing these problems." This was why our objective was to put the facts on the table not only to force the DoD to properly care for their veterans this time around, but also to stop the same diversionary patterns in the future. With the help of top lawmakers such as Senators Daschle and Dodd, plus Representatives Shays and Gephardt, we held out hope that we might alter the paradigm of how the DoD deals with current military members' and veterans' issues. The paradoxical reality was that if we could accomplish this goal, DoD medical abuses might cease, and this would preclude many future veteran health issues that required oversight.

Through a series of email exchanges after our meeting, Daschle's and Gephardt's staffs kept us apprised about a letter their bosses sent to Secretary Rumsfeld concerning the vaccine:

> From: [redacted]@daschle.senate.gov ([redacted])
> To. TRempfer@aol.com
> Major Rempfer—thanks for the email. We have had several conversations with Mr. Perot to keep him apprised of what we are thinking. we informed him that we are working on a joint letter from Sen. Daschle and Rep. Gephardt to the Secretary of Defense on this issue. In addition, we have had a number of conversations with others on this issue, including Sen. Dodd's office, GAO, Connecticut attorney general's office, and the senate veterans affairs committee. We plan to meet again with Rep. Gephardt's staff to determine how to proceed beyond the letter to Rumsfeld . . . attached is a copy of the letter that was sent by Sen. Daschle and Rep. Gephardt . . .

The Democratic leadership wrote Rumsfeld on June 21st, 2001:

> Dear Secretary Rumsfeld:
> As you continue your strategic review of the Defense Department's strategy and policies, we write to express our interest in and concern about reports regarding the Pentagon's continued use of an anthrax vaccine on our military personnel. . . . we are troubled by several reports and actions that raise questions about whether continuing to

> administer the current anthrax vaccine is in the best interests of our personnel. . . . we have been made aware of the fact that a number of military and civilian personnel have come to believe that the vaccination given to our military personnel is neither safe nor effective for its intended use. As you know, the Food and Drug Administration's underlying prerequisite for the approval of any vaccine or drug is that it be proved safe and effective for its intended use.
>
> . . . In light of the questions surrounding the safety and effectiveness of the vaccine and the inability of the vaccine manufacturer to obtain FDA approval, many have asked why the Pentagon has not halted the vaccination program for all military personnel until these outstanding questions are satisfactorily addressed.
>
> . . . the Defense Department has already taken disciplinary action against a number of people . . . although a number of questions about the safety and effectiveness of the anthrax vaccination would appear to remain unanswered, the Pentagon has made it clear that it does not intend to revisit either the punishments already meted out or its decision to discipline those who act similarly in the future.
>
> . . . we also have an obligation to our military personnel to ensure that we do nothing to increase the risks they already face for their decision to serve their country. We would welcome your response to the issues raised here. In addition to your written response, our staff is prepared to discuss this matter with appropriate Defense Department officials.

Unfortunately, the deliberative process between Congress and the DoD was interrupted by the subsequent anthrax letter attacks. We never found out if Daschle's and Gephardt's offices ever received a response to their letter to Secretary Rumsfeld. But, thanks to Mr. Perot, we also attempted to help the new presidential administration in early 2001. The *Military Times* cover story was delivered to the White House, and Mr. Perot provided our briefing book to the White House senior advisor to the president, Mr. Karl Rove. We knew Mr. Rove tasked Deputy Secretary of Defense Wolfowitz to "examine the issues of both Gulf War Syndrome and the Anthrax vaccine and how they can be dealt with." Rove added, "They are political problems for us." He was right, and fortunately tasked the president's civilian appointees in the DoD to examine the issue. Deputy Secretary Wolfowitz in turn tasked two undersecretaries to investigate the vaccine and provide recommendations to Defense Secretary Rumsfeld.

Mr. Rove called while Russ and I were in Washington for Mr. Perot's meetings. Between congressional office visits, Russ and I took a needed break for lunch at a Capitol Hill diner within a couple blocks of the capitol. When Mr. Rove called our cell phone, I left the noisy restaurant for the quieter lobby. Rove told me he needed to know what was going on with the anthrax vaccine. I listened intently and then frankly explained that my colleague Major Dingle and I had gone in a different direction than the chain of command. I explained how we came to Colonel John Boyd's ethical "fork in the road" and what we were "doing." I cautioned that we were concerned about the DoD's blatant violation of the law over our troop's health rights.

Mr. Rove informed me that they had an under secretary of defense for acquisitions reviewing the program. This thrilled me because I knew the man appointed to this post was "Pete" Aldridge. The Honorable Mr. Aldridge was the same official who served in the previous elder Bush administration as the secretary of the Air Force. He was the dignitary who gave the keynote address at my 1987 academy graduation. He was the man who planted the seed in the minds of nearly a thousand newly commissioned officers in 1987: "If a regulation or policy doesn't make sense, get involved; propose a revision that does contribute to your mission."

Under secretary for personnel, Dr. David Chu, was the other President Bush appointee that we later learned was tasked to "get involved" and "propose a revision" for the anthrax vaccine program. Mr. Rove asked me to send his office our research materials again and recited his secretary's email address. I eagerly emailed his office that night when I returned home. I sent the briefing book, hoping that I remembered the email address. Mr. Rove's secretary replied:

Subj: Re: email test - per request of Mr. Rove

Date: 5/24/01 7:16:03 AM Eastern Daylight Time

From: @who.eop.gov

To: TRempfer@aol.com

File: emailtes.zip (1279 bytes) DL Time (TCP/IP): < 1 minute

Got your email.

Karl Rove's secretary confirming White House office receipt of briefing book email.

I tried not to barrage Mr. Rove. Content with the "Got your email" confirmation, I had summarized that our "research has uncovered two major areas where the anthrax vaccine program breaks the law—the first is the violation of service member's prior consent rights under 10 USC §1107, absent a waiver of those rights by the President—and the second deals with the unapproved manufacturing process changes, which 'adulterates' the product, making it subject to seizure and destruction by federal authorities." I included our Connecticut attorney general's letters to the secretary of defense and the FDA requesting they "cease and desist the illegal conduct." I concluded by saying, "It is my ardent desire that service members receive worthy vaccines and protection in the future and the current anthrax vaccine does not pass the test."

During one of my later conversations with Mr. Rove's secretary, she informed me that our information was being passed on to the National Security Council (NSC), specifically, Mr. Stephen Hadley, deputy national security advisor. This encouraged me because during the election, Mr. Hadley spoke about the anthrax vaccine on behalf of the president, mirroring our concerns. In the fall of 2000, Mr. Hadley responded to a *PBS NewsHour* question:

> The vaccination program is a very serious issue. Maintaining the trust and confidence of our men and women in uniform is critical to the future of our armed forces. Some months ago, Governor Bush called for the Commander-in-Chief and our military leaders to be very mindful of the concerns of our men and women in uniform and their families about the vaccine, and called for the government to do more to address their concerns. Hopefully, the current administration will respond.[21]

The former assistant secretary of defense in the Reagan administration, Lawrence Korb, also weighed in: "The Anthrax Immunization program was a disaster from its inception. It should have been voluntary, not mandatory, and should not have been started until there was much more evidence that it was needed and safe. It certainly had a terrible impact on morale." Colonel David Hackworth, a prominent, decorated Vietnam veterans advocate, also responded:

> Yes. Trust is a very essential. . . . And the basic problem is that trust has just disappeared from the people at the bottom because they've seen again and again misuses of power. The anthrax thing is a very clear example. It's a vaccine that no one has confidence in, that's being shoved down the throat of those people on active duty and those reserve units that are deployed to an area where anthrax might be used. There are so many questions about it—not only among the ranks, but also among Congress and the scientific community.

I was encouraged with the uniformity of concern over the implementation of the anthrax vaccine program. It was good to know these feelings existed within the ranks of the people that were now in power, and I was more encouraged when Mr. Rove's secretary gave me a phone number for the NSC. I followed up, as I always did. On July 23rd, 2001, one of Mr. Hadley's men called me back, and I shared with him some thoughts on a solution in terms of an executive order. He told me that an executive order might be premature. He also cautioned that the NSC might not be the right organization to get involved because they focused on "macro" issues.

I emailed some solutions, but I never had any further communications with the NSC. Within a couple of months, I received emails and attached memos from a reporter that explained details of the review process within the DoD in an article published by the *Washington Post*.[22] Someone at the DoD leaked a memo by the head of the joint chiefs of staff, General Henry H. Shelton. General John Abizaid, director of the Pentagon's J-5, the directorate for strategy and policy, prepared the memo, written on August 30th, 2001, and referenced the August 10th recommendations to Defense Secretary Rumsfeld from Under Secretaries Aldridge and Chu.

The *Post* reported, "Language in the memo supports long-standing rumors that they may recommend canceling the contract with BioPort and developing a new facility for the military's vaccine needs" [Memorandum from undersecretaries recommending to minimize the anthrax vaccine at Appendix R]. The DoD did not make the recommendations public officially, but the article hinted the memo by Shelton was a preemptory move to preclude the cancellation of the anthrax vaccine. The Shelton memo to Rumsfeld forestalled the imminent report on the future anthrax vaccine funding. Shelton was adamant that the anthrax vaccine was the "centerpiece of our defense" [CJCS memo dubbing anthrax vaccine a "centerpiece" at Appendix S]. It appeared that the Pentagon's OODA loop wagons were circling tightly.

The "centerpiece" assertions were odd to read, because at the time, the vaccination program had ground to a virtual halt due to the non-validation of the manufacturing line—they had run out of vaccine. The article went on to contend that "sources" viewed the memo as an "effort to derail the undersecretaries' review." The article documented that "Pentagon leaders also are apparently engaged in a heated battle among themselves over the vaccine." It was clear that military officials at the highest levels, spanning several presidential administrations, stood to be extremely embarrassed if the "centerpiece" they'd staked their reputations on was cancelled.

With the help of Mr. Perot, I contacted the reporter who in turn graciously provided the Aldridge-Chu recommendations memo to Rumsfeld, as well as General Shelton's memo. From the source documents, it was clear that the new civilian leadership was prudently recommending "minimum level" use of the vaccine and recommended procurement of "bio-detectors and stockpiles of antibiotics" instead of their "centerpiece." The civilian leaders went one step further by recommending the "comprehensive review of doctrinal positions," the development of a "national long-range vaccine," as well as a "coherent institutional process to assess and prioritize biological threats and approve the use of associated countermeasures."

The recommendations corresponded with my first congressional testimony on policy and doctrine. DoD leaders apparently concurred with our concerns that the anthrax vaccine policy resulted from a less-than-"coherent" process to counter a threat that was known to be treatable with simple antibiotics. Someone evidently listened to Mr. Perot's warnings, our inputs, and our testimonials that appeared to be wholly consistent with the new recommendations to the SecDef.

Mr. Perot gave us access to both the reporters and the many high-level government officials. Without Mr. Perot, none of the meetings, networking, and access to key documents were possible. The documents helped us to understand that our participation in government worked and the anthrax vaccine program was on the verge of cancellation. For now, the experiences perhaps did not impact the outcome, but they illuminated the dysfunctions. Even though the government knew the truth, forces unbeknownst to us at that time labored in the laboratories and spin shops to halt ours and Mr. Perot's progress in stopping the mandate.

Behind the scenes, at the highest levels, some senior military leaders objected to the civilian military leadership's recommendations to minimize the program and develop a new vaccine. Unfortunately, all the progress would be incredulously halted within weeks, on the heels of 9/11, with the anthrax letter attacks. The anthrax letter attacks frightened the nation, abruptly terminated the Perot-Rove-inspired intellectually honest review, resuscitated the anthrax vaccine program, and ensured the negative patterns would continue for many years to come.

In the midst of this setback, Mr. Perot sent me a magnificent sword, called King Arthur's Excalibur. He sent one to Russ too when he first fell ill. The philanthropist gentleman wanted to thank us for our efforts, even though we could not have done any of it without him—or without the Perot factor. The engraving included a Winston Churchill quote, which defined our common commitment. Thanks to our champion, and despite the diversions and distractions resulting from the anthrax letter attacks, I would always remember and internalize Churchill's motivational quote on Perot's Excalibur sword plaque: "never give in, never give in, never, never, never!" Mr. Perot's and Winston Churchill's sentiments symbolized the emotive force sustaining our marathon theme and provided the motivation to power through the troubling miles ahead.

Sword with Winston Churchill quote, "never give in, never give in, never, never, never!"

MILE 13: Anthrax Letter Attacks (2001)

The number thirteen is superstitiously associated with bad luck. Mile 13 in our marathon was associated with the bad luck of the untimely anthrax letter attacks. The anthrax letter attacks underhandedly undermined much of the progress made to date with respect to challenging the illegalities of the anthrax vaccine program. In response, we had to use the DoD's principles of war—the principle of "Maneuver." As fighter pilots, we knew we had to overcome the setback.

Sun Tzu wrote, "The difficulty of tactical maneuvering consists in turning the devious into the direct, and misfortune into gain." One year earlier, Russ and I collaborated with the Justice Department by reporting to the government what we believed were illegal and unapproved manufacturing changes to the anthrax vaccine by the manufacturer. We initiated a False Claims Act suit, or Qui Tam, on behalf of the United States of America against BioPort Corporation. The suit alleged BioPort, the manufacturer and distributor of the anthrax vaccine, was responsible for false claims to the United States government. The basis for such a lawsuit[1] arose from our belief that the anthrax vaccine sold after 1990 to the United States government was "nonconforming," or manufactured in violation of the Federal Food, Drug, and Cosmetic Act.

Before we went to the Department of Justice (DoJ), we reported our concerns to a congressional committee, which in turn requested an official inquiry by the GAO. Ultimately, the GAO validated our concerns about the unapproved manufacturing changes we discovered and published their findings.[2] The GAO commented, "As our testimony today reports, in the case of the anthrax vaccine, the Michigan facility did not notify FDA of a number of changes made in the manufacturing process in the early 1990s and no specific studies were undertaken to confirm that vaccine quality was not affected." The GAO investigation, initiated based on Russ's analysis and discoveries, confirmed the decade-old unapproved alterations. The adverse implications of the changes were never studied in relation to their association with Gulf War Illness maladies or to help that population of veterans in their decades-long ultramarathons for proper medical care.

We also provided similar data to our state's attorney general, Richard Blumenthal. The *Hartford Courant* published stories on May 19th and 22nd, 2000, detailing our attorney general's investigations. Excerpts included, "State Attorney General Richard Blumenthal is calling for an inquiry into whether the anthrax vaccine being used to protect state National Guard troops from biological warfare complies with state and federal laws." The reporting by our dogged journalist friend Dennie Williams identified problems with lingering questions about the safety and legality of the drug's manufacturing process and whether the filters and fermentation systems used for the harvesting of anthrax vaccine were approved properly by the FDA.

Our retired regulator colleague, Sammie Young, commented for one story saying, "Based on my reading of the 1997 and 1998 inspection reports . . . this is a company that is completely out of control, and they should not be producing medicinal products for human use." The "adulteration" allegations brought forth in the False Claims Act filing were serious and well-researched [supportive GAO report excerpts at Appendix T]. We effectively suspected the manufacturer of fraudulently selling a product not produced in accordance with the Food, Drug, and Cosmetic Act or federal acquisition regulations. Though the court dismissed our allegations procedurally, due to our not being the original source of the evidence, the facts and GAO verified the unapproved manufacturing changes to the vaccine at the core of the fraud allegations.[3]

After a year of investigation, the Department of Justice (DoJ) did not intervene on behalf of the United States. Instead, the DoJ explained to us over the phone, just after 9/11 and after the anthrax letter attacks, that they did not want to deny the DoD a supply of anthrax vaccine. Over the year prior, the DoJ gave us the opposite impression. In explaining their non-intervention, they appealed to us during the conference call that we needed to understand who their "client" was. I took notes as they spoke. The DoJ attorney asked, "Who is my client?" He lectured, "That is the essential question." He conceded, "BioPort cut corners," but added a qualification, "There are violations, and there are violations." While admitting the manufacturer "didn't follow FDA regulations," he said, "the DoD was clearly aware of what was going on at BioPort."

Russ and I scratched our heads during and after the conference call. It was clear who this DoJ attorney believed was their "client." The actual client was the US taxpayer who was footing all the bills. Citizens most definitely were not "aware of what was going on at BioPort." Instead, the DoJ protected its clients in the executive branch, the DoD, and the FDA. They were all on the same team and they circled the wagons instead of pursuing the fraud case. The DoJ lawyer confessed that even if the manufacturer were found guilty of the fraud, the Treasury Department—another executive branch agency—would pay all of the bills due to the DoD's "intimate" participation in the violations. He explained that any recovery would be a "circular exercise" of money in and out of the US treasury. The DoJ representative told us this was "not desirable," and that the DoJ didn't want to be the reason troops "stopped getting the vaccine."

The DoD's fear tactics made everyone afraid to stand in the way of efforts to protect the troops. Despite the admitted "violations," Russ got the DoJ attorney to acknowledge over the phone that the decision was a "political" one, a "cost benefit calculation." The DoJ didn't want to be perceived as "derailing the program." The DoJ investigators discovered the DoD was deeply involved in this dilemma, even in the manufacturing realm—they chose to protect their "client." The DoJ and courts employed similar quibbling rationales of "government knowledge" and sovereign immunity claims in order to quell False Claim Act cases during the COVID-era.

Needless to say, in addition to ending the DoJ's involvement in the False Claims Act suit, the first round of anthrax letter attacks in

September of 2001 also derailed the internal Perot-inspired Rove-Wolfowitz-Rumsfeld review of the anthrax vaccine program. The letter attacks, within a week of the September 11th, 2001, tragedy, suspiciously capitalized on the 9/11 fears in order to revive the anthrax vaccine program. Less than a week after the 9/11 attacks, I received a one-liner from Mr. Rove's secretary stating, "Karl has advised me that this issue will be handled by the DoD. Thank you."[4] This was also the day prior to the first anthrax letter mailings.

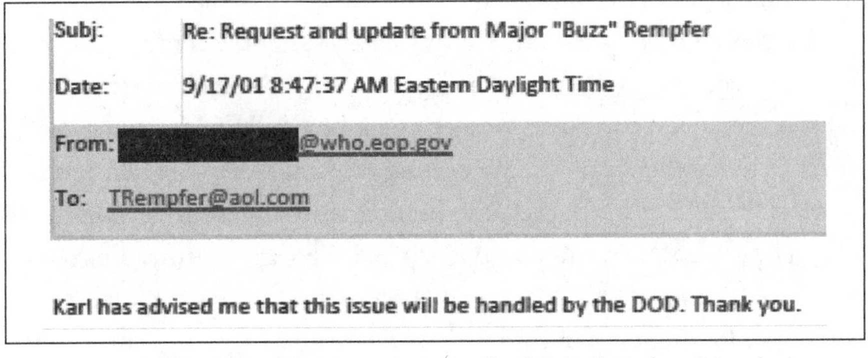

Subj: Re: Request and update from Major "Buzz" Rempfer

Date: 9/17/01 8:47:37 AM Eastern Daylight Time

From: [redacted]@who.eop.gov

To: TRempfer@aol.com

Karl has advised me that this issue will be handled by the DOD. Thank you.

Email from White House just prior to the first anthrax letter attack.

The anthrax letter attacks proved to be a pivotal turning point for the DoD's anthrax vaccine "problem," as Rove termed it. The attacks resulted in the expedited re-approval of the manufacturer. Early on, the FBI profiled the criminal who sent the anthrax letters as having a "scientific background," adding that they had "likely taken appropriate protective steps to ensure their own safety, which may include the use of an Anthrax vaccination or antibiotics." The FBI knew that the anthrax vaccine was restricted to the realms of the DoD or government research.

The FBI must have found the anthrax mailer's warning to "TAKE PENACILIN" [sic] intriguing. If the perpetrator really were an Islamic fundamentalist or terrorist, it was highly unlikely they would warn recipients to take antibiotics. Plus, prior to the letter attacks, knowledge about efficacy of simple antibiotics was limited to Army anthrax researchers. It seemed clear no one intended on killing anyone, but instead the perpetrator sought to strike fear amidst the post-September 11th environment.

The FBI believed that "the targeted victims . . . are probably very important to the offender. They may have been the focus of previous expressions of contempt." The FBI profiling should have pointed agents to the contempt against lawmakers who were investigating the DoD's anthrax vaccine—an obvious motive. After 10,000 interviews, 6,000 subpoenas, and 80 searches, we expected the FBI to investigate this evident link.[5]

The White House responded vaguely when asked about the source: "All indications are that the source of the anthrax is domestic. And I can't give you any more specific information than that. That's part of what the FBI is actively reviewing. And I just can't go beyond that.[6]

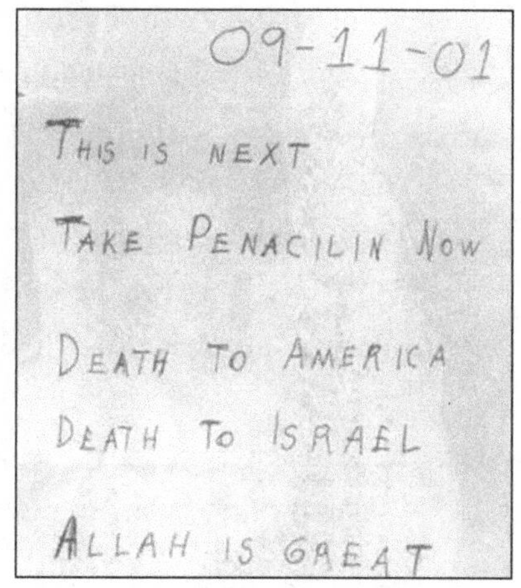

09-11-01

THIS IS NEXT

TAKE PENACILIN NOW

DEATH TO AMERICA

DEATH TO ISRAEL

ALLAH IS GREAT

Anthrax letter warning "TAKE PENACILIN [sic] NOW" from the FBI website.

Each revelation added to our concerns due to our awareness of the US Army motive to get the anthrax program resumed. Notably, Army involvement was a documented common denominator for the only two anthrax epidemics in US history. The first epidemic occurred in 1957 during an Army anthrax vaccine trial in a textile mill in Manchester, New Hampshire.[7] Four Americans died.[8] The second epidemic in 2001 killed five from highly lethal anthrax spores. The spores were eventually genetically linked to a US Army lab.[9] The fifth and final innocent victim, Ottilie W. Lundgren, lived in our home state. The multiple deaths four decades apart from lethal spores were troubling. The proximity of the US Army to all of these events warranted investigation to shed light on the coincidences, patterns, M.O., and the full truth.

Despite previous concerns expressed by the president and other key political figures, and likely ignorant of the US Army's historical pattern of military medical mistakes with vaccines, our national leaders capitulated to the new threat. Leaders' concerns prior to the attacks included:

President George W. Bush: The Defense Department's Anthrax Immunization Program has raised numerous health concerns and caused fear among the individuals whose lives it touches. I don't feel the current administration's anthrax immunization program has taken into account the effect of this program on the soldiers in our military and their families. Under my administration, soldiers, and their families will be taken into consideration.[10]

Presidential Candidate Al Gore: I feel the concerns are genuine . . . based on the concerns I have heard from military personnel directly; I think we are justified in taking a closer look.[11]

Presidential Candidate John McCain: I think that there should be a pause . . . and I would get the best scientific and medical people together and make a better argument than they've made.[12]

After the letter attacks, the "health concerns" were ignored. The "closer look" was abandoned. No "pause" occurred. The stalled anthrax vaccine program instead accelerated due to the letter attacks. During this time it became apparent that even senior DoD leaders, when asked by the media if the anthrax vaccine would continue to be used, were noticeably uncomfortable. Despite the negative reviews, the DoD blindly endorsed what they internally recognized as an inadequate vaccine—a bad order, bad cause, and bad means by otherwise good people. Confusing news transcripts from the DoD revealed the awkward exchanges with the media:

Q: Before September 11, the Pentagon was wrestling with the decision about whether to invest more money into BioPort, the nation's sole maker of anthrax vaccine, which so far has had difficulty producing FDA-certified vaccine. Are you now prepared to put—invest whatever it takes in that facility to produce a vaccine, or are you inclined to go in some other direction?

Secretary Rumsfeld: The answer is, I think, that I would have to talk to Dr. David Chu and Pete Aldridge . . . it has been something that I—preceded us, me, here as a problem for this department . . . those problems that have not been worked out.[13]

This DoD presser confirmed our knowledge of the internal DoD review, Aldridge and Chu's work, and also the knowledge of the problems. Media exchanges added to the confusion:

> **Q:** I wonder if you can give us an update on the Pentagon's anthrax vaccine?
> **Secretary Rumsfeld:** We're going to try and save it. There have been other efforts that have failed over a period of years. And it may or may not be savable, but I met this morning with Pete Aldridge and David Chu, and we discussed this at some length. And they or their representatives are going to be meeting with people from HHS and Secretary Thompson's office and try to fashion some sort of an arrangement whereby we give one more crack at getting the job done with that outfit. It's the only outfit that—in this country that has anything underway, and it's not very well underway, as you point out. We're trying to fashion a way that the—it's a combination of things, but they have not been approved by the FDA, as I understand it. They do not have what looks to be—well, I shouldn't be characterizing a private entity that way, but things have not been going swimmingly for them. And what we're trying to do is figure out a way where we might get some help so that they might improve their performance.[14]

Reading those quotes, I wondered how the thousands of troops must have felt hearing or seeing those words after being forced out over the anthrax vaccine, after the years of quality control problems, the non-finalized vaccine license, the unapproved manufacturing processes, the failure to get the vaccine properly approved for the investigational use against inhaled anthrax, and the lack of Gulf War Illness study. I wondered how others felt about the sole manufacturer getting "one more crack at getting the job done." Compared to those extensive concerns, the pithy responses from the top DoD official were less than a ringing endorsement:

> **Q:** Are you taking the anthrax vaccine, Mr. Secretary?
> **Secretary Rumsfeld:** No.

Q: You're not being inoculated; you're not taking a series of tests.
Secretary Rumsfeld: No. No.
Q: All right. No vaccine.
Secretary Rumsfeld: No, no, no.[15]
Q: Mr. Secretary, have you been vaccinated against anthrax?
Secretary Rumsfeld: No . . . Have you?[16]

Our defense secretary possessed unmatched skill in dodging tough questions:

Q: Mr. Secretary, I have a question to you on the anthrax vaccination program . . . a couple of weeks ago you said you were going to try to give BioPort one more chance. What did you have in mind?
Secretary Rumsfeld: I would rather have you talk to Pete Aldridge about what we have in mind. It involved, as I think I indicated at the time, some complex negotiations and discussions about how they might find a way forward that would be useful and constructive from the standpoint of the Department of Defense.[17]

The media sensed the discomfort with the issue, but did not relent or accept indecision:

Q: Have you reached a decision yet on the future of the anthrax-vaccine program?
Secretary Rumsfeld: No.
Q: You have not. Any sense when you will?
Secretary Rumsfeld: Torie can get you an answer to that from Pete Aldridge and David Chu. I just happen not to be current . . . I should've said, "Not to my recollection." I think they are reopening that company.[18]

Sensing the confusion and discontent within the Pentagon over the vaccine, we pressed on in our attempts to educate more senior leaders. We began with the new official that Mr. Perot previously introduced Russ to at the congressional hearing—a former pharmaceutical executive himself. The new assistant secretary of defense for health affairs, Dr. William Winkenwerder, invited us to the Pentagon to brief our concerns. Amazingly, the anthrax vaccine agency director received a promotion and became the doctor's assistant. He was the same officer that authored the first Anthrax Vaccine Agency newsletter with his infamous distraction: "Much of the hand wringing and bizarre allegations about the vaccine is coming from a vocal minority of people who think the 'field' is where a farmer works and 'Gortex' is one of the Power Rangers . . . For those who have had to fight for it, freedom has a special flavor the protected will never know."[19]

Beyond the unprofessional, diversionary, and derisive tactics, worse yet, this officer now held a position where he tried to block our concerns. Despite the assistant's best attempts to block our meeting, we outflanked him. One of his haughty pre-meeting notes included:

> Maj Rempfer: Thank you for your recent request. The Assistant Secretary of Defense for Health Affairs is not aware of having accepted any offers for a briefing on anthrax, other than those provided by his staff, the AVIP Agency, and numerous other relevant agencies whose charter it is from the Secretary of Defense to plan and execute this program . . . Further, it is not customary for military officers within the Armed Services to make direct requests for an audience with the Assistant Secretary without making such requests formally through their chain of command in their respective Service or OSD component head. Please route future requests accordingly.

I responded to the assistant's attempt to blow me off with the full intention to press forward with the meeting. I politely wrote back to him that it also wasn't "customary" for a DoD military member to mislead our appointed or elected civilian officials. I asserted that we were respectfully requesting an opportunity to set the record straight. A little string pulling later, the assistant responded with our meeting arrangements. Once again we came full circle in our attempts to explain the ever-mounting legal concerns to our leaders. The assistant wrote back:

> Dr. Winkenwerder has agreed to a 45-minute meeting on 14 Jan 02, from 1400–1445. . . . Dr. Winkenwerder would prefer to keep the group to no more than 5 attendees. I trust you have the other information necessary regarding the read-aheads, etc. . . . Regards, US Army Military Assistant, ASD (Health Affairs)

Something was surely going on in the Pentagon. We were full circle, no different than with Tiger Team Alpha, being asked—or at least entertained—for our inputs and concerns over the anthrax vaccine. We only arranged the meeting and were allowed in the building, because entities inside knew something was wrong. The only question would be whether there would be enough courage inside the puzzle palace to stop the "bad order and discipline" once and for all.

Once again in good faith, Russ, JR, Lou, and I gathered our forces and made our way to DC. This time we ventured deep into the DoD's OODA loop, inside the Pentagon. We were given an hour and fifteen minutes with the assistant secretary of defense, but once again found ourselves less than hopeful about the possibilities. We were met and accompanied by a deputy who informed us that she had worked with Mr. Cragin when he was defending the anthrax vaccine program. She proudly dished out the boilerplate script about the threat and protecting the troops. The pompous platitudes worked for the last four years for them, so we understood.

We sat in the meeting, accompanied by the lawyers that were defending the DoD in other legal forays launched by attorney Lou Michels. Plain-clothed guards sat just outside the meeting room in the foyer as though we were the enemy. The meeting ended politely. Not too long after, we discovered we must have stirred the pot because we were asked to follow-up with additional position papers on the program. The acting Assistant Secretary of Defense for Reserve Affairs, Craig "Raven" Duehring, who had engaged me two years earlier after my *Washington*

Post opinion editorial, asked for Russ's, JR's, and my policy inputs before and after meeting. He wrote:

> Buzz, OK, the latest. I have arranged to have a meeting between the Health Affairs folks (perhaps with additional guests) and our Reserve Affairs people on Tuesday afternoon. We intend to hash out, hopefully, once and for all, all the problems associated with the NEW vaccine. I know I have asked for the bottom line before, but I must ask once again. Can you give me the questions that must be answered and the conditions that you feel must be fulfilled before you would accept providing the NEW vaccine to our troops? . . .
>
> For the purposes of this discussion, omit your concerns about the old vaccine and omit the discussion about the people who have already seen their careers damaged by their actions. We can discuss that later. I honestly see these as three separate, but related, issues. I really need your help in addressing the NEW vaccine. I only want to do what is right for our folks, and to that end, I need your help. . . . Thanks very much. I think we are really making some meaningful progress.
>
> Craig W. Duehring, Principal Deputy Assistant Secretary of Defense

It was clear to us what was going on. The DoD was trying to package the new program with a "NEW" vaccine to resell it to the troops. But there was also an element of intellectual honesty at play with their research into some of the regulatory flaws of the vaccine. At one point "Raven" commented and asked, "I think the biggest issue remaining on the table is whether or not the NEW vaccine is experimental. Can you tell me what the FDA regulation is that requires a human field trial? I'd like to read it for myself. Thanks." Once again, we provided all the applicable US Code, Code of Federal Regulations, and Federal Register citations.

As we compiled the data, Russ cynically asked, "If they have a giant staff and a huge budget, why do they need us to provide them documents and citations from government documents?" It became clear the DoD wanted to emphasize that they were using a "new" vaccine, even though it lacked a finalized license. We sensed this was just an extension of the sales job to salvage a fatally flawed program, but we appreciated the chance to give our inputs—just like Tiger Team Alpha. We tried to play ball, but "full circle," we would not compromise on corrections to military records or the requirements of the law—we played hardball, respectfully.

We prepared two documents. The first was an initial position paper identifying how to implement the program legally, with informed consent or by obtaining a presidential waiver, if indeed the DoD wanted to make the program mandatory. Excerpts:

> Sir, Application of the rules prescribed in 10 USC 1107, 21 USC 355, 5 USC 301, 42 USC 289, Executive Order 13139, and DoD Directive 6200.2 in no way preclude the use of the vaccine if this is deemed to be the most viable form of protection. Our SECDEF has two options:
>
> 1. The NEW AVIP can be conducted on a voluntary basis, with soldiers being afforded their "informed consent" required under 10 USC 1107, 21 USC 355, 5 USC 301, 42 USC 289, Executive Order 13139, and DoD Directive 6200.2.
>
> 2. OR—The NEW AVIP can be conducted on a mandatory basis under these same laws, executive orders and DoD Directives if the President of the United States determines that the threat warrants a waiver of our soldier's informed consent.
>
> We appreciate your consideration of our bottom line, and can provide you any supporting documents as we did with Dr. Winkenwerder. . . . The ultimate loyalty at this point is protecting our Commander in Chief from becoming embroiled in the illegalities inherent in the use of the old anthrax vaccine and the old AVIP.
>
> V/Respectfully, Major Thomas L. "Buzz" Rempfer

JR arduously prepared the second lengthy position paper. This was only necessary because the anthrax vaccine top gate guard, the pharmacist, prepared a rebuttal to our initial "bottom line." The pharmacist and the other gate guards ensured the anthrax vaccine program was a "no bad news" initiative and made every attempt to minimize and deny the problems. Whether attrition, adverse reactions, or altering the body of scientific evidence, the "duty" fell to these anthrax vaccine gate guards to confuse our nation's leaders about the earlier, less favorable assessments of the anthrax vaccine. Though the pharmacist was particularly disappointing to witness, over the years we witnessed other DoD gate guards testing the ethical limits as well.

This is why our response to Dr. Winkenwerder's office and the pharmacist gate guard required our "bottom line." A top DoD official specifically requested it and we realized someone had to rebut the gate guard's attempt to debunk our concerns. We held firm about the right way to implement the program and submitted our work once again, four years after the original unanswered Tiger Team Alpha submissions. Full circle, the concerns were dismissed—like us.

Raven, the assistant secretary of defense, told us he did all he could in the months that followed. We appreciated what he tried to do and the faith he showed in our work. As many had done before him, he balanced his loyalty to the chain of command and the political realities against his instincts about the vaccine. He took a risk, as fighter pilots do. Raven gave us one more chance to enter the DoD's loop, to do our duty, but then signed off with this final message:

> Buzz, I have full confidence in this administration and the people who work here. If I didn't, I wouldn't stay around. I believe the current vaccination program is safe and that our people need the protection it provides. Certainly, the anthrax scare of last year should have convinced everyone that the threat is very real.

Raven had done more than any person inside the Pentagon in the previous four years to protect our forces from bad medicine and bad policy. Ultimately though, he went NORDO—meaning "no radio"—the communications were once again cut off. Full circle, we had to gather our intelligence on the pending status of the program from media reports. In one *Washington Post* story on the internal loop of the DoD decisions, a DoD deputy assistant secretary of defense for health affairs, who previously worked on a House Committee that defended the DoD's anthrax vaccine program, commented: "We continue to assess the many factors involved in establishing this policy . . . We do not have a final decision as yet . . . We are very close. We are dotting the i's and crossing the t's . . . When we . . . have [a] decision, we will make an announcement . . . We will not, however, address the internal processes that led to our decision."[20]

The "internal processes" that the DoD "will not, however, address" indicated that they didn't want anyone to know about the debate and dialogue with us. At least Raven's engagement signaled that someone on the inside still cared enough to try. But Russ was right that the DoD as an institution did not have the guts to correct the wrongs. They did not correct what they'd done to a thousand or so troops, probably because to do so would be an embarrassment to a thousand or so generals. To me, Orwellian warnings played out in this parallel universe we observed.

The English novelist Eric Arthur Blair used the pseudonym George Orwell for his iconic novel titled *1984*. He coined "ignorance is bliss" and "war is peace"—paradoxical phrases to describe the control dynamics between the fictional power centers of Oceania, Eastasia, and Eurasia. He warned of deception tools like "doublethink" and "newspeak."[21] Orwell explained the "labyrinthine world of doublethink" like this: "To know and not to know, to be conscious of complete truthfulness while telling carefully constructed lies . . . consciously to induce unconsciousness, and then, once again, to become unconscious of the act of hypnosis you had just performed."[22] Fortunately, we were conscious and aware of their labyrinth.

We were not ignorant or blissful because we knew the facts in our war for the truth. We knew active-duty officers worked on Capitol Hill and testified in court-martials wearing civvies. They blocked constituents and gave suspect testimonies. We knew paradoxical doublethink and misleading statements like "not inconsistent" or "not aware" were used concerning the legality and approval of the vaccine. We knew the truth was the DoD's use of the vaccine for inhalation anthrax was not consistent with the vaccine's approval. We understood the dangers Orwell warned us about. In our intellectual war against anthrax vaccine, doublethink and newspeak were alive and well. We refused to unconsciously be victims of their hypnosis.

My compatriot Enzo Marchese coined another unique term that spoke to the deceitful nature of the gate guard's misinformation campaign. His expression, "dupe-speak," was fitting for the duplicitous gate guards, but was not meant as an insult to the targeted masses. Any insult was directly the result of the perpetrators who treated our citizens as though they were dupes in their attempts to deceive the media, public, commanders, and troops alike. Yes, "dupe-speak" was apropos, and the devious dupe-loop was too dangerous to not dispute and try to disrupt.

Worse still, officers like the pharmacist gate guard were empowered and promoted to defend the policy while most senior officials feigned ignorance of the problems. We were witnessing Orwellian warnings being played out within the Pentagon's anthrax vaccine program. We analyzed the paradox of their education campaign rhetoric versus the pre-1997 science and documents. We knew this wasn't the military we joined as we saw real illnesses and punishments occur. As we worked to expose the flaws, the gate guards more vehemently protected their program. The dupe-looper-in-chief pharmacist added to the dupe-speak, saying, "In aggregate, what they show is anthrax vaccine has a side effect profile similar as that of other vaccines."[23]

"In aggregate" appeared to be false based on the new information showing a hundred-fold increase in adverse reaction rates, possible birth defects, and suspect deaths. The pharmacist gate guard spoke with unchecked impunity. He added, "We have conducted eighteen human safety studies—short and long term—retrospective and prospective." The unknowing reporter, reader, and officials had no idea what this meant or if it was true. Russ analyzed the referenced post-anthrax-vaccine-program studies, compared to the preprogram admissions about a "paucity of data," as described by the Institute of Medicine. Most of the studies were initiated after the anthrax vaccine program began and included record reviews with little to no transparency.

To us it was clear—the pharmacist and his fellow gate guards were saying whatever was necessary to quell concerns and make people go away. But the troops could not escape the abuses. The gate guards repeated the patterns of delay, denial, and abandonment that marked the syphilis testing, nuclear testing, and Agent Orange travesties. The pharmacist added, "Lots of people are confusing, 'it happened after vaccination' with 'it happened because of vaccination.' . . . If people are getting sick," the pharmacist declared, "It is not due to the vaccine."[24] A similar 'correlation does not equal causation' retort reemerged years later as a COVID-shot denial tactic.

Denial was the M.O. and delay was the objective. The pharmacist gate guard was just an emboldened pawn conducting an ill-fated mission. Medical doctors weren't quoted, and the DoD allowed a pharmacist to mislead our troops, citizens, media, and elected officials. And now the DoD had the anthrax letter attack fear to push the program's revival. Though troops fell ill due to the vaccine, sometimes in large numbers or "pockets" as the DoD referred to them, our leaders looked the other way because the anthrax letter attacks justified their threat as real. The answer about what to do with the vaccine program—the same program Rumsfeld admitted hadn't gone "swimmingly"—came from the former Air Force leader who spoke at my academy graduation.

"Pete" Aldridge told my academy graduation class in 1987 to make sure "if a regulation or policy doesn't make sense, get involved; propose a revision that does contribute to your mission." He and Under Secretary Chu wrote the pre-September 11th recommendations to minimize the use of the vaccine, procure antibiotics and bio-detectors, and develop a coherent process for dealing with these defense measures in the future. At that point, I knew that Aldridge came to the same conclusions. As fighter pilots, we'd say we were all reading from the same "lineup card," the lingo of fighters for our mission plan. If members of our formation were not looking at the right lineup card, it could lead to in-flight disaster. Sadly, I remember reading the DoD news briefing transcripts, realizing the "lineup card" had changed:

> **Q:** Can you tell us where you're at on the anthrax-vaccination program?
> **Aldridge:** Yes. As you know, the—my favorite topic [chuckles] the BioPort facility was, in fact, approved by FDA for production. It is now producing vaccine. We're trying to get the contract in place to—because we actually had a stop-work contract—a contract which we actually stopped work on. It is now being renegotiated to get back on the contract.[25]

It was our "favorite topic" too, but for Russ, me, and thousands of others, it didn't invoke chuckles. It was indeed decided within a few more months. Whereas Secretary of Defense Cohen signed off on the original mandate, Rumsfeld did not put his name on the newly relaunched program. We didn't blame him. Instead, the deputy secretary of defense, Paul Wolfowitz, directed the resumption.[26] The USAF restarted its vaccinations.[27] A Pentagon spokesperson preemptively referenced the "threat" in public announcements when addressing any anticipated resistance: "We're not expecting that because of events last year and the actual use of anthrax."

We weren't expecting it either. The nation traveled full circle, and even to war over 9/11 in Afghanistan and worries about weapons of mass destruction (WMD) in Iraq—including suspicions of anthrax. Ironically, the WMD never materialized, and the anthrax letters turned out to be an insider attack. Regardless, the anthrax vaccine was again used as part of a "medical Maginot Line," no different from the 1991 first Gulf War, or in 1998 following the expulsion of the United Nations Weapons Inspection teams. I thought about the cliché, attributable to Sir Winston Churchill, that "a nation that forgets its past is doomed to repeat it." I thought about our nation, our history, and the full-circle nature of the patterns of deceit and M.O. we encountered.

Though frustrated, due to my belief that the government was fooled by the anthrax letter attacks, I hoped that in the years to come, the legislature and the executive branches of our government would similarly come full circle, as Senator Daschle and President George Bush did with Agent Orange. Especially considering the Department of Veterans

Affairs (DVA) study of Gulf War veterans revealed that the DVA had data linking anthrax vaccine to Gulf War Illness (GWI), which had not been released to the public, we hoped in time the truth would come out.

The DVA's own committee, tasked to investigate the anthrax vaccine and Gulf War illness, acknowledged, "Substantial questions remain about the possible contribution of vaccines, including the anthrax vaccine, to chronic ill health experienced by veterans of the 1991 Gulf War." The DVA committee also pointed out that "recommendations were enacted into law in the Force Health Protection statute, PL 105–85," but that "counterparts at the Department of Defense [must] ensure that these laws are implemented."[28] In other words, internally the DVA appeared to know that unpublished studies linked Gulf War Illness to the anthrax vaccine, but it was too damaging. The DVA more likely than not buried the studies to stymie the Gulf War veteran population's ultramarathons for answers.

The anthrax letter attacks became a too-convenient excuse to bury the inquiry altogether. The anthrax letter attack criminal diversions by domestic actors led directly to a war and the resuscitation of the anthrax vaccine program. Even after, the facts also showed the FDA failed to properly finalize the anthrax vaccine licensing rule when the agency allowed the manufacturer to reopen after the letter attacks. The new product labeling documented a momentous increase in the number of systemic adverse reactions to the anthrax vaccine.[29] One section of the post-letter attack anthrax vaccine labeling documented an adverse reaction increase of 175 times, which was 17,500 percent higher than previously experienced. When the anthrax vaccine program was first announced, the FDA-published rate was 0.2 percent versus the new rate of 5 to 35 percent. The truth was in the fine print.

Despite the troubling culture that allowed the entire anthrax vaccine debacle to occur and then be resuscitated after the letter attacks, there were also indicators that administration officials were still quietly listening. Mr. Rove's boss, President Bush, made the development of a new anthrax vaccine a national priority. Within days of the anthrax vaccine manufacturer's re-approval, and the revelations on the new product label,

President Bush stated in his 2002 State of the Union Address: "The nation will develop vaccines to fight anthrax and other deadly diseases."[30] A White House press release on February 5th, 2002, discussed the "acquisition of the next-generation anthrax vaccine."[31] The president reiterated his call "to quickly make available effective vaccines and treatments against agents like anthrax" in his 2003 State of the Union Address.[32] Though the presidential advisor's secretary had informed me that, "this issue will be handled by the DoD. Thank you," it was evident oversight continued. Our nation and our DoD needed a new vaccine.[33] The president ordered its development. Sadly, as of 2024, over twenty-five years later, the DoD never acquired the approved next-generation anthrax vaccine.

Suffice it to say, our nation was hoodwinked by the anthrax letter attacks. In August 2001, our government was on the verge of stopping the shots, resurveying judgements, and we hoped our leaders might assert civilian control over the military by correcting the wrongs. Maybe they would even do something about the dupe-loop by the DoD and their dupe-speak tactics. We hoped they would halt the discharges, denial, delay, abandonment, and abuse of power and discretion. Our crew didn't want to see our DoD doom itself by repeating historical mistakes and never breaking the cycle. We were proud to serve and to try to stop the confidence-shaking uniform deceit. Unfortunately, everything changed after the 9/11 tragedy with the subsequent anthrax letter attack deceptions, seemingly designed to derail the vaccine oversight and promote war.

Little did we know that after the first anthrax letters in September of 2001 that more anthrax mailings were yet to arrive. Remarkably, these disappointing setbacks served to refuel us as much as our sporadic victories. We maneuvered, adjusted our strategy, and devised new tactics to turn the "misfortune into gain." Russ and I went back to the drawing board. Russ, the intellectual heavyweight in our unyielding campaign, had one more brilliant idea to encourage our leaders to hold the institutions of our government accountable.

MILE 14: Citizen Petition 1 (2001)

Undeterred, Russ went to work on our next mechanism for redress, a "Citizen Petition." This new tool derived its power under the auspices of section 10.30 of Title 21, which governed the Federal Food, Drug, and Cosmetic Act. The petition allowed citizens to contest their government's bad behavior. This new instrument required the government to answer the mail.

Russ was also an airline captain. As he traveled around the country, he visited university medical libraries in order to further our research on the anthrax vaccine. He searched for the "truth in the fine print" at the University of Connecticut Health Center, Duke University's Medical Center Library, the University of Texas Health Science Center at San Antonio, the University of Washington's Health Sciences Libraries, the California State University Library in Long Beach, and even the FDA's Reading Room in Washington, DC. If our DoD leaders had only emulated Russ's thoroughness, the entire catastrophe would never have happened.

On our home turf as well, I joined Russ at the Connecticut State library, which held a repository of government records. Russ taught me how to research these federal archives and the Federal Register. He showed me the original FDA entries on the anthrax vaccine. We realized that the FDA had never finalized the licensing process on the anthrax vaccine as required by law. The state librarians assisted Russ in confirming there was never a final ruling issued for the anthrax vaccine beyond the FDA's "proposed rule" on Friday the 13th of December 1985.

The FDA's published proposed ruling confirmed the vaccine used by the DoD in their mandatory force-wide vaccination program had "not been employed in a controlled field trial" as required by law. The federal document established, "No meaningful assessment of its value against inhalation anthrax is possible due to its low incidence." This supported our contention that the vaccine's use for inhaled anthrax was investigational, making the mandatory program illegal absent a presidential waiver of a troop's right of prior consent. Russ had unearthed the most important key absurdity in the dilemma—the vaccine license had not been finalized by the agency responsible for its regulation, the FDA. This is when Russ uncovered the buried truth that the FDA did not technically ever license the vaccine, contrary to the DoD's boilerplate script.

The petition was essentially Russ's rewrite of our False Claims Act suit, but a direct means of asking the FDA Commissioner to correct regulatory mistakes. The fact that the FDA had never finalized the anthrax vaccine license was our main "hook" for the petition. Either nobody realized this fact that Russ discovered in the libraries around the country, or they knew it and covered it up. We also cited FDA policy which barred the government from procuring drug products from a manufacturer suffering from regulatory violations, noncompliance with current good manufacturing practices, and a notice of intent to revoke a manufacturer's license. The petition asked the FDA Commissioner to "issue a final rule" on the anthrax vaccine in the Federal Register, since the vaccine at that point remained technically unlicensed by the FDA.

Immediately following the September 11th tragedies, the anthrax-contaminated mail discovered in the US postal system created nationwide fear. Understandably, lawmakers and citizens asked why the government didn't have an available vaccine. Behind this backdrop, we observed fear-mongering of the threat being used by the DoD and its supporters to influence an expedited approval of the anthrax vaccine's non-validated manufacturing line. So, we cast another spear in our multi-azimuth attack through this legal mechanism with our citizen petition. My eternally researching flight lead, Russ, included the FDA policy barring the government from procuring drugs from a manufacturer with BioPort's regulatory record. The policy stated:

> CGMP [current good manufacturing practice] deficiencies supporting a regulatory action also support decisions regarding non-approval of drug marketing applications, government purchasing contracts. . . . Therefore, the issuance of a warning letter or initiation of other regulatory action based upon CGMP deficiencies must be accompanied by disapproval of any pending drug marketing application, or government contract for a product produced under the same deficiencies.

In plain language, the government "must" have disapproved contracts with BioPort because they were selling the DoD vaccine produced during periods of "deficiencies" after receiving "warning" letters and "other regulatory action." The FDA issued a warning letter to the anthrax vaccine manufacturer on August 31st, 1995, and a Notice of Intent to Revoke letter on March 11th, 1997.[1] The FDA's rules barred purchase of the anthrax vaccine according to this policy—Section 400.200 of their own compliance policy guides, or CPG 7132.12. Russ's and my petition simply asked for the very enforcement apparently required by FDA's own rules.

I flew to Washington, DC, and filed the petition with the FDA's Dockets Management Branch on October 15th, 2001. I planned to file the petition on the same day that BioPort initially advertised they intended on filing their application to recommence manufacturing. Our goal was to force the FDA to "do it right" and properly finalize the license, especially if they intended on reopening the plant. The former FDA Chief of Biologics of the Center for Biologics, Evaluation, and Research, Mr. Sammie Young, helped me with my mission. He kindly provided ground transportation from Baltimore-Washington International Airport to FDA HQs in Rockville, Maryland.

Sammie was more than just my ride that day. He assisted us with court filings in the years that followed in an arduous process of judicial review challenging the anthrax vaccine's uncompleted license. Upon arriving at the FDA, the clerks realized our petition dealt with anthrax vaccine, so they swiftly summoned a supervisor from a back room. Fortunately, I had my former FDA regulator wingman in my formation. Instead of an official supervisory blow off, we were met with a warm greeting. "Sammie, it's good to see you again!" Thanks to Sammie, the skids were greased, and we filed the petition. The dockets management branch officially time stamped our petition—#8163 on October 15th, 2001, at 0951 hours.

Former FDA executive and USAF Colonel Sammie Young.

Our sixty-page petition requested that the FDA revoke the license and follow its own rules based upon court precedent.[2] The lengthy history of manufacturing problems, including failing to report equipment and process changes to the FDA, appeared to meet the threshold for license revocation. This was without even mentioning the DoD's "intimate" extra-regulatory involvements. Our petition's cover listed our filing authority under the Federal Food, Drug, and Cosmetic Act, and asked that the Commissioner take the following administrative actions:

1. Issue a final rule on the drug category placement of anthrax vaccine. . . .
2. Declare as adulterated all stockpiles of anthrax vaccine. . . .
3. Enforce FDA Compliance Policy Guide, Section 400.200. . . .
4. Revoke the anthrax vaccine license (#1260) held by BioPort Corporation.

After filing the citizen petition, Sammie dropped me off back at the airport. I placed calls to congressional offices, including Senator Daschle's. I always kept our Capitol Hill allies informed with each step in our multi-axis attack. Daschle's office answered, but I was placed on hold. Soon after, the receptionist in the Democratic leadership's office came back on the phone and informed me that my staffer contact was "unavailable." I asked to leave a message. I was again put on hold. When the secretary came back on the line she shocked me by informing me that they were evacuating the office because they had just opened an "anthrax package." She hung up. I stood in the airport concourse in complete disbelief. I flew home, stunned to have witnessed in real-time the second round of anthrax mailings before the breaking event hit the news.

We never heard from Daschle's office again. Senator Daschle's office went completely "NORDO" (no radio) in the fall of 2001. In addition to going incommunicado, their investigation ceased. The senator even recommended the vaccine for his staff after previously uncovering its serious safety, efficacy, and legality problems. "One aide to Senate Majority Leader Tom Daschle said he took the shot because of the military doctors advising him and his colleagues."[3] Considering Daschle's investigation of the vaccine, we were surprised to read that he stated, "I support that decision" to use the vaccine. Despite the fact that Daschle's office was in

Anthrax letter and envelope opened in Senator Daschle's office.

the midst of the highest-level inquiry to date, DoD advisors successfully scared and convinced him and his staff to take the very vaccine they were investigating "because of the unknowns, because of the uncertainty."[4] In the end, I imagine they felt very manipulated and used by the deadly scheme.

JR often referred to the vaccine policy issue as the "stinky elephant at the cocktail party." Many knew it stunk, but few would acknowledge it, and almost no one dared to call out the façade of force protection. One lawmaker who did smell something was also the sole medically trained doctor-turned-senator on Capitol Hill, Bill Frist. CNN captured his skepticism over the vaccine.[5] Soon after the anthrax attacks, Senator Frist said, "The vaccine is a dated vaccine, it's an old vaccine. There are very real and potentially serious side effects from the vaccine and anyone who elects to receive the vaccine needs to be made aware of that." We agreed.

Senator Frist continued, "I do not recommend widespread inoculation for people with the vaccine in the Hart Building. . . . There are too many side effects and if there is limited chance of exposure the side effects would far outweigh any potential advantage." Again, we agreed. While some acknowledged the risks for civilian staff, others recommended it despite knowing the problems. But no one apparently felt our troops deserved "to be made aware of" the "very real and potentially serious side effects from the vaccine." Nor did our troops qualify to be able to "elect" to receive the vaccine, despite the law technically guaranteeing this explicit equal right. The double standard for our nation's troops, the hypocrisy, was hard to ignore.

After the anthrax mailings, it made "common sense" that the political paralysis was then exploited to sucker staffers into taking the vaccine. It seemed that a bad cause through bad means by otherwise good people ruled the day. A logical question would be: Why was the vaccine necessary when antibiotics successfully treated those exposed? Of the approximately ten-thousand postal workers and congressional staff offered the vaccine with informed consent, only about 1 percent actually submitted to the shot—and none of the 99 percent who refused contracted inhalation anthrax infection. Though the vaccine was not required, the successful attack broke vaccine hesitancy and squelched any remaining opposition or oversight.

Case in point: Daschle's office went off the grid and was AWOL (absent without leave). We had witnessed that once people submitted to the vaccine, they stopped questioning it further. This happened throughout the military, and the phenomenon reoccurred with COVID-19. With opposition dissolved, BioPort simultaneously filed for the approval of their new manufacturing line, and the DoD pushed the vaccine on Congress. A *Washington Post* article documented that "the Capitol Hill physician, John F. Eisold, had strongly urged about seventy people on Capitol Hill to receive the vaccine."[6] The article added, "Federal health officials say the vaccine is safe even though it has not received final approval from the Food and Drug Administration."

What was missing in the article was the explanation that Eisold was also a Navy admiral.[7] He placated congressional concerns as he advocated they take the vaccine.[8] The admiral emailed members and their staffs that "recent misleading news reports regarding the anthrax vaccination of Capitol Hill workers have been circulating and causing great concern," but without giving examples. Eisold recommended the vaccine for Daschle's staff, advising, "The vaccine will be given as part of a CDC research protocol and will require obtaining informed consent from all participants, since the vaccine is not FDA approved for post exposure prophylaxis." Incredibly, they were offering informed consent to civilian staffers for a use of the vaccine that was "not FDA approved"—the same legal right requested by our troops since the anthrax vaccine was not technically licensed by the FDA or approved for inhalation anthrax.

Predictably, two DoD generals also used the letter attacks to justify continued use of the vaccine. In a letter to the FDA, these DoD generals capitalized on the "situation" by stating, "The anthrax attacks in October 2001 illustrated the risk of an unprotected population in an environment contaminated with a biological warfare agent."[9] A congressional staffer for Senator Jeff Bingaman explained, "When anthrax hit Daschle's office, they had NIAID [the National Institute of Allergy and Infectious Diseases] telling them the anthrax vaccine was unnecessary considering the staff were taking antibiotics." He clarified, the "DoD offered them the anthrax vaccine and around thirty staffers took it at the recommendation of their staff leadership." He wrote, "Therefore, they are wedded to their decision and recommendation to staff, believe their experience was a good one, and are grateful that DoD circumvented NIAID."[10]

The DoD officials seized the opportunity to push the vaccine and get the original office investigating the program to accept the vaccine and terminate their critical inquiry. Later, Senator Bingaman and his staff knew something was wrong with the way it all went down, and passed a resolution expressing the sense of the Senate. The resolution made four main points, including that the "Secretary of Defense should reconsider the mandatory nature of the anthrax and smallpox vaccine immunization program, pending the development of new and better vaccines." The resolution added that the "Board for Correction of Military Records should reconsider adverse actions already taken or intended to be taken against service members for refusing to accept the anthrax vaccine," that the intelligence community should "reevaluate the threat of anthrax," and that the "Secretary of Veterans Affairs should assess those adverse events being reported with respect to the anthrax vaccine."[11]

It was surreal to witness the second round of anthrax attacks after visiting Washington to deliver the petition. It was more unsettling to have been on the phone with the office that was helping us at the very moment the attack occurred. Realizing the coincidence might be suspect, I contacted the FBI. I figured FBI might be curious why I was on the phone with Senator Daschle's office when they opened the anthrax letter, or why I was in Washington. The FBI should have been curious why Daschle wrote Rumsfeld. The FBI even advertised, "If you have credible information that might help, please contact the FBI immediately."

So, that's what I did. I wanted to ensure the FBI was aware of our work with Daschle's office, his staff's investigation of the anthrax vaccine policy, and my ill-timed phone call to the senator's staff. Officially the FBI, with the exception of one runaround email, never directly contacted me. On that occasion, the FBI's Operations Center recommended

I contact the FBI's New York division even though that office was no longer responsible for the case:

> Dear Mr. Rempfer: Thank you for contacting the FBI's Internet Tip Line. In your message, you advised that you have previously provided information regarding Anthrax to the New York Division. . . . Accordingly, you should contact the New York division and request an update. . . .

The FBI's apparent blindness to my inputs, while publicly asking for information, added to the repeated patterns of willful ignorance by government officials. Since the FBI's investigation languished for almost a decade, it was unlikely they ever investigated the similar statistically relevant pocket of deaths during the 1957 US Army anthrax vaccine trial in Manchester, New Hampshire. Perhaps the FBI retrospectively tested to see if those spores were also linked to highly lethal US military stockpiles as some scientists surmised based on predictive models involving biological warfare attacks on civilians—as was the case in 2001.[12, 13] Regardless, our objective was to work with any government officials who listened. Despite the FBI's blinders, we still had a job to do in righting the wrongs associated with the defunct anthrax vaccine program.

Within a year, the FDA responded to our petition and significantly acknowledged they had "not issued a final order on AVA's [anthrax vaccine adsorbed] classification"[14] or license, but would do so. The conflicts of interest and circularity became clearer still in FDA's response:

> DoD has been significantly involved in developing the formulation and manufacturing process of all three versions of the anthrax vaccine: The DoD vaccine, the Merck vaccine, and MDPH's vaccine. **DoD's continuous involvement with, and intimate knowledge** of, the formulation and manufacturing processes of all of these versions of the anthrax vaccine provide a foundation for a determination

> that BioPort's anthrax vaccine is comparable to the original DoD vaccine. . . .
>
> DoD was involved in developing the three versions of the anthrax vaccine and had knowledge of the manufacturing processes of each version, DoD is thus similar to a manufacturer that made manufacturing changes to its product as contemplated by FDA's Comparability Guidance.[15]

In other words, like the DoJ and the FBI, the FDA looked beyond the manufacturer's failures to comply with the law because of the DoD's "intimate" associations with the violations. Interestingly, the FDA literally threw the DoD under with the admissions. The explanation of the circular relationship provided perspective on how the DoJ justified its political decision to drop our False Claims Act suit. The FDA's acknowledgment of the never-finalized license included the assurance that they would fix it seventeen years after the original anthrax-vaccine-proposed rule. It was hard to believe that the FDA-DoD duo forgot to finalize the license. Notably, the FDA's response proved the DoD's "FDA approved" soundbite for anthrax vaccine was fictitious.

The FDA response to our citizen petition, admitting the agency had "not issued a final order on AVA's [the anthrax vaccine's] classification" or license but would do so, served as an important vindication, fatal admission and severe chink in the anthrax vaccine program armor. More determined than ever to fight on against these absurdities that defied "common sense," we passed the lead to our legal colleagues. We needed our attorney colleagues, the JAGs, to help force the judiciary to hold the DoD and FDA accountable. It appeared that only the federal courts would have the power to judge the FDA's negligence and the DoD's "continuous involvement with, and intimate knowledge" of the anthrax vaccine.

Through the citizen petition, we forced the FDA to do their job. With the FBI, we tried to assist them in doing their job, but came to realize the bureau was politically paralyzed and intellectually broken, as many Americans would come to understand in the decades ahead.

Had the FBI listened, they could have wrapped up their anthrax letter attack investigation in months instead of dragging it out for the next decade. The bureau deliberately ran their investigation like a marathon instead of a sprint, losing most spectators along the way as a result and by design.

It made no sense—so we got involved. This was our duty according to the inspirational words by former US Air Force Secretary Aldridge from my academy graduation. Overcoming the political paralysis of derelict executive branch agencies, and collaborating on forcing judicial review through the courts, became our next azimuth of attack.

MILE 15: Judicial Review (2003 to 2005)

In September 1995, the US Army developed a plan to obtain FDA approval for proper licensing of the anthrax vaccine against aerosolized or inhalation anthrax. The plan's text included the admission that "This vaccine is not licensed for aerosol exposure expected in a biological warfare environment."[1] The Pentagon office in charge of the anthrax vaccine, the JPOBD, held a meeting in October 1995 to discuss the plan for mass vaccination mandates.

The minutes revealed the Army knew FDA approval was required for inhalation anthrax, that the efficacy tests used to license the vaccine were for a different vaccine, and that there was no scientific data to support a change in licensing. Meeting minutes stated, "A meeting was held on 20 Oct 1995 to discuss the process for modifying the MDPH [Michigan Department of Public Health] anthrax vaccine license to indicate a reduced number of injections and to expand the indication to include protection against aerosol challenge of spores."[2] Officials admitted, "the original series of six doses was established in the 1950's for an anthrax vaccine similar to but not identical with the MDPH vaccine." "Studies of vaccine effectiveness in humans working in tanneries showed protection against cutaneous disease, but there was insufficient data to demonstrate protection against inhalation disease" [meeting minutes at Appendix U].

The JPOBD meeting minutes captured the office's director quoting George Washington. He stated that the DoD's position should be: "Soldiers are citizens first." He cautioned that whatever studies were formulated, they had to be done with this concept in mind. He understood that our troops had the same constitutional rights as other citizens. In 1995, it seemed that the generals agreed with us! Not only did the DoD know the anthrax vaccine was experimental, but they understood the legal requirements. This general was reminding everyone of the very same constitutional lessons that General McGinnis taught us. After our meetings with General McGinnis, we could not understand why the DoD conveniently overlooked everything they recorded in their memos and meeting minutes. This kind of behavior seemed inconsistent with the vision of the Founding Fathers, as well as the requirements of our oath of office.

In September 1996, an investigational new drug application [FDA forms at Appendix V] was prepared by the DoD and submitted by the manufacturer to the FDA following an Army, Joint Staff, and Office of the Secretary of Defense review process. This meant everyone was aware of the necessity to obtain an FDA approval for a licensed indication for inhalation anthrax prior to mass vaccinations. By late 1997, the same month Cohen announced the program, and without FDA approval of the investigational new drug application, a JPOBD report noted, "Anthrax and Smallpox are the only licensed vaccines that are useful for the biological defense program, but they are not licensed for a biological defense indication."[3]

The law, 10 USC § 1107, was passed by Congress in 1998, the same year the anthrax manufacturer shut down for quality control deviations, and the same year the vaccine program started. The president signed the law, which directly impacted the DoD's investigational use of anthrax vaccine. The statute included "the requirement that the member provide prior consent to receive the drug in accordance with the prior consent requirement imposed under the Federal Food, Drug, and Cosmetic Act." That right could "be waived only by the President."[4] Instead, the DoD ignored the law, violated prior consent rights, and shattered the oath of office:

> I do solemnly swear that I will support and defend the Constitution of the United States against all enemies, foreign and domestic; that I will

> bear true faith and allegiance to the same; that I take this obligation freely, without any mental reservation or purpose of evasion; and that I will well and faithfully discharge the duties of the office on which I am about to enter. So help me God.

Everyone ignored the oath of office and the law, not to mention the FDA regulations on investigational drugs and good manufacturing practices. I realized that "judicial review" was the next course of action now that other checks and balances failed to provide civilian oversight over the DoD. I maintained my faith in the system and let judicial review work, no matter how long.

Judicial review was basically the form of oversight envisioned by the framers of the Constitution to ensure the activities of the legislative, executive, and various agencies of our government comported with the requirements of the Constitution. Some would contend that an absence of proper judicial review, and excessive concentration of unchecked power in the executive branch of government, led to totalitarian regimes prior to World War II in Germany, Italy, and Japan.[5] This, of course, was precisely why Hamilton, Madison, and their fellow Founding Fathers devised a system where separation of powers, checks and balances, and even our duty of flying high cover as citizen soldiers might preclude such a fate. As pilots and operators, the menacing anthrax vaccine mandate simply defied logic and common sense.

The failures of the DoD to ensure the vaccine license was finalized, to ensure the product was not adulterated in compliance with the Food, Drug, and Cosmetic Act, to ensure the drug was not used for an experimental application without prior consent or the proper presidential waivers, and to ensure that the vaccine was properly studied as a possible cause of Gulf War Illness were simply logical, common sense exercises. It was a slap in the face to the Gulf War veteran population and the ultramarathons for justice and proper treatment they already endured.

To not ensure compliance and to not study the effects was no different than taking a broken jet airborne—you simply do not execute such a mission. I realized this was why we all so vehemently objected to the manner in which the DoD executed their anthrax vaccine operation. It made no sense at all. I reaffirmed that none of it added up, especially the willfully ignorant way the DoD ignored our initial on target Tiger Team Alpha investigation findings. The DoD continued to feign ignorance in order to avoid a duel with their own officers and the truth.

Over time we hoped that the judicial review process would forthrightly settle the complex issues encircling the anthrax vaccine dilemma. One encouraging example was in February 2003 when a US District Court judge ruled in a case concerning a US Army soldier's anthrax vaccine refusal. A fellow academy graduate, Steven Masiello, brought the case on behalf of the wrongfully discharged soldier. The judge's ruling subtly cautioned the DoD:

> The Court is not passing on the merits of the anthrax program. The plaintiff has raised significant questions about that program. If the Court were reviewing the program, the Court would be very concerned about the question that the plaintiff has raised. Title 10 United States Code Section 1107 provides that whenever the Secretary of Defense requests a member of the armed forces to receive an investigational new drug, the Secretary must provide a member with notice about the investigational nature of the drug and require the member's consent. . . .
>
> There have been no tests showing that the vaccine is effective at protecting human beings from exposure to inhalation anthrax, although animal studies by the Army exist. The Court will not substitute its opinion for that of the Army, but it will not review the matter. And its ruling today should not be understood as an approval of what the military is doing in this case. The military will be held accountable to the public if it is using its own soldiers as guinea pigs to determine whether the anthrax vaccine has long-term health consequences and whether it protects against airborne anthrax. Those decisions are, as I said, decisions that are committed to the Executive Branch of the Government. The Court neither approves nor disapproves of those decisions, because it is not the function of the Court to do that. Those decisions will be debated, and ultimately the Executive Branch will be held accountable

> to the public for those decisions. And that is the way the system of government works.[6]

Though the court clearly chose to not interfere with the DoD's program, the court also specified it was "not passing on the merits of the anthrax program." Steadfast, Masiello did not allow the court to evade its constitutional requirement of "judicial review." Based on the proximity of imminent warfare against Iraq, and the ongoing war on terror, the court attempted to dodge its constitutional responsibility. Determined, Masiello appealed to the court:

> The District Court, recognizing the merits of Ms. Barber's claims, noted that the military, in violation of regulatory, statutory, and constitutional law, was vaccinating members of the Army with an IND without obtaining their informed consent. The District Court's dismissal on justiciability grounds was based on "the current status that we find ourselves," but "should not be understood as an approval of what the military is doing in this case." The trial court further noted that the Army might be using its uniformed personnel as "guinea pigs." Contrary to the District Court's ruling, regulatory, statutory, and constitutional rights do not disappear, or become non-justiciable, simply because we are at war.[7]

Though every veteran, soldier, commander, and citizen surely supported our DoD in defending our national security interests at home and overseas, all agreed that regulatory, statutory, and constitutional requirements should not be abrogated. Nor was it good government or precedent to allow overt violations of the law by the DoD to be excused simply because our troops faced hostilities overseas. Though judicial deference to the DoD was commonplace and steeped in legal precedent, it was my conclusion that it was not due in this case. The issues at question were at play for many years preceding any conflict or war, and would remain contested long after hostilities subsided. Even the courts grew tired of the emergency and war excuse.

The DoD seemed to take advantage of such deference and in fact used it time and time again to avoid judicial review—essentially perpetuating the very alleged abuses of power and discretion at issue. Examples abound from a variety of DoD briefings to courts on legal matters related to challenges of the anthrax vaccine program. The DoD built upon such noninterference precedents by quoting past judicial deference. A case concerning the illegal experimental use of the anthrax vaccine, brought by fellow academy graduate John "Lou" Michels and attorney Mark Zaid, revealed past precedents cited by the DoD in their calculated motions to manipulate the courts into dismissing the case.[8] The DoD appealed to the court for noninterference, writing:

> First, the military justice system must remain free from undue interference, because "the military is a 'specialized society separate from civilian society' with 'laws and traditions of its own developed during its long history.'" [Councilman, 420 USC] at 757 (quoting Parker v. Levy) . . .
>
> Second, Congress sought to balance the competing interests in military preparedness and fairness to service members charged with military offenses, by "creat[ing] an integrated system of military courts and review procedures . . . it must be assumed that the military court system will vindicate servicemen's constitutional rights."

I thought about this troubling language. Despite the precedents, I did not think the Founding Fathers actually wanted the military to be a "specialized society separate from civilian society." It seemed un-American and potentially dangerous. The quote from retired Admiral Stanley Arthur, who commanded US naval forces during the Gulf War, spoke to this quandary:

> The armed forces are no longer representative of the people they serve. More and more, enlisted as well as officers are beginning to feel that they are special, better than the society they serve. This is not healthy in an armed force serving a democracy.[9]

What happens when the assumption that "the military court system will vindicate servicemen's constitutional rights" proves incorrect, and instead, the US military courts abuse the "deference" naively granted to it by the federal judiciary? What if the DoD violates our troops' constitutional rights, their prior consent rights, and further violates their constitutional Fifth Amendment right of due process by denying them a defense in court? The DoD did this and then barricaded its abuses by arguing against court intervention and by citing past precedent. The troubling circularity attempted to block the safeguard of judicial intervention. The DoD pleaded:

> Federal courts have a long and uninterrupted history of judicial deference to the military in its handling of internal military matters.
>
> See, e.g., *United States v. Stanley,* . . . Rejecting analysis that "would call into question military discipline and decision-making" and "would itself require judicial inquiry into, and hence intrusion upon, military matters"
>
> *Chappell v. Wallace,* . . . "Courts are ill equipped to determine the impact upon discipline that any particular intrusion upon military authority might have."
>
> *Brown v. Glines,* . . . "The special character of the military requires civilian authorities to accord military commander some flexibility in dealing with matters that affect internal discipline and morale."

Interestingly, the DoD cited the "laws and traditions of its own developed during its long history," as well as its "uninterrupted history of judicial deference to the military in its handling of internal military matters." I wondered if "interrupting" the "history of judicial deference" was not only in order, but overdue. Certainly, the DoD was only worthy of "due respect to the autonomous military judicial system" if it didn't abuse its power and discretion and its troops' rights. The plea by the military for the federal courts to stay out of their business required judicial review. Court "analysis" that called "into question military discipline and decision-making" and "intrusion upon military matters" became necessary in the case of the anthrax vaccine dilemma in order to halt the historic violations of the law.

It's reasonable that the "special character of the military requires civilian authorities to accord military commanders some flexibility in dealing with matters that affect internal discipline and morale." But nowhere is it written that this flexibility is absolute. With anthrax vaccine, the "civilian control of the Legislative and Executive Branches" was breached by invasive DoD dupe-looping and manipulation. At some point, the judiciary must step in to restore the oversight, "checks and balances" and "separation of powers" as the Framers' intended. Contemplating the judge's words, I concluded "that is the way the system of government works."

Just as Masiello continued to appeal to the federal courts pro bono on principle, so too would our gallant attorneys, Lou Michels and Mark Zaid. In a new path advocated by the federal judge in the Masiello case, Michels and Zaid refiled their case under the Administrative Procedures Act (APA) against Defense Secretary Rumsfeld on behalf of "similarly situated soldiers" irreparably harmed by the anthrax vaccine policy. The Administrative Procedures Act ostensibly circumvented convoluted government immunity defenses utilized by the DoD defendants. The case was filed in the United States District Court for the District of Columbia, requesting that a federal judge properly declare the anthrax vaccine an experimental drug.

A separate motion was also filed seeking a temporary restraining order and preliminary injunction against the DoD defendants to prevent further anthrax inoculations without prior consent, or without a presidential waiver according to the law. Specific aspects of the suit cited the FDA's failure to properly finalize the anthrax vaccine license, as proven by our citizen petition response from the FDA. The case included the anthrax vaccine's experimental use for inhalation anthrax and the DoD's deviation from anthrax vaccine license dosing requirements. The DoD's response utilized the urgency of hostilities to attempt to excuse the lawbreaking:

> With our military forces currently engaged in armed hostilities in which the threat of biological and chemical weapons is a constant and

> substantial reality, plaintiffs ask this Court to grant extraordinary and unprecedented relief. They seek to undermine a key component of military readiness and defense against battlefield use of biological weapons—the Anthrax Vaccination . . .

The DoD also employed the same tactic of citing past precedent of judicial deference to implore the court not to interfere with the alleged illegal activities:

> First, this Court lacks jurisdiction to hear plaintiffs' claims, which are nothing short of an attack on decisions of the military chain-of-command regarding military readiness and the conduct of battle. As the Supreme Court has held: "Civilian courts must, at the very least, hesitate long before entertaining a suit which asks the court to tamper with the established relationship between enlisted military personnel and their superior officers," because "that relationship is at the heart of the necessarily unique structure of the Military Establishment." *Chappell v. Wallace*, 462 US 296, 300 (1983).

I agreed that to "hesitate" before intervening was warranted, but to "never" intervene was not. The DoD continued by trying to exclude Administrative Procedure Act review based on "military authority exercised in the field in time of war or in occupied territory."[10] But the DoD could not explain the reality that their program began far before a "time of war" and stretched far beyond the confines of "occupied territory." The DoD would plea that "instead, the military justice system affords the military plaintiffs their proper remedy for claims that an order to submit to inoculation is unlawful." But the DoD failed to explain that in every instance that our troops attempted to get the arguments or concerns about the illegality of the program addressed within the "military justice system," the DoD denied the admissibility of the evidence.

Dale Saran, the US Marine Corps JAG who defended anthrax vaccine refusers in the Pacific and who helped us with our *Military Times* editorial board brief, commented on the mockery inherent in the system.

The federal courts, at times, held that their turf was not the place to hear the arguments. The DoD maintained that court-martials weren't the place for the arguments. Dale concluded that, "Evidently nowhere can this issue be heard" due to the DoD's unhindered circular successes in blocking the legal debate over the anthrax vaccine mandate.

Other cases were brought on behalf of troops with documented injuries from the vaccine. In March of 2003, a federal tort claim lawsuit for 114 injured troops was filed against the manufacturer with specific counts including: negligence, breach of warranties, breach of the right to be treated with essential human dignity, strict products liability, fraud, deprivation of civil rights, as well as a spouse's loss of assistance, companionship, and consortium. Though the DoD-funded defense would once again contend that the plaintiffs were attempting to question the DoD anthrax vaccine program, the legal challenge on behalf of harmed troops and their families was really about harms and the laws of our country. We all hoped that the manufacturer could no longer hide behind the DoD or its fortress of legal immunity and indemnification. The attorneys for the sick and deceased military members wrote:

> This case is about whether a private corporation and its predecessors can be held liable for selling the DOD an experimental and defective vaccine knowing it would be used on soldiers without their informed consent. It is about whether these defendants, engaged in a for-profit private enterprise, are entitled to immunity if they have produced and manufactured a vaccine in violation of numerous federal regulations and standards.[11]

All these cases were filed because what the DoD was doing didn't make sense, didn't add up, and defied logic. We realized the DoD and its surrogates could hide their bad behavior behind legal motions, but in time their abuse of power and discretion would hopefully catch up to them. The truth did catch up with the DoD in a 2003 preliminary injunction that affirmed, "*Anthrax Vaccine Adsorbed ("AVA") is an experimental drug unlicensed for its present use and that the AVIP violates federal law (10*

USC § 1107)." The court's injunction "ORDERED that the Motion for a Preliminary Injunction is GRANTED. In the absence of a presidential waiver, defendants are enjoined from inoculating service members without their consent."

The court explained, "The apparent change in position from the December 1985 proposed rule and the cryptic use of a double negative (i.e., 'it is not inconsistent'), fail to persuade this Court that the view expressed in the 1997 letter is the FDA's formal opinion." The judge cited our FDA citizen petition noting, "In a response to a citizen petition dated August 2002, the FDA's Associate Commissioner of Policy noted that the FDA still has yet to finalize the rule proposed in the December 13, 1985, Federal Register." As a result, the judge ruled, "**This Court is persuaded that AVA [the anthrax vaccine] is an investigational drug** and a drug being used for an unapproved purpose. As a result of this status, the **DoD is in violation of 10 USC § 1107**, Executive Order 13139, and DoD Directive 6200.2," and concluded, "the Court finds that the plaintiffs meet the requirements for a Preliminary Injunction." Excerpt from judge's ruling:

case law. In sum, because the record is devoid of an FDA decision on the investigational status of AVA, this Court must determine AVA's status for itself. This Court is persuaded that AVA is an investigational drug and a drug being used for an unapproved purpose. As a result of this status, the DoD is in violation of 10 U.S.C. § 1107, Executive Order 13139, and DoD Directive 6200.2. Thus, because the plaintiffs are likely to

Case 1:03-cv-00707-EGS Document 18 Filed 12/22/2003 Page 33 of 34

The women and men of our armed forces put their lives on the line every day to preserve and safeguard the freedoms that all Americans cherish and enjoy. Absent an informed consent or presidential waiver, the United States cannot demand that members of the armed forces also serve as guinea pigs for experimental drugs.

Excerpts from Judge Emmet Sullivan's federal court ruling.

By not allowing the DoD to treat our troops as "guinea pigs," the same judge, Emmet G. Sullivan, for the United States District Court for DC followed up almost a year later with a 2004 permanent injunction. He ruled, "Accordingly, **the involuntary anthrax vaccination program, as applied to all persons, is rendered illegal** absent informed consent or a Presidential waiver; and it is further ORDERED that the Defendants' Motion for Summary Judgment is DENIED."

Yet the most important part, from a precedent perspective, included subsequent precise language from the informed consent section of the Code of Federal Regulations (CFR) about "no penalty or loss of benefits."[12] In the DoD's attempt to overcome the Judge Sullivan ruling, the military assured the court that they would continue the anthrax vaccine program strictly on a voluntary basis. Continued use of anthrax vaccine ensured "**no penalty or loss of entitlement**." This was the first application of a new law under 21 USC §360bbb-3[13] and the PREP Act[14] called an Emergency Use Authorization (EUA). The law guaranteed informed consent.

It is presciently instructive to reflect back on this first application of EUA law for the anthrax vaccine. That EUA was a seminal event. Between 2003 and 2007, the government lost in court over the anthrax vaccine and their position was ruled "**not substantially justified**" in the *Doe v. Rumsfeld* litigation. The declaration of the vaccine and mandate as investigational, experimental, unlicensed, and illegal precipitated a workaround with the first ever application of the EUA law. The groundbreaking EUA application unequivocally affirmed that the vaccine's use was strictly voluntary for the members of the armed forces, with no punishment and no adverse effects on employment. The precedent application of the EUA law included the exact same statutory language that exists today, including the word "consequences." The unique ruling allowed use of the EUA anthrax vaccine "**on a voluntary basis**;" meanwhile the optional wording of the law had not changed, and the Federal Register entry memorialized the precedent:[15]

> **Refusal may not be grounds for any disciplinary action** under the Uniform Code of Military Justice. **Refusal may not be grounds for**

> **any adverse personnel action**. Nor would either military or civilian personnel be considered non-deployable or processed for separation based on refusal of anthrax vaccination. **There may be no penalty or loss of entitlement for refusing anthrax vaccination**.

Bottom line up front, or BLUF, as military folks like to say, the service members challenging the anthrax vaccine program were the "prevailing party" because the government's position was "not substantially justified." As a result, the government paid our attorney fees. Years later, the DoD agreed to pay $1.8 million to attorneys in COVID mandate cases, too.[16]

CONCLUSION

The Court concludes that plaintiffs are entitled to fees and costs for litigating this action, including on appeal, because plaintiffs are the prevailing party and the government's position was not substantially justified. Plaintiffs' request for

Federal court monetary award ruling for plaintiffs after the DoD lost the case.

The court explained that the Equal Access to Justice Act (EAJA) provided that a prevailing party in a non-tort suit against the United States was entitled to fees and expenses unless the government's position was "substantially justified." The court found that "given the unreasonableness of the agencies' initial position before the lawsuit, and their sharp changes in December 2003—issuing a final order after eighteen years that contradicted the proposed order—the Court concludes that the government was not substantially justified in this case." Our attorneys in the case against the anthrax vaccine were reimbursed approximately $230,000.

The court would not have ordered the government to pay, and the government would not have paid, were it not for the fact that they broke the law. The government lost because the anthrax vaccine was not properly licensed by the FDA, and therefore the mandatory anthrax vaccine program was patently illegal before 2006. The final ruling ordering payment protected the government from further admissions of guilt or liability for their violations of the law. But the precedent, and the fact that the case assured EUA use of unapproved medical products carried "no penalty or loss of entitlement," incurred vast implications in the future with COVID.

MILE 16:
Amerithrax
(2001 to 2010)

We enjoyed a great deal of positive progress through our interactions with Attorney General Blumenthal, Mr. Perot, Mr. Rove, Senator Daschle, Representative Gephardt, and with the DoD's recommendations to "minimize" use of the existing anthrax vaccine and develop a modern product. So, the anthrax letter attacks seemed improbably coincidental on the heels of the 9/11 tragedy. My suspicions that someone involved with the anthrax vaccine program perpetuated the letter attacks motivated me to contact the FBI. Within days of the second anthrax attacks, I contacted an FBI agent from the New York office who was also a USAF reservist.

The New York FBI office was originally tasked to investigate the bio-incident. That agent informally kept me apprised and confirmed my communications. I reported the illegal and investigational use of the vaccine by the DoD, suspected illegal adulteration of the product by the Army, and provided an email copy of the Senator Daschle letter to SecDef Rumsfeld questioning the punishments "meted out." The FBI agent seemed empathetic and assured me I would be interviewed. Soon after, the investigation, dubbed Amerithrax, was stripped from the NY office.

<<Your first e-mail was brought to the SAC in Charge of the Anthrax Investigation in the FBI- NYO and has a copy to be analyized for pertinent facts to the case by the Task Force.>>

NY Office FBI agent confirmation of tips, November 10th, 2001.

Despite never hearing back from the FBI officially, I persisted with my tips, especially after the FBI officially named a US Army anthrax vaccine scientist as a person of interest. One of my FBI tips stated, "We've discovered that the cultural M.O. of the US Army has been to disparage and divert attention from their own misconduct and the Hatfill scapegoat theory would fit this pattern. . . . Opposition to the US Army anthrax vaccine program found a similar attack against officers questioning the mandatory vaccinations vs. an analysis of the issues and facts related to a non-validated manufacturer." Years before the FBI named this person of interest, my tips suggested the FBI investigate the US Army scientists responsible for the vaccine, with the access, motive, and a history of anthrax vaccine deceit. I documented the FBI's reply to my tips.

HOME |PRIVACY NOTICE |LINKS |CONTACT US |SITE MAP |SEARCH

Your tip has been successfully submitted.

The FBI's automated email confirmation of received tip.

I told the FBI, "I was in DC, on the phone with Senator Daschle's office, on the 15th of October 2001, at the exact moment they opened the anthrax letter. I was discussing a briefing I had provided his office pertaining to the anthrax vaccine, and the subsequent letter Senator Daschle had written to SecDef Rumsfeld in opposition to the DoD's mandatory inoculation program." I added that "perspective on M.O. and motive are items I may be able to offer your investigation." My inputs covered obvious areas to investigate given Senator Daschle's role.

For many years, the FBI seemed uninterested, or at least did not verify that I was on target with my suspicions on motive. I also included the DoD Inspector General Defense Criminal Investigative Service's (DCIS) documented referrals of our allegations to the FBI about our tangential complaints related to the vaccine's suspected adulteration and illegal mandate.

My tips detailed, "In addition, it's important that I highlight one additional case for your cross-reference. DoD IG / DCIS case numbers 84142, 79472 and 79473 were also referred to the FBI by the DoD." The cases dealt with our charges of illegal conduct regarding the anthrax vaccine, including alleged adulteration of the vaccine and dubious

HOTLINE CASE CODE S[illegible]ET

Case Number	Status	Origin	Date Received	Date Opened	Priority	Referral
84142	C	L	05/08/2002	05/20/2002	RU	I

Activity	Sub Activity	Allegation	Sub Allegation	Analyst	XRef
DCIS		IMC	SOC	[illegible]	Y

Status Description

Organization	Sub Organization	City	State/Country
DCIS MAFO		ARLINGTON	VA

Subject Name	Grade	Sub Type

Case Description
DOD'S INVOLVEMENT IN THE ADULTERATION AND ILLEGAL USE OF THE ANTHRAX VACCINE; FALSE AND INACCURATE TESTIMONY UNDER OATH BY DOD OFFICIALS CONCERNING THE ANTHRAX VACCINE'S INVESTIGATIONAL STATUS - CROSS REFERENCED WITH 79472 AND 79473 (PI REVIEWED AND DECLINED)

DCIS IS REFERRING COMPLAINT TO FBI AND FDA TO EVALUATE ALLEGATIONS

The DoD IG and Defense Criminal Investigative Service referrals of complaints to the FBI.

testimonies by DoD officials to the Congress. The DoD IG referred the cases to the Defense Criminal Investigative Service (DCIS), who in turn referred it to the FBI on November 20th, 2002, according to a DoD IG hotline tracking data form. I explained, the "cases may be helpful as your investigation continues as a means of reviewing the historical depth of the controversy regarding the anthrax vaccine."[1]

Over the many years that the FBI investigation languished into a murky marathon of negligence and skullduggery, I persistently ensured the bureau was aware of the precipitating events. I highlighted my "17 SEPT 01 communiques from the White House Office (Karl Rove's secretary), one day before the anthrax mailings, as well as his memo to the DoD in the spring of 2001 that the anthrax vaccine was a 'political problem.'" I also reinforced the disturbing reality that "Senator Daschle's investigation of the Anthrax Vaccine program, which began in the summer of 2001, was effectively halted by the anthrax attacks on his office as US Army officers offered the very protection he was investigating."[2]

The FBI investigation seemingly stalled, despite the fact that a White House spokesman revealed in the first year of the inquiry about the suspected "domestic" source of the anthrax spores.[3] Officials affirmed that the spores more likely than not originated from one of three US Army anthrax spore stockpiles where the Ames strain was stored—one being the Fort Detrick laboratory. It was not until 2010 that the FBI definitively concluded the anthrax letter attacks against Senator Daschle and others were most likely committed by a US Army scientist whose motive was to save his "failing" anthrax vaccine program by creating a "scenario where people all of a sudden realize the need to have this vaccine"[4, 5] [FBI communications at Appendix W].

2. Motive. According to his e-mails and statements to friends, in the months leading up to the anthrax attacks in the fall of 2001, Dr. Ivins was under intense personal and professional pressure. The anthrax vaccine program to which he had devoted his entire career of more than 20 years was failing. The anthrax vaccines were receiving criticism in several scientific circles, because of both potency problems and allegations that the anthrax vaccine contributed to Gulf War Syndrome. Short of some major breakthrough or intervention, he feared that the vaccine research program was going to be discontinued. Following the anthrax attacks, however, his program was suddenly rejuvenated.

MR. TAYLOR: The other question you have, Dr. Ivins is a troubled individual, particularly so at that time. He's very concerned, according to the evidence, that this vaccination program he's been working on may come to an end. He's also very concerned that some have been criticizing and blaming that vaccination program in connection with illnesses suffered by soldiers from, I think, the first Gulf War. So that was going on, according to the evidence, in his mind at that particular time.

With respect to motive, I'll point again to -- with respect to the motive, the troubled nature of Dr. Ivins. And a possible motive is his concern about the end of the vaccination program. And the concerns had been raised, and one theory is that by launching these attacks, he creates a situation, a scenario, where people all of a sudden realize the need to have this vaccine.

Regarding motive, the FBI affirmed the intent of the anthrax letter attacks—they "rejuvenated" the "failing" anthrax vaccine program.

Though vindicating to witness that the FBI report ultimately made their findings on point with my original allegations and tips, it remained disturbing to me that the joint DoJ and FBI report actually concealed the facts related to Senator Daschle's inquiry into the anthrax vaccine. This vital revelation existed nowhere in the report. The FBI must have known the American people would certainly be intrigued that one of the targets of the anthrax letters actually questioned the anthrax vaccine directly to the US secretary of defense just prior to the letter attacks. The American people may have been further intrigued that the FBI was made aware of this connection and possible motive within days of those attacks, almost ten years earlier.

This might add perspective as to the FBI's own possible motive in concealing these facts, and instead they disingenuously reported to the American people that the "Review of Correspondence to Senator Daschle" implied no one had any "inkling as to why they had been specifically targeted." Instead, the FBI suspiciously "determined that none of these had any ostensible connection to the anthrax mailings."[6] In the final report, the FBI never told the American people about any of the tips, which possibly signaled, at a minimum, an incompetently bungled, decades-long delay, or at worst a calculated cover-up. The FBI apparently never pursued the DoD IG DCIS-referred complaints about serious alleged "criminal allegations."[7] The FBI appeared to ignore the unresolved DoD IG and DCIS whistleblowing, referred to their public corruption squad, and didn't address the allegations embedded within these complaints.

The government's final report also provided zero perspective on the many years of congressional and GAO investigations that revealed how and why the anthrax letter attack gate guard's program failed.[8] The FBI glossed over the vaccine's failures at the core of the motive, such as the systemic adverse reaction rates being reported at "a level more than a hundred times higher than the 0.2 percent published in the product insert." The FBI provided no analysis of the pressures the scientist was under due to the alleged illegal "change in the composition of the vaccine from the vaccine originally approved in 1970."[9] The FBI ignored all the other parties who "did not notify FDA of a number of changes made in the manufacturing process in the early 1990s," and never considered that "no specific studies were undertaken to confirm that vaccine quality was not affected."[10] Beyond a doubt, the FBI's findings left many stones unturned.

While briefly mentioning Gulf War Illnesses, the FBI never followed up on the alleged illegal activity that "found up to a hundredfold increase in the protective antigen levels in lots produced after the filter change that year," nor did they investigate the DoD researchers who "hypothesized that the filter change altered the composition of the vaccine by increasing the level of protective antigen in the finished product." The FBI made no attempt to evaluate the impact on our service member's health due to the unapproved "change in the filter from ceramic to nylon."[11] The GAO had noted "marked variances" in their report on the elevated adverse reaction rates. The GAO reported, "Such marked variances from the product insert data suggest the possibility of change in the composition of the vaccine from the vaccine originally approved in 1970."[12] The FBI failed to report these facts to the American people. Being candid in the final report would have revealed more motives for a prolonged investigation and for hiding the truth.

Soon after the attacks, the GAO questioned whether terrorist entities could "overcome the major technological and operational challenges to effectively and successfully weaponize and deliver a biological warfare agent to cause mass casualties." The GAO remained one of the few watchdog agencies to quantify the "attendant publicity" lavished by DoD leaders "about the importance of the threat." GAO reports analyzed the DoD's anthrax vaccine program and the unsolved anthrax attacks that appeared to justify resuscitating the program. The GAO reaffirmed that after initiation of combat hostilities in Iraq, "Intelligence assessments have not changed since 1990 for chemical and biological warfare threats on the battlefield or by terrorists." The GAO added, "It would be very difficult for a terrorist to overcome major technical and operational challenges to effectively and successfully weaponize and deliver a biological warfare agent."[13] GAO's findings, and their message to lawmakers, appeared to hint at an inside job.

Before and after this period, dauntless examples of lawmaker willingness to challenge the uncomfortable or inconvenient truths did occur on other issues. Senator Dianne Feinstein diligently exposed wartime interrogation techniques, viewed by many as torture. That word, torture, is uncomfortable and her committee's conclusions were inconvenient. Despite consecutive presidents wanting to "turn the page,"[14] Senator Feinstein didn't allow it. She didn't accept the willfully ignorant rationalizations simply because it was easier to just "turn the page." That conviction and adamancy was not evident in the anthrax letter attack investigation. The half measures in pinning down the motive back to the anthrax vaccine program helped, but the apparent lack of full accountability missed the target. Once again, there was an accountability body count of one, without any institutional culpability—it was easier to simply "turn the page."

Due to the FBI's attempts to obscure their own investigative negligence, the American people never comprehended the full implications of what occurred. The American people were not told the uncomfortable, inconvenient truths. The FBI had tips pointing to Army scientists within weeks of the anthrax letter attacks, but did not investigate the leads for almost a decade. When the FBI did report to the American people, they left out any admission and all evidence related to the attack on Senator Daschle that the bureau knew was directly related to motive.

Despite these obvious omissions and diversions, no one in the government ever resurveyed the punishments that Senator Daschle asked SecDef Rumsfeld to review. No one ever resurveyed the resuscitation of the anthrax vaccine program after the FDA finally licensed the vaccine in 2005. No one questioned DoD assertions that "the anthrax attacks in October 2001 illustrated the risk." Too much time had passed, perhaps by design. Everyone wanted to "turn the page." No one wanted to go back and resurvey or analyze any of the breakdowns.

The Amerithrax investigation suffered from many of the same dysfunctional themes identified in our own struggle. The decade-long investigation alone proved that unsettling point. Bureaucratic diversions and investigative misdirects also resulted in a settlement in excess of $5.8 million to a wrongfully accused non-military scientist.[15] Some questioned the lone-wolf scenario based on the evidence and complexity of the attacks. Though I was agnostic about the identity of the perpetrator, I firmly believed that the FBI got the motive correct regarding the objective to save the anthrax vaccine program—the crux of my tips on motive to the FBI.

Given the FBI's damning conclusions, it was disconcerting for the bureau to not in turn inform the American people about the obvious Senator Daschle attack links to motive. If the letter attacker's motive was to save the anthrax vaccine program, and Senator Daschle was the only entity attacked who questioned that program and the punishments meted out to our troops, the FBI was remiss to overlook the nexus linking the letter attack victims to motive.

Some journalists advocate an unorthodox and contrarian method, or heterodoxy, when analyzing cases such as these.[16] By looking for the opaque or hidden evidence, such as the ignored Daschle-Rumsfeld letter from the anthrax case, deeper motives are better understood. The FBI potentially hid these facts to obscure their investigative misdirects and to protect the DoD. Had they revealed the existence and relevance of the Daschle-Rumsfeld communications, it would have bolstered the FBI's findings and the DoJ's Amerithrax case. Omitting and obscuring such critical evidence weakened the conclusions and its credibility.

Albert Einstein explained this kind of behavioral phenomenon saying, "The world we have made as a result of the level of thinking we have done thus far creates problems we cannot solve at the same level of thinking at which we created them."[17] By design, that is the level of thinking you get when you install a former military man, Robert Mueller, as director of the FBI just days before the anthrax letter attacks. Then he retires a decade-plus later—shortly after the Amerithrax investigation's half-measured conclusions. On Director Mueller's watch, the nation's citizenry never learned the full story. Like many before and after, Mueller was the top FBI gate guard in charge of those who curated the evidence to ensure we never learned the inconvenient, opaque, and hidden truths. A different level of thinking, a heterodoxic approach, was required.

Einstein's hypothesis predicted the FBI's and the DoD's incapability to correct the errors due to their bureaucratic cultures and political "level of thinking." The FBI's failure to assess the motive and the damage caused by the anthrax vaccine program ensured no one ever questioned its resurrection or corrected the punishments meted out. The FBI's level of thinking would have surely disappointed Einstein's "passionate sense of social justice and social responsibility."[18]

MILE 17:
Learning to Overcome (2006 to 2015)

Personally, and professionally, I carried on. I volunteered for a variety of assignments and was finally able to get back into operational flying in support of our Air Force. After being grounded, the chief of the USAF Reserves promulgated guidance to block any aviators that did not accept the anthrax vaccine from returning to flying. Fortuitously, I executed a flanking maneuver by volunteering for flying duty where pilots were desperately needed—drones.

The chief's directive seemed vindictive, so I waited it out. They apparently forgot the memo based on the subsequent needs of the service. The memo blocking pilots from returning to flying duty read, "Accession to the Selected Reserve will be denied to any individual known to have previously separated in order to avoid the anthrax vaccine." According to the military at the time, "Accession is contrary to the best interest of the Air Force Reserve," and "denial of reassignment or accession will be based on actual knowledge of anthrax vaccine avoidance as the underlying reason for the earlier reassignment or separation." There was no question the vaccine was the only underlying reason I was ordered to resign. Luckily for me, they "turned that page."

Curiously, the memo stated, "When the anthrax vaccine becomes available, applications for accession or reassignment of individuals known to have avoided the anthrax vaccine will be considered on a case-by-case basis, provided they are willing to be vaccinated" [memo at Appendix X]. That meant at the time they formulated the policy, the Air Force was well aware the vaccine stockpiles ran out due to the manufacturer's invalidated status. Enforcement of the policy more likely than not became complicated by the interim federal court rulings that blocked the mandate. Over the years, I moved military units several times and fortunately never got caught by this policy. Maybe I just got lucky outflanking them. But more likely, it was because I worked for some really good folks. And I don't mean to thumb my nose at the guidance of those in command. Instead, I'm grateful for the commanders on the ladder's lower rungs—the heroes who allowed me to serve. Thankfully, I avoided the spite of some due to the leadership of others.

My new flying specialty found me in unmanned aerial vehicles or drones. The USAF later tagged them as Remotely Piloted Aircraft or RPAs. Most of my fighter pilot colleagues dreaded the thought of flying unmanned aircraft from a console, but for me, drones became the ticket to serving my country again in a meaningful mission. For my final ten years of duty, I flew reconnaissance missions to protect America's troops. I can honestly say in flying those missions, I contributed more to the troops on the ground than I ever did flying F-16s, A-10s, or F-117s.

That was the great thing about drones. You actually had the endurance to really get to know the people you served. Compared to most fighters with less than an hour of time on station, drones spent dozens of hours continuously overhead, talking to ground units on the radio, and typing to them via chats. We were there before they needed our help, providing valuable, persistent reconnaissance. When the proverbial stuff hit the fan, we were there providing mutual support as their eyes in the sky. After things calmed down, we

Lt Col Rempfer and fellow crewmember with a General Atomics MQ-9 Remotely Piloted Aircraft (RPA).

were still there. Twenty-plus hour sorties gave us the staying power to be effective while burning extraordinarily little gas.

One of my favorite stories I shared during briefings around the country about drone capabilities and synergies came in 2008 when our US Marines surged in Afghanistan. One night I reported for my midnight shift and received my briefing about the supported unit. I recognized the unit designation as the same outfit where my son's high school friend served. I received the changeover briefing from the pilot and sensor operator currently flying the drone, and I took control of the mission. I checked in with the ground controller on the radio and right off the bat confirmed the unit we supported was the same one as our young Marine's. I passed along his name and asked if by any chance the Marine on the radio knew our young Marine.

The ground controller directed us to zoom out with our camera and scan several hundred meters south along a riverbank to a walled compound. I called "contact." He talked our eyes onto a line of troops in sleeping bags along the north side of the compound wall. I called "contact" again. He said, "Your boy's in one of those sleeping bags." As luck would have it, the next night we supported the same group of Marines. This time my boy was not far from the radio. The controller said, "he wants to say 'hi' to you." He gave a burst from his squad automatic weapon.

Fortunately, my son's childhood friend made it home safely and came to visit us soon thereafter. He told the tale from his perspective, and how grateful he and the other Marines were to have Uncle Buzz overhead keeping an eye on them. The memory brings chills and tears to my eyes to this day. Had I not continued to serve—had I capitulated to the efforts to get rid of me ten years earlier—I would not have been able to fly high cover for this young Marine and his comrades. I would not have been able to overfly my own son's forward operating base when he deployed a few years later—embarrassing him with annoying calls on the radio from his dad in the skies above. I was grateful to serve for almost a decade providing overwatch for my boys!

Rempfer honoring his boys as an Armed Forces Communications Association Panelist.[1]

So yes, I was grateful to not only be back flying for my country, but more importantly to be able to serve in such a valuable mission to protect our troops on the ground. During those years I also tried to stay engaged intellectually by arduously accomplishing the professional military education (PME) our service prescribed for us. One of my hero bosses once mentored me, "Do your PME, Buzz, so they don't have an excuse to not promote you!" I followed his advice. Not only did I do all my required PME, but I also pursued my master's degree. After unyieldingly applying to the Naval Postgraduate School several times, the school finally offered me a scholarship for an eighteen-month master's degree program in security studies.

During my master's program, I initially intended to author my thesis on a topic related to unmanned aircraft or civil support. Ironically, the Department of Justice funded the program through the Department of Homeland Security (DHS). My classmates were all government professionals from DHS, Secret Service, Customs and Border Protection, and the FBI. The timing of our class coincided with the FBI's announcement that they finally homed in on their anthrax letter attack suspect, the US Army scientist from the Fort Detrick lab. I wrote one of my master's papers on this development in the case. I then chose to do my thesis on the subject matter after the government wasted another half billion dollars of taxpayer funds for additional anthrax vaccine for the Strategic National Stockpile in 2008, on top of the billion plus since 9/11.

It defied common sense that a program on the verge of cancellation just before 9/11, but resurrected due to the anthrax letter attacks, would get even more money even after the FBI disclosed the motive of the attack was to save the failing program. And now they were pumping billions more into the unnecessary vaccine, but this time to stockpile it for civilians through the very organization that was funding my graduate-level master's degree. The DoD's original purchase price per dose rose from $2.26 in 1998 to $29.91 ten years later, a 1,235 percent increase. The Obama administration purchased another $911 million in 2016 for the known inadequate vaccine to protect from a noncommunicable threat, and which hadn't infected anyone until the US Army scientist apparently unleashed it to save the anthrax vaccine program.[2]

My thesis advisors initially cautioned me on the thesis topic, but ultimately supported my decision to proceed. Something was wrong—they knew it. The dupe-speak and dupe-loops we experienced in all the years fighting the anthrax vaccine program differed diametrically from the academic freedom and intellectual inquiry permitted by the Naval Postgraduate School's Center for Homeland Defense and Security. The premise of my thesis focused on warning DHS and the Department of Health and Human Services (DHHS) about the troubling DoD history with anthrax vaccine, rebranded as BioThrax®. Particularly in light of the FBI admissions about the vaccine's resuscitation following the post-9/11 anthrax murders, I sensed the vast allocations of funds for more BioThrax for civilians was a political stunt to assure confidence in the vaccine.

I titled the thesis *Anthrax Vaccine as a Component of the Strategic National Stockpile: A Dilemma for Homeland Security*. The DoD archived the work on the Defense Technical Information Center website.[3] The abstract explained how past problems with the Department of Defense anthrax vaccine impacted the Department of Homeland Security and the Department of Health and Human Service's policy following the 2001 anthrax letter attacks. Those departments included the old anthrax vaccine in the Strategic National Stockpile (SNS). The thesis, and a Homeland Security Affairs Journal[4] article, explored the Department of Defense's experience with the vaccine, enumerating past safety, efficacy, regulatory, and legal problems. The thesis suggested public health policy alternative courses of action, including the use of antibiotics and development of a new vaccine. These recommendations literally could be cut and pasted from Defense Secretary Rumsfeld's recommendations by his undersecretaries from August 2001.

I worked hard on the thesis, incorporating a "quadrangulation" technique that utilized four research methodologies: a literature review, case study, program evaluation, and gap analysis. The multi-prism approach evaluated peer-reviewed and published literature sources which demonstrated the politically driven evolving record on safety, efficacy, regulatory, and legal issues. The methodological "quadrangulation" of the thesis effectively "boxed in" the root facts on anthrax vaccine to hopefully effect future policy decisions by documenting the troubling historical patterns. The literature review was my favorite section. It was really in honor of my old buddy Russ, who was the first to discover the intellectually shady fact that the DoD scientists attempted to alter the literary record at the same time the anthrax vaccine program began.

This first methodology tackled the available literature and revealed a shift in the literary record around 1998. This was the same time that the DoD announced plans for mandatory anthrax immunization of the armed forces. Literature was generally negative about the vaccine prior to this point, but it shifted in 1998 to an overall pro-vaccine stance. The highly politicized atmosphere surrounding the DoD mandatory inoculations helped to explain the disingenuous evolution of the literature. As Perot discovered, the DoD scientists and researchers rewrote their articles and published new "science" to "conform with policy" during a time when top leaders attempted to assure its citizens and soldiers that it could protect the troops in the Middle East in the midst of weapons-of-mass-destruction (WMD) verification problems with the Iraq regime.

The alleged illegal, unapproved changes to the vaccine's manufacturing process, potentially improving the vaccine, might also explain a shift in the professional assessments. The preponderance of supportive literature on anthrax vaccine emanated primarily from government and military sources. Civilian reviews generally critiqued the vaccine negatively before and after 1998. It was interesting to note, most literature on the vaccine emanated from a small group of government-affiliated scientists. Even those authors' vaccine reviews shifted from generally negative during the pre-policy period to generally positive in the post-policy pronouncement era.

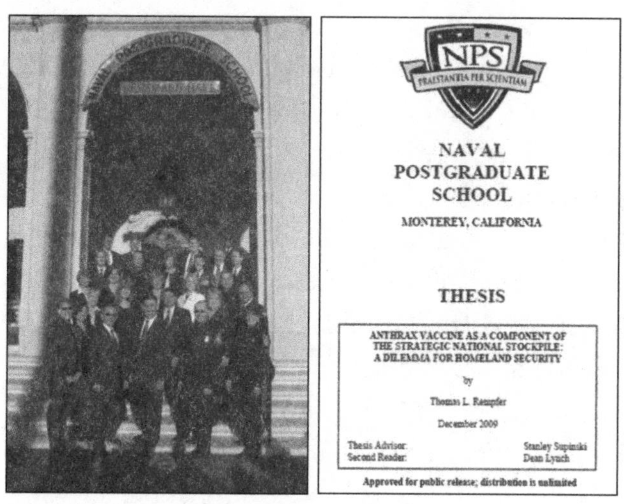

NAVAL
POSTGRADUATE
SCHOOL
MONTEREY, CALIFORNIA

THESIS

ANTHRAX VACCINE AS A COMPONENT OF
THE STRATEGIC NATIONAL STOCKPILE:
A DILEMMA FOR HOMELAND SECURITY

by

Thomas L. Rempfer

December 2009

Thesis Advisor: Stanley Supinski
Second Reader: Dean Lynch

Approved for public release; distribution is unlimited

Naval Postgraduate School class and thesis.

The paper earned the school's Outstanding Thesis Award.[5] The validation meant a great deal. I loved Monterey, California, my professors, my classmates, and the experience. The Naval Postgraduate School's Center for Homeland Security also published an article after the thesis' publication titled "Silo thinking in vaccine stockpiling persists."[6] The Center's conclusion echoed my reality, "Lt. Col. Thomas Rempfer's course paper in Technology for Homeland Security may not have translated into an applicable policy or practice in the field. It did, however, expose the 'silo' thinking that plagues many agencies that are tasked with homeland security responsibilities in the United States." The commentary added this substantiating note, "Significantly, the FBI also recently finalized its findings, corroborating Rempfer's research."

Almost a decade after publishing the thesis, its conclusions and recommendations were closely mirrored in the main tenets of the United States' 2018 National Biodefense Strategy announced by the president.[7] Maybe someone read my thesis! The first ever US Biodefense Strategy remarkably resembled my thesis conclusion recommendations point for point. The thesis suggested PSDs/PPDs (Presidential Study/Policy Directives) and the president implemented the same actions via the new administration's unique lexicon and policy terms as NSPDs/NSPMs (National Security Presidential Directives Memoranda). The thesis suggested a Biodefense Czar under DHS, and the president's vision implemented the same action of independent oversight under DHHS. The thesis suggested a Biodefense Commission, and the president's vision instituted a cabinet-level Biodefense Steering Committee. The thesis cited anthrax vaccine and the letter attacks as the premise for the recommended biodefense strategy, and the president's directive specifically cited the anthrax letter attacks in his call for "accountability."[8] In other words, the president ensured civilian control of the biodefense enterprise, which was the fundamental thrust of the thesis' recommendations.

The Naval Postgraduate School's Center for Homeland Defense and Security also invited me back to present a talk about how government and military agencies can harness the positive contributions of whistleblowers.[9] The main point of my talk was to keep whistleblowing from happening in the first place by listening to your people rather than retaliating against them. Of course, it is not that simple in complex situations, but leaders always benefit by erring on the side of prudence and caution versus retaliatory reactions to a potential whistleblowing scenario.

NAVAL
POSTRAGUATE
SCHOOL CHDS
ALUMNI
SHORT TALKS
-- WHISTLE-
Discover how leaders can harness the energy and positive contributions of potential whistle-blowers.

Rempfer and family at NPS graduation and subsequent "TED"-style talk.

In my whistleblower talk, I outlined key points based on my experiences that could be valuable to others similarly situated in the future—both supervisors and subordinates. The monumental objective is to change the paradigm and leverage potential whistleblowers as valued members of any organization. No doubt the entire potential whistleblowing process is incredibly stressful for both leaders and employees. And that's the whole point: to make it standard operating procedure (SOP) and to shield the concerned employee, versus transforming that person into someone to shun and squash. Most institutions encourage their people to blow the whistle if necessary, but when they do, everyone knows they get crushed. The laws support whistleblowers, but are rarely enforced or enacted in order to protect this class of personnel. Instead, leaders should encourage dissent, respect it when professionally received, and protect their whistleblowers. Disagreement is okay and dialogue is good, but it should never evolve into retaliation and reprisal. Indeed, it's a two-way street and the whistleblower must follow the rules as well to ensure their complaint is legitimate after exhausting available avenues of redress.

The talk to the gracious audience emphasized the points above, plus advocated "Four Cs" to better understand the whistleblowing process: Concerns, Complaint, Consequences, and Corrections. Concerns marked the initial step, and ideally the last. Once articulated, leaders take action to address and resolve the dispute. If not, a Complaint may be necessary. Leaders should guide employees through this process, but regrettably rarely do. If Consequences result from the Concerns expressed, particularly if in response to the allegations themselves, then it should result in Corrections to remedy any injustices. It is important to note that in any professional-realm disagreement, the concerned employee must be vigilant to avoid supplying any ammunition that the policy gate guards might use against them. Perceived misconduct in any whistleblowing scenario creates a classic switcheroo that derails the core Complaint. I always told my kids—remember, "No ammo!" Do not let anyone change the subject—give them "no ammo" to use.

Of course, if the Concerns were important enough to highlight in the first place, then one must be "unyielding" by professionally demanding resolution. Being respectful and reasonable throughout any dispute process is essential. Respecting and trusting the process, regardless of the outcome, must steer your conduct and expectations. We will review these points at the end of this book about my journey, because it is perhaps the most valuable takeaway. Be "unyielding" to prevail, but you also must be realistic, grateful, trusting, and humble throughout the process.

These learning points proved that my naive idealism never lost its vector and fuel. My days at the Naval Postgraduate School rejuvenated and recharged my faith in academia, the military, the government, and humankind. My professors and classmates became my heroes. My wife was eternally grateful that I had someone else to talk to about this topic other than her!

MILE 18: How to Make You Whole (2011 to 2022)

My Walter Mitty and Forrest Gump qualities probably propelled me fatefully into these predicaments. The raw optimism embodied by those characteristics similarly kept me from ever giving up. I did sprinkle a little realism into my expectations over the years. What I learned was that even when you win, don't count on any headlines, and expect minimal or no accountability. To get "made whole," or to right the wrongdoing, expect a long, unyielding, but vital journey.

Over the years, we made our points and gained many victories, yet forces beyond our control often marginalized the win for the ultimate outcome. I came to understand that it is not just about the win. It is about behaving like a champion. I had my fair number of champions that looked out for me and made sure I was not discarded. So, while this entire story might seem depressing and disheartening at times, I chose to try to emulate the champions that backed me up and protected me through all the struggles. Colonels Russ Dingle, Kevin Wear, Ron Moore, Juan Gaud, Lee Pritchard, Gregg Davies, Brett Buras, Randy Russell, Scott Keller, Alan Buck, Mike Kavanaugh, and Austin Moore are examples of my stalwart champions, to name a few. These heroes and fearless allies are the ones I remembered—they far outnumbered the less heroic actors.

Along the road, we earned hard-fought successes in our attempts to set the record straight. One of the earliest related to our forced discharges from our A-10 unit. After our dismissals, we turned our efforts toward our own former guard unit in an attempt to collect the documents which would set the record straight on our own forced expulsions from the guard. Though we attempted to always focus on the federal-level problems, we committed to correcting the misrepresentations made by the DoD over our own involuntary resignations. Ultimately, the state's Freedom of Information Commission concurred with us that we were "ordered" to resign, despite the National Guard's misrepresentations and objections. Excerpts included:[1]

> FIC2000–303—FREEDOM OF INFORMATION COMMISSION—OF THE STATE OF CONNECTICUT—In the Matter . . . the Connecticut National Guard was in the process of addressing the controversial issue of an anthrax vaccine program, that the complainant herein was an officer in the Guard during such time, that he was assigned by the Guard to research the vaccine, that he decided not to take the vaccine, and that he ultimately resigned his commission as a Guard officer as a result of the anthrax vaccine controversy, **after being ordered to so resign**.

Central to our larger concerns about the shoddy implementation of the anthrax vaccine program, we also knew the proper procedures were not followed in our own expulsions. We were never granted one iota of due process. The military compounded this restriction of our fundamental rights by violating Freedom of Information Act laws when we attempted to obtain records documenting our ordered resignations. Additional FOIA Commission findings included:[2]

> FIC2000–304—FREEDOM OF INFORMATION COMMISSION OF THE STATE OF CONNECTICUT—It is found that the respondents do not contend that the requested records that they maintain are exempt from mandatory disclosure. It is also found that they simply ignored the complainant's requests in this matter, and that such inaction resulted in the respondent's violations of the complainant's rights under the FOI Act, both in letter and spirit. It is further found that such violations were without reasonable grounds.

The *Hartford Courant* documented the resulting sanctions levied on the National Guard leaders for their violations of the statutes in an article titled "National Guard Is Fined." The article stated, "The state Freedom of Information Commission on Wednesday fined the Connecticut National Guard and the state's adjutant general for failing to give two Air National Guard pilots information about the military's mandatory anthrax inoculation program." The willful withholding of the records followed a pattern of obstruction of justice mimicked by most every DoD organization we engaged for proper due process and redress over many years.

In another *Hartford Courant* article, Senator Christopher Dodd inquired on our behalf and met with similar obstruction. The article elaborated, "US Senator Christopher Dodd is questioning how the Pentagon's Inspector General handled a complaint by two Connecticut Air Force Reserve majors that two high-ranking Army officers gave false or misleading testimony."[3] The complaints were not investigated and were summarily dismissed when the DoD controlled the mechanisms of the investigation. Russ and I realized that just as the original questions about the anthrax vaccine were ignored, so was our right of due process and for accountability.

We also helped draft state-level legislation concerning the use of experimental vaccines.[4] Our goal was to put into place state-level provisions in the Connecticut statutes that mirrored protections under federal laws if, or when, the federal government failed to enforce its laws (for example, Title 10, Section 1107). Representative Shays also joined in the call for the correction of military records for any troops punished over the anthrax vaccine. In a press release, he wrote that the disciplinary actions should be reversed "against those whose legitimate health concerns prompted a decision to opt out of the program." According to a *Hartford Courant* article, "He urged Cohen to review all of the several hundred disciplinary actions against those resisting anthrax inoculations. He cautioned, 'Without an assured supply of vaccine, continuing to order any more soldiers, sailors, airmen, and Marines to start the shots they may never finish constitutes military malfeasance and medical malpractice.'"[5]

These efforts precipitated the affirmative ruling from the DC Federal District Court. An article discussed the ruling that linked back to our forced constructive discharges from our pilot positions. The headline read, "Judge advances anthrax vaccine refusal case," and "Pentagon must reconsider exonerating two military pilots discharged after resisting inoculations prior to FDA approval." The article explained that, according to a federal court, the military corrections board did "not accurately describe the outcome of the Doe litigation." Judge James Robertson stated, "Contrary to the board's conclusion, the plaintiffs in the Doe litigation clearly prevailed. To base denial of Rempfer's constructive discharge and compensatory relief claims on the fiction that the Doe plaintiffs lost would be arbitrary and capricious." The judge advanced the same argument in posthumously supporting Russ's parallel claim, concluding, "Taken as a whole, Judge Sullivan's decisions in *Doe v. Rumsfeld* conclude that, prior to the FDA's December 2005 rulemaking, it was a violation of federal law for military personnel to be subjected to involuntary [anthrax] inoculation because the vaccine was neither the subject of a presidential waiver nor licensed for use against inhalation anthrax."[6]

Despite the efforts and rulings, ultimately the government never corrected our records. One court would direct reconsideration, but the next would find the military review and repeated denials sufficiently addressed the court's earlier concerns. Though frustrating, the goal to correct records continued over the many years that followed. It was not until July 25th, 2018, following multiple appeals to the president of the United States, that the DoD was directed to publish record-correction guidance for the military correction boards. They were directed to consider record corrections based on past errors, inequities, and injustices [memo at Appendix Y].

On October 30th, 2019, the correction board finally came through and fully corrected the service record for one of the previously convicted and jailed Marines. Sergeant James D. Muhammad successfully received a discharge upgrade to fully honorable, restoration of his E-5 rank with the reversal of his demotion to E-1, back pay, and the award of a good conduct medal. The *Military Times* reported on the unprecedented development with an article titled "Troops Who Refused the Anthrax Vaccine Paid a High Price."[7] A tandem article, titled "In the Shadow of Anthrax,"[8] depicted the precarious parallels of the anthrax vaccine mandate and the looming COVID shot mandate facing our troops [article excerpts at Appendix Z].

Military Times' "In Anthrax's Shadow" cover story.

Sergeant Muhammad, the ever-vigilant warrior for justice, published his own opinion editorial explaining, "I offer my story in the hope that I will not be the last and so the burden for the 1,000 or more similarly punished troops might be lightened as we continue to press for broad based legislatively mandated relief."[9] Shortly thereafter, one of James's Marine colleagues, Ocean Rose, received similar favorable consideration. But up to a thousand troops' cases languished. Based on those ignored injustices, James and I lobbied elected representatives and submitted draft legislation to Congress to encourage the legislature to finish the task. We titled the bill "The Lt Col Russ Dingle Memorial Anthrax Vaccine Justice Act (MAVJA)." An updated bill included COVID mandate harms: the Memorial Anthrax and COVID Vaccine Justice Act (MACJA).[10, 11]

Sgt James D. Muhammad.

In both the correction cases and the proposed legislation, a key justification included the concept of equitable treatment. Sergeant Muhammad requested that I submit an affidavit for his application making this very point on equity. My explanation to the correction board included the fact that "I refused anthrax vaccine in 1998, was not punished through any judicial or non-judicial means, and served until 2015, wherein I reached mandatory retirement, with active-duty benefits, and was discharged under fully honorable conditions." I explained that "there are literally hundreds of service members that were similarly treated, as opposed to the extreme disciplinary processes and penalties that James Muhammad underwent." I made the case that the point of Sergeant Muhammad's "appeal is not to re-adjudicate the core judgments in this case, but instead to ask for clemency based on a reflection of undeniable inequities of justice represented by James Muhammad's case when compared to others in our armed forces." The board apparently agreed, and Sergeant James Muhammad received full justice after fifteen years.

In my own tangential substantiated reprisal case from 2011 (explained in Mile 1), a decade later, the Air Force directed record corrections after being ordered to do so by the SecDef. They also expunged references related to the substantiated illegal firing, which took another two years.

> SUBJECT: Correction of Military Records - BC-2012-03031
>
> In accordance with the Memorandum for the Chief of Staff, AFBCMR BC-2012-03031, dated 23 Oct 13, your records will be corrected to expunge any reference to your improper release from the Arizona Air National Guard. Having complied with the BCMR directive, NGB has reviewed our records and expunged any and all reference to your improper release.

National Guard Bureau directive to expunge records.

During the intervening review, the IG overruled attempts to excuse the suspected cheating by the chain of command. They also dismissed the general's insinuation about blocking me from assuming command due to my professional dissent years earlier over the anthrax vaccine. Though the hiring process was wholly unrelated to anthrax vaccine, the general who fired me attempted to connect the two issues—command selection reprisal and prior anthrax vaccine refusal. The Pentagon provided me with a transcript of the general's testimony that revealed the failed diversion. As I read it, I had to wonder why he hired me in the first place. Before he hired me, I disclosed the anthrax vaccine history and he complimented me on my candor. This was the same general officer who personally complimented my successful operational duties in the past.

```
I KNEW you were the right guy for our unit!  Keep up the great work (like last night)
```

Positive feedback email from leadership prior to being fired for challenging hiring violations.

It's important when attempting to correct records to put laudatory comments on the record to "lead turn," or preempt, future efforts by the DoD to disparage its opponents. In my case, even the DoD IG determined there were no "comments that reflected poorly on performance."[12] From the transcript, the general also complimented me extensively in his testimony saying, "He's very smart. He's brilliant, in fact. I hired him. He's that smart. And that's just because I hired him. He's just—he's very smart." He added, "And he is a brilliant guy," and that "in about 99 percent of every category you look at, Col Rempfer mops the floor with everybody. He is that good." Even after firing me over the commander position dispute, he stated, "He would be a great commander," and "Col Rempfer is the best."

Maybe at that point the general knew he was being called on the flying carpet for the hiring violations, so perhaps he was trying to smooth things over and distance himself from the alleged cheating. He continued by stating, "Col Rempfer was very successful in the things we asked him to do. . . . We hired him initially as a pilot, we had given him other responsibilities. He was successful at everything we gave him." The general verified there was only solid performance in my past duties by saying, "He's given a very special assignment—something we're going to build, because he's just brilliant."[13] None of it made any sense, and that's probably why the reprisal and corrections cases were ultimately decided firmly in my favor.

The final official argument by the correction board as to why they did not initially support a command credit, or a special promotion board, came down to what they referred to as "intangible considerations." Fortunately, the SecDef's office did not buy into the dupe-speak or enter their dupe-loop. The only "tangible" facts proved that the chain of command's actions were arbitrary, capricious, and most importantly—they broke the rules. Fortunately, the SecDef's office discarded the diversionary excuses for the suspected cheating, as well as the immaterial attempts to excuse the substantiated reprisal. They ordered a credit for command in my records.

The final ruling by the SecDef's office not only directed the command credit inclusion in my professional records, but also consideration by a retroactive promotion board. Apparently, the SecDef's office had never reversed a USAF reprisal remedy ruling, but did so twice in my case. Success with the command credit led to a 2018 Special Selection Board (SSB) to consider my promotion for the next rank. The USAF waited three more years, but finally an action officer at the Pentagon emailed me the good news about the promotion's belated processing.

> Colonel Rempfer, At long last, please see the attached notification of your advancement to Colonel on the Reserve Retired List. Congratulations and thank you for your patience with this process. You will need to use the attached document for a new ID card which reflects the rank of Colonel.

The Air Force also sent me a gracious memo announcing, "It is my pleasure to inform you of advancement to colonel," and "Congratulations, this achievement attests to your ability and performance in the Air Force

Reserve. You can be justifiably proud" [memo at Appendix AA]. The staff officer also provided a letter by the secretary of the Air Force approving the promotion, and affirming "The board recommended Lieutenant Colonel Rempfer, Thomas L. for promotion to the grade of colonel . . . I approve Lieutenant Colonel Rempfer's advancement on the Reserve retired list." Ten years after the substantiated reprisal and three years after selection for promotion, the Air Force inked the long delayed correction and promotion. While disheartening that it took so long, I appreciated the recognition and remedy following the hard-fought case.

The laws governing Special Selection Boards, 10 USC §628 and §14502 state, "A person who is appointed to the next higher grade as the result of the recommendation of a special selection board convened under this section shall, upon that appointment, have the same date of rank, the same effective date for the pay and allowances of that grade, and the same position on the active-duty list as he would have had if he had been recommended for promotion to that grade by the board which should have considered, or which did consider, him."[14] In my case, the Air Force did not ensure senate confirmation or the promotion's retroactivity. The law seemed crystal clear, and the Air Force's original notification to me about the expected process and retroactive date of rank was as well. In their 2018 letter advising of the promotion board, they specifically wrote, "If selected your date of rank will be 01 June 2012, as if the original board selected you" [letter at Appendix BB]. The lack of USAF coordination for senate consideration effectively blocked the retroactive date of rank and the entitlements guaranteed by law.

In the end, the all-too-familiar patterns encompassed a disappointing combination of the time-tested tactics of disparage, discharge, dismiss, denial, delay, and if they lose, marginalize. It was plain to see. The Air Force fought back against the promotion opportunity for many years, no different than the command opportunity the chain of command blocked me from assuming. Once I earned the rank, the service effectively negated any substantive value by ceremonially processing the promotion without retroactivity. Since it was about principle, I remained grateful the USAF promoted me in retirement. Then again, I earned it and I served the time, so as Mr. Perot's gift, the King Arthur Excalibur sword, and the Churchill quote, inspire, "never give in, never give in, never, never, never!" Hopefully, in time, they'll fix the date—it's the law.

Like many of the struggles documented in these pages, with delayed justice or none at all, the systemic problem with these ordeals lies within the perplexing reality that a case can be dragged out for ten years, or more. It becomes another marathon for the pursuit of justice. In the end, that very delay negated the equity otherwise protected and intended by the law. Such an outcome was morally and ethically wrong on two levels. First, the obstruction meant many of the wronged personnel never entered the marathon or simply gave up along the way. Second, the system should not allow bureaucratic delays, perhaps by design, to evade full justice remedies.

The poetic irony of the outcome meant I was ceremoniously promoted only due to the substantiated misconduct of some senior leaders that didn't want me to assume command or be promoted in the first place. I did earn it, but none of the positive corrections or remedies ever would have happened absent the violations. Though not as fully designed, the system did work. I was content knowing I never compromised my values to receive the promotion. I never stabbed anyone in the back. I always fought for the rights of those I supervised. When all was said and done, I had no regrets. I was proud of my conduct and expected the same from our US Air Force.

MILE 19:
Part Two
Pandemic
(2019 to 2023)

The pandemic started in 2019, and COVID-shot mandates launched in 2021. The world-shaking events marked the beginning of another marathon before the first one finished. I first learned about the severe acute respiratory syndrome coronavirus 2 (SARS-COV-2) pandemic near ground zero on a layover in Shanghai, China. The nomenclature for the virus originated from an earlier coronavirus responsible for a SARS outbreak in 2003. As the world later learned, the subsequent SARS-COV-2 or COVID pandemic was likely created within the vicinity of the Wuhan Institute of Virology (WIV) in China. By the time it was all over, the pandemic proved less deadly than originally feared, but more divisive from a societal perspective due to controversial mandatory public health measures.

I loved flying to China, and touring her cities on layovers. I dodged traffic while jogging on their busy streets, hunted bargains at my Chinese friends' shops, and learned just enough Mandarin phrases to not seem like an American cliché. My last trip to China in January 2020 was eerily quiet. No one was on the streets, no shops were open, and there was literally no one to talk to. Our flying into China stopped, except for cargo runs primarily to transport medical supplies. Once the pandemic started, we did our jobs and the flying reminded me of missions from my military days. This time it was to provide essential goods and supplies to the homefront.

We managed the risks as essential workers and followed all the rules. Not too long into the pandemic, the patterns, similarities, and déjà vu moments began. What we experienced across twenty-plus years of combating deceptions and groupthink related to the anthrax vaccine suddenly resurfaced. The COVID public health handlers appeared to dig out the old propaganda playbook. They picked up where they left off, forgetting the rules in FDA law and Hippocrates's lessons. In 2021 alone, our government outlaid $1 billion for a publicity campaign to prematurely advertise COVID inoculations as safe and effective through the nation's top media outlets.[1]

Considering the pandemic was ongoing, and science on the shot's safety and efficacy was evolving, the government's one-sided promotion seemed imprudently rash. Our past experiences with anthrax vaccine directly related to the pandemic, with remarkable commonalities in the patterns of bad government behavior. The taxpayer-funded educational propaganda campaign was just one example. Three other troubling issues reemerged. The first problematic recurring theme related to why we allowed scientists to play with petri dishes in the first place, without the strictest of controls. The second troublesome recurrence related to bad behaviors by bad actors who relied on ad hominem attacks against any who questioned the counterfeit narratives. They encouraged the crushing of any doctor or citizen who challenged the medical establishment's prescriptions related to the third recurrence—illegal medical mandates. This mile in the journey reflects on these patterns, which played out predictably and disappointingly during the pandemic.

An early example of government efforts to combat information that didn't fit their counterfeit narrative, and to aggressively create counter-disinformation, leaked through FOIA disclosures from the NIH. The NIH director, Francis Collins, recommended a "devastating published take down" of the Great Barrington Declaration.[2] That initiative from a group of infectious disease epidemiologists and public health scientists, including a Stanford Nobel Prize winner, expressed "grave concerns about the damaging physical and mental health impacts of the prevailing COVID-19 policies." The group's letter garnered almost one million signatures and recommended "Focused Protection." Their key goal included minimizing social harms related to arbitrary universal lockdowns, with the objective of protecting vulnerable members of society.

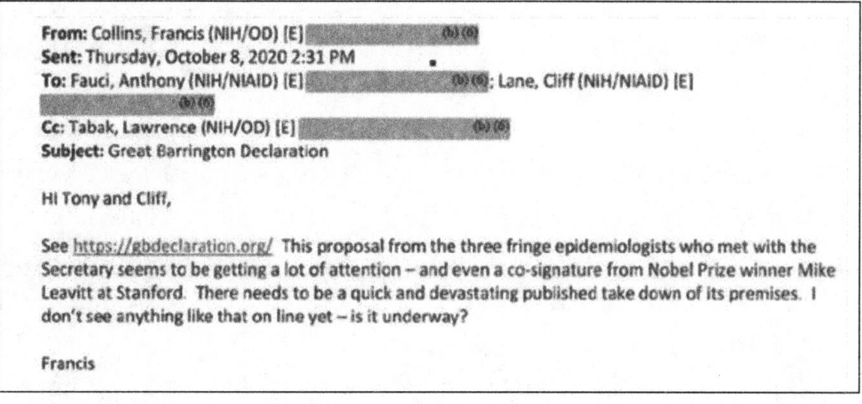

From: Collins, Francis (NIH/OD) [E] (b)(6)
Sent: Thursday, October 8, 2020 2:31 PM
To: Fauci, Anthony (NIH/NIAID) [E] (b)(6); Lane, Cliff (NIH/NIAID) [E] (b)(6)
Cc: Tabak, Lawrence (NIH/OD) [E] (b)(6)
Subject: Great Barrington Declaration

Hi Tony and Cliff,

See https://gbdeclaration.org/ This proposal from the three fringe epidemiologists who met with the Secretary seems to be getting a lot of attention – and even a co-signature from Nobel Prize winner Mike Leavitt at Stanford. There needs to be a quick and devastating published take down of its premises. I don't see anything like that on line yet – is it underway?

Francis

NIH Director Dr. Francis Collins's "devastating published take down" email.

Dr. Collins assaulted the credibility of participating epidemiologists as "fringe" and expressed concern over the fact that a Nobel-Prize-winning Stanford chemist cosigned the declaration.[3] Rather than embracing the diversity of thought or evaluating the validity of the declaration's recommendations, he attacked the messengers and advocated a "devastating published takedown of its premises." Enzo's term, "dupe-loop," applied to these deceptive directives that descended from the very top of the American government's public health ranks. Somewhere along the way Collins forgot about those he served and his true mission. He pushed a public relations campaign to protect controversial policy rather than serving as a champion of intellectual diversity and discourse at a time when America required measured leadership.

During the pandemic era, the same disturbing patterns we witnessed with anthrax vaccine resurfaced. Doctrinal departures defined the patterns. We learned our government offshored and outsourced biologic research to foreign laboratories in Wuhan, China, and even in Ukraine.[4] The very CB Taboo warned about during my 1980s Air Force Academy training, and reflected on with my 1999 congressional testimony, rapidly infected and impacted the entire world.

The CB Taboo (Chemical and Biological Warfare Taboo) against dangerous biological research also emerged as ammo from Russia. While government officials admitted the existence of the debated laboratories, they quibbled that labs were not for "biological weapons."[5] But did the distinction matter? The issue wasn't if we funded bioweapon labs versus bioresearch labs. Wuhan was likely not a bioweapon lab, but a worldwide pandemic still potentially originated from a lab where US actors funded research. The problem was that we were messing with petri dishes again at the risk of doctrinal common sense. Instead of serving as an example, many perceived that defensive biological research resulted in proliferation and harm to our nation's credibility in the realms of restraint and compliance.

During the pandemic inquiry on the virus's origins, reputable figures suggested the plausibility of an accidental lab leak. One academic was Francis Boyle, PhD, a Harvard-trained lawyer and bioweapons expert. Boyle's expertise with the Biological Weapons Convention of 1972 precipitated his assistance in drafting the Weapons Anti-Terrorism Act, which was ultimately signed by President George Bush Sr. in 1989. Boyle's Wuhan lab-leak warnings preceded his cautions about US-Ukraine biological research.[6] The US government ultimately acknowledged Wuhan lab collaboration, and the DoD later admitted to "providing support to 46 peaceful Ukrainian laboratories."[7, 8] Given the history with anthrax vaccine, and the patterns of "intimate" US involvement with perilous petri dishes, the revelations justifiably jarred our trust.

Another lab-leak theorist was Dr. Robert Redfield, a former CDC director. In a CNN interview he stated, "I still think the most likely etiology of this pathology in Wuhan was from a laboratory." He added, without implying "intentionality," that "I do not believe this somehow came from a bat to a human and at that moment in time that the virus came to the human, [and] became one of the most infectious viruses that we know in humanity for human-to-human transmission. Normally, when a pathogen goes from a zoonotic to human, it takes a while for it to figure out how to become more and more efficient."[9] Redfield charged that the public health official's emails proved they "knew" the truth about gain-of-function, or the tinkering with petri dishes to create a more infectious virus. Regarding origins, he alleged they "misled congress" in a "cover-up" to obscure that the "virus was, in fact, a consequence of science."[10] Boyle's and Redfield's unorthodox and scholarly positions on COVID origins defied the script.

Regardless, the subsequent derision of experts like Dr. Boyle and Dr. Redfield mirrored what we witnessed throughout the anthrax vaccine ordeal. Despite being vindicated after many years regarding the anthrax vaccine illegalities, we absorbed the clichéd allegations of internet misinformation long before it became a part of the American lexicon. Suppression of discussion and disparagement resurfaced regarding COVID origins, the necessity of mRNA (messenger ribonucleic acid) injections, and the canceling of doctors. Our nation's leaders wholly ignored President Eisenhower's prophetic warning about industrial complex capture of government and regulatory agencies.[11] Collectively, our society re-erected that era's "pillar of shame" with cold war-McCarthyism bad behaviors. Once again, our citizens were wrongfully blacklisted, while the government and their surrogates perpetuated the "biggest cover up in medical history."[12, 13]

In both dilemmas, the official dupe-speak was sacrosanct, yet appeared politically biased and of questionable trustworthiness. The dupe-loop and debate over the legality and necessity of COVID inoculations mirrored our anthrax experiences too. This time the government, in league with mainstream and social media, squelched debate and judicial review over complicated scientific and legal issues, ostensibly to combat vaccine hesitancy. The gate guard's bad behavior resembled McCarthyistic orthodoxy with its many obliging shills. More ominous still, the similarities in the dupe-loops of the lab leak in 2001 and the suspected lab leak origins of the pandemic-era COVID inoculations witnessed the gate guards impacting every element of the event-decision loops. In both cases, they possibly created the threats, hyped the threats, promoted vaccines to remedy the threats, and impugned or punished any who questioned their threats or their countermeasures—gate guards operated in all segments of the dupe-loops.[14]

From 1997 to the 2000s, citizens and soldiers witnessed the same circular groupthink as the policies and rhetoric changed. Before the

DoD changed the rhetoric, and when they did answer the questions, they agreed that the anthrax vaccine had a "higher than desirable rate of reactogenicity," a "lack of strong enough efficacy against the aerosol route of exposure,"[15] was "unsatisfactory for several reasons,"[16] and possessed "unusually hazardous risks associated with the potential for adverse reactions."[17] "Evidence in experimental animals that the vaccine may be less effective against some strains of anthrax"[18] meant the vaccine was "not licensed for a biological defense indication"[19] [excerpts from *Vaccines* textbook at Appendix CC]. At least with anthrax vaccine, the media and many officials actually analyzed the dichotomy. But during the pandemic, most of the public health officials and the media read from the boilerplate script. Anything contrary to that script was deemed misinformation. The groupthink was as contagious as the virus, and equally as dangerous as the doctrinal shift with biological research proliferation.

Our position during the anthrax vaccine dilemma was the same as the DoD's before the vaccine loyalty test. We reviewed the documents collected over the years by the Gulf War Illness advocates that came before us. When the questions we asked could not be answered, we found the answers ourselves. The documents demonstrated that the truth and the facts were changed so that the anthrax vaccine "education campaign" would conform to policy. The questions couldn't be answered because the military and government experts that understood the complexities of the issue conveniently forgot the CB Taboo—COVID was no different regarding policy and origins.

Our troops, parents, and citizens observing the incompetence and negligence realized the experts failed to be the experts. They forgot or did not know what the documents said. They failed to be the "circuit breakers" when the questions couldn't be answered. During the pandemic, the same behavioral dysfunctions and loyalty tests infested public health requirements where accepting a mandatory single dose of an mRNA injection, regardless of regimen or immune efficacy, became the litmus test.[20] Wary troops and concerned citizens did their own homework and rooted out the inconsistencies, often with information from alternative media, scientists, and courageous doctors who dared to practice medicine and live up to their oaths.

Because we swore to our own oath of office, we empathized with those sworn to the Hippocratic oath: "You do solemnly swear . . . That you will be loyal to the Profession of Medicine and just and generous to its members." But what we observed during the pandemic didn't include some of those loyal to the profession of medicine being "generous to its members." Many of the institutional medical figures put daggers in the backs of their fellows. Their oath insisted that their fellowship stay "aloof from wrong, from corruption" and to exercise their "art solely for the cure of your patients." It commanded, "Give no drug, perform no operation, for a criminal purpose." During the anthrax vaccine controversy, the mandate unequivocally violated the law, and therefore was arguably criminal. Then again with the pandemic, illicit mandates of unapproved medical products and the withholding of treatments once again appeared to challenge both the laws governing medicines and the oaths guiding the medical profession.

During the anthrax vaccine years, we suffered through these anomalies. A deputy assistant secretary of defense for health operations policy testified, "This is not a medical program—it is a line commanders' program to prevent combat casualties."[21] As a result, any who dared to question the rhetoric and policies were systematically discharged. The message was sent loud and clear for the doctors to stay out of the debate, as it was relegated to one of "good order and discipline" and combat readiness. This political reality made the military medical community's tacit acquiescence of the "commanders' program" seem to be an endorsement. The reality was that the DoD deliberately and systematically carved doctors out of the operation.

Two decades ago, Russ Dingle painstakingly chronicled the entire recorded history of the anthrax vaccine, including the laws governing biologics. He built a library chronicling every tangentially related published medical article and research paper going back to the 1940s, numbering over seven hundred titles. He retrieved the entire recorded history of the regulations governing vaccines and all their myriad iterations over the last fifty years. Through the Freedom of Information Act, Russ obtained all the officially recorded documents from the FDA and many more from agencies within the DoD. Gulf War vets and organizations shared their files. It was a mountain of material, much of it mundane in tone and arcane in content. But when he sifted and sorted it all out, the true nature of the anthrax vaccine became quite clear and quite simple. The anthrax vaccine was unsatisfactory. Period. So, why were most doctors AWOL then and now?

Interestingly, we observed only a small group defending the DoD's anthrax vaccine program groupthink. They appeared to be very short-lived on the scene of defending what turned out to be an illegal program. As fighter pilots, we referred to such temporary engagements in battle as a "high-speed pass." You come in, guns blazing, and depart the fight as quickly as possible. Those spreading and defending the vaccine disappeared after a quick high speed pass.

One of the nation's most renowned psychologists, Dr. Philip Zimbardo, spent his life analyzing groupthink scenarios such as these, beginning with his Stanford prison experiment in 1971. Dr. Zimbardo taught my Naval Postgraduate School psychology class and graded my papers related to the parallels between the anthrax vaccine dilemma and his life's work. In my thesis, based on his teachings, I wrote about the hermeneutics of psychology. This lens allowed an evaluation of the ethical breaches associated with the anthrax vaccine, beyond individual sociopathic deviations. It permitted focus on the institutional and situational contexts. In Dr. Zimbardo's book, *The Lucifer Effect*, he evaluated the controversies surrounding prison abuse at Abu Ghraib in Iraq. This example helped us to place the idiosyncratic social behaviors in perspective by altering the common notion of a "bad apple" and considering larger forces.

More apropos, Dr. Zimbardo considered the prospect of a "bad barrel," or "the idea that the social setting and the system contaminate the individual, rather than the other way around." The "bad barrel" concept acknowledged the potentially "corrosive influence of powerful situational forces."[22] At Abu Ghraib, otherwise good people did bad things while other otherwise good people watched. But these incidents were not isolated, from a historic standpoint. For other historic examples in the medical realm, one needs to look no further than presidential apologies fifty years after syphilis testing, forty years after radiation testing, and government acknowledgments about the toxic effects of Agent Orange thirty years too late. Each of these examples highlighted the institutional or systemic nature of abuse and the situational forces leveraged for rationalization. The situational forces sprang back to life during the anthrax vaccine controversy, and again with divisive medical mandates during the COVID pandemic.

In evaluating the anthrax vaccine and COVID shot controversies, I found the psychological vantage point valuable to explain the perceived breakdowns. Dr. Zimbardo's explanations of the "psychological causes behind such disturbing metamorphoses," where people collectively commit wrongs, helped me to better compartmentalize the bad behaviors. Just as addictive tendencies and denials have medical explanations, the tendency to succumb to situational forces was a known aspect of bad human behavior. The prism of psychology helped critical thinkers to understand why others bought, sold, and defended the immoral schemes. Intellectually, it allowed us to move from anger and frustration over the abnormalities to a place where we instead identified the bad human behaviors and endeavored to curb them over time.

Like the Abu Ghraib controversy, my analysis in the anthrax vaccine thesis—and again with the parallels from the pandemic—adopted the "premise that ordinary people, even good ones, can be seduced, recruited, initiated into behaving in evil ways under the sway of powerful systematic and situational forces."[23] It also proved valuable to consider some of the subtopic psychological concepts that seemed to apply, such as confirmation cognitive biases. From writing my thesis in 2009, and into the next decade, this was largely an esoteric concept. Yet during the pandemic and the commensurate informational struggles, more and more people recognized the dangers of situational ethics and cognitive biases. Both sides made the charge.

Cognitive "confirmation bias" involved the intriguing ability of humans to confirm and rationalize what they want to believe and discount evidence to the contrary. Cognitive bias included the government's fixation on the "Maslow's hammer"[24] of universal mRNA shots during the pandemic—even more so than with the limited military-centric anthrax vaccine dispute. With both the 2001 anthrax endemic and the 2020s COVID pandemic, the government hammer of vaccination became the tool, and everyone, troops and citizens alike, were pounded.

The second cognitive concept referred to as "probability neglect" appeared to directly apply where people "subjectively overestimate the probability of highly undesirable but objectively rare outcomes." Obviously, with an anthrax attack, "intense negative emotions are involved," and our "attention is captured by the dreaded outcome," regardless of the "relatively small chance of the threat actually occurring." Probability neglect "is an important contributor to sustaining disproportionate fears,"[25] and was especially ripe for abuse in a pandemic scenario.

A third cognitive concept called "negativity bias" also suggested that "human beings are much more powerfully influenced by negative than positive information."[26] With anthrax vaccine and the pandemic, a fascinating psychological dilemma appeared where probability neglect and negativity bias manifested in multidimensional forms of fear and behavioral manipulations promoted under the endless renewal of government emergency powers.[27] Officials leveraged these powers to protect drug makers and paralyzed citizens into obedient compliance.

Confirmation bias, probability neglect, and negativity bias all appeared to apply and connected the anthrax vaccine and COVID shot controversies on an intellectual level. Ironically, in the case of the anthrax attacks, it was the government that hyped the threat to justify use of the vaccine before and after the government insider anthrax attacks. Even though the FBI confirmed that a government scientist committed the attacks to save his program, and to revive use of the failing vaccine, federal investigators similarly succumbed to these situational forces by not overcoming biases by surmounting the "evil of inaction."[28] Despite the FBI findings, no one resuscitated a review of the program or renewed the call for corrections due to the illegalities.

To reverse these unhealthy congnitive behavioral trends, Dr. Zimbardo advocated the encouragement of "integrity system heroes" where a dimension of humility should exist in organizations to facilitate expressions of minority opinion. This environment replaces egocentrism with sociocentrism in order to overcome the "root causes" of inherent situational evils over time. "Chronicity" is a term used by Dr. Zimbardo to describe the healing power of time.[29] We saw it over and over again where human beings seemed incapable of correcting their wrongs, and then only future generations and Presidents appeared willing to do so once all the perpetrators were gone. Chronicity, healing over time, may be the key as it has been in the past.

Dr. Phillip Zimbardo's model of heroism, called the motivational and decisional framework, advocated engagement, sacrifice, active gallantry, and passive fortitude. To an extent the FBI did exhibit some gallantry by finally pinpointing the self-appointed perpetrator and his motive to save the failing anthrax vaccine program, but the bureau and military failed to follow through, extract accountability, and correct the wrongs. I understood it. That was a part of the human condition, and that's precisely why we must analyze the lapses to hopefully do better in the future. This is the long-term "quest" where "collective heroism" accrues over time.[30] Russ, Dr. Buck, and our wingman, Mr. Perot, were examples of the "integrity system heroes" from our story. Similar integrity system crusaders inevitably emerged during the pandemic era as well.

The story of Dr. Buck was the preface for the larger story of the pandemic era two decades later where good doctors were silenced or never spoke up. Those that did were crushed, this time by the institutional sovereigns over the profession of medicine. The psychology of bad human behavior and institutional coercions we witnessed more than twenty years ago were just a teaser for what occurred again with the pandemic, this time on a worldwide scale. Anthrax never rose to the level of a pandemic, and the controversy was less than endemic. The two isolated epidemics followed the Army testing of the DoD's anthrax vaccine in Manchester, New Hampshire in 1957, and then again after the Army scientist mailed anthrax letters in 2001 to save the failing anthrax vaccine program. Those epidemics were hints of the patterns of bad behavior that contaminated the country in 2020 and beyond. Our earlier, less lethal circumstances as a more isolated, captive audience, with considerable media and congressional support, allowed at least a fighting chance to separate fact from myth. The pandemic era posed a more difficult task.

As the COVID pandemic dragged on, many Americans felt similarly betrayed by the apparent disregard of legal rights, as well as the brazen dismissal of the obviously critical and legitimate issue of natural immunity. The bad or negligent human behavior was not new. My thesis work on social psychology helped me to compartmentalize the disturbing reality of how "good people can be induced, seduced, and initiated into behaving in evil ways."[31] In our historic case, the DoD procured an experimental vaccine with known problems and illegally ordered the vaccine's use on troops under the threat of punishment. The DoD knew that the vaccine suffered significant limitations and the mandate violated the nation's laws. Social psychologists warned us of such a phenomenon where, "With numbing regularity, good people were seen to knuckle under the demands of authority and perform actions that were callous and severe."[32]

During the pandemic, our leaders, practitioners, and citizens knuckled under the demands of authority. But, in fairness, at the same time the world was dealing with an infectious disease that spread rapidly, killed, and struck fear across the globe. Whereas with anthrax we witnessed a rare non-communicable threat, this time the global scale of infectiousness caught the attention of every human being on the planet. This time the manipulations of human behavior required collaboration and collusion, but at the same time the coordinated coercive mandates awakened open-minded people worldwide. What they did to Dr. Buck then now victimized multitudes of doctors, nurses, and scientists nationwide during the pandemic.

Knowing that vaccines inevitably would be developed and propelled as the next phase of the pandemic response, I treaded cautiously, behind the scenes, and tried to check my biases. Most doctors and scientists, at least initially, did so as well, but the pandemic provided fertile ground for a rebirth of the bad human behaviors—bad causes and bad means by otherwise good people. Everything was happening again. It was just all too familiar.

Then came the variants: Alpha, Delta, Omicron, BA.2/4/5, XBB, EG.5, Pirola, and JN.1. The longer it lasted the more our immune systems and trust were challenged with "off target effects" that potentially created "frameshifting" through "mistranslation events" that "decrease efficacy or increase toxicity."[33] Despite unknown effects from the inoculations, or the suspected "glitch" from "slippery sequences" of mRNA, officials promoted the shots as a one-size-fit-all panacea without considerating the ever-increasing proof of immune dysfunction. As the science evolved, citizens naturally questioned the glitches and slippery motives as well.

The motives became suspect because the previous taboos conceivably and predictably created the virus, albeit hopefully accidentally. Well-intentioned experts created the vaccines, while less well-meaning public officials forced the vaccines on others through mandates without regard to the legal limitations or medical necessity. Full circle, the bad human behaviors birthed a new cognitive taboo of politicized medicine to restrain the practice of medicine and to force the injections on people worldwide. Legitimate questions remained regarding origins, risks, necessity, and legality, which justifiably tested the propriety of the onerous mandates.

An esteemed cardiologist published a peer-reviewed paper stressing the need to address these unanswered ethical-based questions. Dr. Aseem Malhotra respectfully urged a suspension of COVID shot mandates due to a rise in cardiac arrests linked to mRNA vaccines. The doctor warned of "unprecedented harms" related to the vaccines and asserted recipients were more likely to get injured by the vaccine than by the disease itself. He emphasized a return to strict pharmalogical "evidence-based" practice of medicine.[34] The doctor injected a healthy warning that "politicizing medical discussion is an exercise in intellectual bankruptcy."[35] Our experience across two decades matched this doctor's new awareness of the unhealthy patterns.

Unfortunately, during the pandemic reasonable answers to legitimate questions and reciprocity of respect seemed elusive. What began as an opportunity to unite everyone toward a common goal devolved into the bad human behaviors described in Dr. Zimbardo's warnings. Situational ethics and groupthink prevailed. People worldwide—vaccinated and unvaccinated—served as the tools of manipulation by being pitted against one another. Literally every level of society morphed into a global COVID Stanford prison experiment. The challenge ahead required never repeating the psychological manipulations and instead managing the delicate balance of rights and policy so paramount to the best interests of trust in public health. Evidence of progress was slow, but glimmers of progress equated to reversing the negative trends.

As future generations grade the bad behaviors of pandemic era politics, our society will likely be embarrassed by the report card. A society should not be proud of harming our troops, citizens, or colleagues. The harms were medical, social, professional, and intellectual. There is no greater good that justifies such bad behaviors. The same human nature tendencies must also be kept in check to not serve up vengeance as a response to the harms in pursuit of justice.

As future leaders and scholars cooperatively evaluate the root causes of the behavioral and societal breakdowns, it will be vital to identify the effects from the foremost bad actors during the pandemic era in order to guard against a recurrence and extract accountability.

MILE 20:
Fauci Effect (1984 to 2022)

Anthony S. Fauci, M.D., became director of NIAID about the time I entered military service, almost forty years ago. His half century tenure meant he was around during the anthrax vaccine disaster, but he did nothing to alter its illegal trajectory. He aptly fits into the gate guard box. His compartment of responsibilities was to oversee, research, prevent, diagnose, and treat infectious diseases. Based on his failure to act in the anthrax vaccine debacle, his much debated role in the worldwide pandemic was unsurprising. The doctor should have inspired confidence and trust, but instead his bad effects ranked him as a chief bad actor during COVID. Whereas the "Fauci effect"[1] initially promoted the altruistic idea that Fauci epitomized integrity in science, history will show the antithesis—the "Fauci effect" will most likely be synonymous with "untruthful"[2] disinformation and a momentous loss of trust and credibility in our public health institutions.

American lawmaker and doctor, Senator Rand Paul, is one elected official who attempted to hold Dr. Fauci accountable for allegedly lying to Congress.[3] He alleged, "Fauci misled the country about the origins of the Covid pandemic ['likely the product of gain-of-function research at the Wuhan lab'] and shut down scientific dissent."[4] The senator questioned Fauci's funding for gain-of-function research at the Wuhan laboratory in China. He challenged Fauci's effort to discredit and divert attention from the lab leak theory and his apparent misstatements to the US Congress. Though a verdict on COVID origins remained elusive, FOIA documents from the Departments of State and Interior indicated a lab leak was "most likely" based on circumstantial evidence and indications that scientists sought to create viruses with unnatural or synthetic man-made "furin cleavage" features.[5]

As of June 2022, even the World Health Organization (WHO) Scientific Advisory Group for the Origins of Novel Pathogens (SAGO) expressed that they were "open" to investigations about the "laboratory as a pathway of SARS-CoV-2 into the human population." The WHO group recommended inquiries into "any and all scientific evidence that becomes available in the future to allow for comprehensive testing of all reasonable hypotheses."[6]

Objectively documenting the enigma of Dr. Fauci's bad behaviors required me to reflect on what we witnessed during our earlier war for the truth. Fauci's inaction in our battle, and his ascendance after the anthrax letter attacks, provided perspective on his role during COVID.[7] Senator Rand Paul's duel with Fauci as a top public health figure went straight to the root issue of whether or not Dr. Fauci funded research that led to the pandemic. It appeared suspicious primarily because Fauci attempted to divert attention on origins by disparaging alternative theories and theorists. While those tactics were not indictable in and of themselves, his evasive testimonials understandably eroded American confidence. Dr. Fauci's legacy will most likely be that he appeared to flout CB Taboo doctrinal precedents and oversaw sponsorship of dangerous offshore research that allowed bad actors to tinker with petri dishes—then covered it up.

This was a repetitive theme throughout the anthrax vaccine dispute as well. Officials suspiciously obfuscated the truth, but were never held to account. In the end, it's conceivable the same fate of getting off the hook scot-free awaits Fauci. History may not be so kind in dismissing his role or the waffling and quibbling. Without accountability lessons won't be learned. The light bulb must be illuminated for the American people to gain awareness if health officials violated the public's fragile trust. Compared to our anthrax vaccine timeline, the pandemic learning curve and revelations on COVID origins moved at light speed. In contrast, the "glacial pace"[8] of truth awareness for the anthrax vaccine and the anthrax letter attacks obstructed the lessons learned.

To the issue of the COVID origins, even the terminology caused confusion—probably by design. The allegations surrounding gain-of-function research, or the increased infectiousness of a virus, was technically referred to as research on "enhanced potential pandemic pathogens."[9] The National Institute for Health (NIH) justified this genre of research as a "compelling public health need and conducted in very high biosecurity laboratories." The NIH supported "research that may be reasonably anticipated to create, transfer or use potential pandemic pathogens resulting from the enhancement of a pathogen's transmissibility and/or virulence in humans." The NIH abbreviated enhanced potential pandemic pathogen research with the acronym ePPP. They acknowledged it "requires strict oversight and may only be conducted with appropriate biosafety and biosecurity measures." The NIH added, "research is inherently risky and requires strict oversight," but that "the risk of not doing this type of research and not being prepared for the next pandemic is also high." In contrast to Fauci quibbling about the confusing terms,[10] the NIH admitted "ePPP research is a type of so called "gain-of-function" (GOF) research."[11]

Evidence showed that early on Dr. Fauci attempted to dismiss the allegations of a lab leak by insisting the COVID virus' "mutations" were "totally consistent with a jump of a species from an animal to a human." The suspect part came with later disclosures that Fauci may have "deliberately suppressed" the lab leak theory. He apparently coordinated articles debunking the lab leak theory in science journals. Later, email exchanges showed he received a congratulatory note from EcoHealth Alliance's Peter Daszak. Daszak turned out to be an instrumental actor in the muddied dispute along with an NIH-funded University of North Carolina researcher, Ralph Baric, who co-authored papers on gain-of-function efforts as early as 2015.[12, 13, 14] Reference Daszak's email to Fauci, he wrote, "From my perspective, your comments are brave, and coming from your trusted voice, will help dispel the myths being spun around the virus' origins."[15]

Through a medical journal, *The Lancet*, in February 2020, Daszak jointly published this statement: "We stand together to strongly condemn conspiracy theories suggesting that COVID-19 does not have a natural origin."[16] The authors doubled down in July 2021 in another *Lancet* article titled "Science, not speculation, is essential to determine how SARS-CoV-2 reached humans." They softened their ridicule of questions about virus origins from "conspiracy theory" to "speculation," but reaffirmed their conclusion that "the virus evolved in nature."[17] Months later, other scientists published a *Lancet* rebuttal claiming, "There is no direct support for the natural origin of SARS-CoV-2, and a laboratory-related accident is plausible."[18]

Even Fauci had received preemptive lab leak warnings. As early as January 2020, scientist Kristian Andersen explained that "phylogenetic analyses" were "completely off," that there were "unusual features of the virus." He emailed that "one has to look really closely at all the sequences to see that some of the features (potentially) look engineered." Andersen's email warned that their conclusions "find the genome inconsistent with expectations from evolutionary theory" [email at Appendix DD]. Andersen later recanted his cautions, but those online social media posts were subsequently excised from the information sphere.[19]

Congress recorded the government's obstruction of efforts to isolate the origins of COVID, plus the fact that Andersen co-authored the infamous paper titled "The proximal origin of SARS-CoV-2," which was "prompted" and cited by Fauci to dispute the lab leak.[20] Just prior to a $8.9 million NIAID grant,[21] Anderson reversed by concluding, "we do not believe that any type of laboratory-based scenario is plausible."[22] Despite the reversals, diversionary "canards," and attempts to frame expertise as "synonymous with science," reputable independent journalists and congressional investigators called out the suspected "lab leak coverup by Dr. Fauci."[23]

In retrospect, the revelations were very damaging. At a minimum, the communiques created the appearance that Fauci tried to hide the warnings and downplayed the possibility that the NIH funding for research that Daszak funneled to the Wuhan lab could have resulted in the pandemic.[24] It worked for a while, until exposed. Subsequently, the NIH admitted that the lab "did indeed enhance a bat coronavirus to become potentially more infectious to humans," something the institute referred to as an "unexpected result" at the lab in Wuhan. NIH laid blame on Daszak as their surrogate for not reporting the "unexpected result." Daszak's EcoHealth Alliance disputed the allegation that they violated their NIH grant by claiming the company reported the "data" (the "unexpected result") in their "year four report in April 2018."[25]

A *Lancet* task force looking into the origins of the pandemic shut down due to concerns about conflicts of interest by Daszak as one of its members. Task force chair Jeffrey Sachs of Columbia University in New York made the call. Daszak's EcoHealth Alliance subcontracted $600,000 in US government funded research to the Wuhan Institute of Virology, which Sachs deemed as "disqualifying conflicts of interest."[26] Daszak's apparent lack of forthrightness resulted in the task force's closure. Daszak's prior affirmation to *Lancet* stated, "I have no conflicts of interest." The fine print revealed he collaborated with the Wuhan lab's Dr. Shi Zhengli by funding her work through NIH grants. Daszak's literary efforts to discount the lab leak theory lacked credibility. The *British Medical Journal* quoted Sachs' conclusion that "it is clear that the NIH co-funded research at the WIV [Wuhan Institute of Virology] deserves scrutiny under the hypothesis of a laboratory-related release of the virus."[27] Eventually Dr. Sachs concluded a lab leak was the most likely hypothesis. He also explained the dirty underbelly of NIH grants that led to disturbing behavior by compromised scientists where they concocted the counterfeit narrative about natural evolution to cover-up the more likely lab leak explanation.[28]

Who knew what to believe? Curious Americans—no, the entire world—deserved to know the truth. In May 2020, *National Geographic* quoted Fauci stating he was "very strongly leaning toward this could not have been artificially or deliberately manipulated."[29] But Jeffrey D. Sachs, in June 2021, insisted that "both hypotheses—natural zoonosis and research-related infection—are viable at this stage of the investigation." He added, "Natural zoonotic events are inevitable."[30] Even if a naturally occurring event, Sachs' final *Lancet* report suggested that the "research-related incident" possibly occurred at the Wuhan Institute of

Virology lab.[31] The British Parliament was also briefed in late 2021 that the "'most likely origin' of COVID-19 was a leak from the Wuhan lab, and that the virus may in fact be genetically engineered."[32]

That news came during a briefing for the Science and Technology Select Committee by Dr. Alina Chan, with specialties in gene therapy[33] and cell engineering from MIT and Harvard. She added the "virus has a unique feature, called the furin cleavage site, and without this feature, there is no way this would be causing this pandemic." Dr. Chan explained that leaked documents related to Wuhan and EcoHealth Alliance showed a connection between the NIH funded work and "inserting novel furin cleavage sites." She mocked that they tried "to put horns on horses," and then magically "at the end of 2019 a unicorn turns up in Wuhan city."

The more you looked into it the more it matched our twenty-plus years of witnessing the obfuscations by government officials—so it does lend credence to the possible conclusion that Fauci and company had something to hide. On point, other scientific journals reported the "highly unusual" furin site similarities to earlier COVID shot patents.[34] Like anthrax vaccine, something was amiss with the literary record as unpublished and non-peer reviewed studies presented the opposite conclusion that the virus likely originated from animals in the Wuhan Market.[35] But, none of the alternative theories addressed the original concerns about the "unusual features of the virus" that made it "look engineered" and "inconsistent with expectations from evolutionary theory." And none of this explained how the first ill patients were workers at the Wuhan lab.[36] Time will tell, and someone, someday will hopefully tell us the truth.

As with the requirement for candor on the lab leak possibility, similarly the NIAID's Director Fauci had a moral and ethical obligation to use his considerable influence to impact the anthrax vaccine after the anthrax lab leak. He publicly addressed the matter on many occasions and at any time he could have used his authority to rectify the legal violations before the courts were compelled to do so in 2003 and 2004. He didn't. He didn't when he and I discussed the controversy in the same PBS broadcast with Betty Ann Bowser shortly after the anthrax attacks. Instead Fauci advocated for the vaccine to address the threat, even though it was not properly licensed.

> **Betty Ann Bowser:** The rash of anthrax scares and deaths have prompted health officials to recommend that lab workers and others at risk of exposure to anthrax be inoculated against the deadly disease. Dr. Anthony Fauci of the National Institutes of Health explained the reasoning behind that.
> **Dr. Anthony Fauci:** That in fact is classical public health strategy, when you have a vaccine for a particular disease and you're going to have health care people, laboratory people, or even people like firemen or others who are going to go into a potential disaster zone, that those are the individuals that you want to be protected because of the repeated exposure . . .

Fauci dodged the issue of the problems with the vaccine, but fortunately PBS highlighted the problematic reality of the concerns. No one at the time realized the motive for the attacks in the first place was to save the failing program by an Army scientist. The aftermath found Fauci at the helm of the nation's biodefense apparatus that embellished the letter attack bioincident "threat" to promote the vaccine and secure lavish budgets to advance the new biodefense enterprise. The patterns were significant since from 2021 to 2023 the government leveraged the COVID crisis to rapidly field, push, and then mandate unapproved countermeasures. The PBS interview documented the similar rush to field a countermeasure based on a bioincident.

> **Betty Ann Bowser:** Still, the perception that the vaccine wasn't safe persisted. Hundreds of pilots in Air and National Guard units around the country refused to take the shots. For some, like Major Tom Rempfer, a pilot with the Connecticut National Guard, that refusal meant the end of a military career . . . Ultimately if civilians are going to need protection against anthrax, the answer may be found in a new

vaccine. Dr. Fauci thinks the events of September 11 will speed that process.

Dr. Anthony Fauci: In usual times, that is a process that takes years and years. It likely will take more than a year, or maybe two now, but I can tell you the amount of time that it's going to take given the urgency of the situation to translate from an observation that's basic science to a usable product, if you were, is going to be markedly truncated because of the urgency . . .

Betty Ann Bowser: Fauci said a new vaccine would clear up all the nagging questions about the old vaccine's safety. Meanwhile, because of BioPort's problems, there isn't enough vaccine in the Pentagon's stockpile to continue a mass inoculation program.[37]

At least Fauci advocated for a new vaccine, something the military and government knew they need since 1985—almost four decades later the new vaccine still did not exist. Fauci would also later testify supportively that antibiotics, not the vaccine, remained as the recommended "first line of defense" against anthrax. All of this begged the question as to why the NIAID permitted the outlay of billions of tax payer funds for a national stockpile of a vaccine that was not even required. They knew simple antibiotics worked and they needed a new vaccine. In testimony to the Committee on Homeland Security years later Fauci affirmed:[38]

Dr. Fauci: The best approach towards anthrax is antimicrobial therapy.

Though the PBS interviewer implied my career ended, it did not. I even returned to flying. My fate was tied to my mindset as a fighter pilot and operator. We followed the book, the rules, and the checklist. We didn't take broken airplanes airborne. We didn't make others do so. We didn't present hopes or beliefs as facts. That's what separated our approach from that of Dr. Fauci. He had not been an operational practicing doctor for most of his career. Perhaps he did not think of the implications of pushing a vaccine on hesitant operational practitioners concerned about necessity, safety, and efficacy. No one ever questioned my hesitancy about flying broken jets. Prove it works and I'll fly it. Prove a vaccine works and I'll take it. Operators get that logic. The DoD's Code of Conduct helped to logically explain why we fought back against illegal mandates. The code insisted I not "take part in any action which might be harmful to my comrades," that I "obey lawful orders," and "resist by all means available."[39]

Two decades ago, the DoD foisted the non-operational mentality and anthrax vaccine on the troops. In the pandemic era Fauci and others imposed this mandate mentality and their medicine on the hesitant operators of the world ranging from airline crews to truck drivers to health care workers. The bureaucratic disability suffered by Dr. Fauci, and others in the public health establishment, was that they seemed insensitive—and that's putting it charitably—to the interests of the very working people they served. Instead, Fauci insisted that once you "make it difficult for people in their lives, they lose their ideological bullshit, and they get vaccinated."[40] This mindset lacked the operational empathy to understand the hesitancy of ordinary Americans.

While some supported Dr. Fauci's leadership style, others lost confidence with his bureaucratic oscillations and governmental dupe-speak. Dupe-speak defined my disagreement with Fauci's ill-advisements of the American people. It is likely the brunt of why Senator Rand Paul challenged Fauci's dupe-speak and alleged that the origins of COVID was "the biggest coverup in the history of science."[41] Senator Paul knew, as Orwell warned in his essay to English compatriots during World War II, that "History is written by the winners."[42] The senator, and other operational practitioners, shunned the possibility that the Fauci dupe-speak might prevail. Reasonably, it is one thing to offer a vaccine's protection to the operators and workers of your country, but it is another level of arrogance to force it on them, especially when not trusted.

Fauci was not alone in his attempts to reconstruct, hide, or obscure the truth. It is truly fascinating how deeply you had to dig through media search results to find the quotes where Fauci first advised against mandates, then flip-flopped. The quotes are there, but the social media machinery choked the results and truth to hide the hypocrisy. A columnist for *The*

Hill identified the internet suppression of queries by revealing, "You'll find that many of the answers are buried, dating to early 2020, or simply impossible to find." He asked, "Why is that? Shouldn't we Americans be allowed to look up such information and then make judgments ourselves?"[43] Operators asked the same questions and all we wanted was to find the straight answers.

If you dug long enough you could find them, such as Fauci's earliest responses on the prospect of mandates for COVID inoculations. His response was, "No, definitely not." Fauci assured us, "You don't want to mandate and try and force anyone to take a vaccine. We've never done that. You can mandate certain groups of people, like health care workers, but for the general population you can't." Fauci allayed our concerns saying, "We don't want to be mandating from the federal government to the general population. It would be unenforceable and not appropriate."[44] Fauci insisted, "They have the right to refuse a vaccine. I don't think you need a contingency plan. If someone refuses the vaccine in the general public, then there's nothing you can do about that. You cannot force someone to take a vaccine."[45] The media and the tech giants jointly owned responsibility for the lack of reporting and information suppression regarding historic data points such as these, but the gate guards shared culpability.

Another example came from government relationships with the media through the Trusted News Initiative (TNI).[46] This Pravda-like collaboration complicated the search for information. The initiative's partners marched seemingly without question to the shifting political winds by advertising, tacitly authorizing, approving, and promoting health policy, while feigning ignorance about the counter-evidence.[47] The unprecedented $1 billion paid to the media to promote the COVID inoculations for the government-pharmaceutical league should have sounded alarms.[48] Reports indicated that Pfizer even indirectly funded fact checking, which added to the conflict of interest concerns.[49] Similar worry revolved around legal discoveries documenting that NIH scientists, including Fauci, received more than $350 million from industry royalties.[50] Revelations about drug industry money through user fees permeating the budgets of the world's regulators exacerbated the problem.[51]

Given the enormous financial incentives, supervisors in the military, government, and business world faced significant professional jeopardy if they did not capitulate to the heavily funded misinformation counterfeit narrative.

In addition to the evident media-tech-government collusion and the disconcerting financial conflicts, Fauci embraced his deity status, the "Fauci effect" accolades, and represented himself as the embodiment of "science."[52] Was it any surprise that the "Fauci effect" further revealed itself when the politics of vaccination mandates changed one year into the pandemic? No. In an interview with CNN's Jake Tapper, Fauci's response evolved to, "I have been of this opinion, and I remain of that opinion that I do believe at the local level, Jake, there should be more mandates." Unlike his prior quotes against mandates, or science demonstrating the benefits of natural immunity, the updated quotes promoting mandates were readily available at the top of the search engine results. The new dupe-speak became, "There really should be" mandates. Fauci even encouraged the FDA to hasten vaccine approval to facilitate "a lot more mandates." Of course, "This is serious business, so I am in favor of that," Fauci asserted.[53]

It's important to note that Fauci didn't fly solo in his flip-flopping or waffling advocacy for the vaccine mandates. One of the gate guards from decades earlier, the now retired Army pharmacist and Merck employee, also weighed in from the bleachers as another non-operational expert. Regarding the push for children's vaccinations, the old gate guard assured parents, "There is a need, there is a benefit, the safety is reasonable,"[54] This tepid endorsement may not have been reassuring to the moms and dads, though. The pharmacist's venture into the COVID shot discussion only reminded me of his, and other gate guard's, writings about the anthrax vaccine over the previous decades. In both eras they influenced the scientific record, after the fact, to conform to the latest political-medical objectives and policy. These examples illustrated the patterns that connected our anthrax ethical dilemma to the scarily identical pandemic era attempts to alter public perceptions about relevant science through suppression or revision.

Years earlier we witnessed similar trends where researchers from the original 1957 US Army clinical trial for anthrax vaccine published an update on the anthrax vaccine for a chapter in the medical textbook *Vaccines*. Prior to their rewrite, the vaccine was long considered to suffer from safety and efficacy deficits.[55] [1989 DoD letter to Senator John Glenn at Appendix EE]. Even the previous editions of the very *Vaccines* textbook previously documented the "unsatisfactory" nature of the product due to its unknown purity, undefined nature, undesirable constituents, and efficacy issues.[56, 57] However, by 2008 the scientist gate guards assured us "It can be prescribed with the confidence commensurate with dozens of human safety studies and experience in 1.8 million recent vaccines."[58] Of course, those millions of humans were our troops who were illegally forced to submit to the vaccine.

The literary and testimonial smokescreens by Fauci and Daszak, as the gate guards muddying the COVID origins, resembled the methods from our dilemma when dealing with the anthrax vaccine gate guards who hid or altered the truth as the politics of medicine dictated. Their academically suspect efforts blocked intellectual inquiry. Despite the smokescreens, the parallels between the two controversies deserved reexamination. Part of that examination should consider another late in the debate assessment by a longtime DoD biodefense official, Dr. Robert Kadlec. In describing Fauci's behaviors in "directing people away" from the lab leak hypothesis, he used phrases to describe the ploy: it was an "information operation" that resulted in a "cover-up," a "colossal misdirection," while adding other descriptors such as "denial and deception."[59]

With COVID, DoD documents also lent credence to these assessments. The not for profit organization Project Veritas revealed military documents that suggested a lab leak link. Though the DoD and the Defense Advanced Research Projects Agency (DARPA) took credit for a role in the development of COVID-19 vaccines, there was a glimmer of oversight as they rejected participation in GOF/ePPP experiments over safety and gain-of-function moratorium concerns.[60 61] Regardless, the "lifting"[62] of the moratorium on certain gain-of-function research just prior to the presidential transition required further analysis since the accepted risks involved dangerous research while restricting alternative therapies. The DoD's foresight was laudable, but that Americans had to learn from Project Veritas about our government's central role was not.

The DoD documents raised concerns about the COVID inoculations and asserted "SARS-CoV-2 is an American-created recombinant bat vaccine, or it's precursor virus," with EcoHealth Alliance as the creator at the Wuhan Institute of Virology.[63] A complaint that revealed the documents warned that the "DoD rejected the program proposal because vaccines would be ineffective and because the spike proteins being inserted into the variants were deemed too dangerous (gain-of-function)." The author added, "The DoD now mandates vaccines that copy the spike protein previously deemed too dangerous." Like a genuine operator, the whistleblower offered, "To me, and to those who informed my analysis, this situation meets no-go or abort criteria with regards to the vaccines until the toxicity of the spike protein can be investigated."

Other military data released from the Defense Medical Epidemiology Database (DMED) spurred Senator Ron Johnson in February 2022 to write to Defense Secretary Lloyd J. Austin III about significant suspected increases in adverse reaction rates allegedly due to the vaccines. The senator probed if "diagnoses for neurological issues increased ten times from a five-year average of 82,000 to 863,000 in 2021." He also questioned whether "DMED data showing registered diagnoses of myocarditis had been removed from the database."[64] Senator Mike Crapo joined Johnson in efforts to oppose vaccine mandates imposed on our nation's troops and workers.[65]

Particularly given the post-facto—now publicly recognized—attempts by Fauci and others to divert attention from the lab leak conclusion, the hypothesis gained strength. Another whistleblower, Dr. Andrew Huff, worked with Daszak at EcoHealth Alliance and questioned the gain-of-function research. The insider pointed to Fauci for the COVID cover-up, suspected US intelligence agency involvement in the precursor events, and asserted the "manmade pandemic" was due to EcoHealth Alliance's "development of the agent SARS-COV2."[66, 67, 68] The crucial emergence of an internal whistleblower increased the need for comprehensive investigations.

The cover-up worked initially, but over time Americans and the world community read the documents, decoded the dupe-speak, and dissembled the dupe-loop. After observing gate guards for a while my gut pointed me to an Occam's razor answer, where the most obvious explanation is likely. The probability that a research-related infection originated in the Wuhan lab, where Fauci funded research through NIH and Daszak, was the most obvious and likely conclusion. China's parallel military research complicated the effort to uncover the ground truth or shared blame as the lab leak hypothesis moved from conspiracy theory to the most plausible explanation.[69] Given the controversy, it was inexplicable that the US government funded Daszak's research with $826,277 before the pandemic and allocated $653,392 more in 2022.[70, 71]

At a minimum, Fauci did not appear to have been forthright with the American people or Senator Rand Paul—despite being forewarned about the "unusual features of the virus." Colleagues warned Fauci about the sequences and genome that "(potentially) look engineered," and that they were "inconsistent with expectations from evolutionary theory." Even if simple neglect, Fauci did not apparently warn administration officials or the American people or about the laboratory origin concerns, but instead chose to preemptively sidetrack the pertinent body of medical literature and challenge concerned scientists and lawmakers. If the whistleblower warnings and the non-zoonotic hypothesis were true, Fauci deserved an investigation into his conduct and the apparent failure to investigate the underlying cause of the virus' origins.

When deposed as a witness regarding pandemic controversies, Fauci repeated, more than 250 times that he didn't know, didn't recall, or didn't remember. When asked about the work that he funded in Wuhan, he dodged questions, testifying that he was "completely in the dark about the totality of the work."[72] Whether connected to the pandemic origins or not, his omissions and deflections appeared to challenge his core duty to candidly protect public health and the public's trust.[73]

Beyond the quibbling about a lab leak versus a natural pathway, Fauci's past statements challenged his credibility. Fauci declared, "one of the encouraging aspects about the efficacy of the vaccine . . . if you do get infected, the chances are that you're going to be without symptoms, and the chances are very likely that you'll not be able to transmit it to other people." He added, "when people are vaccinated, they can feel safe that they are not going to get infected, whether they're outdoors or indoors." Fauci's "bottom line" goal was "to get people to appreciate you get vaccinated and you're really quite safe from getting infected." Fauci parsed, "There will always be breakthrough infections, but given the denominator of people who are vaccinated, that's a very, very rare event . . . it really is a big . . . endorsement for . . . getting vaccinated."[74] Fauci assured people they would not die by saying, "all three vaccines have proven 100 percent effective."[75]

Even the manufacturer apparently affirmed that the vaccines were never tested to block transmission.[76] In contrast, Fauci insisted that breakthrough infections in the vaccinated were "extremely unlikely," with a "very, very low likelihood—that they're going to transmit it." Fauci incorrectly asserted that the vaccinated are "dead ends"—"preventing the spread of the virus throughout the community"—"the virus is not going to go anywhere."[77] Fauci insisted that if vaccinated, "The risk is extremely low of getting infected, of getting sick, or of transmitting it to anybody else, full stop."[78] Later, Fauci literally served as a human case study for verifying vaccine policy pronouncement contradictions based on his own recurring infections after being inoculated and double-boosting.[79] In the end, breakthroughs turned out NOT to be a "very, very rare event"—NOT with a "very, very low likelihood." No, we realized they were very, very common—Fauci was very, very wrong—Fauci was not science after all, "full stop."

Fortunately, future historians, scientists, doctors, policymakers, legislators, citizens, students, and own our children will have ample opportunities to analyze the "Fauci effect." We'll let them decide. Until then, my goal was to document the arguably negative influence Fauci and others had on science, on a healthy scientific "dialectic," and on trustworthy public health policy. The historic patterns documented by Fauci's bad behaviors in both dilemmas, and the mentalities of gate guards maneuvering and manipulating throughout the anthrax vaccine and COVID shot controversies, demanded reflection. In the future, guardrails must be erected to protect against the bad behaviors and accountability injected to arrest the unhealthy patterns by the bad actors.

MILE 21: Citizen Petition 2 (2021)

The rollout of COVID mandates merged the dilemmas. The intersection between the anthrax vaccine dilemma and the COVID countermeasure mandates inevitably focused on the same fundamentals—safety problems, questionable efficacy, and fake FDA approvals. One way we extracted accountability during the anthrax vaccine travesty was through our citizen petition.

Our 2001 petition proved that the anthrax vaccine license was never finalized. Russ and his librarian helpers verified this fact from the get-go. The FDA agreed, but ignored it, so the federal courts deemed the entire enterprise illegal. One of our colleagues from the anthrax vaccine effort, Dr. Meryl Nass, working with the Children's Health Defense (CHD) Scientific Advisory Committee, asked for assistance in drafting a citizen petition for the COVID shots. I held great respect for CHD's founder, Robert F. Kennedy, Jr. He sent a gracious message years earlier, motivating me in our aligned missions—of course, I could not decline Dr. Nass' request.

> Dear Buzz, Your experience is profound testimony to the power of this cartel and to obliterate our national values, human rights, and public health in pursuit of profits and power. They have helped transform our political landscapes and public health dialogue into a Kafkaesque nightmare. The thing that gives me most hope, in this battle against indomitable power, is finding courageous souls like yours beside me in the foxhole. My uncle, President John F. Kennedy, observed that moral

> courage is a commodity far rarer than physical courage. I'm delighted to meet a template for both virtues! See you on the barricades.

RFK Jr. is the son of the late attorney general and senator, Robert F. Kennedy.[1] Championing human rights and exposing corruption courses through his blood. He, like us, reluctantly waded into the swamp of vaccine oversight. Once he did, he discovered the disturbing cottage industry and the revolving door between government officials, pharmaceutical company top management, and their boards of directors. He acted by creating his non-profit, emulating the lessons from his book *American Values* where he discusses his father's reflection on the stoics. From that book, RFK Jr. embodies the hope that "in an absurd world, the acceptance of pain, accompanied by the commitment to struggle, transforms the most common men into heroes and provides the most tragic hero peace and contentment." Throughout his life of advocacy, RFK Jr. steadfastly emulated the unconquerable task of Sisyphus, condemned to push a rock up a hill for eternity. He wrote, "In applying our shoulder to the stone, we give order to a chaotic universe."[2]

From a Sisyphus-like moment in his own life, he tells the story of when he was dragged kicking and screaming into critically analyzing vaccines, initially due to his focus on helping parents with children suffering from vaccine-based toxic mercury exposures. Like Russ and I, RFK Jr. was never anti-vaccine as our own and our kid's shot records amply prove. Instead, we were anti-illegal-mandate. We were process-oriented citizens who asked our government to listen and to follow the regulatory regimes and rules meant to ensure safe and effective vaccines . We only got smart on the issues once our careers and families were threatened, just as RFK Jr. became an expert when helping his relatives and their friends fight for their medical rights. The COVID shot mandate controversy became particularly relevant to his nonprofit considering half of polled parents were not confident about employing novel EUA technologies on children.[3]

Despite RFK Jr.'s open-minded entrance into what should have been healthy intellectual and academic inquiry, early in the COVID shot controversy the mainstream and social media labeled him in the top tier of anti-vaccine misinformation agents, the "Disinformation Dozen."[4] Digital disparagement tactics do not tend to deter a champion freedom fighter, but instead merely served to energize his intuition that he was on the right track. Like us, RFK Jr. simply wanted the government to follow its own rules when it came to pharmaceutical products, and we were proud to follow his courageous lead. RFK Jr.'s role in the 2020s was analogous to Mr. Perot as a champion and defender from our earlier dilemma in the 2000s. Without influential and honorable allies such as these men, challenging the dis-mal-information would have been insurmountable.

As a democrat, from a family of lifelong democrats, RFK Jr.'s involvement was a litmus test across the political spectrum in the highly polarized vaccine debate. RFK Jr. was a needed replacement on the front lines of the fight against mandates. His photographic memory and grasp of the facts also filled a gap vacated by this story's previous hero, Mr. Ross Perot. RFK Jr. could recall applicable laws and studies, as well as name the names of those responsible for the regulatory capture at FDA. He identified the "negative leadership" origins of the problem, just as Mr. Perot did with the Bush administration on anthrax vaccine. Russ would have been proud of RFK Jr.'s leadership, and we still had our faithful friend JR to provided mutual support as well.

Engaging the propriety of the vaccines this time around troubled me. The angst visited every home and hearth in America. This was a pandemic. Anthrax wasn't. This time the threat was a communicable disease. Anthrax wasn't. Public health officials attempted to field a tool to address the threat. So why the resistance? Why did we, like Sisyphus, join forces to push against the insurmountable task, but without regret? Four dimensions explained the predicament.

First, the novel technologies in the vaccines raised concerns. Second, the mandates on the population struck unharmoniously against fundamental beliefs in liberty and medical freedom for those citizens who were unwilling to barter inalienable rights for a transactional work permit. Third, government officials once again misinterpreted statutes and broke the law in order to coerce those citizens and sucker businesses to mandate compliance. Fourth and finally, when hesitancy emerged, officials sowed division and fostered contempt to bully the population.

Some shrugged off the term "hesitancy" as insulting, perceiving it as "patronizing and presumptuous."[5] I wasn't worried so much about the terminology as I was wary of the tactics, the underlying science, and the policymakers. The actual hesitancy more aptly applied to our citizen's reluctance to trust the NIH, FDA, and CDC. Because the policy choices warranted cautious deliberation by those agencies, it was reasonable to expect that a citizen petition would help compel them to follow their own rules. Doing so might preclude the very vaccine hesitancy concerns the government paradoxically created through its history of violating health rights.

From the get-go, I promised my wife to keep my role in the background this time around. She endured decades of my pacing and listening to me refine and reiterate my talking points. I promised my wife a different strategy this time, one that preserved my unyielding endurance and her marathon of patience. In the beginning, I stayed behind the scenes with my help. Dr. Nass put me to work. One of Dr. Nass' longtime colleagues and a holocaust survivor, Vera Sharav, joined our crew.[6] Vera personally understood the "darkest of dark" times and sagely warned us when "you have to stop the train." Collectively our group churned out a citizen petition to contest the FDA's actions on COVID inoculations. RFK Jr. and Dr. Nass were our "petitioners."

We cited all the appropriate legal authorities in asking the FDA to issue, amend, revoke, or refrain from taking several administrative actions related to emergency use authorizations for COVID inoculations.[7] The core mechanism came from Russ's 2001 example, leveraging 21 CFR §10.30.[8] That regulation allowed any citizen to request the Commissioner of Food and Drugs to issue, amend, or revoke a regulation or order, or take or refrain from taking any other form of administrative action. The highlights of our new petition included a plea to revoke or refrain from approving any future EUA for any COVID shot where the risks of serious adverse events or deaths outweighed the benefits of the vaccination. We asked the FDA to review the evolving VAERS safety surveillance data, amend EUA advertisements, and review the extent to which vaccine benefits and risks were unknown, all in accordance with 21 USC §360bbb-3.[9]

Given the evolving efficacy data, we recommended the FDA amend efficacy indications on all digital and written guidance related to EUA COVID inoculations as "probably" effective or "possibly" effective in accordance with the labeling guidelines published under 21 C.F.R. §201.200(e)(1).[10] Though nuanced, we insisted the FDA decisively distinguish the EUA "authorized and allowed" regulatory status for Americans, as opposed to the incorrect and misleading "approved, safe, and effective" language repeated widely. Most importantly, RFK Jr. and Dr. Nass' petition presciently warned against the testing of COVID shot on minors, a salient vision unfortunately wholly ignored by federal, state, and local public health establishments.

Equally visionary, and similarly disregarded, the petition suggested advanced advisements to the DoD regarding potential mandates. Reminders about the legal requirements and the "option to accept or refuse" the vaccines with "no penalty or loss of entitlement" proved to be another disregarded effort to save jobs and preserve military readiness. We emphasized the 2005 DoD EUA[11] precedent for anthrax vaccine that guaranteed the optional or voluntary nature of the inoculations, as did 10 USC §1107a. That law's genesis was the 2004 National Defense Authorization Act (NDAA)[12] and a parallel civilian law, 21 USC §360bbb-3(e)(1)(a)(ii)(III).[13]

We cited CDC VAERS data,[14] which already reflected historically unprecedented adverse events. The underreported data demonstrated the adverse events and deaths far exceeded all deaths from all vaccines over the entire thirty years of government tracking data—over 1.6 million adverse events, almost 37,000 deaths, and close to 69,000 disabilities and counting.[15] The dramatic statistics grew worse considering reporting to VAERS passive database included only 1 to 13 percent of known adverse reactions. Earlier government efforts to evaluate VAERS reliability affirmed the underreporting.[16] Extrapolating these raw numbers over time caused alarm.[17] Since the thresholds already exceeded historic red flag limits, we asked the FDA to revoke the EUAs where the risks of serious adverse event or death outweighed the benefits.

Though we did not highlight it in our petition, other international databases showed similarly alarming statistics, such as the worldwide database managed by the WHO Collaborating Centre for International Drug Monitoring in Uppsala, Sweden. The WHO's collaborative drug monitoring organization recorded over five million potential side effect reports as of December 2023.[18] A preprint extrapolation study also revealed an "excess risk of serious adverse events" for the Pfizer and Moderna mRNA COVID inoculations.[19] The findings warranted scrutiny in light of the rapidly accelerating global vaccination numbers, equating to fourteen billion doses in over 70 percent of the world population—about six billion people.[20]

As time wore on the original concerns grew worse as excess deaths increased, with allegations the COVID inoculations represented a "rapacious" versus a noble lie.[21] The science indicated the "mRNA vacs [vaccines] dramatically increase inflammation on the endothelium and T cell infiltration of cardiac muscle and may account for the observations of increased thrombosis, cardiomyopathy, and other vascular events following vaccination."[22] Studies found risk following mRNA-based COVID injection was "strongly associated with a serious adverse safety signal of myocarditis," was "223 times higher" compared to the "average of all vaccines combined for the past 30 years," and "represented a 2500% increase in the absolute number of reports in the first year of the campaign when comparing historical values prior to 2021."[23] Still other studies documented "hidden cardiotoxic effects."[24, 25, 26, 27] Finally, alarmingly, autopsy findings indicated death due to cardiac failure "can be a potentially lethal complication following mRNA-based anti-SARS-CoV-2 vaccination" and other autopsy studies found "a high likelihood of a causal link between COVID-19 vaccines and death from myocarditis"[28] These findings added context as to why autopsies were discouraged or omitted altogether.[29]

DoD affiliated doctors, including the anthrax vaccine truth-teller, Dr. Renata Engler, acknowledged that heart inflammation problems were coincident with COVID vaccination in previously healthy troops.[30] One hypothesis suggested the shots triggered "mRNA COVID-19 vaccine-induced myocarditis and potential increase in sudden deaths among elite athletes."[31] Studies hypothesized that mRNA vaccines-induced myocardial inflammation, particularly for male athletes that may have been triggered by an adrenaline induced adverse reaction,[32] with a "low prevalence" of myocarditis associated with actual COVID-19 infection.[33] As evidence mounted about the shots potentially affecting heart function adversely,[34] brave top athletes refused to be casualties, objected to the mandates, and spoke out about the unapproved shots.[35]

Some predicted surveillance will "clarify the long-term effects of these experimental drugs and allow us to better assess the true risk/benefit ratio of these novel technologies."[36] A former BlackRock managing director, Edward Dowd, estimated a 23 percent increase in excess deaths, and 1.36 million disabilities in younger vaccinated populations coincident with 2021 COVID vaccinations and boosters.[37, 38] Dowd recommended a fraud investigation related to clinical trial data and alleged FDA complicity.[39] He claimed the "the rate of change is the smoking gun."[40] The insurance industry Society of Actuaries published data that corroborated Dowd's theory.[41] Over time, the cautions advised in our petition gained validation through UK data showing a 20 percent increase in excess deaths and increased rates of hospitalization and death in the vaccinated populations, which exceeded that of unvaccinated citizens by 200 percent or more.[42] Data from government statistics verified all-cause mortality up 100 to 300 percent.[43] Canadian data also suggested higher death risks for vaccinated versus unvaccinated individuals.[44]

Our petition warnings to the FDA preceded widespread awareness of the growing "vaccine breakthrough case" phenomenon where vaccinated subjects still fell ill and transmitted the virus. CDC marginalized this reality as a "small percentage," but over time breakthroughs constituted the majority of the spread.[45, 46] The draft petition insisted that drug labeling comply with regulations and to adopt the FDA's "probably" or "possibly effective"[47] indications based on breakthrough cases. Yet while admitting waning efficacy, Fauci and others twisted the evolving science into justifying boosters by claiming, "The proof of the pudding will be after you get people vaccinated and boosted, and we have a greater durability of protection that doesn't wane as easily."[48] But that didn't happen, and in fact the "proof" they failed to acknowledge was the fact that natural immunity protection was "higher than that conferred by vaccination."[49]

As the months dragged on, and the COVID shot efficacy waned, increased breakthrough cases and death statistics resulted in the government altering the long acknowledged definition of a vaccine. According to the CDC, the definition of the term "vaccine" changed from producing sterilizing "immunity" to instead providing merely an immune "response" or "protection."[50] Many were shocked to witness the reactionary new definitions from the CDC and by the FDA due to the scientific realities of breakthrough infections. COVID-19 "vaccines" did not live up to the previous definition, so they just changed it for these "medical countermeasures," or MCMs according to the FDA and CDC websites and documents—making science conform to policy.[51]

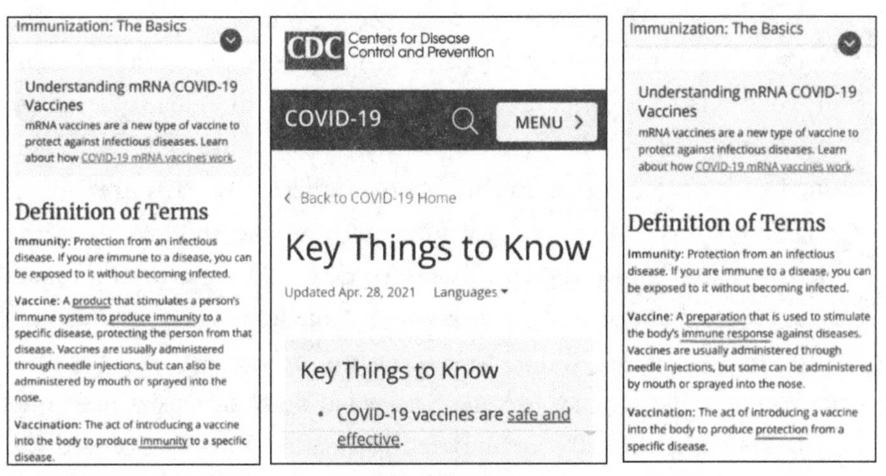

CDC's pre-COVID definition of "produce immunity," vs. "immune response" and "protection".

We also called the FDA's attention to the important misbranding rules to ensure the government, doctors, and providers didn't confuse EUA allowed injections with fully approved biologic products in violation of 21 USC §355(a) & §352 & §331(a)(d). We reminded the FDA of their own language for approving an injection as EUA, that "it is reasonable to believe that the known and potential benefits of Pfizer-BioNTech COVID19 Vaccine . . . outweigh its known and potential risks."[52] We asked for the public to receive the same "reasonable to believe" candor on potential benefits versus risks, instead of the FDA's misleading "safe and effective" soundbites.

Our petition hoped to ensure compliance with the same statutorily required "safe and effective" language used internally at the FDA—the same legalese sent to manufacturers—to avoid misleading the American people and nullifying their rights for fully informed consent. We also sought to encourage multi-drug treatments that proved promising for prophylactic and early care. The problematic aspect of alternative therapies involved the regulatory conflict where EUAs for the inoculations required "no adequate, approved, and available alternatives."[53] For the FDA to allow alternative treatments appeared to disallow the COVID injection EUAs.

Many medical professionals and elected officials objected to the disapproval of early treatment with alternative therapies. One example from 2020 related to three US senators asking for FDA clarification on the previously approved EUA for hydroxychloroquine (HCQ).[54] This seemed to be the most politically charged alternative early treatment, but ivermectin (IVM) served a similar purpose and suffered the same criticism as an effective therapy. The purpose of FDA approved off-label use of these products under the practice of medicine, or the repurposing the drugs, was to inhibit the replication of SARS-CoV-2. Studies demonstrated "ivermectin use was associated with decreased mortality in patients with COVID-19 compared to remdesivir,"[55] and proved the "most effective drug candidate," with the "best result," when compared to other therapies.[56] Courts concurred concerning the FDA's overreach, finding, "FDA is not a physician. It has authority to inform, announce, and apprise—but not to endorse, denounce, or advise."[57, 58]

Our petition encouraged FDA to expeditiously evaluate existing research that demonstrated "statistically significant reductions in mortality, time to clinical recovery, and time to viral clearance."[59] At a minimum, the petition asked that off-label use not be discouraged. A renowned cardiologist, Dr. Peter McCullough, surmised that 85 percent of hospitalizations and mortality during the pandemic were avoidable with widespread awareness and support for these sorts of early multidrug treatments. Dr. McCullough was outspoken on the superiority of natural immunity and the lack of medical necessity to vaccinate previously recovered patients.

Dr. McCullough categorized four domains of spike-protein-induced harm from mRNA injections: cardiovascular, neurologic, thrombotic, and immunologic; while other scientists referred to the pathogenic toxicity of the mRNA products as "Spikeopathy." His concerns about the "crisis of compassion," the need for a "self-check in the house of medicine," and the "great controversy" of "grand deception" ailing modern medicine that echoed our petition's points for the FDA to resurvey.[60] Dr. McCullough and others eventually called for a global moratorium on COVID-19 mRNA product use due to well-documented serious adverse events that resulted in an "unacceptably high harm-to-reward ratio," as well as unanswered "questions pertaining to causality, residual DNA, and aberrant protein production."[61]

The petition also challenged the FDA to resurvey and compare historic data points such as the mid-1970s swine influenza vaccine experience. "A proven association between the 1976–1977 swine influenza vaccine and Guillain–Barré syndrome (GBS) halted that particular national vaccination campaign"[62] due to adverse events and deaths that barely scratched the reported deaths and GBS related demyelinating disorders reported after COVID shots. An example included a Special Forces Green Beret doctor, Lt Col Peter Chambers.[63] These controversies touched even closer to home when a neighbor suffered demyelination, disability, and death following a COVID-19 injection. Brave military doctors, such as Major Samuel Sigoloff and Lt Col Theresa Long, saw the signals and rang COVID shot alarms along with Dr. Chambers.[64, 65, 66]

A report by the National Academy of Sciences documented key lessons learned from the Swine Flu Vaccination Program to include: overconfidence, preexisting personal agendas, failure to address uncertainties, insufficient questioning of scientific logic, among others.[67] Regarding the halted Swine Flu Vaccination Program, the CDC's Emerging Infectious Diseases concluded, "In 1976, the federal government wisely opted to put protection of the public first."[68] I hoped the FDA might reflect on these historic lessons, listen to physicians like Chambers, Sigoloff, and Long, and place the protection of the public first. Given the suspected unacceptably high adverse reaction rates, FDA should have "wisely opted" to "resurvey its judgments."

We included an action to protect prior consent rights for service members in anticipation of future FDA approvals. The troop's protections for medical choice actually exceeded that of the general population due to the precise language in the law written for them based on the poor record of Gulf War experimentation. An exception included circumstances where the president waived consent requirements.[69] According to 10 USC §1107(f) the president may order such a waiver if the president determines in writing that obtaining consent is not in the interests of national security in connection with a member's participation in a "particular military operation."

Like Title 21 for civilians, the military law for EUA inoculations was codified as 10 USC §1107a.[70] The §1107a language was similar to §1107(f) in order to ensure our troops were granted prior consent and had the "option to accept or refuse administration of a product." Prudently, no presidential waivers of informed consent under 10 USC §1107(f) or 10 USC §1107a occurred. Precariously, if the president had signed a prior consent waiver for the troops such an action would have undermined the propriety of subsequent FDA approvals for COVID shots. Given the later military mandate, a presidential waiver could have legally justified a shot mandate and signaled that the DoD actually learned lessons from the anthrax vaccine experience.

Regardless, over the early months of COVID-19 shot availability, I found myself feeling proud of military leaders for exercising restraint in not prematurely forcing the injection mandate on the troops. Our nation's troops are as intelligent as they are courageous, and certainly don't qualify as a vulnerable population. My hope was the nation's military leaders would honor and respect their troop's prior consent rights—but the DoD's legally compliant policy didn't last.

Just as the military initially honored the statutes governing mandates of unapproved biologics, EUA law was crystal clear that COVID injections for all citizens should have been optional as well in accordance with 21 C.F.R. §360bbb-3(e)(1)(a).[71] Yet throughout America, and in literally all sectors of society, from schools, businesses, government and industry, coercive tactics became the order of the day to first encourage, incentivize, and threaten, then to make COVID vaccination a condition of employment, education, or day to day life activities.

It is unlikely most Americans would be content with mandates for EUA injections if the government candidly informed them—as required by law—that the EUA COVID shots were authorized for emergency use only, and technically were, in fact, "unapproved medical products."[72] A citizen's right to participate voluntarily in a medical trial, without penalty or loss of benefits, was concurrently defined in 21 CFR §50.25 and 45 CFR §46, the regulations covering the "Protection of Human Subjects" and "Informed Consent."[73, 74] That regulation defined "Basic elements of informed consent," and made clear that "participation is voluntary, that refusal to participate will involve no penalty or loss of benefits to which the subject is otherwise entitled, and that the subject may discontinue participation at any time without penalty or loss of benefits to which the subject is otherwise entitled." Additional elements of informed consent included a medically related caveat about the "consequences of a subject's decision to withdraw from the research and procedures for orderly termination of participation by the subject." Regardless of whether the EUA use of COVID shots was considered research, the context of "consequences" was relevant.

21 CFR §50.25, from the Code of Federal Regulations regarding informed consent, bolstered the reality that EUA products could not be mandated due to the commonality of the language between 21 CFR

§50.25 and the EUA law under 21 USC §360bbb-3. The wording was unmistakable. Yet the equally voluntary context in the regulation, the statue, and the Federal Register was ignored or omitted from government opinions on the matter. The regulation and statute both included the optional choice or voluntary nature of participation, as well as "consequences," as did the DoD EUA precedent for anthrax vaccine in the Federal Register.[75]

"Consequences" in the CFR was specifically related to "a subject's decision to withdraw" from the research or clinical trial requiring informed consent, effectively a medical consequence. The informed consent CFR specifically stated that "participation is voluntary," and the language matched the DoD 2005 EUA precedent for anthrax vaccine that upheld the "no penalty" intent of the law. This is literally the only relevant application of the word "consequences" outside of 21 USC §360bbb-3, and its context had nothing to do with loss of job or loss of entitlements and benefits. In actuality, the law and precedent barred such punitive actions. Though the Cures Act provided waivers for informed consent, nothing overrode optionality or permitted mandates.[76]

The government unmistakably ignored and omitted this historically, statutorily complete, and technically accurate interpretation of the word "consequences" from the CFR's. Instead, they utilized that singular undefined term, "consequences," from the EUA statute to leverage loss of livelihood as a mechanism to achieve increased acceptance of unapproved EUA products. The EUA law for unapproved products affirms voluntary conditions by stating, "the option to accept or refuse administration of the product." The statute adds mention, but unlike the CFR, without clarification about the "the consequences, if any, of refusing administration of the product." Based on this undefined caveat, the original intent of the voluntary nature of the EUA law was contorted to support mandates. The entire mandate scheme hinged on an improperly interpreted word—"consequences." The misinterpretation equated to patently illegal EUA mandates.

The vagueness of the statute regarding this singular unclear word resulted in misinterpretations that caused vast negative implications and job loss nationwide. Mandate legality misinterpretations patently contradicted the CFR language on informed consent, as well as the DoD voluntary EUA precedent from 2005. Nowhere in Title 21 covering an EUA is job loss or loss of entitlement allowed, but instead it is specifically forbidden in the CFR on informed consent. The government's and the DoD's position unconscionably defied the first ever EUA precedent with the DoD for anthrax vaccine in 2005 that guaranteed "no penalty or loss of entitlements."[77] DoJ counsel even advised that if the DOD imposed a mandate they "should seek a presidential waiver before it imposes a vaccination requirement."[78] DoD brazenly imposed a mandate, but sought no waiver, most likely because to do so defied civilian mandate legality.

Even a Supreme Court of the United States precedent, *Jacobson v. Massachusetts*, 197 US 11 (1905), was often referred to as "settled law" with respect to mandates.[79] In reality, it applied to state police powers under the enumerated authorities of the Constitution. That case only affirmed a small fine levied as penalty for noncompliance with an immunizing smallpox vaccine. As with 21 USC §360bbb-3, nowhere in the precedent case was a loss of livelihood included as a penalty. Inexplicably by 2021, "penalty and loss of benefits" became the legal standard and misinterpretation that ignored over a century of human subject protection.[80]

In filing the citizen petition we understood that, at various levels and sectors of our society, vaccine mandates survived judicial scrutiny. Nevertheless, mandates were never intended for EUA unapproved medical products, despite some unchecked state, local, private entity, or subordinate federal agency attempts to contort the law on this matter. Further, the Preemption Doctrine[81] or Supremacy Clause of the US Constitution, Article VI., §2,[82] supported the law's "option to accept or refuse" consent requirement for EUA products as superseding all lower-level mandates. This is likely why federal officials initially prudently shunned mandates.

The option to refuse EUA COVID injections is codified in the law, and early on President Joseph Biden supported the same by stating, "I don't think it should be mandatory. I wouldn't demand it to be mandatory."[83] Based on this, we assumed the federal government would concur with our petition's request to inform the American people of their legal rights and the option to accept or refuse administration of unapproved COVID injections. The law ensured no adverse work, educational or other non-health consequences, per 21 USC §360bbb-3(e)(1)(a)(ii)(III).[84] We thought the FDA would support, defend, and uphold the federal laws that govern biologics.

Russ would be proud to know we inserted his intellectual discoveries in the petition. Russ always pounded, so we reminded the FDA, that their mission is "protecting the public health by ensuring the safety, efficacy, and security of human and veterinary drugs, biological products."[85] Russ was big on mission adherence. We also clipped some themes from Russ's earlier works of art from the 2001 citizen petition by including his teaching points about President Roosevelt's signing of the Federal Food, Drug, and Cosmetic Act. It closed many drug safety and efficacy legal loopholes, and forever altered the landscape of biologic consumer protection in America.[86]

Russ always hammered the importance of the 1962 Harris-Kefauver amendment,[87] which set in motion the regulatory standards for licensure with proven efficacy. A 1972 mandated review of biologics sought to ensure proof of efficacy and to make sure that vaccines were not misbranded. I did my best to continue Russ's work by pointing out the dilution or adulteration of drug law safety and efficacy requirements, particularly with less stringent standards for EUA products.[88] The historical

Russell E. Dingle with his wife, Jane, and daughters, Megan, and Emma.

footnotes required reflection, and so the petition provided these reminders to the FDA. Russ's greatest contribution was his emphasis on the FDA's Preamble: "The importance to the American Public of safe and effective vaccines cannot be understated."[89]

The new petition echoed Russ's themes and his legacy: "The FDA erred with the anthrax vaccine, and it took a Citizen Petition and federal court decision to make the FDA comply with the FDCA . . . With this Petition, we look forward to the FDA's appropriate, tough regulatory action to bring its COVID vaccine regulations and guidance into line with federal law."

Russ's yeoman work two decades earlier helped RFK Jr.'s team to insist that the FDA ensure "American public receives only safe and effective vaccines." The petition warned the FDA that "most Americans are not aware of the strict compliance requirements for EUA COVID vaccines, nor do they know that these biologics are 'investigational' or 'unapproved medical products.'" The misinformation fell squarely on the FDA's shoulders due to the failures to "provide and enforce accurate public messaging." Our bottom line warned the FDA that "reversing this trend is imperative; the FDA must comply with law." We emphasized, "Acting on this Citizen Petition will enhance the FDA's credibility with the public." I knew Russ would be proud of us for following his lead, and we all hoped to be proud of the FDA as well.

Aside from the petition, RFK Jr. battled on, writing his exposé on Dr. Fauci and the "bewildering cataclysm that began in 2020" with the pandemic. His book, *The Real Anthony Fauci*, described a "collective pivot" by the establishment away from regulatory due diligence and stringent oversight. RFK Jr.'s father and his uncle, Senator Teddy Kennedy, ensured this establishment had the regulatory and legal tools to protect the public health from bad medicine. But these powers used their helm to "generate fear," "promote obedience," "discourage critical thinking," and undertake "mass public health experiments." The book unfolds the "well-worn tactics of rising totalitarianism," "manipulation of science," and "vilification of dissent," with the hope that someone with influence will reverse the "collective pivot" and the bad behaviors.

RFK Jr. held Fauci to account since he was the one at the controls when the government and its proxies "orchestrated gaslighting" and marginalization through "draconian diktats." This life-long democrat and public health advocate was a worthy adversary. With his nonprofit, this petition, and books, RFK Jr. joined the multitudes of courageous doctors, scientists, and nurses worldwide who exhibited the fortitude to stand up to the injustices and unhealthy orthodoxies. His example served as a light of hope against the "indentured servants on Capitol Hill" acting as "sock puppets" for the very industry they swore to oversee.[90] RFK Jr.'s revelations about the upturned world we witnessed required strong words in telling truth to power.[91]

Ultimately, the FDA ignored our petition in 2001 only to later suffer multiple adverse federal court rulings declaring the anthrax vaccine mandate illegal. Likewise, the FDA ignored RFK Jr.'s and Dr. Nass' petition warnings in 2021 only to later have all mandates similarly halted and rescinded due to their illegalities and unpopularity. Our teams filed both petitions out of a sense of citizen duty as the law encourages. Like 2001, the government should have listened to RFK Jr. in 2021 rather than laying siege to our rights. Had the FDA resurveyed their judgments it would have spared significant national embarrassment and an enormous loss of public trust.

We were lucky to have RFK Jr. as our champion in the pandemic—an all too rare American leader willing to join citizens on the "barricades" for medical freedom. Like Perot, RFK Jr.'s 2024 presidential bid represented his commitment to the nation to fortify America's leadership gap, fearlessly speak the unvarnished truth, and dissemble the regulatory capture of government.[92] Russ and Ross Perot would be proud to know this new "integrity system hero" joined our formation in 2021, the same formation those heroes launched two decades earlier.

MILE 22:
Natural Immunity? (2021)

The natural immunity dialogue goes to the core of a germ theory versus terrain theory discussion.[1] Simplistically, germ theory is about external threats and is offensive in nature. Think about it as using external countermeasures, or shots, to fight off germs. Terrain theory is more defensive in nature and how our bodily natural terrain's immune system can protect us. It also speaks to a metaphorical view of the institutional approach of our government and the DoD in mandating countermeasures as opposed to emphasizing the innately natural biological approach to combating threats through the power of our immune systems. In practice both germ theory and terrain theory are useful, as opposed to focusing on one without consideration of the other.

I developed my own anecdotal opinion about natural immunity and terrain theory versus vaccine-centric germ theory "protection" after catching COVID from a vaccinated pilot who suffered from a breakthrough case. In July, after returning home from a cargo trip to Asia, my employer called me for a health check in. The checkup from my chief pilot boss advised me, "You've been exposed." He asked, "How are you feeling?" Coincidentally, that morning I began experiencing mild flu-like symptoms. Though my complaining to my wife began that morning, unfortunately she gave me a nice welcome home kiss the day before, so it wasn't just me at risk.

The supervisor informed me of the suspected exposure and asked me to get tested. I asked him which pilot got sick. He couldn't reveal that private medical information, which I respected. I said, "I think I know." One of the pilots told our crew a couple of days earlier as the trip ended that he didn't feel well, but assured us all he was "vaccinated," so no worries. I didn't worry at the time, but I did subsequently contact that colleague to see if he was indeed the pilot that tested positive. Sure enough, though double Pfizer injected pilot tested positive and transmitted the virus to his wife and me, and now from me to my wife as well. I was grateful to my company for their proactive contact tracing effort, as well as to my fellow pilot for his candor. I tested positive that afternoon, and my wife did as well a couple of days later.

SARS COV 2 RNA (COVID-19), QL, RRT-PCR, RESPIRATORY SPECIMEN 07/20/2021 (#196350)

Report	Result	Ref. Range	Units		Status	Lab
Result	positive					

Positive SAR COV 2 RNA PCR Test.

The days that followed dragged on. After it was over, I explained to my friends and family that it wasn't the worst virus I experienced, but it was absolutely the longest and the most exhausting. As we researched our illnesses and managed our recoveries, what we found fascinatingly missing from our government public health officials at all levels was any emphasis on immune system maintenance. Over the years prior my wife had emerged as a militant vegan, so her indoctrinations—and my desire to be fed—made me sensitive to our health choices as a backdrop to all the subsequent pandemic polemics. The good diet, low to no animal product, was vital. But according to the government, the only thing that seemed to matter was the shots. Not diet, not exercise, not readily available proven alternative therapies, or overall healthy lifestyles, but instead the shot once again became the quick fix panacea that would save us from the virus.

We both figured my wife's recovery would outpace mine, but in fact, we both agreed after it was all said and done that I recovered more quickly. The only difference in our recovery regimen was sleep. I slept eight to ten hours each night and had a three to four hour nap each afternoon. She did chores every day and per her normal routine only gets about seven hours of sleep nightly, and that's only if the doggies permitted the rest. Our anecdotal conclusion found that the regenerative benefits of sleep were very real. Truth be told, it baffled my wife since I like my candy, and I'm only vegan-ish. I value minimizing animal products in my diet, but I'm not a disciplined vegan like my partner. Neither of us suffered any comorbidities, so ultimately our recoveries were squarely in line with statistical expectations. But that doesn't mean anyone should sneeze at the disease. It was serious, and must be dealt with by respecting the threat.

Both of our doctors believed in the benefits of ivermectin and prescribed it. We both continued our pre-exposure prophylactic daily regimens of vitamin C, zinc, quercetin, melatonin, and vitamin D3. Were it not for the healthy diet my wife swears by, I'm not sure what our outcomes might have been. It exhausted us, as marathons do, but we survived. Honestly, you do tend to second-guess yourself about whether or not you should have been vaccinated. We were even reluctant to talk to friends and family about the illness until after we recovered due to the societal pressures and prejudices over the injections. Naturally, the last thing you want is to be one of those cases where the news or social media mockingly broadcasts your un-injected status.

Just as we did not advertise our own health choices in our social media footprint, we also did not chest thump our survival. In fairness, it could have gone either way. We had acquaintances that were not so lucky. In retrospect, I felt fortunate that our own humoral immune response conquered the virus, but I was humbled by the fact that many were not so lucky. For those in vulnerable populations, like two of our parents that lived with us, we didn't argue against the hopeful value of the shots when they chose to get inoculated.

My wife and I kept an eye on the development and approval of other medical alternatives. Protein-based injections, without the mRNA technology, such as Novavax, were on our radar scopes, but were also EUA. Until July 2022, those non-mRNA alternatives remained unauthorized.[2]

Understandably, citizens felt cornered by the mRNA monopoly. Even once allowed under EUA, the option to accept or refuse the unapproved injections, such as Novavax or any of the mRNA unapproved medical products, should have been our choice according to the law. Due to our natural immunity, injection of any of the unapproved shots was not the priority due to our "sustained immune response" that should have lasted for "up to 1 year after primary SARS-CoV-2 infection."[3] But as mid-fifty-year-olds, we kept an open mind in balancing our informed consent rights against the full pallet of medical options.

As a pilot, I also remained cognizant of government aviation medical rules under 14 CFR §61.53, which specifically prohibited operating an aircraft if I suspected any "medical deficiency."[4] This rule required that I guard against the illness, but also against possible and unknown adverse reactions to the available unapproved medical products, such as the EUA injections. The rules warned about taking any "medication" or "treatment for a medical condition that results in the person being unable to meet the requirements for the medical certificate necessary for the pilot operation." Plus, the Federal Aviation Administration's (FAA's) rules advised designated Aviation Medical Examiners (AME's) "to not issue airmen medical certificates to applicants who are using . . . medications FDA (Food and Drug Administration) approved less than 12 months ago." The FAA added that it "requires at least one-year of post-marketing experience with a new drug before consideration for aeromedical certification purposes."[5] The FAA rules impacted my own and many pilot colleague's medical decisions.

The FAA's rules applied to all EUA "vaccines." In addition to ignoring post-marketing surveillance guidance, the FAA appeared to misinform their AME's that the COVID "vaccines" were "approved" instead of the fact that all the available injections were EUA.[6] The FAA's guidance stated, "FDA approved vaccines are acceptable," and therefore simple deductive reasoning made clear that unapproved EUA injections were 'unacceptable.' Only EUA shots were available. Despite the FAA's contradictory guidance, I rechecked my qualitative and quantitative antibodies, and my biases. My SARS-COV-2 IgG and IgM antibody results remained robust and my biases were in check.

The irony of contracting the virus from a "vaccinated" pilot was only to be topped a few months later after being exposed and quarantined due to identical circumstances. Again, my company barred me from flying for a period due to exposure to another vaccinated pilot's breakthrough case. This time around, after the subsequent exposure I tested with even higher quantitative antibodies. It appeared that my natural immunity worked. My doctor, who is a retired USAF F22 pilot and flight surgeon, said, "I do think your natural immunity kicked in."

The F22 pilot and friend, Dr. Jay "Bones" Flottmann, as well as my local primary care nurse practitioner, made me feel lucky to receive care from medical professionals that didn't get caught up in the politicization of medicine and groupthink. Both prescribed ivermectin and both were firm believers in medical choice. The upshot was, we experienced multiple exposures from breakthrough cases and developed robust natural immunity. Why mess with millions of years of human evolution and jeopardize our natural immunity with injections that the FDA agreed were unapproved medical products authorized under an emergency? Multiple studies also affirmed this conclusion, as well as the fact that "people who have recovered from COVID-19 might not benefit from COVID-19 vaccination," due to "increased adverse events following vaccination."[7]

Other advocacy groups publicly took a more firm and critical approach, particularly on the natural immunity and mandate issues. Based on the evasiveness and lack of transparency by the CDC and FDA on these controversial subjects, many groups cast strong language to describe the government-medical-biopharma syndicate. Terms included: authoritarianism, patriarchal arrogance and abuse, and paternalistic nihilism. One group, ICAN, the Informed Consent Action Network, struck out particularly hard against the CDC when the "vaccinated" breakthrough cases escaped all restrictions compared to the unvaccinated naturally immune. ICAN insisted CDC lift discrimination against the naturally immune and forced disclosure of the CDC's V-safe[8] data.[9]

ICAN provided the CDC with sixty-plus studies. The studies demonstrated natural immunity as longer lasting and superior to EUA mRNA shot immunity. The CDC produced one study in comparison that ICAN charged to be irrelevant, particularly considering the weight of the studies supported the robust power of natural immunity. Natural immunity was proven to be durable, hopefully preventing infection and transmission.[10] In contrast, EUA shot immunity waned. Recipients suffered recuring infection and became transmission vectors for the disease.

ICAN was not alone. CHD and other nonprofit and health organizations chimed in. The Brownstone Institute in particular began coalescing the peer reviewed published literature on the topic. They began by citing 79, then 81, then 128, then 150, and finally over 160 studies showing the advantages of natural immunity over the COVID injection's protection.[11] [12] Another reputable organization, the American Association of Physicians and Surgeons (AAPS), joined these organizations by offering alternative viewpoints other than the government's prescription, as well as guidance on multidrug early home treatment entirely missing from the CDC's remedies.[13] In contrast, the CDC appeared to deliberately withhold data that didn't fit what critical observers perceived to be a politicized pro-COVID-injection counterfeit narrative.

For many months, the CDC withheld data on infections and recovery. Finally in November 2021, the center published updated estimates of aggregate infections. The number of 146 million Americans equated to 44 percent of the population.[14] Between people in America partially (81.4 percent) or fully (69.5 percent)[15] vaccinated, and those that previously recovered from the virus, the nation would very soon be at or over herd immunity levels. The superiority of natural immunity deserved study according to Dr. Marty Makary from the Johns Hopkins School of Medicine, citing an Israeli study showing acquired immunity to be twenty-seven-times more robust.[16]

The longer the pandemic lasted the more this data inevitably favored herd immunity and the statistical weight natural immunity added to that equation. A San Francisco study added credence to the body of evidence favoring natural immunity by stating, "Overall, fully vaccinated cases were significantly more likely than unvaccinated cases to be infected by resistant variants." The study noted, "In contrast to previous studies, we found that vaccine breakthrough infections are more likely to be caused by immunity-evading variants as compared to unvaccinated infections." The study also focused on the immune escape related concerns by concluding, "The predominance of immune-evading variants among breakthrough cases indicates selective pressure for immune-resistant variants locally over time in the vaccinated population concurrent with ongoing viral circulation in the community."

The San Francisco study also warned, "Our longitudinal antibody analyses show that vaccine breakthrough cases are generally associated with low or undetectable qualitative and neutralizing antibody levels in response to vaccination."[17] The diminishing neutralizing antibody findings created concern regarding longer-term efficacy of the inoculations and their boosters. Even the World Health Organization questioned booster efficacy based on the "evolution of the virus" in questioning that "a vaccination strategy based on repeated booster doses of the original vaccine composition is unlikely to be appropriate or sustainable."[18] These findings supported other studies revealing the shot's "negative" efficacy,[19] that "immune function among vaccinated individuals . . . was lower than that among the unvaccinated individuals," and that "booster shots could adversely affect the immune response."[20] [21]These immune erosion and antibody evasion anomalies required review prior to mandates.[22] Other troubling signals, such as DNA repair impairment, suppression of innate immunity, and immune imprinting demanded more study.[23] [24]

Beyond scientific concerns, lawsuits filed by state attorney generals against the injection imperatives claimed the government exceeded its constitutional authorities and specifically raised the natural immunity dispute. One filing declared, "Defendants . . . have not addressed . . . why natural immunity should not be considered an adequate alternative to vaccination." The state attorney generals argued, "Nor would it be possible to reasonably explain away this omission; by all indications, natural immunity confers superior resistance to COVID-19 than any of the currently available vaccines, and one in three Americans had COVID by the end of 2020."

The suit cited *Science* author Meredith Wadman's analysis that concluded, "Having SARS-CoV-2 once confers much greater immunity than a vaccine—but vaccination remains vital." The author referenced the Israeli study that established natural immunity provided better protection against the Delta variant of coronavirus than vaccinated protection.[25] I always like to go to the actual study, which concluded, "This study demonstrated that natural immunity confers longer lasting and stronger protection against infection, symptomatic disease and hospitalization caused by the Delta variant of SARS-CoV-2, compared to the BNT162b2 two-dose vaccine-induced immunity." But objectively, it's important to note they found, "Individuals who were both previously infected with SARS-CoV-2 and given a single dose of the vaccine gained additional protection against the Delta variant."[26] Regarding variants, other studies also showed that "hybrid immunity" from prior infection and boosters "conferred the strongest protection."[27]

The anthrax vaccine litigation and scientific debate went on for years, as will the analysis of the evolving science related to COVID and the Operation Warp Speed EUA unapproved injections. Almost forty years have lapsed since the government acknowledged the old anthrax vaccine's inadequacies and, as of 2024, we still don't have a new vaccine for anthrax. None of this will be decided definitively in the near-term. If anthrax experience provides a predictive template, we may not know much more in the decades to follow. Therefore, it remains essential that the body of evolving science, or even the public policy pronouncements, are not scrubbed.

A prime example of scrubbing related to quotes from one of the president's early speeches. Internet searches suppressed results. Fact-check websites listed the quotes, but tempered the context. The president stated, "This is a simple, basic proposition. If you're vaccinated, you're not going to be hospitalized, you're not going to be in an ICU unit, and you are not going to die."[28] "You're not going to—you're not going to get COVID if you have these vaccinations."[29] The president promoted an unhealthy schism, stating the virus was "much more transmissible and more deadly in terms of non-unvaccinated people." In fact, the opposite was true for unvaccinated recovered people. Regardless, the incorrect messaging laid a foundation for dividing the population over coercive mandates with high-level ridicule of the unvaccinated.

The president had company with ignoring the issue of natural immunity. Dr. Fauci answered questions for CNN's Sanjay Gupta on the same subject. Gupta asked, "I get calls all the time, people say; I've already had COVID, I'm protected. And now the study says maybe even more protected than the vaccine alone. Should they also get the vaccine? How do you make the case to them?" Fauci replied, "You know, that's a really good point, Sanjay. I don't have a really firm answer for you on that."[30] Ironically, years earlier Fauci had a firmer response when interviewed by C-SPAN. He unequivocally stated vaccination was unnecessary following an infection from a virus. Dr. Fauci said, "The best vaccination is to get infected," and "the most potent vaccination is getting infected."[31] The issue deserved evaluation by the government, especially the DoD since military regulations explicitly exempted redundant vaccinations where prior immunity existed. Like anthrax vaccine, our view matched the government's old position.

We tried to not take the politicization of medicine personally. We recovered. We were immune. Though we were wounded to hear the president proclaim, "This is a pandemic of the unvaccinated," we tried to not take it out of context. The president provided more context with, "And it's caused by the fact that despite America having an unprecedented and successful vaccination program, despite the fact that for almost five months free vaccines have been available in 80 thousand different locations, we still have nearly eighty million Americans who have failed to get the shot."[32] Instead of slighting the president, I doubted the competence and objectivity of his advisors that would prepare or recommend incorrect and spiteful soundbites.

A *Lancet* article, titled "Stigmatising the unvaccinated is not justified," critiqued this level of thinking as well. The author schooled that "people who are vaccinated have a lower risk of severe disease but are still a relevant part of the pandemic. It is therefore wrong and dangerous to speak of a pandemic of the unvaccinated." He called on "high-level officials and scientists to stop the inappropriate stigmatisation of unvaccinated people, who include our patients, colleagues, and other fellow citizens, and to put extra effort into bringing society together."[33]

Framing the hesitancy or wariness pejoratively as a failure seemed unfair. From a messaging standpoint, especially from our president, I viewed it as incomplete and harmful. I wished he expressed it all differently because the pejorative depictions certainly didn't motivate us to get inoculated after recovering—and our doctors didn't recommend it. In fact, newly emerging science asserted, "Individuals who have had SARS-CoV-2 infection are unlikely to benefit from COVID-19 vaccination."[34] Moreover, new studies revealed that "for the first time, our study links prior COVID-19 illness with an increased incidence of vaccination side effects."[35] Months later, data from the Cleveland Clinic indicated that more doses of vaccine directly corresponded to increased cumulative incidence of COVID-19 infection. Study findings stated: "The higher the number of vaccines previously received, the higher the risk of contracting COVID-19." Another analysis demonstrated "those not 'up-to-date' on COVID-19 vaccination had a lower risk of COVID-19 than those 'up-to-date,"[36] and that zero doses surprisingly corresponded to the lowest incidence of infection.[37]

As the months wore on, and the variants became less threatening, leading countries such as Iceland withdrew their COVID restrictions altogether. Ministry of Health infectious disease officials declared that "Widespread societal resistance to COVID-19 is the main route out of the epidemic." That "widespread social resistance" phrase was code for acquiring natural immunity. They added, "To achieve this, as many people as possible need to be infected with the virus as the vaccines are not enough"—"previous infection is at least as high, if not higher than that provided by two-dose vaccination."[38] The Danish authorities also altered recommendations for injections to only "people aged 50 years and over as well as selected risk groups."[39]

Across the world, common sense and the validity of terrain theory appeared to prevail over time. Vulnerable populations were prioritized. Evidence poured in pointing to potential immune system harm and negative efficacy from the vaccines. Logically, but belatedly, natural immunity finally figured back into the calculus and the science. Unfortunately, it took longer for the science, law, and common sense to protect against the illegal unapproved shot mandates.

MILE 23:
Federal Mandates (2021 to 2023)

By mid-2021, it was safe to assume that anyone who trusted the injections, the FDA, or the CDC most likely voluntarily received one or two shots. Nevertheless, after the August 2021 Pfizer shot FDA "approval," the president announced multiple federal-level mandates.[1, 2] The COVID injection mandates symbolized "the wall" in our multi-decade marathons against illegal mandates of unapproved medical products. A wall in a marathon is that point where complete exhaustion saps all your energy, yet you must go on. Millions of Americans hit the wall—COVID shot mandates. The mandates targeted large businesses with more than one-hundred employees, as well as our nation's healthcare workers, federal employees, federal contractors, and the DoD. Many in these groups of critically thinking citizens and soldiers fought back, breaking through the metaphorical wall in the marathons against illegal mandates.

The harassing mandates directed their fire on the segment of the population that already made their choice. In consultation with their primary care providers, those citizens determined the injections were medically unnecessary or unwise. Mandates seemed vindictive, malevolent, oppressive, and were aimed squarely at the consent rights of a singular subset of the American people. The government's rhetoric supported the perception of a vendetta as the president couched his mandate commencement speech with the targeted theme of "Vaccinating the Unvaccinated." President Biden unapologetically advised all Americans, "If you want to work with the federal government and do business with us, get vaccinated. If you want to do business with the federal government, vaccinate your workforce." He proclaimed, "The vaccine has FDA approval." This was an oversized statement, and did not include two of the EUA products made by Moderna and Johnson & Johnson (J&J).

Ironically, J&J's manufacturing was subcontracted to Emergent BioSolutions (EBS), the rebranded name for the old anthrax vaccine manufacturer. EBS was in the money until a half billion doses of J&J COVID shots were ruined through contamination.[3] Ultimately, the FDA shuttered EBS's COVID injection operations for J&J due to more violations.[4] The FDA limited use of the J&J shot in April 2021, and again in May 2022, due to blood clot risks.[5] The shelved product expired and the CDC ordered the destruction of 525 million unwanted EBS doses.[6]

Separate from the J&J rift, litigation was underway revolving around the available Pfizer BioNTech product that was only "allowed" or "authorized" by a simultaneously re-issued FDA EUA. In the footnotes, and buried deep within the documents and websites, the FDA told the truth in the fine print about continued EUA for Pfizer as opposed to the "approval" mirage for an unavailable "approved" Pfizer product marketed as Comirnaty. In truth, "approved" doses never became available in the US during the public health emergency, but the staff apparently forgot to brief the president. High-level government sources confirmed that no approved shots were available. The sources confirmed that the newly approved product was "not even in the mix." This was likely due to the fact the EUA injections were liability free and "covered persons" were indemnified.[7] The approved shot was not. It made business sense to provide only EUA shots as long as the government kept reissuing EUAs for the indemnified products. The drug makers also strong-armed secretive agreements with purchasers that acknowledged "the long-term effects and efficacy of the Vaccine are not currently known and that there may be adverse effects of the Vaccine that are not currently known."

Months later, the FDA's magic wand also re-waived liability for Moderna's EUA shots, while simultaneously approving an unavailable version marketed as Spikevax. Like the Pfizer-BioNTech COVID injection, the FDA reissued the EUA for Moderna's original shot.[8] Circularly again, the FDA republished the EUAs since the approved product was not actually available. If the FDA rescinded the EUAs, the manufacturers would lose their liability shield. According to the FDA's reissued emergency use authorizations the "EUA would remain in place for the Pfizer-BioNTech COVID-19 Vaccine for the previously-authorized indication and uses."[9] The whole scheme was a swindling of the American people to goad the populace into accepting the existing injections based on the illusion of FDA approvals. But why should American workers, children, and our troops accept the risks, face penalties, and suffer consequences if the manufacturer does not? Answer—they should not according to 21 USC §360bbb-3—the law.

The president paternalistically scolded, "We've been patient, but our patience is wearing thin. And your refusal has cost all of us. So, please, do the right thing." Biden admonished, "But just don't take it from me; listen to the voices of unvaccinated Americans who are lying in hospital beds, taking their final breaths, saying, 'If only I had gotten vaccinated.' 'If only.'"[10] Months later the president repeated a scary scripted warning, "For unvaccinated, we are looking at a winter of severe illness and death—if you're unvaccinated—for themselves, their families, and the hospitals they'll soon overwhelm." He added, "If you're vaccinated and you had your booster shot, you're protected from severe illness and death—period . . . booster shots work."[11]

Ultimately, the dire predictions and the optimistic product endorsements didn't pan out. The president seemed to pit his people against one another saying, "For the vast majority of you who have gotten vaccinated, I understand your anger at those who haven't gotten vaccinated." Though the negative messaging was disappointing to witness, by the State of the Union address the following year the president's message changed to a healthier tone: "Let's use this moment to reset. So, stop looking at COVID as a partisan dividing line . . . Let's stop seeing each other as enemies and start seeing each other for who we are: fellow Americans."[12] With that, I agreed.

Average citizens weren't the only ones getting stern warnings. The president threatened, "If these governors won't help us beat the pandemic, I'll use my power as president to get them out of the way."[13] My reaction when I heard the speech made me remember the other things the president probably needed to get "out of the way." I promptly searched and found the CDC and FDA published information and illustrations that explicitly assured the American public that the government would not impose vaccine mandates. They were still online. So, I took screenshots.

"Frequently asked questions" and "Myths and Facts" that contradicted the directives still existed online as of the mandate announcement. Soon after, they disappeared from the digital sphere. The previous government position clearly stated: "The federal government does not mandate (require) vaccination for individuals." Another stated: "No. The federal government does not mandate (require) vaccination for people." The government added that the "CDC does not maintain or monitor a person's vaccination records." The old position matched our position.

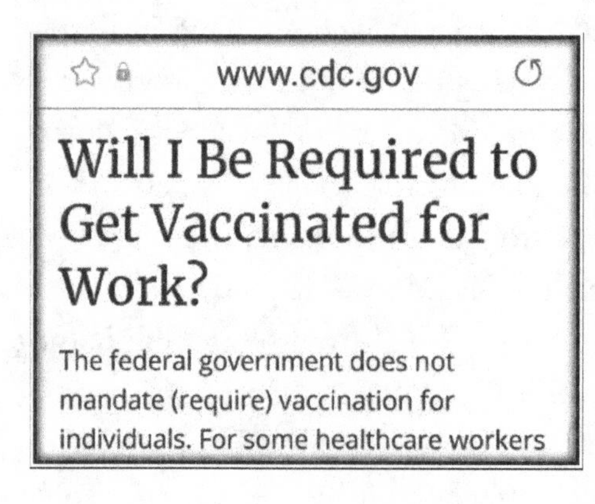

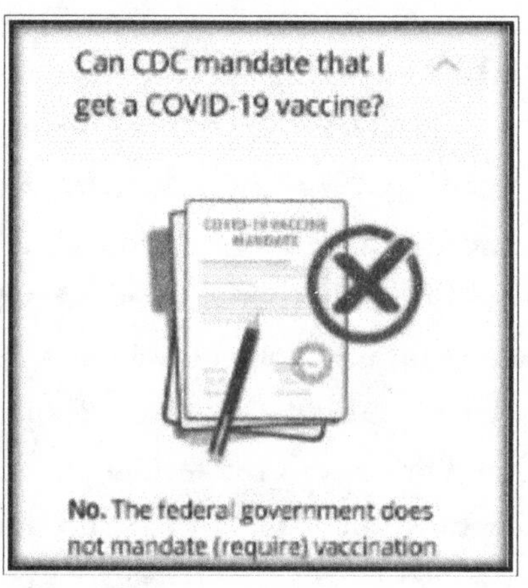

"The federal government does not mandate (require) vaccination".

The CDC's questions and answers and "Myths and Facts" disappeared. They got them "out of the way." I'm not sure if the executive branch actually forgot about them, or thought people would not remember the government's earlier formal position. Perhaps the top-level messaging was so hurried that the bureaucrats at lower levels didn't get the word to scrub the websites of the previous anti-mandate policies. What I did sense, in reflecting back on my earliest education, was that our leaders veered away from the founding principles with this unprecedented direction. They diverted from consent, from the law, and their previous plan.

My brother and I enjoyed great discussions about the nation's foundational emphasis on consent during the prior election cycle. He reminded me that the consent of the people is rooted in the Declaration of Independence. From the beginning the Founders declared, "We hold these truths to be self-evident, that all men are created equal, that they are endowed by their Creator with certain unalienable Rights, that among these are Life, Liberty and the pursuit of Happiness. That to secure these rights, Governments are instituted among Men, deriving their just powers from the **consent** of the governed."[14] Logically, consent seemed inconsistent with mandates. The mandates felt fundamentally wrong and the even the White House agreed until they changed their message. At its core, consent was constitutionally rooted, but it was also affirmed in the laws governing the "option to accept or refuse" unapproved medical products. Our leaders forgot all of this, and ignored the fact that abridging consent might not settle well with the "governed."

Though it sounds dramatic, the issues drove to the core of the Constitution. American's consent rights, the "option to accept or refuse" unapproved medical products, were protected as were other enumerated rights under the First Amendment of the Bill of Rights. That article prohibited "abridging the freedom of speech, or of the press; or the right of the people peaceably to assemble." It guaranteed the right to "petition the Government for a redress of grievances."[15] The Fifth and Fourteenth Amendments also protected all American's rights for "due process of law" and "equal protection of the laws." This equated to the option to refuse unwanted medical procedures.[16] This was one of those scenarios where our freedoms were at risk and redress of grievances must be exercised. That's what we attempted to do with both our 2001 and 2021 FDA citizen petitions. The courts also took on cases for "Life, Liberty and the pursuit of Happiness."

The first judicial redress of grievances over mandates occurred when the Fifth Circuit Court of Appeals acted decisively in halting the federal government's attempt to mandate COVID injections through an Occupational Safety and Health Administration (OSHA) Emergency Temporary Standard. This was the COVID shot mandate on large employers. The court cited "grave statutory and constitutional issues with the Mandate." The court recognized the ignored issues by acknowledging the "naturally immune unvaccinated worker" being "at less risk than an unvaccinated worker who has never had the virus." The court agreed, "the Mandate fails almost completely to address, or even respond to, much of this reality and common sense."

The judges reminded the government that "occupational safety administrations do not make health policy," and that in attempting to do so, "OSHA runs afoul of the statute from which it draws its power and, likely, violates the constitutional structure that safeguards our collective liberty."[17] Since everyone forgot the prudence of the past, the judges reminded the agency of its previously stated policy position: "Health in general is an intensely personal matter," and that "OSHA prefers to encourage rather than try to force by governmental coercion, employee cooperation in [a] vaccination program."[18] Again, our position matched former OSHA policy.

The OSHA mandate debate rapidly accelerated to the Supreme Court. I listened with interest to the high court's unprecedented hearing regarding the OSHA mandate on January 7th, 2022.[19] My observations included the pro-injunction justice's subtle calm and impartiality.[20] Remarkably, I did not witness any discussion during the hearing about the new science revealing waning efficacy, or the fact that the injected contract and transmit COVID—and no discussion occurred about natural immunity. Significantly at the time, scientific evidence exposed the Omicron variant breakthrough rate with an "88% likelihood to escape current vaccines."[21] [22]

The plaintiff's counsel cautioned the high court, "The agency cannot pursue that laudable goal unlawfully." The attorneys advised the justices against permitting the mandates simply because they might believe this "illegal action will lead to good effects, and so we will allow that to happen." Fascinatingly, the SCOTUS hearing included no discussion with respect to the 2005 anthrax vaccine EUA precedent that assured "no penalty or loss of entitlement." And plaintiffs did not argue the Title 21 protections that guaranteed EUA inoculations were optional under the

Food, Drug, and Cosmetic Act. Patiently, I awaited the SCOTUS ruling with optimism.

To my relief, large business employees won! Ultimately, up to one hundred million Americans remained employed due to the Supreme Court's rejection of the OSHA mandate.[23] It was academic for the justices since "OSHA has never before imposed such a mandate. Nor has Congress." The high court affirmed that the mandate exceeded agency statutory authority. The court struck down the order based on their conclusion that "this is no 'everyday exercise of federal power,'" but a "significant encroachment into the lives—and health—of a vast number of employees." The "vaccine mandate is strikingly unlike the workplace regulations that OSHA has typically imposed . . . A vaccination, after all, 'cannot be undone at the end of the workday.'"

The SCOTUS justices pronounced that the doctrines of "nondelegation" ensure "democratic accountability by preventing Congress from intentionally delegating its legislative powers to unelected officials." They explained that the "major questions"[24] doctrine "serves a similar function by guarding against unintentional, oblique, or otherwise unlikely delegations of the legislative power." The justices concluded, "Both doctrines prevent 'government by bureaucracy supplanting government by the people.'" The justices added, "The question before us is not how to respond to the pandemic, but who holds the power to do so. The answer is clear: Under the law as it stands today, that power rests with the States and Congress, not OSHA."

The Supreme Court did "not impugn the intentions behind the agency's mandate." Instead, they sought to "only discharge" their "duty to enforce the law's demands when it comes to the question who may govern the lives of 84 million Americans." Graciously, the affirming justices wrote, "Respecting those demands may be trying in times of stress," but absent judicial interventions "declarations of emergencies would never end and the liberties our Constitution's separation of powers seeks to preserve would amount to little." The high court concluded, "Although Congress has indisputably given OSHA the power to regulate occupational dangers, it has not given that agency the power to regulate public health more broadly." At the end of the day, the majority halted the "encroachment into the lives—and health" for all Americans.

Following the Supreme Court ruling, the executive branch withdrew the OSHA mandate and published an opinion in the Federal Register "strongly encouraging vaccination."[25] The US Court of Appeals for the Sixth Circuit dismissed the case as moot on February 18th, 2022.[26] The government's encouragement meant strictly optional shots at large employers as intended by law for unapproved EUA medical products[27]—we scored a win thanks to the original lawsuit filed by the National Federation of Independent Business and the Ohio plaintiffs.

Aside from the OSHA mandate, a case from the Southern District of Texas struck down the federal employee mandate. That court ruled, "The motion is GRANTED as to Executive Order 14043,[28] enjoining the government from implementing or enforcing the Executive Order until the case is resolved on the merits."[29] The judge insisted, "This case is not about whether folks should get vaccinated against COVID-19—the court believes they should." The judge clarified, "It is not even about the federal government's power, exercised properly, to mandate vaccination of its employees. It is instead about whether the president can, with the stroke of a pen and without the input of Congress, require millions of federal employees to undergo a medical procedure as a condition of their employment." The court concluded the overreach was "bridge too far."[30] Review of the federal employee mandate occurred in the US Court of Appeals for the Fifth Circuit with the court initially declining to lift the injunction,[31] then lifting it on administrative versus meritorious grounds,[32] and then finally the court reaffirmed the ban.[33] Pending review, the government prudently chose not to resume the mandate.[34]

Apart from OSHA and federal civilian employee's mandates, within weeks of the distinct federal contractor mandate announcement over half the state attorney generals filed suits. The states grounded their case in the US Constitution and Tenth Amendment, which protected federalism and reserved promulgation of public health policy to the states. The filings cited the principle that "the powers not delegated by the

Constitution to the United States or prohibited by it to the States are reserved to the States." Mandates provided examples of reserved powers, so the argument held the federal mandate was "an unconstitutional exercise of authority and must be invalidated." The suits also cited the failure to "provide an exemption to persons with natural immunity to COVID-19, even though natural immunity exists and is at least as effective as vaccination in preventing re-infection, transmission, and severe health outcomes."[35]

The federal contractor mandate disallowed natural immunity exemptions, but fortunately they were not necessary. The United States District Court for the Southern District of Georgia ruled to strike down the federal contractor mandate promulgated under Executive Order 14042.[36] Like other courts, it found the case was not about injection efficacy and acknowledged the seriousness of the pandemic. The court reminded us that "even in times of crisis this Court must preserve the rule of law and ensure that all branches of government act within the bounds of their constitutionally granted authorities." Citing the Supreme Court, the judge recognized the public's interest in combating the viruses spread, yet "that interest does not permit the government to 'act unlawfully even in pursuit of desirable ends.'" The unlawful part related to how the "President exceeded the authorization given to him by Congress." Thus, the court granted the injunction.

Again, in a sigh of relief for millions of Americans, the court acknowledged the great deference afforded to the executive branch, but cautioned that the Congress must also "speak clearly" on the exercise of power with "vast economic and political significance"—such deference is not "a blank check for the President to fill in at his will." The court, as with the earlier OSHA ruling, objected to the promulgation of "such a wide and sweeping public health regulation as mandatory vaccination for all federal contractors and subcontractors." They argued enjoining EO 14042 "would, essentially, do nothing more than maintain the status quo" where "entities will still be free to encourage their employees to get vaccinated, and the employees will still be free to choose to be vaccinated."[37] That sounded reasonable to me. Fortunately, six separate federal court injunctions impacting fourteen states agreed with this logic and the law.[38]

Further judicial review regarding the federal contractor mandate by the US Court of Appeals for the Fifth, Sixth, and Eleventh Circuits found the mandate was not "anchored" contextually in the law, nor the historical implementation of the Property Act. Courts questioned "the imposition of an irreversible medical procedure without precedent in the history of the Property Act's application." They argued the government sought to "usurp" constitutional roles "by doing something that it has no traditional prerogative to do—deploy the Property Act to mandate an irreversible medical procedure." Ultimately, the Sixth Circuit mirrored the Eleventh by declaring the "intrusion upon traditional state prerogatives raises serious constitutional concerns under federalism principles and the Tenth Amendment." Citing "intangible harms" in the form of "invasions of state sovereignty and coerced compliance with irreversible vaccinations," the court expressed concern that the mandate "requires vaccination everywhere and all the time," despite not being "'anchored' to the statutory text, nor is it even 'anchored' to the work of federal contractors."

The court concluded, "The government has not made the 'strong showing' required to justify" its mandate since the "Property Act likely confers no authority upon the president to order the imposition of the contractor mandate." The court ruled to "DENY the government's requested stay" of the injunction.[39] The Fifth Circuit added, "The President asks this Court to ratify an exercise of proprietary authority that would permit him to unilaterally impose a healthcare decision on one-fifth of all employees in the United States. We decline to do so. Thus, we AFFIRM the preliminary injunction issued by the district court."[40] Eventually, the government yielded to the court orders and injunctions by declaring, "the Federal Government will take no action to implement or enforce Executive Order 14042"[41]—we won that round too thanks to dozens of state attorney generals.

So no, it was not a constitutional crisis after all. Though the founding documents do say, "That whenever any Form of Government becomes destructive of these ends, it is the Right of the People to alter or to abolish it, and to institute new Government," the good news was that we were not at that point, had never gone there, and would not in this circumstance. Despite how disheartening it was to see the mandates announced, I compartmentalized it. It marked another fateful step in the journey—just like the anthrax vaccine—the system was tested. In time, judicial oversight and elections overruled the mandates convincingly. What remained was correcting the damage. In accordance with my understanding of Thomas Paine, I predicted that good people with good means would ultimately also ensure the good cause of record corrections.

During the anthrax vaccine dilemma, the institutions we swore an oath to defend misled us for over six years. The courts finally set the record straight by 2007. The DoD and FDA were "not substantially justified."[42] That final outcome cannot be undone. With the COVID injection approvals, the federal mandates, and the confusing regulatory actions and spin, the exact same M.O. resurfaced. Was it possible that the federal government was not telling the truth and was "not substantially justified" again? Even the approval fanfare was a ruse since the advertised "approved" doses never materialized in the US marketplace throughout the entirety of the pandemic. Fortunately, favorable court rulings, consent, and common sense prevailed in the end.

With all the pandemic inertia related to the injections, I sensed it was initially well-intentioned—even if not well thought out. Unfortunately, the subsequent bullying and coercion tainted the good intentions, something also reminiscent of the anthrax vaccine dilemma. The bullying with COVID shot mandates added a fascinating dimension. With the anthrax vaccine it was in your face. Your commanders bullied you, coerced you, or kicked you out, and they enforced it with the hammer of the Uniform Code of Military Justice. This time mandates emanated from the executive branch, using bullying state and local authorities, plus private entities, to dictate policies that breeched constitutional limits. Compared to the anthrax vaccine coercion, the players in these new mandates used the full pallet of bureaucratic trickery.

In contrast to the 2005 DoD and FDA EUA precedent event that ensured "no penalty or loss of entitlement," the government's new interpretation of 21 USC §360bbb-3 argued it supported mandates at the risk of job loss. They did so by misinterpreting the law's wording—"consequences"—to support a preferred context. The government's new interpretation conflicted with both a literal read of the law and precedent. The interpretation was articulated in a memo by the acting assistant attorney general and the DoJ Office of Legal Counsel. It opined as to whether the Food, Drug, and Cosmetic Act prohibited entities from requiring the use of a vaccine subject to an Emergency Use Authorization. The DoJ's misinterpretation of the precedent and statute concluded the law "does not prohibit public or private entities from imposing vaccination requirements, even when the only vaccines available are those authorized under EUAs."[43]

But the lawyer writing the memo of advisement never told anyone, or maybe didn't know, that the actual context of "consequences" in the law for informed consent under 21 CFR §50.25 related to opting out of clinical trials. The intent was, "If you drop out of this clinical study there could be medical consequences." But nowhere did anyone envision a mandated EUA unapproved drug threatening employment. Obviously, if a person dropped out of a voluntary clinical trial there is no job jeopardy. The 2005 DoD EUA precedent supported this fact, and the CDC website pre-announcement also supported this understanding—as did the EUA law.

Unfortunately, whether with mandates for the military, federal employees, federal contractors, health care workers, or large businesses, the government opted to ignore the original intent, wording, and precedent implementation of the EUA statute where "no penalty" could be imposed on an individual for choosing the option to refuse an unapproved EUA medical product. This flawed and rash decision relied upon the foundational misinterpretation of that one word—"consequences"—and of course the DoJ forgot to mention the informed consent statute context in their interpretation. Instead of honoring the consent context within the law, or at least not misinterpreting it, the DoJ cleverly concocted a concept called "secondary consequences."

The DoJ attorney proposed the peril of secondary consequences or "exclusion from certain desirable activities," such as school and employment. They simply made it up, just like the historic fabrication that the FDA had licensed the anthrax vaccine properly until being overruled by a federal court in 2003. The law granted none of it. It didn't even begin to say it. Officials misinterpreted a tortuous fabrication of secondary consequences to support discretionary consequences. They omitted precedent and independent government analysis that concluded the EUA law affirms the "right to decline the vaccine."[44] The disingenuous quibbling about "consequences," and conveniently forgetting that the government previously agreed there could be "no penalty or loss of entitlement," didn't bode well for the administration's strategy.

Central to the hoodwinking of the America, and perhaps misleading the president, was the fact that the FDA approved a license for a new, unavailable, Comirnaty shot just prior to the mandates. They maintained it was the same injection as the existing stockpiled Pfizer-BioNTech shot, but the truth in the fine print proved otherwise. Behind the FDA's con, documents confirmed the "two formulations differ," that they were not "biosimilar," that they were not "interchangeable,"[45] and that the "the products are legally distinct with certain differences."[46]

Plus, existing EUA stockpile labeling clearly stated "Pfizer-BioNTech," not the new "Comirnaty." Car dealers can't sell a 2020 model as a 2021—regulatory rules work the same way. Manufacturer labeling cannot deceive the public. The FDA's new spin conceded that testing "demonstrated that the modified formulation is analytically comparable to the original formulation." "Comparable," "modified," none of it was permitted as interchangeable by the Federal Food, Drug, and Cosmetic Act from a regulatory perspective. The gate guards sought to confuse everyone since it was allowed medically, but voluntarily, under the practice of medicine.

Driving the licensing point home from our story many years ago, the DoD's General Blanck played similar word games to confuse US senators. On April 13th, 2000, in testimony to the Senate Armed Service Committee, Senator Pat Roberts explicitly asked General Blanck about the investigational new drug application to get the vaccine approved for inhalation anthrax.

> **Sen. Roberts:** General Blanck, the annual Congressionally mandated chemical and biological defense program report to Congress submitted on March 15, 2000, states: "The Department submitted data to the FDA last year to license the vaccine to provide protection against aerosol exposure to anthrax." My question is why is the Department seeking a license for the vaccine when the license for the anthrax vaccine has existed since 1970?
> **Gen. Blanck: It is really for the facility, not for the vaccine per se**.
> **Sen. Roberts:** Oh, I see, okay. All right. That clears that up.[47]

Just as in 2000, licensing matters were confusing, as were the terms experimental and investigational for biologic products. The truth was that the medical products were unapproved. The gate guards in both eras told the truth "per se," just enough twisting of the truth to get the overseers off their backs or to give the senators the answers they wanted to hear. Because the Comirnaty license is for a facility, and since that facility had not bottled, labeled, or shipped the new Comirnaty, the existing EUA stockpiles were not licensed "per se." The available EUA COVID mRNA shots being given to the troops were not licensed, "per se," just like anthrax vaccine, since the license is "really for the facility." In the senator's words, I hope "that clears that up." This historic clarification meant the government was implementing illegal mandates using unapproved medical products at the peril of our troops again, and now our citizens, too.

Our colleague, JR, obtained Blanck's senate testimony on the anthrax vaccine. JR, Russ, and I saw through the diversionary quibbling. At the time, the general utilized the confusing explanation to divert attention from the DoD's investigational new drug application for anthrax vaccine—one of the smoking guns that ultimately proved the DoD was mandating the shot illegally. The same legal logic directly applied to COVID injections since licensed facilities did not apparently provide "approved" shots to our troops prior to the mandate deadlines. The available EUA Pfizer BioNTech injection was not licensed or approved, so the government's mandates risked the same fate as the anthrax vaccine order—patently illegal—"not substantially justified." The vials mandated for use on our troops and citizens were plainly labeled, "For use under Emergency Use Authorization." The medical professionals were supposed to protect our troops and citizens from these confusions, and supposed to ensure they received informed consent and prior consent. They did not speak up or blow the whistle on this gargantuan fraud despite their medicine cabinets being exclusively stocked with EUA unapproved shots.

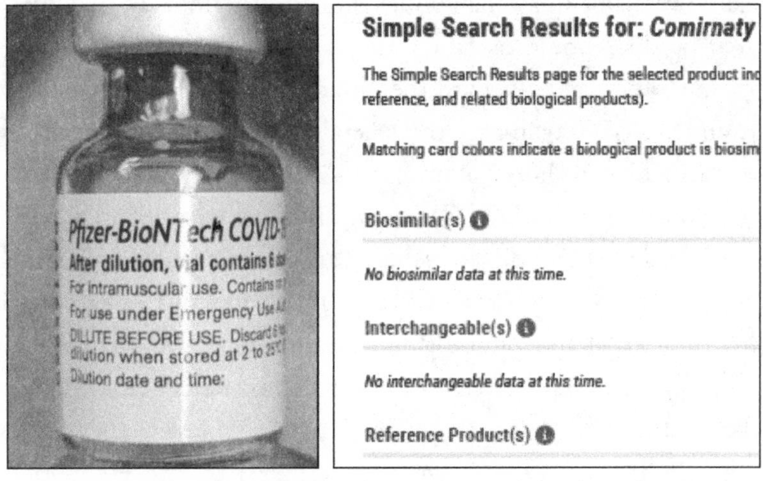

DoD doses of Pfizer-BioNTech & FDA Purple book confirming "no interchangeable data".

Logically, if the FDA was correct to advise the president that mandates were legally permissible, then why did the DoD not impose a mandate until the controversial "approval" for Comirnaty was announced in late August 2021? This requires a two-part answer: First, again, the only available doses of the Pfizer BioNTech EUA injection stated clearly on the label, "For use under Emergency Use Authorization," The DoD knew they could not mandate those products. Second, unfortunately, the DoD took the bait with the "approval" news. The DoD didn't mandate COVID injections earlier because they knew they could not mandate EUA products by law. The enormity of the fraud expanded when the EUA for the unapproved Pfizer product was subtly extended at the very same time that the unavailable licensed Comirnaty product was approved. Since 21 USC §360bbb-3 and 10 USC §1107a both affirmed EUA products were optional, the requirements for mandatory citizen and soldier inoculations appeared to be patently illegal since approved Comirnaty was unavailable when mandated and while the refusers were sanctioned.

The distinct mandate for DoD's military members resulted in the federal courts strictly scrutinizing the issue of interchangeability. The court explicitly agreed with our interpretation and rejected the DoD's extrajudicial claim that the Pfizer-BioNTech shot was interchangeable with Comirnaty—the FDA approved, but unavailable, injection. It was unlikely that the FDA approved product would ever be available due to liability problems. The liability question may be behind the "approval" racket perpetuated on the people and troops. Regardless, the federal court in the Northern District of Florida in *Doe v. Austin* upheld that "the DoD cannot mandate vaccines that only have an EUA." The legal distinction applied to all available doses and cited the DoD directive that guaranteed: "mandatory vaccination against COVID-19 will only use COVID-19 vaccines that receive full licensure from the Food and Drug Administration (FDA), in accordance with FDA-approved labeling and guidance"—meaning not EUA injections.

The SecDef also assured everyone that "I will seek the President's approval" if licensed doses were not available. In the end, no FDA-approved shots were ever made available by the DoD throughout the mandate deadlines and no waiver of prior consent was requested.[48]

> Mandatory vaccination against COVID-19 will only use COVID-19 vaccines that receive full licensure from the Food and Drug Administration (FDA), in accordance with FDA-approved labeling and guidance. Service members voluntarily immunized with a COVID-19 vaccine under FDA Emergency Use Authorization or World Health Organization Emergency Use Listing

SecDef memo: "only use COVID-19 vaccines that receive full licensure"—"FDA-approved".

> Consistent with FDA guidance, DoD health care providers will use both the Pfizer-BioNTech COVID-19 vaccine and the Comirnaty COVID-19 vaccine interchangeably for the purpose of vaccinating Service members in accordance with Secretary of Defense Memorandum, "Mandatory Coronavirus Disease 2019 Vaccination of Department of Defense Service Members," August 24, 2021.

DoD memorandum on 'interchangeable' use products, in conflict with SecDef memo.

The memos proved the DoD wasn't following the SecDef's directive. The court affirmed that members of the armed forces have a "statutory right to refuse," according to 10 USC §1107a, and that the "DoD guidance documents explicitly say only FDA-licensed COVID-19 vaccines are mandated." Those injections did not exist throughout the mandate's deadlines. Without a regulatory basis, DoD defendants contended, "once the FDA licensed Comirnaty, all EUA-labeled vials essentially became Comirnaty, even if not so labeled." About a year after the mandate, the DoD quibbled, "Comirnaty-labeled vaccine is in fact available for DoD to order as of today's date (May 2022)."[49] The DoD argued "10 USC §1107a does not apply," but the court didn't buy it. The judge admonished, "The DoD's interpretation of §1107a is unconvincing," because "FDA licensure does not retroactively apply to vials shipped before BLA [biologic license amendment] approval," and affirmed that "drugs mandated for military personnel be actually BLA-approved, not merely chemically similar to a BLA-approved drug."

The judge added that "the distinction is more than mere labeling: to be BLA compliant, the drug must be produced at approved facilities." The judge could have added "per se" for effect based on past anthrax vaccine testimony. The judge advised, the "DoD cannot rely on the FDA to find that the two drugs are legally identical for § 1107a purposes." For the first time a federal court broke through the ruse about interchangeability as "most plausibly interpreted as a factual, medical claim rather than a regulatory claim." The court recognized under the practice of medicine that a doctor could interchangeably use a COVID injection for medical purposes, but that practitioner's discretion did not translate into forced inoculations from a regulatory basis.

These interim rulings allowed the courts, just like in the anthrax vaccine *Doe v. Rumsfeld* case, to set the record straight. While the non-liable media and uninformed medical surrogates spread misinformation to the American people to perpetuate the ruse that the available shot was approved and interchangeable, in contrast, the manufacturer told the unadulterated truth in the months that followed the mandate announcement. The manufacturer published plainly, "The Pfizer-BioNTech

COVID-19 vaccine has not been approved or licensed by the US Food and Drug Administration (FDA), but has been authorized for emergency use by FDA under an Emergency Use Authorization or EUA to prevent Coronavirus Disease 2019 (COVID-19) for use in individuals 16 years of age and older."[50]

The manufacturer told the truth because not doing so was punishable under section 502 of the Federal Food, Drug, and Cosmetic Act's "misbranding" and advertising laws.[51] The DoD reinforced that "mandatory vaccination against COVID-19 will only use COVID-19 vaccines that receive full licensure from the FDA in accordance with FDA-approved labeling and guidance," but also stated, "Pfizer-BioNTech/COMIRNATY® has the same formulation and can be used interchangeably with the EUA PBS-buffer Pfizer-BioNTech COVID-19 vaccine."[52] While true in the practice of medicine, it was not so under the regulatory regime. In the years to come our courts will likely have to settle the interchangeability of "approved" versus EUA shots.

In addition to the federal suits challenging the DoD's mandate, Oklahoma Guard Adjutant General Thomas Mancino pushed back to protect his troops. The general informed the DoD his state would not enforce the mandate, nor punish his people.[53] The DoD didn't miss a step by threatening to cut off funding or activating the state's troops to federal service to provide the legal authority to see the mandate through to completion.[54] The DoD won that round, but the readiness concerns over the injections were contagious. Florida's top military leader, Major General James O. Eifert, warned the DoD that the shot mandate seriously threatened military readiness, and that forcing America's troops to get the injection endangered national security.[55]

Additional state efforts included Texas' ban on private employer vaccine mandates and a suit against Pfizer for a "scheme of serial misrepresentations and deceptive trade practices."[56] Florida Governor Ron DeSantis and Surgeon General Joseph Ladapo also pushed back by publishing health alerts about mRNA vaccine safety. They advised against shots for males under age forty due to an "84% increase in the relative incidence of cardiac-related death."[57] [58] Florida challenged the "poor safety profile" for mRNA products such as the VAERS-based 4,400 percent increase in life-threatening conditions.[59] They warned citizens about "negative effectiveness" for the COVID vaccines where "vaccinated individuals developed an increased risk for infection."[60] Ultimately, Florida's leaders called for a "Halt in the Use of COVID-19 mRNA Vaccines."[61]

Other scientists contended federal authorities "minimized" or "ignored" safety signals related to cardiac events that impacted .1 percent of high school students.[62] [63] With over 25,000 high schools in America, averaging 1,000 students, this equated to 25,000 serious heart issues. Once public, the ignored VAERS data warned not only of stroke, but of "clear safety signals for death and a range of highly concerning thromboembolic, cardiac, neurological, hemorrhagic, hematological, immune-system and menstrual adverse events (AEs) among US adults."[64], [65], [66]

Even though we didn't enjoy such independent scientific analysis or high-level support from government and military leadership or scientists during our era of questioning anthrax vaccine mandates, this new legal front benefited from the tireless and seasoned experience by our same retired military staff judge advocates. Attorneys Dale Saran and Lou Michels, who fought for us in decades past, were in formation fighting the new mandates.[67] [68] These judicial knights served as the corporate knowledge. They understood the M.O., and reminded the courts about the past regulatory missteps by the DoD and the FDA. Their litigation against the DoD put an end to the anthrax vaccine program illegalities and was the reason the EUA law was invented in the first place back in 2005. The EUA law sidestepped the injunction won in federal court.

Saran's and Michel's new suits to protect service members from the COVID mandates focused on the lack of available approved injections, meaning our troop's prior consents rights should have been honored under EUA law.[69] The attorneys, along with civilian attorneys such as R. Davis Younts, Todd Callender, Andrew Schlafly, and many others, focused on religious exemptions and medical exemptions for natural immunity based on the provisions of multi-service regulation AR 40–562. The exemption processes were new compared to our anthrax vaccine experience. Back then no one could claim natural immunity to anthrax. This time hundreds of thousands of exemption requests threw a wrench in the DoD's OODA loop (Observation; Orientation; Decision; Action). The thousands of plaintiffs represented by Dale, Lou, and the other attorney's courageously symbolized America's best and brightest—critical thinkers who charged that the DoD "violated its own rules." Their core allegations echoed our charges against the illegal anthrax vaccine.[70]

Our contacts at the *Military Times* explained the natural immunity regulation in their print editions by stating, "Army Regulation 40–562 presumptively exempts from any vaccination requirement for a service member that the military knows has had a documented previous infection.[71] The regulation instructed military medical personnel to "ensure patients are evaluated for preexisting immunity." One would think obligatory words like "ensure" might mean something. The exemptions included a "MI" code for medical immunity, explained as, "Evidence of immunity (for example, by serologic antibody test); documented previous infection." The regulation, dated before SARS-COV-2, logically included new diseases—especially for COVID with the vast availability of serologic antibody tests. The purpose was to avoid unnecessary medical procedures. This was an axiom in medicine before 2021.

Dale and Lou dutifully carried the legal guidon, making both the natural immunity and core illegality argument for our troops.[72] Regardless of the patent illegality of mandating a EUA products on our military members, the DoD indisputably bypassed natural immunity exceptions codified in their own regulations. Had the DoD followed the law, listened to their own troops, continued to exhibit restraint with the mandate, or had they followed their own immunity exemption regulations, the multi-azimuth litigation efforts would have been unnecessary.

Personally, though not initially engaged defensively against mandated COVID shots, I realized it would be smart to be prepared to outflank the threat. Unlike some airlines and most healthcare systems, my employer did not preemptively impose a mandate. By encouraging and approving exemptions, my company and pilot union prudently protected medical choice between employees and their doctors. In contrast, the cases against United Airlines related to suits over preemptive company-wide mandates. Whereas much of the media depicted the legal activity as supporting the mandates, the courts made clear their rulings only upheld the propriety of unpaid leaves. A judge ruled the case was "not about the constitutionality or efficacy of vaccine mandates promulgated by the government or private entities." Later court rulings held that the mandate represented "ongoing coercion" and resulted in "irreparable harm"[73] due to income loss.

These first airline rulings included language about how the judge was "disturbed" by United's seemingly thoughtless approach to employee concerns about taking the injections. The judge's opinions expressed that "United's mandate thus reflects an apathy, if not antipathy, for many of its employees' concerns and a dearth of toleration for those expressing diversity of thought."[74] "The Court is disturbed by United's seemingly calloused approach to its employees' deeply personal concerns with injecting a foreign substance into their bodies." United's management condescendingly cautioned employees if they "all of the sudden decided I'm really religious," and to be "very careful"—"you're putting your job on the line."[75] Later Supreme Court rulings hinted that United could have faced legal jeopardy for not providing employees reasonable accommodations if there was no undue hardship for the company to do so, while instead appearing to convey animosity in constraining their employee's religious rights.[76]

Once the federal mandates threatened our troops, I collaborated with other dutiful veterans to assist the noble efforts of the Truth for Health Foundation.[77] I also advised my airline colleagues as they publicly challenged the mandates. I was proud to provide professional and dispassionate fact-based soundbites for my Shakespearian "band of brothers." My experience hopefully served as a rational and moderate voice amidst the craziness. We flew in formation, providing mutual support, fighting against what the court labeled as "irreparable harm."

Another old anthrax vaccine era friend, Lieutenant Colonel Dave "Iceflyer" Panzera, stood tall as a lead plaintiff in another case challenging the mandates. His call sign, Iceflyer, owed its origins to his years of flying ski-equipped LC-130 Hercules on the ice in Antarctica while deployed for his unit from Scotia, New York. This was not Dave's first time out in the cold fighting for his and others' health rights. Dave provided courageous testimony on September 29th, 1999, to the 106th Congress in a hearing on the "Impact of the Anthrax Vaccine Program on Reserve and National Guard Units."[78] He assisted with our advisements to the GAO in the report on attrition. He was the first to publicly debunk the DoD's false assertions of widespread use of the military anthrax vaccine by civilian veterinarians. Iceflyer and I were honored years earlier to sing for Russ's family at his memorial service in 2005. As retired military members serving as airline pilots, together we faced another health hurdle threat. Russ would be proud of Dave for leading the charge in one of the suits against the illegal federal contractor mandate.

My own next hurdle and flanking maneuver involved requesting a mandate exemption from my employer. I decided that outflanking and outlasting the looming threat of mandates was the best near-term tactic. Notwithstanding some initial turbulence, my company wisely dodged the "mass formation psychosis"[79] of mandates by opening up the exemption escape hatch. In the end, I was grateful that my company managed the dilemma, prevented the psychosis, and dodged the onerous mandates. That is not to say that the entire ordeal didn't stress me out. I witnessed considerable anxiety throughout our industry due to veiled threats of termination, but in the end, fortunately, my company approved all of my colleagues and my exemption requests. My stress level went down dramatically thanks to my airline's ultimate prudence.

My first exemption request was medical. I also submitted a Title 7 request, under the religious category. They approved that one too. My Title 7 civil rights request was strictly based on firmly held moral and ethical beliefs. I stated I was wary of the current vaccine technology, rising to the level of religious beliefs. My medical exemption was based strictly on natural immunity. I attached published studies showing no added benefit from COVID inoculation for the naturally immune and how it was "associated with increased risk of severe side effects."[80] My medical exemption was a risk-benefit calculation. I followed rules, trusted the process, and protected myself from anyone changing the subject to one of non-compliance.

Sticking with the processes, being grateful, and complying with any reasonable accommodation required as a contingency for the exemption was the right flight plan for me. At the end of the day, my company directed accommodations included masking, social distancing, and declaring "fit for duty" before each flight segment—the same requirements we performed for the preceding two years. For me, the accommodations were not discriminatory and seemed reasonable. Respecting other citizens' fears of the virus was also a worthy and humble approach. Reciprocity in these attitudes offered hope that others might respect our natural immunities and that the collective public health community might return to that forgotten medical standard.

Mandates troubled me, as they did a significant portion of society. The difficult reality was that even when federal mandates waned, some local, state, and private entity vaccine mandates were still grounded by legal precedent. Vaccine mandates in society were legal for FDA approved products, but were explicitly illegal for EUAs. The White House knew this fact. Regardless of the federal-level limitations, they conveniently forgot their old answer: "Can we mandate vaccines across the country? No, that's not a role that the federal government, I think, even has the power to make."[81] Checks and balances prevailed in refreshing their memories.

The White House was correct from the beginning regarding the limits of its power. The Supreme Court concurred with the original position of the president. Our system of government did not issue the presidency a blank check. Fortunately, the federal courts also checked the president's powers with the mandates in order to maintain the status quo for millions of impacted federal contractors. Similarly in the OSHA ruling, courts upheld the authentic authorities reserved to the states pertaining to enumerated police powers and emergencies. No one ever intended those authorities to upend everyday life and livelihoods.

The legal fights were far from over to reverse the damage. In the interim, our nation's leaders must resurvey their judgments and avoid abusing emergency powers with overreaching mandates that invade the lives of their citizens. Even after the president declared "the pandemic is over,"[82] the emergency powers and manufacturer liability shields continued.[83] One of my pilot training friends referred to emergency powers and mandate constructs as "carpet bombing a threat when you really need a precision guided munition." His astute analogy challenged excessive public health responses, incongruent with modern medical operations and laws. My friend's metaphor spoke for the need to implement measured solutions. The rights of "Life, Liberty and the pursuit of Happiness" required such continual and restrained contemplation.

As fate dictated, the president revoked all the remaining mandates.[84] The government opted to 'turn the page,' titling the president's Executive Order as "Moving Beyond COVID-19 Vaccination Requirements for Federal Workers." The steps in the judicial, regulatory, and bureaucratic dance would surely continue for years to come based on the damage done.[85] The obsolete mandates against variants using rapidly outdated monovalent injections made no sense. The CDC deauthorized the original EUA injections and replaced them with EUA-only bivalent boosters, lacking any parallel, but unavailable, FDA-approved product.[86]

Fatefully, we—the moms, the dads, concerned citizens all, businesspersons, lawyers, lawmakers, and the troops—won—case closed. Those on the front lines shaped fate by warily navigating the obstacles, just as pilots fatefully do. Mandates were in full retreat, with the final beachhead being the historical account. Too bad no one listened to our citizens or their petitions.

The iconic aviation novel *Fate is the Hunter* cast its shadow on our journey and these themes. One of my longtime flying friends recalled the book's romantic resemblance to our saga as we crossed an ocean together. I updated him on our battle for medical freedom. He found it intriguing that mandates were haunting me, like fate, and recalled the book's relevance. Twice in my flying career, mandates fatefully hunted me down—"it appeared fate was on a rampage."

In our story, at least initially, the mandate was the hunter. Like the undaunted aviators in Ernest Gann's novel, we as pilots used our skills "to avoid hazard" and leveraged the "many unexplored crevasses in our reservoir of knowledge" to fly around the threats. The repetitive mandates, fomented by the likes of Fauci and others in this tale, not so skilled in the art of unyielding truth telling, might inevitably be the ones "victimized by fate, perhaps in revenge for their constant teasing of its power." In this latest case, the government, as with the anthrax vaccine, chose to force their policy—good or bad—on pilots and free people. Both are breeds that do not accept that which fate allows them to out-maneuver—to judiciously escape the threat.

As Gann writes, "If fate was bound to defeat us, then it should have chosen fewer and more subtle harassments."[87] That was their fateful and fatal error—the mandates. The pilot psyche confidently knows you can beat both fate and mandates. Just as we outgunned them before, the collective courage of citizens returned the rampage again. Our mission's wingman flying high cover was the Supreme Court. This time in the pattern, multitudes of American heroes served as the hunters, stalking the hazards and harassments of the mandates. Illegal mandates with unapproved medical products did not defeat us thanks to judicial common sense.

In accomplishing this mission, and for any that followed, I hoped and trusted that author Ernest Gann, and the fateful pilots from his stories, would appreciate this take on fate. As well, James Salter, and his tragic characters in Cleve Connell and Billy Hunter, would hopefully all be proud that we applied the lessons they told about the mission, courage, and humility.

MILE 24: www.Hoping4Justice.org

The Hoping 4 Justice effort, and website, began after federal courts ruled the mandatory anthrax vaccine program illegal, and particularly once the FBI found that "saving" the failing anthrax vaccine program motivated the anthrax letter perpetrator to commit the crimes. After the DoD failed to resurvey its judgements and review the punishments meted out, a letter campaign began to respectfully request the president to direct record corrections. To date, over twenty years later, you can count on one hand the number of successful record corrections. Our nation, our presidents, and our citizens can do better. The more than one thousand punished troops whose service records remain blemished deserved better. COVID-era refusers merit justice too.

As Army Surgeon General Ronald Blanck acknowledged, "I think it speaks to the undercurrent of distrust of the government and the military . . . Agent Orange. Nuclear tests in the '50s. People say, 'How can you say this is safe?' Clearly, we have a credibility problem."[1] Hoping 4 Justice's goal targets fixing that credibility problem by correcting the records for our troops. Halting and decisively correcting the vastly larger COVID injustices also stands as a priority in order to not repeat the patterns. General Blanck might even agree after all the years.

The corrections effort began by assisting an Air Force NCO, Richard "Blaidd" Mischke, with an anthrax vaccine disciplinary and demotion case in the late 1990s. Like Blaidd, the lead champion in that endeavor was a true Air Force hero, the late Colonel George "Bud" Day, USAF retired. I was proud to help as his uncertified pro bono paralegal. Colonel Day was a Medal of Honor and Air Force Cross recipient from the Vietnam War. Those are the nation's highest honors for gallantry and intrepidity at the risk of life, above and beyond the call of duty. When Blaidd received an honorable discharge after refusing anthrax vaccine, Colonel Day aggressively tried to get the Air Force to expunge a non-judicial punishment and demotion. Even after the anthrax vaccine illegalities became a matter of legal history, the Air Force would not budge.

Colonel Day helped me, too. Once fired from my position over the later substantiated reprisal and blocked command opportunity in 2011, he authored a letter to my senior leaders to attempt to stop them from making the wrong decision with my termination [Col Bud Day letter at Appendix FF]. Colonel Day's cautions asked the general to "review the overall facts of this case, and reverse this flawed decision that has the potential to bring a huge amount of criticism and ill will." At the time, again, the military did not budge. We fought unyieldingly to get the record repaired and restored. Colonel Day championed my corrections case before he passed in 2013. I am eternally grateful for the help from this true American hero—Colonel Bud Day.

Due to the unfinished business of anthrax vaccine record corrections, one of the Marines I helped find justice returned the favor. James Muhammad assisted the Hoping 4 Justice effort by chronicling his corrections case on our website. The site includes his application for corrections so that it might help others find justice as well. Hoping 4 Justice's request is simple, logical, and reasonable. Reasonableness is central to our mission to correct records for our fellow service members. We encouraged our lawmakers and elected leaders to be the same, without regard to political inconvenience or the potential uncomfortableness of the task. Lawmakers should not turn the page or look the other way. A summary of the Presidential Appeal lists four main points:

1. Service members must accept immunizations that are properly approved, safe, and effective.
2. Senior leaders must ensure immunizations comply with government regulatory standards.
3. If the rules are not followed, the bottom of the chain of command should not be punished.

4. Since this happened from 1998 to 2005, it is common sense and just to reverse punishments.

Any service member sanctioned by the DoD for refusing to be immunized with the anthrax vaccine, during the years the program was deemed illegal, can proactively fill out a form DD-149 and submit the application for correction of military records to their respective service correction board. It is simple. James's DD149 and supporting materials serve as the template.

The Hoping4Justice.org website organization includes on the home page the directions and a template to apply for corrections. The website includes tabs for the Letters to the Presidents, Supportive Documents, Naval Postgraduate School Thesis, Contact info, Opinions, and an About section dedicated to Lieutenant Colonel Russ Dingle. The COVID tab utilizes the same logic concerning the EUA unapproved medical products illegally mandated on our troops in 2021. This section includes legal citations, the "no penalty or loss of entitlement" Federal Register precedent, plus corroborating FDA documents and evidence.[2]

A draft bill to Congress is also on the website. While limited traction occurred with the legislative initiative, it still always feels good to make the effort and participate in government. James and I are well aware none of this may happen until after we are gone, as with harm from nuclear testing, Agent Orange, and other toxic exposures. But that does not mean you don't give it a shot. The draft bill serves as a template to correct the COVID mandate adverse personnel actions in the future. Like anthrax vaccine, the mission for COVID mandates logically follows:

1. Service members have the prior consent right to accept or refuse EUA immunizations.
2. Senior leaders must ensure immunizations comply with government regulatory standards.
3. If the rules are not followed, the bottom of the chain of command should not be punished.

4. It is common sense to halt and reverse any penalty related to EUA COVID product refusal.

Hoping 4 Justice's primary premise stands firmly aligned with the military rules under the Manual for Courts-Martial. Article 90 (10 USC 890) of the manual explains the violation military members allegedly committed with anthrax vaccine refusal: "Willfully disobeying superior commissioned officer." While the military enjoys the "inference of lawfulness," wherein an order is "inferred to be lawful, and it is disobeyed at the peril of the subordinate, this inference does not apply to a patently illegal order."[3] This paradox defined the anthrax vaccine order. The DoD's mandate wrongly enjoyed an inference of legality, and therefore military courts disallowed evidence to challenge the lawfulness. Yet in the end, the courts ruled the mandate was illegal and affirmed the DoD was "not substantially justified." The same patent illegality logic applied to EUA COVID injections, so long as approved shots remained unavailable, or unless the President waived service member prior consent rights.[4] Since the DoD executed a patently illegal EUA COVID shot mandate, the institution did so at its peril.

The historic, patently illegal status of the anthrax vaccine mandate justified the reversing of the past and ongoing injustices. This was why the military fully restored records, rank, pay, medals, and upgraded to honorable the discharges for two US Marine Corps member's records to date. One thousand or more records still require corrections. At least with COVID injection refusals and discharges early on, the National Defense Authorization Act, section 736, ensured honorable discharges characterizations for "discharges solely on the basis of failure to obey lawful order to receive COVID-19 vaccine."[5] Unsurprisingly, the DoD gate guards snuck in a "general under honorable" discharge option, which was less than fully honorable. As a result, many service records require corrections, because discharged troops given general under honorable discharges lost earned education benefits. This was not consistent with the 2005 DoD EUA precedent explicitly affirming no penalty, no loss of entitlements, and no loss of benefits. Elected representatives saw through this vindictive injustice and proposed a bill for parity.[6]

The premise of Hoping 4 Justice and Unyielding relies on learned observations of the DoD's historic, cultural inability to admit errors for too many decades. The DoD appears to care more about near-term public relations than adherence to the law or the rights and of its personnel. With the COVID injection dilemma being on a much larger scale, we witnessed the same behaviors magnified and replicating throughout our government and all sectors of society. Lacking a visionary perspective of the impact of such conduct on the long-term credibility of the military institution, the DoD left punishments uncorrected for up to a thousand troops during the anthrax vaccine dilemma. The DoD repeated the same patterns for any that questioned COVID shot imperatives, resorting to less than honorable tactics by threatening general discharges.

In our original case, after grounding and discharge as A-10 pilots, we found our own justice by stiff-arming the policy and continuing to serve. Our former leaders were embarrassed for sidelining their fellows only to later be proven wrong. They tried to forget entirely what they did. George Orwell's essay, *Revising History*, sent a salient warning to his English compatriots during World War II about the dangers of letting totalitarian regimes win and rewrite history, warning us against allowing leaders to recast the truth to cover-up their errors.[7] We found parallels in the COVID-era controversies. The DoD's past attempt to cover-up more negligible losses due to the illegal anthrax vaccine provided only predictive commentary about the manipulation of the actual vastly larger COVID illnesses and much larger refusal numbers.

It seemed like then, as again with COVID, we lived in George Orwell's novel, *1984*. In *1984*, Orwell's "newspeak," what we unyielding observers relabeled as "dupe-speak," was "designed to diminish the range of thought."[8] Dupe-speak was the modern lexicon of the dangerous Big Brother vernacular. Without a solid foundation of facts to fall back on, the DoD made it up as they went along. Whether it was a closed plant being "renovated," or a vaccine "we believe" is effective, or the "routine" use among veterinarians, or a "safe" vaccine with no proof of long-term safety, or a review by an "expert" who admittedly was not an expert, and finally the disingenuous distorted definition of a "refusal," dupe-speak became the dialect then, and again.

Early examples of the DoD revising history included the attrition caused by the anthrax vaccine.[9] The DoD representatives were not straightforward with Congress about the personnel costs. Such candor would have increased the scrutiny of the fatally flawed policy. The same sense of underreporting with COVID injection adverse events fit this pattern. Congress' realization of this M.O. motivated passage of Section 751, Management of Anthrax Vaccine Immunization Program, in the National Defense Authorization Act for fiscal year 2001. Congress knew that PR prevailed over straightforwardness and therefore forced a modification of Title 10, United States Code, section 1178(a). The new law required that "the Secretary of each military department shall establish a system for tracking, recording, and reporting separations of members of the armed forces under the Secretary's jurisdiction that result from procedures initiated as a result of a refusal to participate in the anthrax vaccine immunization program."[10]

Unfortunately, the DoD's tactics, or negligence in reporting, did not change even with passage of a law to compel candor. In response to congressional inquiries over non-compliance with this law, the DoD admitted the agency had "not fully complied with the law to consolidate the information requested and submit a report to Congress . . . Of significance, neither a transmittal letter nor a report could be found for the year 2001 data."[11] The DoD officials also publicly confessed that the department "doesn't collect data on how many people it has charged with refusing the vaccination."[12] Such administrative omissions and public admissions reaffirmed the "no bad news" nature of the program. It was unlikely these were accidental errors of omission.

The tactics of delay only guaranteed that the legacy of the anthrax vaccine will join syphilis and radiation testing, as well as Agent Orange exposure, in the history books due to the documented stonewalling, denial, and omissions. Though the anthrax vaccine era malfunctions took over twenty years to begin to see corrections, those corrections should not stall due to the renewed pandemic era errors. Inevitably, reflections and corrections will happen. Just as those "dark chapters" ultimately resulted in executive-level resurveying of judgments and course corrections, in time courageous leaders from future generations will hopefully deliver overdue justice for our service members and veterans [President Clinton letter at Appendix GG].

At times I worry we do not learn, but these historic examples bolster my confidence in the ability of humanity to correct and reflect on wrongs over time. President Clinton reflected, "Our greatness is measured not only in how we . . . do right but also [in] how we act when we know we've done the wrong thing; how we confront our mistakes, make our apologies, and take action."[13] The president's report on human radiation testing concluded, "The Administration will continue to take steps to open the government's records, raise ethical standards, and right the wrongs of the past." These executive-level steps proved we are capable of doing the right thing.

Just as President Clinton later corrected the wrongs related to syphilis and nuclear testing, George Bush participated in announcing legislation to advance Agent Orange compensation. He decreed, "We are here today to ensure that our nation will ever remember those who defended her, the men and women who stood where duty required them to stand."[14] These subtle 1991 admissions and remembrances represented our nation's top leaders, decades after the fact, trying to do the right thing. Paradoxically, all these proclamations were coincident with the origins of the last medical dilemma—the DoD's use of the anthrax vaccine. Moreover, as history repeats and repeats, we must be vigilant to foresee the fact patterns as they applied to COVID shots.

Over time, with anthrax vaccine, the facts slowly, full circle, came to light and service record corrections followed. Twenty years ago, my family patiently asked, "Why does this continue when the wrongdoing is so clear?" I explained to them that the answer was not solely sinister, but merely a result of less admirable facets of human nature observed continuously over time. Russ taught me this with his cynical idealism. I answered their question with a question: "How likely was it that the dozens of top military leaders, controlling the preeminent military power on earth, might apologize to thousands upon thousands of troops, holding no power, controlling nothing, not even their own rights?" Those in power rarely offered conciliation voluntarily, particularly when there was enormous financial or reputational liability at stake.

Now contemplate the same bad human behavior dynamics on a global scale, involving the entire planetary population in a pandemic, as opposed to the much smaller subsets of those exposed to syphilis, radiation, or Agent Orange. The disappointing reality was that the people in charge were reluctant to admit wrongdoing when those harmed were powerless compared to the exponential power of our leaders who approved and pushed the illegal inoculation mandates, but then looked the other way as their counterfeit narrative collapsed. The courts, congress, media, and businesses were all shy to call out the unprecedented constitutional overreach of the mandates after promoting the shots and failing to provide oversight. Even though the DoD possessed wide latitude and discretion to unconditionally correct the injustices without admitting wrongdoing, it did not do so. The probability of contrition decreased further when the same mistakes reoccurred across multiple administrations and many decades, as with anthrax vaccine.

From 2021 to 2023, the lessons not learned resulted in massive job losses due to illegal mandates for the unapproved COVID shots at magnitudes far greater than during the affirmed illegal anthrax vaccine mandate. In the healthcare career field alone tens of thousands of medical professionals lost their jobs due to the mandates.[15] Beyond those tallied losses, up to a third of complaint healthcare workers resorted to "sick leave after COVID-19 vaccination," and "staff absences increase with each additional dose.[16]

DoD's statistics were worse. By the end of 2021, the US Navy announced over 5,000 active duty and over 3,000 Reserve members remained unvaccinated.[17] The US Marines revealed that up to 10,000 remained unvaccinated in the active force alone,[18] with another 5,000 in the Reserves.[19] The USAF admitted larger potential losses with over 9,000 active and 30,000 Guard and Reserve airmen unvaccinated.[20] Figures for the Army Guard approached 40,000.[21] The US Army reported similarly large numbers with almost 10,000 unvaccinated in the active force.[22] Up to 22,000 in the US Army Reserves remained unvaccinated.[23] The losses were unexpected by DoD leaders that figured they could bully and coerce their military members into compliance.

Amazingly, exemptions for religious accommodation were near zero. The DoD's accelerated inoculation OODA loop had not counted on this legitimate flanking maneuver, but the courts could not deny the legitimacy of exemptions. Blanket denials trended across all branches, resulting in the subsequent injunctions and well-deserved court admonishments.[24] [25]

The aggregate figures were dramatic, with well over 100,000 troops unvaccinated after the majority of the deadlines passed. Once Congress halted the DoD mandate, the final tally was 69,000 refusals and over 8,000 discharged.[26] Lost training costs per service member resulted in a significant loss of investment by the taxpayers, with equal or higher future replacement training and readiness costs in excess of twenty billion dollars. While the DoD later spun the losses and lack of returnees, the reality was an unredeemable loss of trust and disbelief that military leaders would fight for their troop's health rights in the future.[27] The potential for illegally mandating unapproved boosters only exacerbated the uncertainty for recruitment and retention losses, not to mention potentially "cause a net harm" in an otherwise healthy young adult population.[28]

A brave Marine officer highlighted these concerns in a video posted for the attention of the military leadership, even as the defense secretary himself fell ill to COVID twice despite being fully "vaccinated" and double boosted.[29] The Pentagon leader's condition was predictably consistent with the latest studies that confirmed the probability of reinfection increased and "was higher among persons who had received 2 or more doses compared with 1 dose or less of vaccine," and that "surprisingly, 2 or more doses of vaccine were associated with a slightly higher probability of reinfection compared with 1 dose or less."[30] The unanswered questions the courageous Marine asked about vexing safety, efficacy, and attrition concerns echoed our same questions from over twenty years ago. The intrepid officer commented on a loss of trust and how the policy was "corrosive to morale." He described what we witnessed: "Heavy-handedness" in implementation, without "repeal or recourse," causing "deep and dangerous cynicism."[31]

This officer, and other Marine pilots, publicly served as a humble déjà vu about our illegal mandate over two decades ago, except these service members' credentials far exceeded our own.[32] Another Department of the Navy commander, Robert A. Green Jr., provided brave mutual support by publishing a readiness warning and co-authoring an accountability declaration with constitutional underpinnings.[33] His book was titled *Defending the Constitution Behind Enemy Lines*.[34] The attrition these officers exposed far exceeded those from the anthrax vaccine debacle, but still did not account for the future retention and recruitment problems imposed by the loss of trust. I earnestly hoped these courageous officer's examples, and that of tens of thousands of similarly situated military members, would not have required an extra section on the Hoping4Justice.org or UNYIELDING.org websites. This duty should have fallen to the president[35] and defense secretary. All of these officials received exclusively EUA shots, yet still contracted COVID multiple times, to include rebound infections. These facts and their cases begged for a reassessment of the science. Even if they did not possess the insight to resurvey the science, they were still duty-bound to resurvey their judgements and honor the rule of law.

Hoping4Justice.org and UNYIELDING.org would not have been necessary if leaders told the truth from the outset or if their staff fully informed them about the legal ramifications. The troops would not have been lost, and the generals would not have been embarrassed later by their inactions. The Hoping 4 Justice and Unyielding efforts exist solely to right the wrongs and guard against the very threats to liberty and truth of which Orwell warned. Even though a few record corrections occurred over anthrax vaccine, it is a national embarrassment that our government did not correct records for all similarly situated troops, versus only the unyielding few. It was unacceptable to repeat the same mistakes decades later with patently illegal COVID mandates.

The Hoping 4 Justice and Unyielding marathons will endure or be replaced by more suitable safeguards. Those entrusted with power cannot be allowed to use dupe-speak, operate dangerous dupe-loops, or revise history. *Unyielding* hopes to serve as an enduring history lesson to guard against the bad behaviors we witnessed decades ago by identifying the patterns, the evolving science, and the legal rulings related to the COVID shots. The anthrax vaccine travesty impacted two million troops and played out over two decades. Anthrax vaccine damage paled compared to COVID injection nightmares affecting billions worldwide for decades to come.

The Hoping 4 Justice effort for corrections, and the Unyielding effort to halt new violations, hopes to break the pattern. The cliché about repeating bad behaviors, but expecting a different result, is an applicable saying to frame the insanity. The pattern of overt violations of law with illegal mandates for anthrax vaccine, repeated again for unapproved EUA COVID shots, met the cynical definition of irrationality. Restoration of sanity required accountability.

With anthrax vaccine, there was no accountability within the government for not following their own rules. The cycle will only be halted this time around through intellectually honest judicial action and executive-level accountability from fresh leadership that is unencumbered by moral timidity, institutional venality, and a cultural reluctance to admit past errors. Just as I have loved and trusted my nation, our Air Force, and the Academy, Hoping 4 Justice envisions a future where current or new leaders return that trust by demanding the truth, correcting the wrongs, restraining their gate guards, and arresting illegal policies real-time. Our future presidents should not have to set the record straight thirty, forty, or fifty years later.

MILE 25:
Whistleblower 1-2-3

A worthy lesson learned that is valuable to convey from these experiences is that you really don't want to strive to be a whistleblower. Just like being labeled an anti-vaccine person, even when you're not, shaking off the whistleblower stigma is challenging. To counter the counterfeit narratives, you have to clarify professionally and persistently who you actually are—anti-illegal-mandate—our government should be as well. Lessons learned from someone who did not want the label, did not look for the trouble, but fought back, reframed it, and overcame it, may be helpful. Government and military leaders will benefit from these lessons learned as well, simply because they cannot deny, disparage or divert from a genuine narrative—anti-illegal-mandate.

No one who knows me probably imagined their friend ending up pigeonholed in this pejorative box of being either anti-vaccine or a whistleblower. My upbringing, education, experiences, and accomplishments do not conjure images of the whistleblower cliché. I was not the root cause of any of the underlying circumstances that placed me into this box. My decision to accept the branding led me down a tougher path, but one with no regrets. I chose to flip the script, turn it around, and cast it as a positive. My flight plan and methodology may help others similarly situated in the same dilemmas related to COVID injections. It applies to any personal or professional dispute encountered in the workplace, be it military, government, or business.

So, what are my rules, or the Rules of Engagement (ROE), to protect yourself? These rules might preclude whistleblowing altogether or prevent reprisal. Throughout my unyielding journey I learned, executed, and reinforced a simple formula for combating ethical dilemmas:

1. Be professional at all times. Give no one ammo to use against you. If you give it, they will use it. Never respond tersely or disrespectfully in any communications. Be compassionate and dispassionate at all times. Do not wear your uniform or refer to your organization when expressing your professional dissent publicly so your higher-ups cannot accuse you of ethics violations. Win or lose, maintain your professionalism. Principally, ensure integrity guides your conduct and motivations before, during, and after any professional dissent scenarios. In our disputes, leaders tried ineffectively to change the subject or disparage our efforts. Since we never gave them ammo to use against us, they could not change the subject or stop our advance.
2. While being unyielding, also be reasonable, humble, and respectful, even if your adversaries are not. Stick to the facts and rules—their rules. View disagreements from the perspective of the other party. Reasonably compromising did not equate to a loss. It could be a win-win. In our professional struggles, though humility never equated to going away, the only thing we could be accurately accused of was our "unyielding" persistence. The general was right on that, ironically providing both the title and the motivational marathon theme for this book.
3. Avoid challenging discretionary matters, and understand the limitations of your expertise. Be particularly careful not to charge the hill over discretionary issues where the opposite side has the full authority to make the call. Disagree, sure. Give your input, yes. However, discretionary matters should not escalate to whistleblowing scenarios. Reserve whistleblowing to combating rule breaking. Focus on the government's departure from laws, rules, and regulations. Always expect the government to be given wide latitude and discretion in the arenas of agency expertise.
4. When the organization does not follow their own rules—prosecute. In doing so, never give up or give in. The persistent, unyielding spirit, while irritating to the opposition, wins the day

when you will not go away. If the impasse is important enough to surmount in the first place, then follow through. Stick to facts and rules versus interpretations. Arguments must be black and white violations of rules or law that you seek to enforce. Misinterpretations are their mistakes, not yours. We prevailed over the anthrax vaccine because the adversary blatantly broke the law. It took time to make the case, but we won. The resulting EUA precedents guaranteeing "no penalty" created prescient legal arguments for equitable treatment over illegal COVID orders.

5. Make the system work by using the oversight tools your organizations provide. Most offer iterations of quality control guidance, risk management practices, inspectors general safety officers, human resources, compliance reporting, anti-waste, or fraud, abuse, and mismanagement protections. Read the rules and employ them to protect yourself. For DoD personnel, use the DD Form 149 to correct errors, expunge punishments, and remove sneaky adverse discharge reentry codes.[1] While anthrax vaccine policy complaints and DD-149's yielded limited corrective actions from the DoD, the later illegal firing over hiring was a victory due to honest review. A COVID-19 era example included a retaliation complaint by the only military member court-martialed, Lieutenant Mark C. Bashaw.[2] This sole prosecution indicated the DoD was reticent to prosecute over patently illegal mandates, likely to evade redress.
6. Wisely, live to fight another day by maneuvering smartly. Outflank the efforts to get rid of you. The anthrax vaccine boxed us into a corner. We fought back, won, and continued to serve. With COVID shots, my company gave me an escape hatch by granting exemptions. I took the hint, followed the process, and lived to fly another day. Do not die on the hill, but stand your ground. With future mandates, file complaints, exhaust exemptions, and consider litigation.
7. You do not have to be the tip of the spear in every battle—delegate and collaborate. Let others take the lead. Work from

the background to preserve your energy. These battles can sadly ruin careers, relationships, and lives. Survive by eliciting help from colleagues, compassion from family, and avoid tiring those relationships with your unyielding spirit. My outspoken tactics with anthrax vaccine transitioned to more of an "in the background" assistance of the troops, lawyers, and nonprofits during the COVID injection madness, at least initially. Work behind the scenes, below the radar, to preserve energy so that you can be unyielding when required.

8. Constantly cross-check your own personal and professional assumptions, predispositions, preconceptions, and prejudices to avoid the pitfalls of cognitive bias. We are all human and, at times, become victims of our own biases. Recognizing the cognitive bias trap is vital in your effort to remain open-minded and reasonable, even when the other side is not. My family held as diverse opinions on the injections and their mandates as existed in the spectrum of the American public. They helped me to constantly reevaluate and resurvey my own biases, judgements, and strategies, as well as remain cognizant of the limitations of my own knowledge and expertise.
9. Do not quit! Sadly, the system is not designed to help whistleblowers. Most leaders really don't want the squeaky wheel to be successful. Though laws support protecting whistleblowers, nothing in the law demands that whistleblowers receive mentoring or assistance. If those mechanisms existed whistleblowing might not occur in the first place. Whistleblowing is a lonely journey, designed to be a difficult marathon that begs you to quit. Once you enter the race, be wary of accepting voluntary personnel actions, or "plea deals," that forgo rights for redress of grievances. It is unfathomable that a citizen or soldier should have to quit if they're ultimately proven correct. If you blow the whistle, and enter the marathon, never quit, never, never, never!

10. Above all, despite the moral duty to be resolute and unyielding in the professional realm, always capitulate to win at home with your loved ones. They are all that matters in the end. RFK Jr advocated for a strategy of humility—a "posture of surrender."[3] This healthy and humble stance creates serenity in all arenas and will resonate with family, friends, and colleagues.

Now, how about the rules of engagement, or the ROE, for government, military, and business leaders? How do those leaders guard against overreacting or retaliating against an employee? Leaders also need a simple formula to combat ethical dilemmas in their hopefully unyielding journey to protect whistleblowers and inspire healthy cultures for their organizations.

1. As recommended above, be professional, reasonable, humble, compassionate, and respectful at all times. Let your employees do their jobs. Part of that job includes sincerely allowing subordinates to raise legitimate concerns to their leadership. Be receptive versus defensive. Imagine if our commander called a "knock it off" at the outset in 1997. Imagine if the FDA diligently reviewed, adjudicated, and modified government policy based on our 2001 or 2021 citizen petitions when we first warned that the anthrax vaccine and COVID inoculation mandates were illegal. Since the federal courts later affirmed both, imagine the possibilities for saved embarrassment and preserved trust. Imagine if leaders actually listened to their citizens and troops who follow the rules, instructions, and laws and expect the same from their bosses. The first ROE is to not rush adverse actions, reprisal, and retaliation in order to crush dissent.
2. Be cautious not to go to war with your subordinates over non-discretionary matters, or those codified in regulations or law, especially if your employee has a valid grievance. Ensure you provide a wide enough berth, with an open enough mind, in case the worker is correct on disputes that relate to rules. Imagine

if the general reviewing my commander hiring case actually investigated the issues. He may have discovered his officers cheated, forged, and lied in order to block me from assuming command. The general's intervention would have precluded higher-level investigation and the later substantiated reprisal. Meticulously inquire—avoid cover-ups.

3. Follow your organization's guidance for managing the whistleblower process. Make sure you provide protection and a path forward for subordinates that find themselves in conflict with the organization's direction. Steer the employee back on course, or alter the flightpath of the organization if the process highlights improper conduct and practices that veered off the path. In my two circumstances of whistleblowing, I never witnessed any leaders that I complained about ever humbly resurvey their judgments. Regrettably, it was adversarial from beginning to end.
4. Avoid placing your valued employee into a corner by testing their values and ethics. Life and work are tough enough as it is without expending efforts to drive valued, trained assets from their jobs through mandates or ill-conceived policies. Just as I advise workers, live to fight and lead another day by refraining from being the organization's peddler of controversial policy. You do not have to be the test pilot for a poorly planned program or politically driven dogma. Certainly, intervene if your team member is not doing their job, but wisely avoid making an example of someone who believes they are standing up on principle, and may actually be right.
5. Perhaps most to the point—avoid imposing medically unnecessary policies of questionable legality with injections of unknown safety and efficacy. Use your authorities to slow-roll politically motivated health mandates. Protect subordinates from policy that violates ethical, moral, and legal frameworks. Don't place subordinates in the precarious position of having to accept discharge through voluntary personnel actions, without a right

of redress. If the DoD confidently believed in the legality of their mandate, it is unlikely that they would have asked subordinate leaders to subtly manage discharges to prelude corrections by ensuring expulsions were all zero redress plea bargains. Let them do their jobs until the mandates pass. The anthrax vaccine and COVID mandate historic failures both support this patient prudence.

6. Just as your employees, customers, citizens, or troopers may blow the whistle to avoid guilt, regret, and resentment, avoid resenting that person for doing what they believed was their duty. If they prevail, be cautious not to reprise or retaliate punitively. Do not feel wounded that a subordinate questioned your authority. Use it as a learning point. Medical professionals should not take offense at patients for questioning their practice of medicine. Admit when the science changes. Some voices humbly did so in changing their message from "we can't trust the unvaccinated," and that "it needs to be hard for people to remain unvaccinated," to acknowledging a known three to four fold superiority and durability of natural immunity.[4] [5] Leaders must focus on the root sources of false messaging—the CDC and FDA, or the DoD—the entities that transmitted the bad policy and false information down to operational level practitioners. Rather than rationalizing about "protecting the institution," keep focused on the root causes of the dispute instead of retaliating against the whistleblowers.
7. If suggestions number one through six above do not pan out due to situational and institutional forces, stand tall in formation with your subordinate, employee, worker, customer, client, or team member. Join them. FDA guardians should have blown the whistle regarding reports accusing the agency of "grossly inadequate" oversight inspections "endangering public health" regarding COVID injection manufacturing.[6] FDA had a duty to protect Americans from lot-to-lot inconsistencies, hyper-concentration of mRNA, and "batch-dependent safety signal"

> problems.[7] It is conceivable that the immense credibility damage to the medical and government establishments are irreparable in the near term, but individual practitioners can still salvage their reputations. In my own journey, the officers that abused their power, implemented illegal orders, reprised against subordinates, and then tried to cover it all up, could have been unyielding leaders, adhered to their oaths, and altered the course of events—it's never too late.

My humble suggestions above carry less weight than the counsel of Supreme Court Justice Neil M. Gorsuch. He warned, "Today, our Nation faces not a world war but a pandemic. Like wars, though, pandemics often produce demanding new social rules aimed at protecting collective interests—and with those rules can come fear and anger at individuals unable to conform for religious reasons." The judge's opinion described one case as a "cautionary tale," and how it should "remind us that, in the end, it is always the failure to defend the Constitution's promises that leads to this Court's greatest regrets." Judge Gorsuch suggested, "in America, freedom to differ is not supposed to be 'limited to things that do not matter much. That would be a mere shadow of freedom. The test of its substance is the right to differ as to things that touch the heart of the existing order.'" Gorsuch asserted, "The test of this Court's substance lies in its willingness to defend more than the shadow of freedom in the trying times, not just the easy ones." He warned the court against "falling prey once more to the 'judicial impulse to stay out of the way in times of crisis,'" and that the Supreme Court itself "came to recognize that the Constitution is not to be put away in challenging times."[8]

Judge Gorsuch's opinions specifically related to the "dismissal of thousands of medical workers—the very same individuals New York has depended on and praised for their service on the pandemic's front lines over the last twenty-one months." Gorsuch noted the malice by writing, "To add insult to injury, we allow the State to deny these individuals unemployment benefits too." Gorsuch pled, "One can only hope today's ruling will not be the final chapter in this grim story. Cases like this one may serve as cautionary tales for those who follow." Gorsuch asked, "How many more reminders do we need that 'the Constitution is not to be obeyed or disobeyed as the circumstances of a particular crisis . . . may suggest'?"[9]

Later, a New York Supreme Court ruled in favor of its police officers cornered by the mandates. Initially the court found that the "vaccine mandate is invalid to the extent it has been used to impose a new condition of employment," and that employees "were caused to be wrongfully terminated and/or put on leave without pay as a result of non-compliance with the unlawful new condition of employment." The judge ordered police officers in particular to be "reinstated to the status they were as of the date of the wrongful action."[10] New York Supreme Court judges in cases for sanitation workers and teachers ordered reinstatement and back pay for all city employees, finding that the mandate "was not just about safety and public health; it was about compliance." [11] Workers, funded by RFK Jr's Children's Health Defense, prevailed. The New York Healthcare shot mandate was struck down as "null and void" and then rescinded.[12]

Our unyielding journey in whistleblowing is the cautionary tale Justice Gorsuch described, because it all happened before and many of the actors reappeared. The Justice aptly summarized, in another case related to emergency decrees, how we "experienced the greatest intrusions on civil liberties in the peacetime history of this country."[13] Justice Alito joined the dissenting judges in another Supreme Court opinion, and sadly explained that our Navy SEALS were "treated shabbily" by the Navy and the DoD regarding religious exemptions.[14]

Other federal judges concurred by partially enjoining the USAF and USMC shot mandates on religious grounds.[15] The entire population was treated shabbily, and not just by the authors of the mandates. The courts held some responsibility for not promptly hearing the cases brought forth to challenge the fundamental 21 USC §360bbb-3 arguments—the illegality of mandating EUA products that Americans had a legal right to "accept or refuse." Is it any wonder that the federal judges avoided ruling on mandates after they allowed their own court clerks, paralegals, secretaries, and staff to be forced into compliance? It's not—it was shabby neglect.

This is not the final chapter. Though it is a grim story and cautionary tale of our citizenry collectively falling prey to the manipulation of fear by power actors with a singular agenda to vaccinate, regardless of law and necessity. While disquieting to witness the mostly muted judiciary and the profoundly shabby treatment of all citizens, future generations will look back and grimace at the gullibility, lack of accountability, and oath breaking against patients, troops, employees, and fellow doctors. The next generation will have this script and the rules. They must stop the bad causes, bad means, bad behaviors, and bad actors before the emergency is declared.

Most likely, the Supreme Court will never hear most of the cases, but instead their shadow rulings, and the lower courts, will continue to rule in favor of the front line workers and operational practitioners based on tangential issues, such as exemptions, while dodging the embarrassing core Title 10 and 21 violations and patently illegal EUA mandates. Sensing more unfavorable rulings, the bullies may preemptively pat each other on the back for their campaign to overcome vaccine hesitancy. Considering booster uptake was between 1 and 17 percent, the self-congratulation was premature.[16] These realities led the government to wisely withdraw the mandates in order to vacate lower court rulings, to dismiss the cases as moot, and to avoid more adverse rulings.[17] So, despite the judicial brinkmanship, wins were wins. "Hang in there," as our astute mentors encouraged us two decades ago—unyielding spirits proved contagious over time!

In the interim, the suggested checklists for both whistleblowers and leaders are not all-inclusive, nor certainly are the warnings about the gate guards who were emboldened to promote illegal, unwanted, and unnecessary medical procedures. The learning points suggest citizens and soldiers alike must watch for the patterns that we witnessed in our twenty-five years of figuring out how to manage life in the face of extraordinary struggles. Blowing the whistle, when done correctly, is simply doing the right thing in the face of such difficult choices. It is an unyielding duty in line with how our parents, bosses, and nation raised us, and how we raise, motivate, and train those that follow us. I'm confident that these unyielding principles will prevail over time.

If it happens to you, it is important to understand you are most likely not responsible for the root causes of the whistleblowing activity. However, you are responsible for how you conduct yourself during the process. That conduct will be the key to whether or not you are successful in outflanking possible retaliation. Blowing the whistle on bad behavior, bad actors, bad means, bad causes, and patterns of abuse is clearly an unyielding duty performed to uphold the integrity of your professions. You may also inspire courage in those who follow your path.

One of my dear friends, who listened to my whistleblowing stories, asked, "How did you ever make it to retirement?" I explained that while my story may seem disenchanting, military members do stand out in the fortunate ranks of personnel who, if they follow the rules, can still do the right thing, and survive. I am grateful the military somewhat followed the whistleblower laws, allowed me to continue to serve the first time despite anthrax vaccine, and reversed the illegal reprisal the second time over the cheating by the chain of command in the hiring and firing fiascos. With anthrax vaccine, champions within the DoD protected me. With the hiring reprisal, the DoD promoted me as a remedy based on merit. These positive examples prove the DoD has the potential to serve as a benchmark in efforts to protect whistleblowers. All troops should have been treated the same. The DoD can serve as a role model to steer the nation back on course—leading by example and putting the shovels down by protecting whistleblowers.

In line with rules on whistleblowing, additional potentially valuable material included a personal and professional life management model collaboratively developed with my colleagues while serving at Hanscom Air Force Base. Our model used the acronym **HEED**, which stands for **H**ealth, **E**ducation, **E**thics, and **D**ecision-making. The model's Decision-making step ends with the sub-acronym FAST, which stands for Find risks, Analyze solutions, Select a solution, and Track its effects. Explanation of the model in my final mile ends with the FAST decision-making tool. It works by Finding and Analyzing the risks associated with the repeated unhealthy patterns of deception and manipulation, while also Selecting solutions and Tracking their effects.

Had governments HEED'ed this model, it is conceivable the violations and loss of trust associated with both the anthrax vaccine and COVID injection mandates would never have happened. Instead, leaders either reflexively or intentionally contorted inconvenient analysis and publicly available information as the underlying cause of mistrust, terming it an "infodemic."[18]

Hopefully, these Whistleblower 1–2-3 rules will assist all parties to avoid such mistakes. As Supreme Court Justice Gorsuch wrote, "Make no mistake—decisive executive action is sometimes necessary and appropriate. But if emergency decrees promise to solve some problems, they threaten to generate others. And rule by indefinite emergency edict risks leaving all of us with a shell of a democracy and civil liberties just as hollow." Whistleblowers, analysis, and information serve an essential role in balancing executive action against this slippery slope.

Having explained the diligent duty to blow the whistle, and cautioned against governmental adverse reactions to these critical inputs, I will describe in detail the practical model about how we learned to HEED our training. I will also perform some retrospective FAST analysis on the many toxic patterns we observed before I take you across the finish line.

MILE 26:
HEED Your Training

The **HEED** life management model's co-authors included Colonel Juan Gaud, Colonel Lee Pritchard, and Master Sergeant David Martin.[1] The model's principles built upon a risk management article our team published in the USAF Flight Safety magazine in June 2004.[2]

HEED stands for **H**ealth, **E**ducation, **E**thics and **D**ecision-making. HEED offered a template to conduct safety related briefings, and provided our personnel with a simple visual aid for important concepts to consider in the balancing act of their personal and professional lives. The HEED model provided organizations and personnel at every level a visual checklist to reinforce, or "HEED," their USAF training, while also incorporating the Wingman theme.

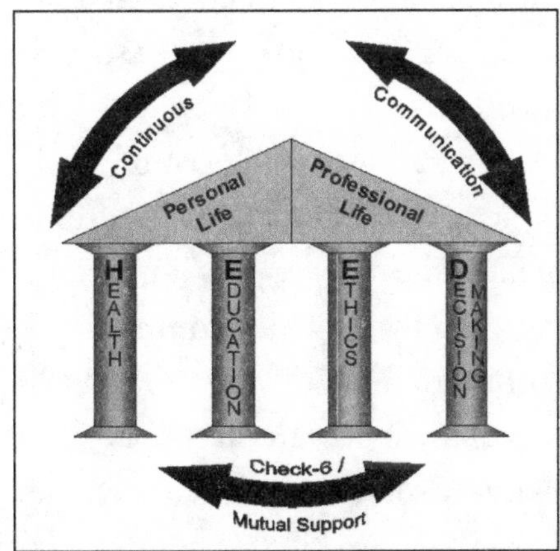

(Remember your boldface: HEED = Health + Education + Ethics + Decision-making; And FAST = 1. Find risks, 2. Analyze solutions, 3. Select solution, and 4. Track effects!

Surrounding the HEED model you find the **3 C's**, which act as pivotal forces for success. These three primary elements of our checklist's cross-check ensured we **C**ontinuously **C**heck-6 for our comrades in arms, our wingman, and **C**ommunicate HEED's pillars in every aspect of our family and work activities. In addition to upholding USAF core competencies and values, a value-added aspect of the HEED model was an emphasis to share the model with family.

HEED and the 3 C's provided a broad-based, memorable framework for daily application by our airmen and their families. Health, Education, Ethics, and Decision-making must be like a pilot's "boldface," or memory items for our service members—a checklist mentality and daily maintenance tool for our troops in all their affairs. HEED is a mindset that reminds us to utilize our common sense on the way out the door to play or work, just like pilots do in being required to carry their checklists and know their boldface cold as they press out on every mission.

We designed HEED as not only a construct for the accomplishment of many of our safety related tasks, but also as crosscheck for the concepts our leaders trained us to internalize. It kept us on the right course personally and professionally, and helped our families and our fellow aviators to do the same. HEED works, no different than pilots cross-checking their attitude and altitude, to ensure we stay on course. A brief review specific HEED pillar application follows:

HEED's first pillar emphasized Health for both our personnel and equipment. Physical health or fitness, as well as mental and spiritual health, were obvious applications. Financial health was also important, as was the health of equipment you, your family, or your teams use.

HEED's second pillar emphasized Education in both the professional and personal realms for both humans and any hardware they operate. It is pivotal that all flight crew and their supervisors routinely checked that their educational aims were on track, and again this was no different than what you would do for your own family members as well.

HEED's third pillar emphasized Ethics—both the values of a person and their work ethic. We did not intend to reinvent the wheel in this realm, but instead to emphasize that the checklist mindset should encompass maxims like the Air Force Core Values. These values should serve as boldface, or memory items: "Integrity, Service, & Excellence!" These thoughts were no different than what our parents taught us—"Honesty, Selflessness, and Hard Work." HEED asked our troops to take the USAF Core Values home to their families. Applied in the home, it's no surprise how quickly children internalize HEED, and even remind us to as well!

HEED's fourth pillar emphasized Decision-making processes. The bottom line is we must always be thinking through our next task, stunt, operation, or recreational activity, constantly weighing risk versus rewards and costs versus benefits. HEED's fourth pillar offered a sub-process Decision-making acronym—**FAST**. FAST simplified decision-making to a memorable format, to inspire increased use. HEED's "FAST" steps include: 1. **F**ind risks, 2. **A**nalyze solutions, 3. **S**elect solution, and 4. **T**rack effects.

The USAF offered the Operational Risk Management Model six-step process. The Air Combat Command offered a three-step simplified process. While both of these decision-making processes were useful, flyers may need a process they can remember on a time-critical basis. Our model's four-step process should not complicate the concept of decision-making, but instead simply emphasized that you must use something.

Whatever you do, keep it simple. Think of HEED as a checklist mentality, a framework, a process of prevention, and food for thought for lead turning risky behavior. By utilizing the enduring USAF theme of taking care of your wingman, you can Continuously Communicate and Check each other's "6 o'clock." The HEED checklist and FAST decision-making cycle, or something similar, provide a simple personal and professional cross-check for your life.

Most aviators and troops never have time to read the hundreds of pages of guidance in USAF Leadership and Development Doctrine. Though this may be a harsh and unwelcome reality, it is true. That is why both citizens and soldiers needed a simple and easy to remember tool—one that all Americans will remember, use, and hopefully pass along to their kids.

Use HEED and FAST, or develop your own checklist for your

organization and your family. Teach your families to live Healthy, get Educated, conduct themselves Ethically, and work hard. Finally, use solid, simple Decision-making, such as FAST, in all you do.

Applying the HEED model and FAST tool to both the anthrax vaccine dilemma and the COVID shot mandates allowed us to Find risks and Analyze threats to the other pillars of health, education, and ethics. The preceding pages documented numerous patterns of behavioral anomalies not normally acceptable in any traditional intellectual realm.

The fact that the behaviors remerged in both the anthrax vaccine dilemma and again with the COVID shot mandates make for case studies worth noting. Identifying the habitual patterns will decrease susceptibility to process breakdowns and psychological manipulations. Examples of observations on the trends and patterns common to both illegal mandate dilemmas follow:

1. Normally governments, their leaders, businesses, and bosses encourage risk management and critical thinking. In the case of vaccine mandates too often responsible officials dissuaded, discredited, or canceled genuine attempts to critically analyze and evaluate risk. This should be the first warning flag that something is wrong—when governments, organizations, supervisors, or their information apparatus squash critical inquiry. The intellectual blackout occurred due to incorrectly framing the suspected Wuhan bioincident as a natural event,[3] while casting contrary views as conspiratorial misinformation. Another practical example involved the subsequent removal of the Governance section from the National Biodefense Strategy. This occurred more likely than not to make the strategy mirror the administration's actual stove piped response and insular silo thinking.[4] These management errors stifled inputs at the highest level during the pandemic. The precipitating failure to properly classify the Wuhan lab leak as a bioincident precluded effective implementation of the national strategy. The resulting revamped strategy memorialized the top-down and stove-piped response

that throttled the truth. Instead of a collaborative decision-making process, they siloed-out the key stakeholder's critical inputs.

2. With both the anthrax and COVID mandates, groupthink tendencies prevailed. With the anthrax vaccine, military doctors blinded themselves by ignoring the controversy and then failed to protect their troops from patently illegal policy. With COVID, some pharmacies and doctors denied lifesaving prescriptions and treatments by blindly following CDC guidance. Some were unfamiliar with published and peer-reviewed scientific and medical journal articles, all while parroting what many Americans perceived to be a highly politicized medicine party line.
3. For both mandate programs, officials promoting the policies at times compromised their integrity in order to protect policy. When writing Karl Rove, Mr. Perot referred to the bad behavior as making "science conform" to political objectives and policy. Mr. Perot identified the "stress theory" advocates, just as we identified the precarious presence of institutional gate guards as a dangerous pattern as well. With the anthrax vaccine, federal courts set the record straight after many years. With COVID shot mandates, the damage was potentially much more immense—so much so that the damage may never be set straight. Future generations' reflections on the misdirected scientific, medical, and moral compasses, plus the bad behaviors, the bad actor gate guards, and the creation of pseudoscience to conform with policy will hopefully make them wise to the incessantly repeating patterns and hopefully keep there from being a next time.
4. Enacting new laws, amending old laws, and misinterpreting existing laws emerged as problematic patterns with both mandates. The Emergency Use Authorization law, under 21 USC §360bbb-3, represented a workaround endpoint for the anthrax vaccine litigation and the intersecting starting point

for COVID injection mandates. True to form, ignoring or misinterpreting the law's requirements mimicked the anthrax vaccine experience. With the anthrax vaccine mandate, the DoD violated 10 USC §1107 and crafted quick-fix, double negative memos misinterpreting the law's requirements until busted by the courts. After the court enjoined the illegal conduct, the EUA law loophole provided a face-saving means to continue vaccinations on a strictly voluntary basis. With COVID shot mandates, the DoD ignored 10 USC §1107a for EUAs, and ignored this first EUA precedent that ensured "no penalty or loss of entitlement." The government rejected EUA law requirements, and the "option to accept or refuse" as the legal standard on authorized, but unapproved, products for all Americans. They replaced precedent and law with faulty misinterpretations of one word: "consequences." As well, the FDA formulated a "framework" to expedite approval or authorization of new vaccine versions, but without the normally required clinical trials required to assess safety and efficacy.[5] Once the public health emergency ended, after being renewed twelve-times,[6] the government rapidly amended the law to permit liability shields and continued use of the EUA products.[7] Be on guard for more quick fixes by public officials. One instance included adding ineffective, unnecessary COVID injections onto the CDC's children's immunization schedule to extend liability protections.[8] [9] Watch out for future alterations of EUA law due to increasing hesitancy.[10]

5. Pitting groups against one another is a classic manipulation tool. National leaders at the highest levels verbalized an unhealthy, divisive "vaccinated" versus unvaccinated dynamic that inappropriately infected every level of our society. Bullying and coercion replaced healthy dialogues with both the anthrax and COVID dilemmas. Our nation's leaders, particularly within the DoD, should agree these conundrums resulted in an absurd waste of time, trust, and human resources when they should have been focusing on legitimate national security threats. We detract

from readiness and the mission when diverted by distrust over imprudent policies.

6. Accelerating vaccinations occurred early in the anthrax vaccine experience in order to get troops inoculated. Once inoculated, people tend to not voice concerns about the vaccines. The acceleration pattern of hurrying and bullying vaccination is a dangerous warning signal, which occurred during both the illegal anthrax vaccine experiment and the COVID pandemic mandates.
7. Fear embellishment drowned out most opposition to the anthrax vaccines, even though it was not a communicable disease, and despite the fact that the threat was unchanged according to the GAO. Fear blinded most people from other concerning facts, like deviations by the manufacturer according to the FDA, known inadequacy by the DoD, and improper licensing. Anthrax vaccine program cancellation appeared imminent until the anthrax letter attacks halted all oversight. The COVID pandemic, involved a communicable disease likely of synthetic origins, but evolved into consecutively less lethal viral variants. The fear inducing techniques during COVID included all the abovementioned concerns, but also included new manipulations of the population based on ultimately unwarranted fears of the unvaccinated. Both bioincidents, and their fear-based responses, spread despite legitimate legal, safety, and efficacy alarms.
8. Misinformation allegations against critics of official policy first materialized during the anthrax vaccine dilemma and then became the norm with COVID mandate controversies. With the COVID pandemic, the government, mainstream media, and social media went further by censoring opposition messaging and actually asking Americans to report those suspected of misinformation.[11] Evolving science and circumstances demanded open critical inquiry, but instead in both mandate scenarios academic freedom was quashed to protect flawed policy and to further the counterfeit dystopian narratives.

Disturbingly with COVID, Congress reported that the Cybersecurity and Infrastructure Security Agency (CISA) under DHS "facilitated the censorship of Americans directly and through third-party intermediaries." Federal court injunctions, though intermittently stayed, affirmed the suspected collusion between the government's media and big tech censorship proxies.[12] Those proxies failed not only in their duty to exert their own constitutional right (freedom of the press), but even worse, they spurned citizen's rights (freedom of speech) by "contacting social media companies for the purpose of urging, encouraging, pressuring . . . the removal, deletion, suppression, or reduction of content containing protected free speech posted on social-media platforms."[13] The Supreme Court shunned such unconstitutional departures, and even used our favorite Orwell metaphors: "Our constitutional tradition stands against the idea that we need Oceania's Ministry of Truth."[14] Apart from unconstitutional collusions, the anthrax and COVID bioincidents included misdirects and misinformation related to their origins, despite our government's involvement at some level in both historic bioincidents: Manchester, Amerithrax, and Wuhan. With anthrax it took a decade for the FBI to admit an insider motive for the lab leak to save the failing anthrax vaccine program. With the COVID bioincident the government and media attempted to derail as conspiracy any suggestion of a lab leak—ultimately deemed by the FBI as the most likely proximal origin of the pandemic. Eventually, oversight by the courts and Congress concluded the government and media "weaponized" misinformation. With anthrax, the bad actors also allowed the proliferation of a weaponized bioagent to rejuvenate a failing military vaccination policy—something only imaginable in a world that sanctioned the blot represented by an Oceania-like Ministry of Truth.

9. In both dilemmas, indemnification and liability waivers for biopharmaceutical companies created the impression of an

immunized cottage industry at taxpayer expense. With both mandates the cornering of captive and unwilling customers added insult to injury. Ironically, the National Childhood Vaccine Injury Act of 1986 gave birth to this new model of protecting vaccine makers from liability.[15] The 1986 Act's advocates, such as Barbara Loe Fisher of the National Vaccine Injury Coalition and others worked arduously to gain its passage in order to secure protections from harmful medical products for America's children.[16] While the law initially accomplished these goals as designed, and positively led to the development of the VAERS, ultimately the law was adulterated. Instead of protecting children, it added the comforting blanket of industry liability shields for manufacturers, even for "design-defects."[17] Similar dilution of the laws applied to 10 USC §1107 where the §1107a EUA Appendix permitted liability-free use of unapproved products along with the DoJ's "consequence" misinterpretations that wiped out consent rights. To add specific historical perspective about nonsensical extensions of indemnification and product liability shields, before Christmas 2022 the government gifted a renewed "PREP Act Declaration for anthrax" through the end of 2027. The declaration was deemed "essential due to the continued national security threat posed by potential exposure to anthrax."[18] Shortly thereafter, the DoD squandered hundreds of millions of taxpayers' dollars for more unnecessary BioThrax vaccine.[19] The "threat posed" conflicted with GAO assessments and the reality that the only anthrax epidemics in America's history were the 1957 Manchester, NH Army vaccine trial and the anthrax letter attacks that the FBI concluded were perpetrated by a US Army scientist with the motive to save the "failing" anthrax vaccine.

10. The unsatisfactory nature of the anthrax vaccine was the official military position for the anthrax vaccine until the policy was announced. Essentially, our original stance was also the government's position, until they changed their official posture

and medical literature. No different initially for COVID injection mandates, the federal government agreed that vaccine mandates were not within their authority, nor prudent. Then the policy changed to include new legal interpretations, which conflicted with EUA law and precedent. Natural immunity, an axiom of science until the pandemic, due to the human bodies' innate sterilizing defense after recovery, was previously accepted as a medical norm until it threatened universal vaccination protection. Despite supportive natural immunity data,[20] and Dr. Fauci's past recommendations,[21] public health officials ignored the inconvenient truths and facts. Like anthrax vaccine, the government's position changed. Officials concocted medical literature, altered "vaccine" definitions, and arrogantly continued gain-of-function research—tainting science to conform with policy.[22]

11. Regulatory shortcuts were the trademark of the anthrax vaccine until caught by the federal courts. Similarly, with COVID injections we witnessed shortcuts with EUA fast tracking, grandstanding about of "FDA approvals," and extralegal interchangeability claims. The FDA and CDC messaging trickery defied their Preamble and mission to "protect the public health," as well as to ensure the "American public receives only safe and effective vaccines." The American public was intentionally confused regarding the interchangeability of "authorized," but unapproved, EUA injections. They were allowed voluntarily in the practice of medicine, as opposed to the Food, Drug, and Cosmetic Act's strict regulatory regime and rules for legally distinct and fully approved products. 17,000 medical experts petitioned to right these wrongs.[23]
12. Crafting snappy, false soundbites with the anthrax vaccine included: "Safe, Effective, and FDA Approved." With COVID shots, the truth was we simply did not know for certain. Safety proved concerning, while efficacy waned. Regarding approval, when the government said the Pfizer injection was

FDA approved, that product, Comirnaty, was not available. The government confirmed, "Pfizer does not plan to produce any product . . . while EUA authorized product is still available and being made available for US distribution."[24] After spending approximately $30 billion,[25] the US government deauthorized the original monovalent and bivalent EUA injections, while providing broad liability immunity through the end of 2024 for "approved" and updated animal testing framework formulas even though the public health emergency had ended.[26]

13. One of the first indications of a problem during both the anthrax vaccine dilemma and the COVID shot mandate controversy was dupe-looping by government policy gate guards that interfered with the practice of medicine and adherence to the Hippocratic oath. With the anthrax controversy, the gate guards pushed an unprecedented doctrinal departure with force-wide inoculations, as well as illegal vaccine mandates that violated consent requirements of our laws and Constitution. With COVID, additional examples included not acknowledging natural immunity and suppressing early treatment or alternative therapies. Such deviations were tantamount to a pilot not following their boldface, checklists, or standard operating procedures. Violations of the Constitution, which is essentially our government's checklist, as well as the dupe-speak by policy gate guards should have set off alarms. They even changed the definitions of vaccines from immunizing to mere protection in order to support policy. Confusing mixtures of legal statuses, whether approved or authorized, further defined the COVID gate guard dupe-loop.[27] These deviations arose due to the politicization of medicine where doctors did not object in unison to government gate guard intrusion into their profession and the practice of medicine.
14. The most disturbing commonality between the two medical mandates was seen in the failure to report problems and adjust policy due to the alarming adverse event profiles. This was

> exacerbated by statistical manipulations to marginalize negative health impacts. The poor safety data for anthrax vaccine pales in comparison to the data manipulations and negative statistics for COVID shots, not to mention the massive denial of injury claims.[28, 29] Be on guard for medical misdirects that imply post-vaccination syndrome and chronic injuries from the shots might be mistakenly attributable to "long covid."[30] The truth regarding the COVID injection adverse events and anomalous neuropathic symptoms will not be known until information suppression ceases and public health officials honestly conduct retrospective surveillance and analysis.

Examples related to those disturbing and common patterns of dismissing adverse events and deaths are important. While deaths related to anthrax vaccine were rare, they did occur. The notable pattern in the two mandate controversies is similar causes of death. Multi-symptom inflammatory syndromes and autoimmune disorders join the two medical dilemmas with vaccines as the common denominators,[31] and of course the commonality of causation denial. Terms like hypersensitivity pneumonia from the anthrax vaccine era seem eerily familiar to the similar inflammatory respiratory illnesses and deaths occurring after COVID injections. Potential attribution blindness with COVID shots was also reminiscent to the same symptom of anthrax vaccine adverse reaction denial.[32] Based on total doses administered, COVID injection injuries will likely overwhelm anthrax vaccine harms by a factor of hundreds or conceivably thousands.

One tragic death from our disaster comes to mind. Army Specialist Rachael Lacy died from pneumonia on April 4th, 2003. Her father, Moses Lacy, said, "The government is covering this up and it is a dog-gone shame."[33] Rachael, studying to be a nurse, was called to active duty at Fort McCoy, Wisconsin. Moses Lacy said his daughter "was a healthy young woman" following her death at the Mayo Clinic in Rochester, Minnesota. He added that the "common denominator is smallpox and anthrax vaccinations." At the time, the DoD acknowledged troops getting pneumonia in the Middle East and at least two deaths, with thirteen placed on respirators.

In a UPI report, Mr. Lacy said, "The Department of Defense is closing their eyes." The Army later held a news conference on the deaths, but tried to deny Rachael's death was connected to the illnesses or deaths of deployed troops. The coroner in the Lacy case, Dr. Eric Pfeifer, reported smallpox and anthrax vaccines "may have" contributed to Rachael's death. The coroner added, "It's just very suspicious in my mind . . . that she's healthy, gets the vaccinations and then dies a couple weeks later." The death certificate listed "post-vaccine" problems.

Astonishingly, one of the gate guards, the pharmacist, denied Rachael's death was related to the anthrax vaccine. Despite the death certificate by Olmsted County Coroner Eric Pfeifer confirming "diffuse alveolar damage" (damage to the lungs), with a contributing condition of recent "smallpox and anthrax vaccination," the pharmacist publicly insisted the case was an "unexplained death,"[34] He added, "Death certificates always conclude something." Later he claimed to UPI that vaccines are "probably not" to blame, that "vaccines have no role," plus a variety of non-committal comments such as "might not," "not at all clear," and that the death of Army Specialist Lacy was "still in the unexplained death program." The pharmacist gate guard confusingly speculated about the finding on her death certificate saying, "Its final contribution has not been finally decided" by CDC.[35] What program? Two years later a government's investigation all but concluded Rachael's death due to the vaccine was "probable."[36] Rachael received five vaccines in one day, so the vaccines may have triggered an autoimmune reaction and death.

USAF Technical Sergeant Sandra L. Larson also died after a sixth anthrax vaccination on June 14th, 2000 at the age of thirty-two. We may never know how many Rachael's and Sandra's there were. Fortunately, Sandra's sister, Nancy Rugo, was her advocate, like Moses was for Rachel. Nancy testified to Congress about Sandra's illness: "She frantically began researching the causes of her condition, and started to suspect vaccine connection."[37] Nancy explained that Sandra died of a "rare blood disease, aplastic anemia, which is considered an autoimmune disease." "Twelve weeks after receiving her shot, she was gone," leaving a child behind.

Nancy Rugo's heartfelt testimony about Sandra's death was followed by yet another October 2000 FDA inspection that cited the manufacturer in eighteen areas, including failing to investigate Sandra's death. *The Lansing State Journal* wrote, "The report also questions the company's handling of a reported June death of a US servicewoman"[38] where she "told family members and the military she believed she was sickened by the vaccine." Nancy Rugo said that her sister "was adamant about it . . . She did a lot of research on it while she was hospitalized."

I think about how lucky I am to have lived the American dream and how important it is to end such avoidable tragedies. We must honor those lost and those who lost loved ones. These honorable young women are just two of the suspected deaths. At least in our challenging times twenty-plus years ago, the mainstream media reported the tragedies, analyzed the death records, and questioned the officials. Who's listening to the millions of families that want answers and hope for justice over liability-free COVID injections? If few are listening, we will not learn.

When fighting the anthrax vaccine, Russ and I HEED'ed our training and professionally placed our analysis and conclusions on our DoD leaders' desks. Our *Army Times* opinion editorials from August 1999 and November 2002 could literally be rewritten today verbatim and republished for COVID shot mandates.[39, 40] It's the same mistakes, same warnings, same patterns, and the same violations of the law. Leaning on aviation metaphors, we explained our analysis of the facts no differently than the preflight of an aircraft prior to a training or combat sortie. "The bottom line was no different: You don't take a broken plane airborne." We cited the FDA's documented violations, the safety, efficacy, and legality concerns previously identified by military officials. "Our findings ran directly counter to the rhetoric promoting the Pentagon's program." We asked leaders "to correct, not repeat, past mistakes." At least then, we had a voice.

Though uniformly ignored at the time by the DoD, a couple of years later the federal court ruled in our favor, arrested the program, and many years later the DoD corrected a handful of records. We implored then, and I entreat future DoD leaders in kind, "Our troops should not be the ones held accountable for past mistakes as our nation moves ahead to ensure that the military and the public are properly protected in the future." The 2021 errors echoed those from twenty-plus years ago. It stains the integrity and credibility of our armed forces to repeat the same lawbreaking. As operationally focused fighter pilots the malfunctions we witnessed, and witnessed again with COVID, were unquestionably avoidable—you do not fly broken jets.

HEED's heuristic techniques provided an analysis continuum to identify patterns and risks resident with both anthrax and COVID. Ignoring the patterns identified in a risk analysis of anthrax vaccine or COVID shot mandates jeopardized the nation's Health. Examples included concerns about "pregnancy complications and menstrual abnormalities" significant enough to warrant calls for a "worldwide moratorium on the use of COVID-19 vaccines in pregnancy." [41] Other experts cautioned about pregnancy and fertility decline,[42, 43] as well as whether or not an mRNA "vaccine induces complex functional reprogramming of innate immune responses."[44]

Additional unknowns about COVID mRNA products suggested risks associated with the lipid nanoparticle (LNP) platform,[45] as well as cancer, stroke, inflammation, clotting, amyloidosis, immune system dysregulation, DNA contamination, encephalitis, myocarditis, retinal vascular occlusion, spike-protein-induced neurotoxicity, musculoskeletal disorders, and multiple sclerosis.[46, 47, 48, 49, 50, 51, 52, 53, 54, 55, 56, 57, 58, 59, 60, 61, 62] Answers to the issues based on the new scientific results did not exist before the rushed mandates and inadequate prior risk assessments.

HEED demands we ask these Health questions and get the answers in the proper order. Possible adverse effects aside, the suppression of Education and intellectual dialogue risks similarly severe Ethical implications, particularly if the integrity of the government's shifting position over time erodes further. As the Decision-making continues, an unyielding FAST analysis of problems and solutions should be encouraged in the spirit of healthy cognitive inquiry. Conducting intellectually honest risk analysis is a mandatory procedure even if "raising safety concerns about the vaccines can be uncomfortable."[63] Employing HEED does not permit expediency or an emergency to supersede a FAST analysis.

In the post-pandemic cleanup and repair process, that is exactly what our Congress finally did, and FAST. In 2022, they voted convincingly to rescind the DoD's COVID shot mandate.[64] Senators also appealed to the DoD to "immediately begin undoing the harm caused."[65] Their appeals mirrored a parallel public messaging effort for record corrections by myself and Sgt James Muhammad.[66] Fortunately a bipartisan majority saw through the flak. Congress determined the risk of the mandate to readiness caused more damage than benefit. The 2023 NDAA halted the "insanity."[67] Congress voted, 'not on our watch.' President Biden signed the bill into law, rescinding the mandate. The DoD dutifully affirmed, "The department will fully comply with the law."[68, 69] In time, the CDC rescinded vaccination cards, those easily forged passports to normality and instruments of illegality,[70] while the Congress continued its unfinished business to secure record corrections and reinstatement for our troops into 2024.[71]

Whereas the judicial branch stopped the anthrax vaccine mandate, the legislature imposed the unprecedented intervention against the COVID mandates, while the CDC attempted to erase the flimsy recordkeeping mechanisms altogether. In reality, the Congress did the DoD a favor by halting the patently illegal mandate prior to more adverse court rulings, like with anthrax vaccine. Congress voted 83–11 in the Senate, 360–80 in the House. Unfortunately, they voted only to repeal, not to repair. So, that was the next mission, like the last, for our harmed troops.[72]

Fortunately, the legions included new voices such as former Vice President Mike Pence. He called the mandate "unconscionable" and advocated for reinstatement and back pay. The DoD's rescission stopped short of that redress.[73] So, we pushed for the DoD to do it right this time. We updated our draft legislation for corrections and focused on adherence to the oath of office, honor code, core values, and Joint Ethics Regulation's requirements for honesty, candor, truthfulness and being straightforward. We emphasized the inexcusable quibbling by the DoD in mandating EUA biological products as interchangeable with non-existent approved shots. This violated the law, defied precedent, and conflicted with the DoD's own mandate memos. Even FDA officials agreed that "absent a license, states cannot require mandatory vaccination," and therefore top officials resigned in protest of the rushed approvals for non-existent products.[74]

Overall, the parallels and patterns between the anthrax and COVID-era abuses required reflection. The mistakes and solutions applied to each of the HEED model pillars. The government had their own models, oversight agencies, branches and departments that could have and should have studied and fixed these malfunctions.

Hopefully, our DoD, the government, and civilian leaders will HEED their training, listen better in the future to those they serve, and display common sense by joining us on the "picket line for medical freedom."[75] They had a duty to perform a requisite retrospective, independent, and objective analysis in order to fix the cultural divide—that chasm that separated those who followed checklists from the leadership that did not.[76]

FINISH LINE—THE DEBRIEF

I told this story about our many missions and marathons because it smoldered inside me for twenty-five years. Throughout the trek my collectivist imprinting gave way to an individualistic sense of right and wrong that my training instructed me to prioritize. While I normally supported the collective good, I had to challenge the bad. The miles turned to years of speaking out and constructively collaborating on citizen petitions to warn the government about the dangers of improperly regulated unapproved medical products. Both marathons proved our petition's warnings correct, but were un-HEED'ed by a government and military unwilling to reevaluate its sunk costs over past misjudgments. The unsettling anthrax vaccine patterns, repeated with COVID mandates, served as the final emotive force to finish the manuscript.

The marathons required documentation and analysis. I ran and wrote them down so my children and their children had something to contemplate from their unyielding father and grandpa. I am so proud that each of those children enlisted to serve our country. Their service made it clear that we raised them in a household of patriotism, duty, and service. I authored this book, as I did my thesis, to attempt to put it to rest—for the last time—so my ever-patient family did not have to endure listening to it anymore. Once the COVID pandemic was over, I hoped this would be my marathon's finish line, but I yieldingly understood the race continues.

As marathons go, even at the finish line, endorphins take over and you get delusional in thinking, "I want to do this again!" That's the timeless tease of marathons—the next one always challenges you. The difference with the marathons against illegal medical mandates is you really would prefer to never enter the race again. Sadly, the marathon metaphor from our anthrax vaccine battle emerged anew with mandates for unapproved and ineffective COVID injections of unknown long-term safety. Though I hoped I could put anthrax vaccine dilemma behind me, the next marathon raced across the world. The familiar arrogance of politics and power during the pandemic met citizen's questions with dismissal, censorship, and the all too familiar dupe-speak. Public health officials played the American people as their pawns with fraudulent messaging about FDA approvals and imposed career-ending mandates on their subjects.

Intelligent people worldwide wondered why the legitimate concerns of the multitudes about mandates and medicine could be flatly rejected. Why bother giving your troops, citizens, or employees critical analysis knowledge and tools, but then punish them upon application? Why challenge people to comprehend problems, but then banish them once they accept the task to critically analyze, synthesize, and evaluate?[1] The answer to me was clear—as Thomas Paine taught—it was the unwise result of bad causes and bad means by otherwise good people.

I was confident the resurgence of such antagonistic trends would not prevail. Instead, I believed an awareness of the abuses would guard our tomorrows. I wasn't so alone this time. We didn't follow the mob. We became the unvaccinated, unhypnotized control group. We recognized bad behaviors and courageously did not comply. As with the anthrax vaccine, we observed patterns that reemerged in the pandemic, even more so based on the breadth, depth, and magnitude of the bad behaviors. Debriefing those patterns, identifying the good and bad competitors, and recording lessons learned from the marathons marked the mission's finish line.

The military trained us when writing and debriefing to make sure you tell them what you're gonna tell them. Check! Then tell them. Check! Then, tell them what you told them. That happens in the debrief that follows. I wrote in the preceding miles about our principles, the government's violations of those principles, and then overcoming injustices by employing those principles. In finishing the race, I'll offer acknowledgements, good and bad, and suggest a pilot's flight plan for fixing the malfunctions so that the failures do not repeat.

As a former military pilot and professional, it was ironic and humbling that I gained my inspiration and courage from many nonmilitary role models. This time around the track my academic inquiry gravitated to Dr. McCullough, RFK Jr., and other outspoken courageous "integrity system heroes."[2] They deserved acknowledgment instead of suppression as did the noble heterodoxy from medical champions I worked with: Drs. Harvey Risch,[3] George Fareed,[4] and Elizabeth Vliet.[5] These doctors revived the missing intellectually honest "dialectic."

There were untold other medical heroes who dared to challenge orthodoxy despite intense groupthink. Whereas we only had Dr. Nass and Dr. Buck decades ago, during the pandemic there was an army of doctors collaborating. They did what our government should have been doing—protecting medical choice. Particularly as science the evolved, and new truths emerged, citizens were intellectually curious to listen to those who refused to be silent. I trusted Russ and JR for the same reasons. They were not scared to question the script, just as the DoD trained us to do.

In contrast, multiple "bandits" emerged in this new but predictable engagement. "Bandit" was a fighter pilot's term for the adversaries we engaged in combat. I lacked confidence in the bandits who were intent on muzzling courageous alternative voices and suspected the long-term intellectual battlefield would not tilt in their favor. The gate guards and bandits who tried to manipulate and steal the truth, such as Dr. Fauci, inevitably fade away. But the "Fauci effect" would not. He announced his retirement in December 2022.[6] The president's words of gratitude to Fauci included that "he has touched all Americans' lives with his work."[7] On that we agreed. Before retiring, Fauci flipped regarding mandates, repetitively parroting a fresh soundbite: "Again, each individual will have to take their own determination of risk."[8] His new prescription translated to individualized risk versus benefit calculations, without penalty or punishment. Fauci's epiphany required prior consent for troops and informed consent for civilians.

Even former White House COVID-19 Coordinator Dr. Deborah Birx admitted she knew that "these vaccines were not going to protect against infection." She added, "I think we overplayed the vaccines."[9] A previous

White House advisor had come to the same conclusion about the litany of "lies."[10] Birx's awareness of the limitations preceded the heavy-handed mandates and dictated a more measured approach—exactly what the law required and citizens expected.

Similar to Birx, the CDC's Director, Dr. Rochelle Walensky, belatedly admitted overly optimistic efficacy assumptions. Early on Walensky said, "vaccinated people do not carry the virus, don't get sick, and that is not just in the clinical trials but it's also in real world data."[11] After the pandemic and five shots, Walensky caught COVID[12] and lamented that she showed "too little caution and too much optimism" regarding the shots. She appeared to blame the "CNN feed" for reporting that the shots were "95% effective."[13] She ultimately admitted efficacy was "waning" and "the truth is science is gray." She seemed to shift blame onto the American public by claiming we mistakenly heard the science was "foolproof." Her backpedaling dismissed the CDC's hyperbole with this tepid excuse: "They wanted to be helpful."

Most of us did not hear our officials relay the science as gray or admit the science was not settled. We heard absolutes—they absolutely threatened our jobs if we did not want their helpfulness or buy into their mandatory unsettled science. Walensky first coined the corrosive "pandemic of the unvaccinated" soundbite. She also encouraged social pressure from family and friends, plus education and counseling to break hesitancy.[14] Like anthrax vaccine, the patronizing messaging and dogmatic indoctrination tarnished public health credibility—it wasn't "helpful."

Their unhelpfulness went further. Through social and mainstream media, cynics of the absolutes were idea-shamed and canceled despite Walensky's admission that "sometimes it takes months and years to actually find out the answer." Yet they mandated that we follow this gray science or else face the consequences—all while the word "consequences" was misinterpreted. In truth, and in the name of accountability, it was the government officials, military brass, and business leaders that should have faced consequences for their hasty mishandling of the law.

No different from anthrax vaccine, the truth surfaced too late, and the overdue excuses tainted our trust in government. After Walensky resigned, the CDC halfheartedly acknowledged natural immunity.[15] The CDC renamed it as "infection-induced immunity," as though they invented the concept. The CDC also ended earlier patient distinctions, "irrespective of their vaccination status," by declaring, "CDC's COVID-19 prevention recommendations no longer differentiate based on a person's vaccination status because breakthrough infections occur."[16] The belated admissions on the evolving science by the gate guard bandits resembled what we identified two decades ago. In both eras they compromised scientific, medical, and regulatory integrity to protect policy. They surreptitiously changed the message as the science evolved.

Officials simultaneously provided equally sly Orwellian answers to straightforward questions. Fauci responded to PBS when questioned, "How close are we to the end of this pandemic?" Fauci's dupe-speak began with, "Well, that's an unanswerable question. And I don't want to be evasive about it." Fauci continued, "We are certainly right now in this country out of the pandemic phase," but provided the caveat about "the global situation, there's no doubt this pandemic is still ongoing."[17] The paradoxical answer left room to turn the emergency back on based on the ongoing "global situation." Orwell's warnings about such paradoxes chillingly resembled Fauci's never-ending pandemic rhetoric, as did PREP Act extensions for indemnity.

So, who do we follow and trust? Is it these bandits who tried to steal the health and trust of the people they served? How do we guarantee future leaders will study the truth in the fine print and provide straight answers? The answers and truths were right in front of us: improper approvals, historically unprecedented adverse reactions, and affirmed illegality of mandates. Our predictive conclusions about the illegality of the anthrax vaccine ultimately earned vindication. Similarly, the judicial review regarding EUA injections and boosters inevitably transitioned toward common sense and justice with more overruled mandates. Considering the injuries to trust, and the rapidly ever-evolving variants and science, the government and military prudently pursued voluntary-only EUA Omicron-adapted booster policies.[18] Good—that was the law.

Adapted bivalent boosters, authorized based on animal bridging studies, tested the intent of the Harris-Kefauver amendment efficacy requirements.[19] [20] Authorization for boosters relied on the anthrax vaccine precedent using the "animal efficacy rule" even though experts had cautioned that correlation of vaccine effectiveness in humans was "fraught with difficulties."[21] The emergency shortcuts challenged the spirit of the FDA's Preamble, which affirmed, "The importance to the American Public of safe and effective vaccines cannot be understated."[22] Whether authorized boosters or "approved" updated shots, efficacy was in doubt. A trustworthy government should have unambiguously admitted the limits and honored informed consent rights. Instead, it butchered trust on a scale that surpassed that of the anthrax vaccine tragedy.[23]

In all those years, and despite the fact that the military and the government was found "not substantially justified" in the illegal implementation of the anthrax vaccine program, no one in the hierarchy of the military or government was held accountable. Leaders never universally corrected the wrongs associated with the thousands of people who were punished. My experience was framed by the reality of those damages and by the fact that the government did not tell the truth then or now, so my mission was to convey the truth this time around—Americans, and their troops, deserved no less. We didn't accept the dupe-speak. We didn't capitulate to the fear, the paradox of control, or the illegally forced compliance. Instead, we enforced the law.

Then, we combatted the bandits with the heroic efforts of wingmen such as Representatives Chris Shays, Dan Burton, Rob Simmons, Nancy Johnson, Ron Paul, Dick Gephardt, plus Senators Chris Dodd, Tom Daschle, Jeff Bingaman, and Richard Blumenthal.[24] During the pandemic, Representative Thomas Massie and others fought to right the scales of justice over the COVID shots in what he termed as a "crime in progress." Massie reminded the Congress that the "vaccine does not stop the spread," educated lawmakers about EUA shot illegality,[25] and sponsored legislation to outlaw gain-of-function research.[26]

Senator Ron Johnson also set the example for other officials by doing yeoman's work as a public servant that was unwilling to idly witness the overt lawbreaking.[27, 28] Senator Johnson articulated the conundrum where "powerful people . . . simply don't want to admit they were wrong and they're going to do everything they can to make sure that they're not proven wrong."[29] Senator Johnson's colleague, former US Army doctor Senator Roger Marshall, exposed the ethics enigma in describing the DoD's shot mandate as an "outrageous and plain mean-spirited vaccine tyranny." He also cosponsored legislation to reinstate fired airline pilots.[30] Florida Governor Ron DeSantis joined the effort by impaneling a grand jury to investigate potentially fraudulent safety and efficacy claims by COVID injection manufacturers.[31] I was proud of these lawmaker champions—unwilling to turn the page or turn their backs on our troops. They were willing to listen, unwilling to ignore the law, and wise to the patterns.

"Provenance" was key to the patterns, at least to a pilot that sought out root causes.[32, 33] The provenance of the anthrax lab leak, perpetrated to save the anthrax vaccine program from cancellation, eventually linked back to a US Army research laboratory.[34] With COVID, provenance about the virus' unnatural or "synthetic fingerprint" added to suspicions about the origins from a SARS-COV-2 lab leak following US funded biological research.[35] Just as with the anthrax letter attacks decades earlier, by using "science—creating a DNA equivalent of a fingerprint," determination of provenance in 2023 was a reasonable expectation.[36] Thus, a Senate report concluded, "the COVID-19 pandemic was, more likely than not, the result of a research-related incident."[37] Congressional committees joined the Department of Energy (DOE)[38] with their "Muddy Waters" report in concluding that a preponderance of circumstantial evidence supported a lab leak based in part on testimonies by the former Director of National Intelligence (DNI), John Radcliff, and the former director of the CDC, Dr. Robert Redfield.[39, 40, 41]

FBI Director Christopher Wray agreed with the DOE and even more definitively concluded: "The FBI has for quite some time now assessed that the origins of the pandemic are most likely a potential lab incident in Wuhan."[42] Testimonials by Harvard, Stanford, and John Hopkins doctors went further and revealed challenges in determining provenance by pointing out the most substantial misinformation during the pandemic came from the government.[43] No different than initial misdirects with anthrax attack misinformation about Iraq WMD, the suspected COVID origins coverup complicated the search for provenance and the ground truth.

Assuming provenance remained paramount, it was reasonable to unyieldingly demand that our government officials fully examine the science, patterns, and origins. Since leaders again attempted to "turn the page," applied the "we may never know" boilerplate script to muddy the truth, then we must demand future leaders establish better controls. The COVID pandemic proved that those entrusted with the biodefense enterprise could not be trusted to comply with protocols, conventions, and treaties—so, take away their petri dishes—the problem's root cause.

Beyond appealing to current and future leaders, advocates of any conduct that deviates from exclusively peaceful, legal, process oriented, intellectual resolution warrants a stern warning. Do not take the bait—do not give them ammo. Cool heads must prevail. Opposing parties in this cerebral clash would love nothing more than to yell "insurrection." That would not play out well, so hopefully no one, anywhere, takes that bait. This is a mental marathon. Our minds propel our positions.

The final front will be the courts to settle the damages following the rescission of the mandates. Though one Supreme Court ruling affirmed the large employer mandate was unlawful, others drifted unresolved in the judicial winds by ignoring "historical example[s]" and "fundamental right[s]."[44] Those examples and fundamental rights were established in the law, 21 CFR §50.25, and the anthrax vaccine EUA that affirmed no adverse "consequences" or "punishments" for exercising the "option to accept or refuse." If the SCOTUS rules on all mandates, that word is final pending future shifts in the judicial winds. In the interim, citizens on all sides must vote and elect unyielding leaders to serve as the change agents who ensure consent, a non-politicized judiciary, and a peaceful future.

It is all about consent. The Constitution and Founding Fathers got that right. But beware, I went to school with the same generation now in charge—we were not trained to cast off civilian authority. On the contrary, they trained us to loyally support our civilian authority. I do so in part with my thoughts, thesis, and this book. Though I might be seen as a rebel by some, others may assess my strategy as too passive and forgiving by emphasizing intellectual conflict resolution. My point is this: When the Founders said "abolish it"—if the "Government becomes destructive of these ends, it is the Right of the People to alter or to abolish it, and to institute new Government"—be cautious. Though we endured a "long train of abuses and usurpations," it was not the "absolute Despotism" suffered by the Founders.[45] Here, the usurpers were our own DoD and government officials. So, the law, our pens, and our minds served as our swords. Truth and facts were the tools, especially facts proving long-term harms. Patience was the best strategy. Elected leaders must peacefully "abolish" the bad behaviors as they grow aware and weary of the abuses. Their job will be to "throw off" the gate guards, shoot down the bandits, and to "institute" a "new" government with a healthier "level of thinking" to better protects us.

Future leaders must heal the seemingly irreparable divides. The most disturbing byproduct of this quagmire festered in the chasms created within families and between friends. Those forces that pitted our populace against one another are the most egregious antagonists in this story. All sides should see this for what it is—unhealthy. Fomenting stress within families did not help, but the imposition of illegal forced vaccination on citizens and soldiers was the root cause. The fighter pilot in everyone must remain cognizant of root causes. Over time, I am exceptionally confident families will reunite after seeing through the information manipulation and counterfeit narratives, trusting and empathizing with the instincts of their unyielding kin. Family erosions will repair as those units no longer abide being "lied to" by the government.[46]

Rather than falling prey to the divisive stunts, my earnest belief is that all Americans should study and observe the repetitive patterns and M.O.'s we experienced during our anthrax vaccine mandate professional crisis and then were repeated during the pandemic. The suspect soundbites were exactly the same—"safe, effective, FDA approved." Predictable tools again included claims of internet misinformation, anti-vaccine monikers, targeting of the unvaccinated, idea-shaming, coercion, manipulation, cancelation, termination, and allegations of selfishness. Businesses, unions, media, big tech, and health experts fell too quickly for the ruse, providing the bad actors in the government their willing army of shills who colluded by bullying the hesitant simply for being anti-illegal-mandate.

Today's unhealthiness will hopefully effectuate deep introspection tomorrow. When we put the pandemic behind us I hope that citizens find clarity, see the patterns, withdraw misplaced anger aimed at the unvaccinated, and respect the core medical rights affirmed and protected by our laws. Those core inalienable rights, and their enforcement, should have been the "starting line" for both the anthrax vaccine dilemma and the COVID fiascos. Had our government and military leaders championed those rights for our people, instead of abrogating them, these marathons, this finish line, and this pilot's mission would have never been required.

In wrapping up this debrief with lessons learned and suggestions for the next race, it is fair to ask: why should anyone care about a pilot's analysis? I offer four answers: First, pilots adhere to strict safety standards, which are demanded by the public, enforced by the government, and exposed by the media when problems occur. Second, the pilot culture is a generation or two ahead of the public health community and government in operational risk management, threat error management, and debriefing to win. Third, pilots are operational practitioners who follow rules and instructions, adhere to checklists and standard operating procedures, execute boldface decisively when required, and never permit the politicization of our profession or procedures as occurred within the practice of medicine.[47] The pilot culture simply does not permit willful ignorance of rules or willful manipulations and defiance of the law. That's the disconnect or cultural divide, and therefore, fourth, and finally, if bureaucratically minded leaders and all medical professionals better emulated and defended the operational mindset, lived up to codes of conduct, complied with standards, followed oaths, enforced laws, and protected those who do, no one would have ever dared to violate the ultimate checklist, the Constitution.

Pilots follow their checklists. The DoD, government, and businesses don't permit pilots to deviate from their checklists. Pilots do not rush to comply—we go-around when things are not safe and do not look right. When pilots make mistakes, public safety is at risk. As a result, government leaders and citizens require risk assessment continuums and absolute candor in threat and error reporting from pilots. Medical professionals traditionally faced similar scrutiny, but in the case of EUA shots an oversight malfunction occurred. The medical profession as a whole might insist that they had the tools and did similar training, but the reality is that collectively they did not transparently conduct the mandatory debrief despite a major disaster in their operations. Though it's a tough pill to swallow, the medical society must consider the transfer of control that occurred. Many deferred to coercion by "experts" that were simply making it up. Some were hijacked by grants, swayed by politics, silenced by professional fear, or indemnified from liability. Clearly corrosive and unhealthy influences, unheard of in the operational aviation realm, prevailed over diligence and reason.

Aviation analogies were relevant to the medical profession and public health realms when it came to the silencing of common sense safety questions. The government's bandit gate guard's and "Gate Keeper's" attempted to steal the citizenry's voice, to get us to stop noticing, and to stop asking questions. It was tantamount to absurdly asking passengers to board an unsafe flight.[48] Can you imagine forcing passengers to fly on experimental aircraft made by liability-free companies or with zero liability airlines? This was what some in the medical profession allowed with EUA shots at the behest of the government and with the cooperation of the media. The medical community should evolve, adopt pilot-like risk assessment practices, develop and adhere to checklists, avoid regulatory capture, disallow practice of medicine politicization, as well as promote a culture where doctors shield doctors and protect patients from mandates for unapproved medicines. While there are notable exceptions, a cultural evolution within the entire medical community will require rigorous honesty, professional courage, and a new paradigm that humbly transforms the Sisyphean task into a surmountable mission of cultural refocusing.

Change will not occur overnight, but on the backside of the pandemic public health establishments will enjoy a window of opportunity for a self-reflective learning moment—potentially resulting in transformative change and the restoration of public confidence. If the opportunity is ignored, the Overton window will shift, hesitancy will grow further, and the roots of distrust will entrench. If internal change proves elusive, if current leaders endeavor to cover-up errors, versus correcting both culture and policy, then future leaders must transform public health bureaucracies and medical institutions. New leaders must impose strict requirements for the medical community to admit errors, correct course deviations, and prescribe adherence to an untainted checklist mentality similar to pilots. Prudent policy by our government and public health officials required the same guiding principles pilots follow to avoid overcorrecting absolutes with burdensome mandates, supported only by coercive soundbites instead of science.

Many Americans joined our mission and marathon against illegal mandates during the pandemic. Collectively, we must boldly insist our leaders fly in formation and run beside us. If unwilling, future leaders must fly high cover over our government and demand a "pilot" program for public health accountability and transformation. The CDC and FDA may try to lead turn demands for oversight and calls for change to the inevitable merge by declaring belated reorganizations in the name of "accountability, collaboration, communication, and timeliness."[49] Beware: such internal reorganizations, in lieu of genuine reform and external oversight, are a classic dodge tactic by the bad actors designed to control the operation without fixing the malfunctions. If we let the gate guards and bandits "bug out," depart the fight without facing their defeat, they will gloss over their errors. If we let them "turn the page," it will happen again—just as it did after the anthrax vaccine illegalities, the anthrax attacks, the COVID bioincident, and with the illegal COVID shot mandates. If we are unyielding, and maintain situational awareness of the tactics of the gate guards, future leaders can fix these malfunctions.

Though this story covered many miles, at its core it was not about masks, lockdowns, alternative therapies, or even background scientific squabbles over safety and efficacy. While all legitimate issues, supplementing the central thematic process concerns, the gripe against illegal mandates at its core was about the processes and laws our officials violated. If leaders cannot follow the simplest of rules related to fundamental medical rights, how can they be trusted with all the tangential policy decisions related to the consequential issues? This was a story about how to do it right, but, when trusts are violated, how leaders, citizens, and troops should react. I hope that the lessons learned translate to noble reform and tilt windmills for generations.

Our children, and their children, will study and learn from what we did and did not do. Our descendants must not let their government do to their kids what we let our government do to ours. They will do it better since we showed them a path, and because they will study the government's part in the pandemic and its response, from deceptions about origins to hoping we would comply with their mandates of ineffective countermeasures. They will learn the incontrovertible truths that provide perspective as to why the illegal countermeasures required mandates. If the products were safe, effective, and properly approved, mandates would not have been required. Instead, our government broke our trust and the laws our elected representatives put on the books to protect us from becoming another footnote documenting such abuses.

Our kids, those new leaders, will get it. We must trust them. That is the final stretch of the marathon and the final step in the process—letting go—letting the next generation fix what ours did not. When they follow our unyielding example to get across their finish line, they must take away the petri dishes if we did not. They must vanquish the bad behaviors and bad means of the bad actors and gate guards if our generation did not successfully shoot down their bad causes. Lastly, they must hold the bandits accountable in eye of history for the theft and maiming of the health, trust, and innocence of millions of Americans and billions of our fellow human beings.

AFTERWORD

Closing out this story, with the heartfelt hope of no more miles or chapters to chronicle, I reflected again on the novel JR gave me about the fighter pilot from the Korean War. I imagined Cleve Connell, of Salter's *The Hunters*, as he launched out solo across the Korean peninsula. The flight was Cleve's last mission. The bandits Cleve hunted would shoot him down. This time Cleve was all alone, with no mutual support from Billy Hunter or anyone else.

> And Cleve was at lonely peace with himself. He felt as if he had finally passed from youth into a real maturity, one in which he soberly realized the price that had to be paid to abide by the ideals that were once so bright and compelling. The reckoning was dear; but for all that they had cost him, he held them even more fiercely. He had nothing frivolous remaining to believe in then, only an obdurate residue more precious than a handful of diamonds.
>
> . . . he flew north, on his ninety-seventh mission . . . They had overcome him in the end, tenaciously, scissoring past him, taking him down. Their heavy shots had splashed into him, and they had followed all the way, firing as they did, with that contagious passion peculiar to hunters.

Like Cleve, I was at peace. Over the years I too "soberly realized the price that had to be paid to abide by the ideals that were once so bright and compelling." Unlike Cleve, that "contagious passion peculiar to hunters" never defeated or "splashed" us. The enemy we faced cowered against our unyielding resolve and mutual support. We were never alone. Their "heavy shots" took many down, but that shame was on the gate guards and those who failed to protect our citizens. Their shame was vastly outweighed by the honor and courage of those they harmed. The passions peculiar to those who questioned orthodoxies hunted and splashed the gate guards.

Justice over time must fatefully ensure a grateful nation's compassionate correction of all the damage for all the troops and all the people harmed. They survived as the "hunters," and their simple, optimistic sense of courage and hope, though somewhat dashed, soberly prevailed. We prevailed through the facts, with no manipulation, no compromise, and by always enduring against the bad and complicated. Though at times the mandates were the hunters, as Ernest Gann feared was fate, we reversed on those bandits too. They never overcame us in the end. We tenaciously stood our ground and shot them down. Our idealistic lens never lost its focus.

We learned and grew as officers and citizens in the process. It was not about winning the race, but holding the line. They trained us to run races even if we could not win. They trained us to take on the challenge of marathons and "never give in, never, never, never." I'm still known to drift into daydreaming of our time in fighters, blurring memories with our anthrax battle and now the COVID fight, often outgunned but never capitulating to any adversary, ever, ever, ever!

I cannot help but drift into daydreaming about our valiant crew . . .

Russ, Enzo, Dom, and I, a four ship in formation, pushed our throttles up. We tightened our harnesses and prepared for our final engagement. We firewalled our throttles, pushed our sticks forward, our aircraft accelerated under the bending negative g-forces. We broke hard in formation to beam the bandits that appeared on our nose, above the horizon, at twelve o'clock.

Grossly outnumbered, we chaffed and flared with our onboard countermeasure systems to defeat enemy missiles already in the air. We thought and fought strategically. We maneuvered tactically. We dove and jinked. We headed further down toward the earth, masking in the terrain.

"*Talley*," my flight leader called, as I gave him bearing, range and distance to the fight.

"*Visual*," I replied, subordinately affirming his lead and that I had him in sight.

"*Engaged, bandit, nose, three miles, head aspect*," my lead informed and described.

"*Talley, visual, press*," I reaffirmed, letting him know I was there for him in mutual support, as I knew he would always be there for me.

"*Lead's engaged on the southern bandit—two, engage the northern bandit*," number one directed calmly called on the radio as we targeted the bandits in a classic side-side split.

"*Two's engaged northern*," I chimed in obediently, targeting the lead bandit's wingman.

I uncaged my missile and waited for the audio tone growl of my infrared seeker to sound exactly right in my headset. I saw my flight leader pull up. Following, I pulled back on my stick and soared up with my lead. We launched our missiles in unison, pickle buttons pressed.

Woooosh, woooosh, rumbled the rockets. The missiles left our wingtip rails. The final spears were in the air for the multi-azimuth, multi-axis attack against the illegal mandates.

"*Kill, multiple illegal mandates*," twenty-five years apart.

"*Egress now, refuel, and reset*" for the next engagement—then, "*Fight's On*!"

Rempfer from pilot training on the T38 and with family in front of the A-10.

EPILOGUE

Despite some negative experiences over the years, I turned them into positives. I am eternally grateful for the myriad incredible opportunities our citizens and country afforded me. I was lucky. My jinks were timely. I out maneuvered my adversaries. I was fortunate to fly our nation's best airplanes. I paid it back by never quitting and by always performing at my best levels. I used the skills from my training to stay one step ahead of the bad causes, means and otherwise good people. I remained positive on the home front, which hopefully motivated each of my three children to enlist and serve. All in all, it was a good run, with no regrets and much gratitude to my nation and the citizens I served. As an Air Force Academy recruiter to this day, I still believe in what the old advertisements said, "The Armed Forces—A Great Place to Start."[1]

My late colleague, Russ, would agree. He gets the final word with his closing thoughts from the final chapter, "the Charge," from our original 2003 *Uniform Deceit* manuscript. His allegations were specific to the anthrax vaccine dilemma, but his prescient visions applied equally to the illegal unapproved EUA COVID injection mandates and whatever comes next. Thank you Russ. You are sorely missed and well-remembered. This effort was a tribute to you.

The Charge, in the Words of Colonel Russell E. Dingle

What does a soldier do when he realizes that a military order jeopardizes the health of his comrades and is in violation of the law? For Major Tom Rempfer and I, United States Air Force fighter pilots and officers, the answer was easy—we did what they trained us to do. We brought our concerns to our commander and attempted to resolve the issues within our chain of command. We naively thought that the military system would work as it had in every other instance in our collective memory.

What happened, though, rocked the very foundation of our belief in the military and challenged our oath to uphold and defend the Constitution.

The Anthrax Vaccine Immunization Program, or AVIP, arrived at our National Guard base in September 1998. Because of the concerns raised by the unit's members, the commander organized a short notice investigative "Tiger Team" to develop a list of questions for higher headquarters. We were assigned to this team. We developed a short list of well-researched questions for the military headquarters in Washington, DC, and waited for answers. But answers to the questions never arrived. Rather, the order became "Take the shot or get out"! This was the first indication for us that something was terrible awry in our military.

Tom Rempfer and I chose to challenge the inoculation order. Our commanders implored us to enjoy our careers, to not "fall on our swords" over the unanswered anthrax vaccine questions. Instead, we chose to do both—to fall on our swords and, as a result, enjoy what our careers were intended to be: honorable ones. We believed in our duty to serve our nation and our obligation to ensure the armed forces was worthy of the American people it serves. We spent the next five years working through military and government channels in an effort to get the Department of Defense to answer the questions, and eventually to acknowledge their wrongdoing in the implementation and execution of the Anthrax Vaccine Program. What began as an effort to notify our commanders about a safety issue had transformed into a battle for the truth—the truth behind an ambiguous program worth billions of dollars, affecting not only every member of the military but the military institution and the entire country as well.

But before we could get to the truth we had more questions to ask and more answers to find. Why didn't the military answer our questions? Why didn't they realize the harm this vaccination program could incur on our forces? What weren't they telling us? After it became obvious that the military was unwilling to address our concerns we reluctantly engaged our elected officials and the media. They wanted answers as well. If we weren't convinced before that something was terribly wrong in our military, all doubt was removed when the military began their anthrax vaccine education and public relations campaign. Designed not to inform, but to win the hearts and minds of its soldiers, the media, and the public, the DoD had declared war on those who would challenge their vaccination program.

In response to media and congressional inquiries, the DoD accelerated the immunization program, rapidly affecting large numbers of soldiers. Most took the shots believing, as they rightly should, that the military would not purposefully subject them to needless risk. Thankfully most did not suffer any adverse reactions. But others did, and some died. Many of our military leaders were content to blindly enforce the order, while others vociferously but ignorantly defended the flawed shot program. Tom and I showed our bosses the Army's own papers detailing the same issues we were raising, but it didn't matter. We shared our research with our elected representatives and the media. They were concerned, but seemed incapable or unwilling to bring the military to task for their controversial inoculation program.

Using our fighter pilot training and fighter pilots tactics, we set out to find the answers. But this was a new enemy and a new battlefield. Just like we had trained, we worked together and independently in this campaign. We optimized our strengths, with me doing the research—building our weapons; while Tom directed the multi-azimuth campaign—enlisting allies and bringing our weapons to bear. Along the way we realized that this was no longer about finding answers to the anthrax vaccine debacle at all. Rather, we set out to uncover a deception perpetrated upon our soldiers, the media, and the public. Through our work, and the selfless work of others, we determined that the military knowingly used an improperly approved vaccine characterized by unusually high adverse side effects for an untested and unlicensed purpose.

For reasons yet unclear, we had entered an era where absolute loyalty to a policy replaced and violated the oath of office every soldier and appointed official swears to uphold. Careers were being made and lost depending on how you pledged your allegiance to the vaccination program. So were relationships. Alliances were formed. Sides were taken. The stakes and the scale of this new age crept towards a constitutional crisis.

And then the supply of vaccine dried up. The FDA had shut down the vaccine plant and quarantined most of the available supply. By default, the program ended with neither side able to claim victory. But this was merely a lull in the battle for the truth. In 2001, the anthrax letters scared America and the FDA hastily approved the renovated vaccine facility to start production again. It was time to re-engage in battle.

For us, the issue was not about objecting to the military leadership's goal to protect soldiers, rather the issue was how to do it legally and with the best medicine and equipment. To do otherwise creates a permeable smokescreen of protection, leaving America's soldiers more vulnerable than ever before. To ignore the law and the protections it affords all citizens jeopardizes the integrity of the military institution. Loyalty to a poorly thought out force protection policy became a condition of employment in the profession of arms and an indelible stain on America's armed forces. Arrogance replaced humility and service before self at the highest levels. Left unchecked, where would this new age of mandatory fealty take the country?

This behavior within a military, required to adhere to the highest standards and codes of truth and honor, is unacceptable. What has been the cost to America and to our military institution? How many officers must reject their oath to defend and uphold the Constitution in order to prop up this lie? How many soldiers must become casualties? This dilemma provides a template of conduct that cannot be condoned. By studying the successes and failures of both sides in this struggle, leaders today and tomorrow will be guided in their efforts to reform our military, ensuring it is one worthy of the American people they serve. The lessons learned for military leaders is that if you teach your officer corps to 'do the right thing,' to stand up for what they believe, to challenge illegal orders and conduct; and you give them the training and tools to do so, you must contend with the consequences when they execute their duty.

Tough questions must be asked and answered, not ignored. Has our DoD as an institution become unwilling to review their actions? Has our DoD abandoned its soldiers at the bottom of the chain of command in order to preserve the reputations of those at the top? Have we willfully, ignorantly, and blindly followed illegal policy on an institutional level? The DoD's Anthrax Vaccine Immunization Program is a warning sign for American society, reflecting dangerous precedents when our military values blind adherence to bad policy above integrity, military professionalism, and expertise.

Our professional dissent required a reaction by the military leadership, yet they collectively looked the other way. As Harvard scholar Samuel P. Huntington asked in the book *The Soldier and the State*, "What does the military officer do when he is ordered by a statesman to take a measure which is militarily absurd when judged by professional standards?" Huntington answered, "The existence of professional standards justifies military disobedience." The United States Air Force's origin owes similar reverence to the concept of such dissent as memorialized by the words of their inspirational founding father, Brigadier General William "Billy" Mitchell, who criticized the Army's use of airpower in the pre-World War II era. Mitchell's words to the DoD leadership of the day accused them of "Incompetency, criminal negligence, and almost treasonable administration of the National Defense." His charge defines the anthrax vaccine dilemma, and we are merely students of his example and leadership.

Lt Col Russell E. Dingle and the A-10 "Warthog" Thunderbolt II, and his quote eluding to the DoD's "intimate" involvement with anthrax vaccine from Scott Miller's movie, *A Call to Arms.*

I was forced out of the guard based on my refusal to take the anthrax vaccine. I could not ask any of my men to do something I was not willing to do, and I was certainly not willing to take the vaccine based on my belief that the order was illegal and immoral.[2]

—Lt Col Russell E. Dingle, USAF, Retired

APPENDIX A

Adjutant General "Unyielding" Letter Denying Appeal of the Illegal Firing

DEPARTMENTS OF THE ARMY AND AIR FORCE
Joint Force Headquarters-Arizona
5636 East McDowell Road
Phoenix, Arizona 85008-3495

NGAZ-ZA 22 June 2011

MEMORANDUM FOR Lt Col Thomas L. Rempfer, 214th RG/LRE

FROM: The Adjutant General

SUBJECT: Response Regarding the Notice of Appointment Appeal Regarding Involuntary Tour Curtailment

1. Although there is no appeal of a Notice of Appointment, per your request, in the interest of fairness I completed a thorough and objective review. After careful deliberation and review of all the information available to me, I have decided to deny your request for reconsideration and to stand by the notice of appointment.

2. Just to be clear, the notice of appointment was issued under a state National Guard regulation and terminates only your five years of service with the Arizona Air National Guard. It does not affect your status as an Air Force Officer nor does it affect your ability to obtain an active duty or reserve position outside the Arizona National Guard.

3. During my review I took everything into consideration to include the letters of support you provided. Regardless of the fact that it was inappropriate to use military email and it could be perceived as undue command influence to solicit letters of support from subordinates, junior officers, and enlisted personnel - I still read each letter. Although it is clear that you have some friends and supporters, I not only question the method by which these letters were acquired but I also doubt if these friends and supporters know the reasons behind the notice of appointment.

4. You have communicated to me in writing and in person, and you have also communicated to others that you will not accept decisions that do not meet with your satisfaction, even when decisions are made by, and under, appropriate authority. You have indicated in no uncertain terms that you believe members of the wing and state headquarters are derelict and negligent in their duties. You do not trust the Arizona National Guard command or organization, and you believe there may be corruption that obtains results regardless of law, regulation or policies. You have even alleged that this may have created mission risk. I do not believe in restricting anyone from pursuing due process or to use avenues afforded to them, but there is a significant difference between questioning authority and undermining authority.

5. LTC Rempfer, your unyielding demands that decisions must fully meet your satisfaction before you will accept them undermines leadership and shows a lack of respect. I do not need to remind you that this is the military and your comments about dereliction of duties, negligence, and possible corruption are detrimental to good order and discipline.

6. I want to thank you for your five years of service in the Arizona Air National Guard and to wish you the best of luck in your future endeavors.

HUGO E. SALAZAR
Major General, AZ ARNG
The Adjutant General

APPENDIX B

USAF Inspector General Reprisal and Hiring Violations Substantiation

DEPARTMENT OF THE AIR FORCE
WASHINGTON DC

Office of the Secretary

JUN 21 2012

SAF/IGS
1140 Air Force Pentagon
Washington, DC 20330-1140

Lieutenant Colonel Thomas L. Rempfer
6058 S. Jakemp Trail
Tucson, AZ 85747

Dear Colonel Rempfer

Our investigation of your allegation was conducted under the provisions of Title 10, United States Code, Section 1034, "*Protected communications; prohibition of retaliatory personnel actions.*" The investigation substantiated your allegation of reprisal. The IO also found that several "procedural violations" of DEMA Directive 25-6 were committed in the hiring of AGR Vacancy Announcement 2011-091A. The Inspector General of the Air Force has reviewed the report of investigation and approved its findings. Additionally, the Department of Defense Inspector General conducted a thorough review of the report, found that it adequately addressed your allegations, and concurred with its findings.

10 U.S.C. 1552 provides that a military member may request the Air Force Board for Correction of Military Records (AFBCMR) consider an application for correction of his or her military records. You may contact the nearest Military Personnel Flight for assistance in this application. The AFBCMR address is:

Air Force Board for Correction of Military Records (SAF/MRBR)
550-C Street West, Suite 40,
Randolph AFB, TX 78150-4742

We have enclosed a copy of the report of investigation, redacted in accordance with 5 U.S.C. 552. Please note that our office, as with other Department of Defense organizations, must comply with the Privacy Act and the Freedom of Information Act. For this reason, further documentation, including any disciplinary actions that may have been taken, is not releasable in order to protect the privacy rights of subjects and witnesses.

APPENDIX C

SecDef-Level Memo Ordering Command Credit and Promotion Consideration

PERSONNEL AND READINESS

OFFICE OF THE UNDER SECRETARY OF DEFENSE
4000 DEFENSE PENTAGON
WASHINGTON, D.C. 20301-4000

13 SEP 2017

MEMORANDUM FOR ASSISTANT SECRETARY OF THE AIR FORCE FOR MANPOWER AND RESERVE AFFAIRS

SUBJECT: Second Remand of Lieutenant Colonel Thomas Rempfer's Case to Air Force Board for Correction of Military Records

On June 7, 2016, pursuant to the Military Whistleblowers Protection Act, 10 U.S.C. § 1034(h), Lt Col (ret) Thomas Rempfer, appealed the Air Force Board for Correction of Military Records' (AFBCMR) decision in his case. Authority to complete such reviews has been delegated to me.

This is Lt Col Rempfer's second appeal to the Secretary of Defense. Both the current and the previous appeal relate to his nonselection for commander of the 214th Reconnaissance Squadron (214 RS) in 2011. He filed several complaints with the Inspector General's (IG) office regarding that hiring process. Shortly after his nonselection, he was fired by the Arizona Air National Guard (AZANG) and he transferred to the Air Force Reserves. He met several promotion boards for Colonel but was not selected and was subsequently retired in 2015.

In 2012, the IG found Lt Col Rempfer was fired from the AZANG in reprisal for his protected communications. The IG also identified six procedural violations in the selection board for the 214 RS commander position. The IG concluded that the supervising official more likely than not changed some of the board's scoring in order to select someone other than Lt Col Rempfer for the command position. As a result of the IG report, a commander-directed investigation (CDI) was initiated to determine if the supervising official committed criminal misconduct in relation to that hiring process. The CDI rejected the IG conclusions and determined no misconduct occurred.

In the AFBCMR's first decision in 2013, the Board granted Lt Col Rempfer partial relief but rejected his requests for a Secretarially-directed promotion to Colonel, a special selection board (SSB), and command credit. He then appealed the AFBCMR's decision to me. In June 2014, I granted his appeal and remanded the case back to the AFBCMR to reconsider its decision and determine whether an SSB and command credit were warranted. In May 2016, the AFBCMR reconsidered the case, rejected command credit, but granted an SSB. Lt Col Rempfer appealed that decision in June 2016. The SSB met in October 2016 and notified Lt Col Rempfer in December 2016 that he was not selected for promotion. Lt Col Rempfer subsequently submitted additional information for my consideration numerous times between December 2016 and August 2017.

I have reviewed all of the materials submitted by Lt Col Rempfer for this appeal, his first appeal, both decisions by the AFBCMR (the original and reconsideration on remand), and the AFBCMR case file including the full IG Report of Investigation, the CDI, and the IG addendum. I find the evidence is clear and compelling that Applicant was improperly denied command after filing IG complaints about that hiring process. In reaching that conclusion, I find the AFBCMR actions were arbitrary and capricious in this case by failing to consider an important aspect of the problem, its explanation runs counter to the weight of the evidence, and its conclusions are not supported by the substantial weight of the evidence.

The Board correctly notes that to find for Lt Col Rempfer it "would have to conclude that only he should or could have been selected for the position and that his nonselection was a clear violation of applicable policies and procedures." However, the AFBCMR never even mentioned the applicable hiring rules in its decision. Those rules provided that whenever a candidate scores at least 10 percent higher than the other candidates, he must be selected; there is no discretion. The original IG report concluded that the scores were manipulated to reduce the scoring gap to within 10 percent to enable the selecting supervisor to choose a candidate other than Lt Col Rempfer. The AFBCMR never discussed the evidence that would have removed discretion in the selection of commander. The failure of the AFBCMR to discuss and acknowledge the critical rules or the substantiated violations was arbitrary and capricious.

The weight of the evidence clearly and convincingly supports the original IG conclusions. The testimony of Lt Col Kavanaugh is particularly compelling. Lt Col Kavanaugh was a disinterested witness interviewed by the original IG investigation. I find him credible. He could not have known that Lt Col Rempfer scored more than 10 percent higher than other candidates without the selection board having told him as much. Further, his statements were corroborated by one of the members of the selection board. The CDI discounts Lt Col Kavanaugh without any compelling rationale. The evidence shows, however, that at least two board members changed their scores for two candidates and the effect of those changes was to bring the total scores within 10 percent, enabling the supervisor to select a candidate other than Lt Col Rempfer. There is no evidence any of the board members intended to change their scores until the selecting supervisor directed them to reopen the board and relook at their scores. The AFBCMR instead relies upon the CDI, which was not tasked with assessing the impact of the hiring violations, but rather whether the evidence supported charging the selecting supervisor with criminal misconduct. Moreover, the witnesses in the CDI profess to have a better recollection 15 months after the event, when they were interviewed for the CDI, than they did six months after the event, when they were interviewed by the IG. Their professed improved memory is unreliable and does not alter the strength of the evidence the IG relied upon.

Based upon these findings, I direct that the AFBCMR reconsider Lt Col Rempfer's case and make appropriate corrections to credit him with command. I find that a directed promotion is neither legally permissible nor warranted. There is ample dispute in the record as to whether Lt Col Rempfer's performance reports are strong or weak. This may be due to nuances specific to the Guard or Reserve components. The best way to resolve whether Lt Col Rempfer's record was strong enough to support promotion, even with command credit, is to allow his record to be compared against his peers in SSBs. Accordingly, I also direct that Lt Col Rempfer meet SSBs in the Guard and Reserve for each promotion board he met following his nonselection for the 214 RS command position. I am confident the AFBCMR can make whatever corrections are

2

necessary to enable Lt Col Rempfer to fairly compete for promotion with appropriate command credit.

Should the AFBCMR decline to reconsider, grant command credit, or direct the appropriate SSBs, then I direct an appropriate Air Force decisional authority take appropriate action to grant this relief. All other matters pertaining to the correction of Applicant's military records shall remain within the purview of the Department of the Air Force, in accordance with applicable law and Service regulations.

Sincerely,

Virginia S. Penrod
Chief of Staff

cc:
Inspector General of the Department of Defense
Thomas Rempfer, Lt Col (ret), USAF

APPENDIX D

Acting Secretary of the Air Force John P. Roth Promotion Letter

SECRETARY OF THE AIR FORCE
WASHINGTON

0 8 JUN 2021

MEMORANDUM FOR CHIEF OF AIR FORCE RESERVE

SUBJECT: CY18 Air Force Reserve Special Selection Board (SSB), Colonel Select (Convened 15 October 2018)

This board was conducted in accordance with DoD Instructions 1320.11 and 1320.14. The report was reviewed and is in compliance with law, regulation, and instructions, information, and guidelines furnished to the board.

The board recommended Lieutenant Colonel Rempfer, Thomas L. for promotion to the grade of colonel. The board report was approved by the Principal Deputy Under Secretary of Defense (Personnel & Readiness) (PDUSD(P&R)) and the President. The Senate returned his nomination without action.

The Department of Justice and the DoD Office of General Counsel have agreed that retired Reserve Officers are not subject to the appointments clause of the Constitution. Accordingly, Lieutenant Colonel Rempfer may be advanced in grade on the retired list after approval of his SSB without Senate confirmation. Lieutenant Colonel Rempfer's SSB was approved by USD (P&R) on 4 March 2019.

I approve Lieutenant Colonel Rempfer's advancement on the Reserve retired list. The advancement of Lt Col Rempfer on the retired list of the Air Force shall not affect his current retired pay or other benefits from the United States to which Lt Col Rempfer is entitled based upon his military service, or affect any benefits to which any other person is or may become entitled based on such military service.

John P. Roth
Acting

Tabs
Tab 1: Summary Sheet
Tab 2: Memo from OSD, 26 August 2020
Tab 3: Memo from OSD, 10 Feb 2020

cc: ARPC/PB

Prepared by: Lt Col Holly Cirelli, AF/REP, DSN 224-8275, holly.cirelli@us.af.mil

APPENDIX E

1994 Senate Committee on Veterans' Affairs Staff Report, SR 103–97

Senate Report 103–97 revealed that Major General Ronald Blanck (who would become the Army surgeon general) acknowledged a possible link between the anthrax vaccine and Gulf War Illness to Committee investigators:[1]

> "Although anthrax vaccine had been considered approved prior to the Persian Gulf War, it was rarely used. Therefore, its safety, particularly when given to thousands of soldiers in conjunction with other vaccines, is not well established. Anthrax vaccine should continue to be considered as a potential cause for undiagnosed illnesses in Persian Gulf military personnel because many of the support troops received anthrax vaccine, and because the DoD believes that the incidence of undiagnosed illnesses in support troops may be higher than that in combat troops."

The Senate Committee subsequently concluded that:[2]

> "Records of anthrax vaccinations are not suitable to evaluate safety . . . However, the vaccine's effectiveness against inhaled anthrax is unknown. Unfortunately, when anthrax is used as a biological weapon, it is likely to be aerosolized and thus inhaled. Therefore, the efficacy of the vaccine against biological warfare is unknown . . . The vaccine

should therefore be considered investigational when used as a protection against biological warfare."

Following Senator Richard Shelby's investigation of Gulf War Illness in 1994, Congress was aware of the DoD tactics regarding veteran health issue disclosures:[3]

> While I have not yet determined the reason for this apparent aversion to full disclosure by DoD, the staff working on this issue from our committee has been constantly challenged by the Department's evasiveness, inconsistency, and reluctance to work toward a common goal here.
>
> I can only conclude, Mr. President [of the Senate], that when dealing with the Department of Defense on this issue, you have to ask the right question to receive the right answer. I do not believe they understand that we are only seeking the truth in a way to help our veterans.

The DoD's reputation and history of care for its own troops was captured in a staff report prepared for the Committee on Veterans' Affairs in 1994:[4]

> For at least 50 years, DoD has intentionally exposed military personnel to potentially dangerous substances, often in secret.
>
> DoD has repeatedly failed to comply with required ethical standards when using human subjects in military research during war or threat of war.

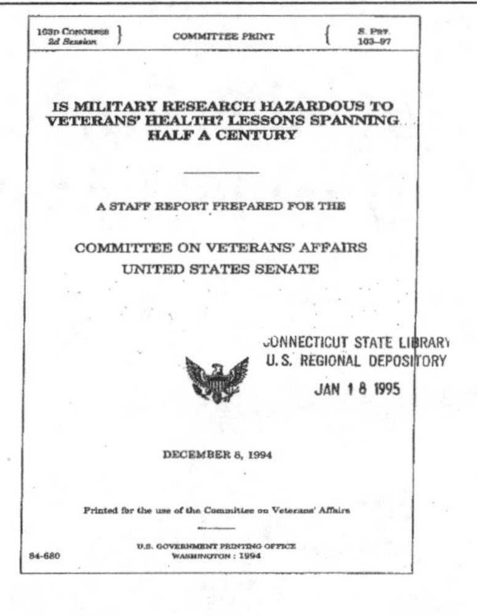

103D CONGRESS 2d Session } COMMITTEE PRINT { S. PRT. 103–97

IS MILITARY RESEARCH HAZARDOUS TO VETERANS' HEALTH? LESSONS SPANNING HALF A CENTURY

A STAFF REPORT PREPARED FOR THE

COMMITTEE ON VETERANS' AFFAIRS
UNITED STATES SENATE

DECEMBER 8, 1994

Printed for the use of the Committee on Veterans' Affairs

84-680 U.S. GOVERNMENT PRINTING OFFICE
WASHINGTON : 1994

Moss' research, but would ensure that the data were provided to DOD.[137]

Although Dr. Moss made no accusations against USDA at the Committee hearing, he has subsequently expressed his views that he lost his job at USDA because of his research findings. He also now reports that his supervisor warned him that he should not discuss his research findings with anyone. Moreover, in an internal USDA memo dated December 30, 1993, Dr. Moss stated that he was advised to "keep quiet."[138] USDA and the Johnson Wax Company are the co-inventors of DEET, an ingredient in most commercially available insecticides, such as Raid.

H. THE SAFETY OF THE BOTULISM VACCINE WAS NOT ESTABLISHED PRIOR TO THE PERSIAN GULF WAR AND REMAINS UNCERTAIN.

At a meeting with DOD officials regarding informed consent in December 1990, the FDA agreed to test the botulinum toxoid (botulism vaccine) for safety.[139] A representative of FDA's Center for Biologics Evaluation and Research explained that the existing supply of the vaccine was nearly 20 years old and consisted of three lots, stored under constant refrigeration. There was concern that the vaccine would break down into toxic products due to prolonged storage. General safety testing was performed by the FDA on all of the lots of botulinum toxoid used in the Persian Gulf; however, the FDA did not complete these tests until January 24, 1991,[140] after the war had started.

While the results of FDA's general safety testing were encouraging, the problem with adverse reactions to the vaccine were not resolved. In her review of the DOD's application for use of the botulism vaccine in the Persian Gulf, an FDA reviewer pointed out that in 1973, the Centers for Disease Control had considered terminating its distribution because of adverse reactions.[141] New lots of the vaccine were manufactured in 1971, but research was not conducted to determine whether the newer lots produced fewer adverse reactions than the older lots.[142]

Since no records were kept for most of the Gulf War soldiers who received the vaccine, there is no new information about the safety of the botulism vaccine resulting from its use by U.S. troops. Therefore, its safety remains unknown.

[137]Correspondence between Secretary Espy and Senator Rockefeller are in Committee files.

[138]Hearing, May 6, 1994; document submitted for the record by Craig Crane.

[139]Minutes of Meeting of the Informed Consent Waiver Review Group (ICWRG), Food and Drug Administration, December 31, 1990.

[140]BBIND 3723, Food and Drug Administration, memorandum from Lawrence A. D'Hoostelaere on "General safety testing of botulinum toxoid," March 2, 1994.

[141]Review by Ann Sutton, Vaccines and Allergenics, DBIND, Food and Drug Administration, to the IND record, November 14, 1990.

[142]Informational material for the use of pentavalent (ABCDE) botulinum toxoid aluminum phosphate adsorbed, U.S. Department of Health and Human Services, Centers for Disease Control, Atlanta, Georgia, Revised May 1982, protocol #392.

I. RECORDS OF ANTHRAX VACCINE ARE NOT SUITABLE TO EVALUATE SAFETY.

Although anthrax vaccine had been considered approved prior to the Persian Gulf War, it was rarely used. Therefore, its safety, particularly when given to thousands of soldiers in conjunction with other vaccines, is not well established. Anthrax vaccine should continue to be considered as a potential cause for undiagnosed illnesses in Persian Gulf military personnel because many of the support troops received anthrax vaccine, and because the DOD believes that the incidence of undiagnosed illnesses in support troops may be higher than that in combat troops.[143]

Unfortunately, medical records and shot records of individuals who served in the Persian Gulf frequently do not report the vaccines they received. In some cases, anthrax was recorded as "Vac-A." However, in many cases, veterans who believe they received anthrax vaccinations did not have them recorded in their medical records. According to testimony received at the Committee hearing on May 6, 1994, vaccines were recorded in separate vaccine records, for soldiers who had such records with them and insisted that the information be recorded.[144]

J. ARMY REGULATIONS EXEMPT INFORMED CONSENT FOR VOLUNTEERS IN SOME TYPES OF MILITARY STUDIES.

Army regulation (AR) 70-25 provides guidelines for the use of volunteers as subjects in military research. Section 3 describes three exemptions whereby military researchers are exempt from the provisions of these protective regulations (the following is a direct quote from the regulation):

a. Research and nonresearch programs, tasks, and tests which may involve inherent occupational hazards to health or exposure of personnel to potentially hazardous situations encountered as part of training or other normal duties, e.g., flight training, jump training, marksmanship training, ranger training, fire drills, gas drills, and handling of explosives.

b. That portion of human factors research which involves normal training or other military duties as part of an experiment, wherein disclosure of experimental conditions to participating personnel would reveal the artificial nature of such conditions and defeat the purpose of the investigation.

[143]Briefing, Maj. Gen. Ron Blanck, Commanding General, Walter Reed Army Hospital, to Committee staff, 414 Russell Senate Office Building, Washington, DC, February 4, 1994.

[144]Hearing, May 6, 1994, testimony of the Rev. Dr. Barry Walker, Persian Gulf War veteran.

APPENDIX F

Federal Register Proposed Rule Excerpts, December 13th, 1985

At the time of the anthrax vaccine mandate, the licensing rule for the vaccine had not been finalized by FDA. Ultimately the federal court found the program illegal due in part to the failure of the FDA to complete the rulemaking. In 1985 the Food and Drug Administration published an expert panel's review of the anthrax vaccine concluding that:

> The vaccine manufactured by the Michigan Department of Public Health has not been employed in a controlled field trial. Brachman employed a similar vaccine prepared by Merck Sharp & Dohme for Fort Detrick in a placebo-controlled field in mills processing imported goat hair . . .
>
> The labeling seems generally adequate. There is a conflict, however, with additional standards for anthrax vaccine. Section 620.24(a) (21CFR620.24(a)) defines a total primary immunizing dose as 3 single doses of 0.5 ml. The labeling defines primary immunization as 6 doses.
>
> Anthrax vaccine . . . efficacy against inhalation anthrax is not well documented.
>
> No meaningful assessment of its value against inhalation anthrax is possible due to its low incidence. [5]

federal register

Friday
December 13, 1985

Part II

Department of Health and Human Services

Food and Drug Administration

21 CFR Part 610

Biological Products; Bacterial Vaccines and Toxoids; Implementation of Efficacy Review; Proposed Rule

---

51058 Federal Register / Vol. 50, No. 240 / Friday, December 13, 1985 / Proposed Rules

(2) Bordt, D. E., J. W. Ahalen, P. A. Boyer, et al., "Poliomyelitis Component in Quadriple Antigen. Controlled Clinical Study of Enhanced Response of Children." *Journal of the American Medical Association.* 174:1166–1169, 1960.

(3) Conner, J. S. and J. F. Speers. "A Comparison between Undesirable Reactions to Extracted Pertussis Antigen and to Whole-Cell Antigen in D.P.T. Combinations." *Journal of Iowa Medical Society.* 53:340–342. 1963.

(4) Weihl, C., H. D. Riley, and J. H. Lapin. "Extracted Pertussis Antigen. A Clinical Appraisal." *American Journal of Diseases of Children.* 106:210–215. 1963.

(4a) BER VOLUME 2033.

(5) McComb, J. A. and M. Z. Trafton. "Immune Responses and Reactions to Diphtheria and Tetanus Toxiods. With Pertussis Vaccine. Aluminum Phosphate Precipitated." *New England Journal of Medicine.* 243:442–444. 1950.

(6) Provenzano, R. W., L. W. Wetterlow, and J. Ipsen. "Pertussis Immunization in Pediatric Practice and in Public Health." *New England Journal of Medicine.* 261:473–478. 1959.

(7) Levine, L., L. Wyman, E. J. Broderick, and J. Ipsen. "A Field Study in Triple Immunization (Diphtheria, Pertussis, Tetanus)." *Journal of Pediatrics.* 57:836–843. 1960.

(8) Tyson, R. M., B. J. Houston, et al., "Measure of Immunologic Responses to an Improved Preparation of Diphtheria and Tetanus Toxoids (Alum-Precipitated Combined with Pertussis Vaccine)." *The Journal of Pediatrics.* 37:357–361. 1950.

(9) BER VOLUME 2005.

(10) Barrete, C. D., E. A. Timm, et al., "Multiple Antigen for Immunization Against Poliomyelitis, Diphtheria, Pertussis and Tetanus." *Journal of the American Medical Association.* 167:103–107. 1958.

(11) BER VOLUME 2001.

(12) BER VOLUME 2002.

(13) BER VOLUME 2099.

(14) BER VOLUME 2016.

Generic Statement

Anthrax Vaccine, Adsorbed

Anthrax is an acute bacterial disease caused by *Bacillus anthracis.* The reservoir is any of several animal species (cattle, sheep, goats, horses, pigs) and the organism produces extremely resistant spores which may persist in soil and contaminate animals or their products. The disease is primarily an occupational hazard for industrial workers who process hides, hair (especially goat), bone meal, and wool, as well as for veterinarians and agricultural workers who may contact infected animals.

Most infections are cutaneous; if untreated they may spread to regional lymph nodes and may cause a fatal septicemia. Primary inhalation and gastrointestinal infections do occur, but with low frequency, and are highly fatal.

Description of Product

Anthrax vaccine is an aluminum hydroxide adsorbed, protective, proteinaceous, antigenic fraction prepared from a nonproteolytic, nonencapsulated mutant of the Vollum strain of *Bacillus anthracis.* It contains no more than 0.83 mg aluminum per 0.5 mL dose, 0.0025 percent benzethonium chloride as a preservative, and 0.0037 percent formaldehyde, which is believed to act as a stabilizer.

The product is tested according to the Public Health Service regulations for biological products and specific additional standards for anthrax vaccine. In addition to tests for general safety and sterility, the product is subjected to a potency assay of its protective activity in guinea pigs, which are challenged with virulent *Bacillus anthracis.*

Indications and Contraindications

Immunization with this vaccine is indicated only for certain occupational groups with risk of uncontrollable or unavoidable exposure to the organism. It is recommended for individuals in industrial settings who come in contact with imported animal hides, furs, wool, hair (especially goat hair), bristles, and bone meal, as well as laboratory workers involved in ongoing studies on the organism.

Contraindications to its use include:

1. A history of clinical anthrax infection which may enhance the risk of severe reactions.

2. Severe systemic reactions with marked chills and fever following a prior injection—in this case further attempts at immunization should be abandoned.

3. The presence of acute respiratory disease or other febrile illnesses in order not to confuse the cause of further fever.

4. Therapy with corticosteroids or other immunosuppressive agents—in this case immunization should be deferred until such therapy is completed. If on long-term therapy, a more intensive immunization schedule should be considered.

Safety

In general, safety of this product is not a major concern, especially considering its very limited distribution and the benefit-to-risk aspects of occupational exposure in those individuals for whom it is indicated. Local reactions are typically mild, with erythema and slight local tenderness for 24 to 48 hours. Some individuals may have more severe local reactions with edema, erythema greater than 5 x 5 cm, induration, local warmth, tenderness, and pruritus. Only a few systemic reactions with marked chills and fever have been recorded. All reactions reported have been self-limited.

Efficacy

The best evidence for the efficacy of anthrax vaccine comes from a placebo-controlled field trial conducted by Brachman (Ref. 1) covering four mills processing raw imported goathair into garment interlinings. The study involved approximately 1,200 mill employees of whom about 40 percent received the vaccine and the remainder received a placebo or nothing. The average yearly incidence of clinical anthrax in this population was 1 percent. During the evaluation period, 26 cases of anthrax occurred. Twenty-one had received no vaccine, four had incomplete immunization and one had complete immunization. Based on analysis of attack rates per 1,000 person-months, the vaccine was calculated to give 93 percent (lower 95 percent confidence limit=65 percent) protection against cutaneous anthrax based on comparison with the control group. Inhalation anthrax occurred too infrequently to assess the protective effect of vaccine against this form of the disease.

The Center for Disease Control has continued to collect data on the occurrence of anthrax in at-risk industrial settings. These data were summarized for the period 1962 to 1974. Twenty-seven cases were identified. Three cases were not mill employees, but worked in or near mills; none of these cases were vaccinated. Twenty-four cases were mill employees: three were partially immunized (one with 1 dose, two with 2 doses); the remainder (89 percent) being unvaccinated. Therefore, no cases have occurred in fully vaccinated subjects while the risk of infection has continued. These observations lend further support to the effectiveness of this product.

Special Problems

Anthrax vaccine poses no serious special problems other than the fact that its efficacy against inhalation anthrax is not well documented. This question is not amenable to study due to the low incidence and sporadic occurrence of the disease. In fact, the industrial setting in which the studies above were conducted is vanishing, precluding any further clinical studies.

In any event, further studies on this vaccine would receive low priority for available funding.

Recommendations

The Panel believes that there is sufficient evidence to conclude that

---

Federal Register / Vol. 50, No. 240 / Friday, December 13, 1985 / Proposed Rules 51059

anthrax vaccine is safe and effective under the limited circumstances for which this vaccine is employed.

Reference

(1) Brachman, P. S., H. Gold, S. A. Plotkin, R. Fekety, M. Wernn, and N. R. Ingraham. "Field Evaluation of a Human Anthrax Vaccine." *American Journal of Public Health.* 52:632–645, 1962.

SPECIFIC PRODUCT REVIEW

Anthrax Vaccine Adsorbed Manufactured by Bureau of Laboratories, Michigan Department of Public Health

1. *Description.* Anthrax vaccine adsorbed is an aluminum hydroxide adsorbed preparation of protective antigen of *Bacillus anthracis.* The product is prepared from a sterile filtrate of a microaerophilic culture of an avirulent, nonproteolytic, nonencapsulated strain. The product contains 0.83 mg of aluminum per single human dose (0.5 mL) and is preserved with 0.0025 percent benzethonium chloride. Not more than 0.0037 percent formaldehyde is added as a stabilizer.

2. *Labeling*—a. *Recommended use/indications.* This product is intended solely for immunization of high-risk of exposure industrial populations such as individuals who contact imported animal hides, furs, bone meal, wool, hair (especially goat hair), and bristles. It is also recommended for laboratory investigators handling the organism. Primary immunization consists of 6 subcutaneous 0.5 mL injections at 0, 2, and 4 weeks and 6, 12, and 18 months. Subsequent boosters at yearly intervals are recommended.

b. *Contraindications.* Prior anthrax infection is an absolute contraindication. Immunization should be avoided in acute respiratory disease or other active infections. Corticosteroid therapy may suppress response. Further immunization should be discontinued in those rare individuals who suffer severe systemic reactions.

3. *Analysis*—a. *Efficacy*—(1) *Animal.* This product meets Federal requirements.

(2) *Human.* The vaccine manufactured by the Michigan Department of Public Health has not been employed in a controlled field trial. A similar vaccine prepared by Merck Sharp & Dohme for Fort Detrick was employed by Brachman (Ref. 1) in a placebo-controlled field trial in mills processing imported goat hair. This vaccine appeared 93 percent protective (lower 95 percent confidence limit=65 percent protective) against cutaneous anthrax. No meaningful assessment of its value against inhalation anthrax is possible due to its low incidence. The Michigan Department of Public Health vaccine is patterned after that of Merck Sharp & Dohme with various minor production changes. It has been distributed by the Center for Disease Control since 1966, first as an investigational new drug and since 1972 as a licensed product. A review of the Center for Disease Control data pertinent to this product for the period 1962 to 1974 in at-risk industrial settings indicates that no cases have occurred in fully immunized workers (see Generic Statement).

b. *Safety*—(1) *Animal.* This product meets Federal requirements.

(2) *Human.* Accumulated data for the Center for Disease Control suggests that this product is fairly well tolerated with the majority of reactions consisting of local erythema and edema. Severe local reactions and systemic reactions are relatively rare.

c. *Benefit/risk ratio.* This vaccine is recommended for a limited high-risk of exposure population along with other industrial safety measures designed to minimize contact with potentially contaminated material. The benefit-to-risk assessment is satisfactory under the prevailing circumstances of use.

d. *Labeling.* The labeling seems generally adequate. There is a conflict, however, with additional standards for anthrax vaccine. Section 620.24(a) (21 CFR 620.24(a)) defines a total primary immunizing dose as 3 single doses of 0.5 mL. The labeling defines primary immunization as 6 doses (0, 2, and 4 weeks plus 6, 12, and 18 months).

4. *Critique.* This product appears to offer significant protection against cutaneous anthrax in fully immunized subjects. This is adequately established by the controlled field trial of the very similar Merck Sharp & Dohme experimental vaccine and by the Center for Disease Control surveillance data conducted on industrial high-risk settings.

5. *Recommendations.* The Panel recommends that this product be placed in Category I and that the appropriate license(s) be continued because there is substantial evidence of safety and effectiveness for this product. Labeling revisions in accordance with this Report are recommended.

Reference

(1) Brachman, P. S., H. Gold, S. A. Plotkin, R. Fekety, M. Wernn, and N. R. Ingraham. "Field Evaluation of a Human Anthrax Vaccine." *American Journal of Public Health.* 52:632–645, 1962.

Generic Statement

BCG Vaccines

Tuberculosis is a communicable disease of world-wide importance caused by *Mycobacterium tuberculosis.* The disease typically involves the lungs but is capable of causing disease in any organ system of the body. The World Health Organization estimates the number of infectious cases of tuberculosis in the world today to be in the range of 15 to 20 million.

Tuberculosis has declined sharply in the United States during the past several decades. United States Public Health Service data indicate that in 1953 there were 84,000 new cases of tuberculosis and 19,700 deaths due to tuberculosis; in 1977 there were only 31,145 new cases and the number of tuberculosis deaths had declined to 3,000. Factors contributing to the observed decline in tuberculosis morbidity and mortality include the gradual increase in socioeconomic level that has characterized the U.S. economy, improved nutrition, the introduction of effective chemotherapy of active tuberculosis, and the increasing use of isoniazid in preventive therapy. There remain, however, localized foci or "pockets" of tuberculosis transmission in the United States, particularly in areas in which preventive medical services are suboptimal or cannot be adequately delivered.

In many other countries, the use of BCG vaccine is credited with a major role in reducing tuberculosis morbidity. BCG vaccination has been the major thrust of the World Health Organization's efforts to control tuberculosis in countries with high rates of transmission of the disease. Although available in the United States, this product has been used but little for the prevention of tuberculosis.

BCG vaccines posed a particular problem for the Panel, owing to the widely disparate results of controlled field trials, and the lack of a reproducible animal model which accurately reflects protective efficacy in humans.

1. *Rationale for vaccination against tuberculosis.* Earlier in this century, a large majority of people became infected with tubercle bacilli as demonstrated by skin test positivity. However, only a small proportion of those who were infected developed overt tuberculous disease. Most people who were infected appeared to have acquired a degree of resistance against developing overt tuberculosis upon subsequent exposure, which, earlier in this century, was frequent and virtually unavoidable.

APPENDIX G
Department of Defense (DoD) Anthrax Vaccine Licensing Amendment

DoD's Joint Project Office for Biological Defense (JPOBD) recognized that the anthrax vaccine was "not licensed for a biological defense indication" based on the fact that the efficacy remained unproven:

Industrial Capabilities Assessment
Summary Report
for the
Production of the Anthrax Vaccine

Preliminary Report

Prepared for the

Joint Program Office for Biological Defense
Falls Church, Virginia

1997

The DoD biological defense vaccine requirement presents a significant production challenge. It may be compared with the U.S. immunization program for public health, which provides protection for 22 diseases. The licensed public health vaccines are approximately equivalent to the diversity of products required for the biological defense program. Again, anthrax and smallpox, considered very effective vaccines, are the only licensed products in the biological defense program, but they are not licensed for a biological defense indication. Although small and fragile, the national vaccine manufacturing infrastructure for U.S. preventive medicine needs is well established. Merck and Company, Inc., has been in the biological manufacturing business for over 100 years.

APPENDIX H

FDA Notice of Intent to Revoke and Inspection Details

The FDA filed a Notice of Intent to Revoke (NOIR) the anthrax vaccine manufacturer's license on March 11th, 1997 (FDA, 1997), for deviations from good manufacturing practices:

CBER

Blood | Therapeutics | Vaccines | Cellular & Gene Therapy | Allergenics | Tissue | Devices

Products | Manufacturers | Health Professionals | Reading Room | Meetings & Workshops | Research | About Us

FDA Warns Michigan Biologic Products Institute of Intention to Revoke Licenses

FDA issued a letter to the Michigan Biologic Products Institute (MBPI), Lansing, Michigan, on March 11, 1997, warning that the agency will initiate steps to revoke MBPI's establishment and product licenses unless immediate action is taken to correct deficiencies at the firm. MBPI is currently licensed to manufacture Diphtheria Toxoid Adsorbed, Tetanus Toxoid Adsorbed, Rabies Vaccine Adsorbed, Antihemophilic Factor (Human), Immune Globulin (Human), Albumin (Human), Anthrax Vaccine Adsorbed, Pertussis Vaccine Adsorbed, Diphtheria & Tetanus Toxoids Adsorbed, and Diphtheria & Tetanus Toxoids & Pertussis Vaccine Adsorbed. MBPI can remain open while it attempts to address the deficiencies cited by FDA.

An FDA inspection of MBPI conducted between November 18 and 27, 1996, documented numerous violations in the following areas: organization and personnel, buildings and facilities, equipment, control of components, drug product containers and closures, production and process controls, laboratory controls, and records and reports. Some examples include:

- failure of the quality control unit to approve or reject all components, drug product containers, closures, in-process materials, packaging material, labeling, and drug products;
- failure to have separate defined areas or other control systems for manufacturing and processing operations;
- failure to assure that equipment used in the manufacture, processing, packing or holding of a drug product is of appropriate design and of adequate size for its intended use and for its cleaning and maintenance;
- failure to properly store and handle components and drug product containers and closures;
- failure to calibrate instruments, apparatus, gauges and recording devices at suitable intervals; and
- failure to record the performance of each step in the manufacture and distribution of products.

Although similar deficiencies have been identified during past inspections, MBPI has failed to make satisfactory corrections. FDA has determined that continuing problems represent a failure to comply with the regulations that safeguard the drug and pharmaceutical industry. However, the agency is not aware of any injuries to recipients of these products as a result of the noted deviations.

In its letter, FDA requires MBPI to submit a written commitment for achieving full compliance within 10 days. MBPI must also submit a comprehensive plan for correcting all deficiencies within 30 days. The action plan must include corrective actions to ensure that the firm's quality assurance unit functions in an adequate, effective and timely manner, including addressing all quality assurance oversight deficiencies, and to conduct a thorough review of all standard operating procedures to achieve compliance with good manufacturing practices as specified in Subchapter C, Parts 210 and 211, Title 21, *Code of Federal Regulations*.

If MBPI fails to correct these deficiencies, FDA may begin the process for revoking the facility's licenses. Under FDA's procedures, MBPI can request a public hearing before an administrative law judge on the proposed revocation of its licenses. The revocation of the licenses would prohibit MBPI from distributing any of its products in interstate commerce.

Form 483 Inspection Report and Notice of Intent to Revoke Excerpts

The agency will initiate steps to revoke MBPI's [Michigan Biologic Products Institute's] establishment and product licenses unless immediate action is taken to correct deficiencies at the firm. MBPI is currently licensed to manufacture . . . Anthrax Vaccine Adsorbed . . . MBPI can remain open while it attempts to address the deficiencies . . .

Although similar deficiencies have been identified during past inspections, MBPI has failed to make satisfactory corrections. FDA has determined that continuing problems represent a failure to comply with the regulations that safeguard the drug and pharmaceutical industry . . . [6]

In its letter, FDA requires MBPI to submit a written commitment for achieving full compliance within 10 days. MBPI must also submit a comprehensive plan for correcting all deficiencies within 30 days.

The manufacturing process for the production of Anthrax Vaccine Adsorbed is not validated.[7]

The observations noted in this FDA-483 are not an exhaustive listing of objectionable conditions. Under the law, your firm is responsible for conducting internal self-audits to identify and correct any and all violations of the GMP [good manufacturing practices] regulation.

APPENDIX I

FDA Inspection Reports Noted Manufacturing Process "Not Validated"

1998 and 1999 FDA "Inspectional Observations" noted on line one that "the manufacturing process for anthrax vaccine is not validated" due to quality control problems—the same years the DoD lanched the mandatory anthrax vaccine immunization program:

DEPARTMENT OF HEALTH AND HUMAN SERVICES PUBLIC HEALTH SERVICE FOOD AND DRUG ADMINISTRATION		DISTRICT ADDRESS AND PHONE NUMBER U.S. Food + Drug Admin 1401 Rockville Pike Rockville, MD 20852 (301) 827-6191	
NAME OF INDIVIDUAL TO WHOM REPORT ISSUED TO: Robert C. Myers, DVM		PERIOD OF INSPECTION 2/4-20/98	C. F. NUMBER 1873446
TITLE OF INDIVIDUAL Director		TYPE ESTABLISHMENT INSPECTED Blood Derivative, Vaccine Mfr	
FIRM NAME Michigan Biologic Products Instit		NAME OF FIRM, BRANCH OR UNIT INSPECTED Same	
STREET ADDRESS 3500 N. Martin Luther King Jr. Blvd.		STREET ADDRESS OF PREMISES INSPECTED Same	
CITY AND STATE (Zip Code) Lansing, MI 48902		CITY AND STATE (Zip Code) Same	

DURING AN INSPECTION OF YOUR FIRM (I) (WE) OBSERVED:

ANTHRAX Vaccine

1. The manufacturing process for Anthrax Vaccine is not validated. For example,

DEPARTMENT OF HEALTH AND HUMAN SERVICES PUBLIC HEALTH SERVICE FOOD AND DRUG ADMINISTRATION	DISTRICT ADDRESS AND PHONE NUMBER CBER, Office of Compliance and Biologics Quality, HFM-605 1401 Rockville Pike Rockville, MD 20852 (301) 827-6191	
NAME OF INDIVIDUAL TO WHOM REPORT ISSUED TO: ^Fuad EL-Hibri	PERIOD OF INSPECTION ^11/15-23/99	C F NUMBER ^
TITLE OF INDIVIDUAL CEO	TYPE ESTABLISHMENT INSPECTED Biologics Manufacturer	
FIRM NAME BioPort Corporation	NAME OF FIRM, BRANCH OR UNIT INSPECTED Same	
STREET ADDRESS ^3500 N. Martin Luther King, Jr. Blvd.	STREET ADDRESS OF PREMISES INSPECTED Same	
CITY AND STATE (Zip Code) ^Lansing, MI 48906	CITY AND STATE (Zip Code) Same	

DURING AN INSPECTION OF YOUR FIRM (I) (WE) OBSERVED:

1. The manufacturing process for the production of Anthrax Vaccine Adsorbed is not validated

APPENDIX J
Indemnification for Anthrax Vaccine

In September 1998, the secretary of the Army granted indemnification to the anthrax vaccine manufacturer, sheltering the company from liability concerns. That same month, private investors, including the former chair of the joint chiefs of staff, Admiral William Crowe, purchased the anthrax vaccine facility and received a multi-million-dollar DoD contract. The documents indemnifying the manufacturer revealed the "**unusually hazardous risks**" associated with the anthrax vaccine. The DoD indemnification letter included:

> The obligation assumed by MBPI under this contract involves unusually hazardous risks associated with the potential for adverse reactions in some recipients and the possibility that the desired immunological effect will not be obtained by all recipients. Although AVA has been extensively tested under the auspices of the Food and Drug Administration, the size of the proposed vaccination program may reveal unforewarned idiosyncratic adverse reactions. Moreover, there is no way to be certain that the pathogen used in tests measuring vaccine efficacy will be sufficiently similar to the pathogen that US forces might encounter to confer immunity. These concerns, coupled with the uncertain and evolving state of product liability law with regard to vaccines, led me to the conclusion that the performance of this contract will subject MBPI to certain **unusually hazardous risks**.[8]

1998 Army Memorandium of Decision to Indemnify

---

transcript begins:

S E C R E T A R Y O F T H E A R M Y

WASHINGTON

September 3, 1998

MEMORANDUM OF DECISION

SUBJECT: Authority Under Public Law 85-804 to Include an Indemnification Clause in Contract DAMD17-91-C1086 with Michigan Biologic Products Institute

Michigan Biologic Products Institute (MBPI), a temporary agency of the State of Michigan, has requested that the Department of the Army include an indemnification clause under Public Law 85-804 (50 U.S.C. 1431-1435) for Anthrax Vaccine Adsorbed (AVA) produced under Contract DAMD17-91-C-1139 with the U.S. Army Medical, Research Aquisition Activity (USAMRAA).

The contract is a firm-fixed effort for AVA production with insurance and facility renovations being provided on a cost reimbursement basis. At present, MBPI is storing over six million doses of AVA at their Lansing, Michigan facility. This vaccine has been accepted by the Government with storage costs identified as a part of the firm-fixed price. MBPI is also responsible under separate contract for the security, testing, labelling and shipping of the Government Material, as required.

The obligation assumed by MBPI under this contract involves unusually hazardous risks associated with the potential for adverse reactions in some recipients and the possibility that the desired immunological effect will not be obtained by all recipients. Although AVA has been extensively tested under the auspices of the Food and Drug Administration, the size of the proposed vaccination program may reveal unforewarned idiosyncratic adverse reactions. Moreover, there is no way to be certain that the pathogen used in tests measuring vaccine efficacy will be sufficiently similar to the pathogen that U.S. forces might encounter to confer immunity. These concerns, coupled with the uncertain and evolving state of product liability law with regard to vaccines, lead me to the conclusion that the performance of this contract will subject MBPI to certain unusually hazardous risks.

The definition of the unusually hazardous risks to which the contract indemnification clause will apply is as follows:

"The risk of adverse reactions, or the failure to confer immunity against anthrax, from the administration to any person of the vaccine

-----
Page 2

manufactured or delivered under this contract. For the purposed of this clause, the phrase "adverse reactions" includes anaphylaxis and any other foreseeable reactions, as well as any unforeseen reactions."

I have considered the availability, cost, and terms of private insurance to cover these risks, as well as the viability of self-insurance, and have concluded that adequate insurance to cover these unusually hazardous risks is not available to the contractor at a reasonable cost. While limited "claims-made" insurance is available to the contractor, the terms and conditins of the insurance are not deemed to be practicable. On the basis of this review, I find that use of an indemnification clause in the contract will facilitate the national defense.

In view of the foregoing and pursuant to the authority vested in me by Public Law 85-804 (50 U.S.C. 1431-1435) and Executive Order 10789, as amended, I hereby authorize USAMRAA to include the indemnification clause set forth at FAR Subpart 52.250-1, together with Alternate I, in Contract No. DAM17-91-C-1139, provided the contract defines unusually hazardous risks precisely as set forth above.

Should it prove necessary in implementing this Memorandum of Decision to incorporate language into the contract to clarify terms found in the indemnification clause, the contracting officer shall not include any such clarifying language without the prior review and approval of the Office of the Assistant Secretary of the Army (Research, Development and Acquisition).

It is not possible to determine the actual or estimated cost to the Government as the result of the use of this indemnification clause, inasmuch as the liability of the Government, if any, will depend upon the occurrence of an incident in the definition of unusually hazardous risks.

The contractual documents executed pursuant to this authorization shall comply with the requirements of FAR Subparts 28.3 and 50.4 as implemented by the Department of Defense and the Department of the Army. This Memorandum

-----

Page 3

of Decision shall be incorporated in its entirety into Contract No. DAMD17-91-C-1139, and shall specifically cross reference FAR 52.250-1. The contracting officer shall not require, and the Army shall not reimburse the contractor for the cost of insurance coverage applicable to the unusually hazardous risks and in excess of the minimum required by FAR Subpart 28.3

/signed/
Louis Caldera

APPENDIX K

Career Immunization Record Documenting Fully Immunized Status

10/28/2017 milConnect: Benefits and Records for DoD Affiliates

Immunizations for Rempfer, Thomas L

View your current and historical immunization records below to assess your military medical readiness.

Action Date	Next Action Date	Action Code	Immunization Type	Dosage Sequence
2014-10-14		I	43 - hepatitis B vaccine, adult dosage	3
2014-10-06		I	149 - influenza, live, intranasal, quadrivalent	24
2014-05-13		I	43 - hepatitis B vaccine, adult dosage	0
2014-04-04		I	43 - hepatitis B vaccine, adult dosage	0
2013-10-23		I	149 - influenza, live, intranasal, quadrivalent	0
2012-12-11		I	111 - influenza virus vaccine, live, attenuated, for intranasal use	22
2012-12-11		I	115 - tetanus toxoid, reduced diphtheria toxoid, and acellular pertussis vaccine, adsorbed	0
2011-10-07		I	140 - Influenza virus vaccine, split virus, for intramuscular use - preservative free	21
2010-11-03		I	111 - influenza virus vaccine, live, attenuated, for intranasal use	1
2010-02-25		I	126 - Novel influenza-H1N1-09, preservative-free, injectable	1
2009-10-14		I	111 - influenza virus vaccine, live, attenuated, for intranasal use	1
2008-10-04		I	15 - influenza virus vaccine, split virus (incl. purified surface antigen)	1
2007-10-26		I	15 - influenza virus vaccine, split virus (incl. purified surface antigen)	1
2006-09-01		I	111 - influenza virus vaccine, live, attenuated, for intranasal use	1
2005-10-31		I	15 - influenza virus vaccine, split virus (incl. purified surface antigen)	1
2005-01-27		I	15 - influenza virus vaccine, split virus (incl. purified surface antigen)	0
2003-11-14		I	15 - influenza virus vaccine, split virus (incl. purified surface antigen)	0
2003-11-14		I	16 - influenza virus vaccine, whole virus	0
2003-05-12		I	9 - tetanus and diphtheria toxoids, adsorbed for adult use	0
2003-05-12		I	96 - tuberculin skin test; purified protein derivative solution, intradermal	0
2002-11-12		I	15 - influenza virus vaccine, split virus (incl. purified surface antigen)	0
2002-11-12		I	16 - influenza virus vaccine, whole virus	0
2001-11-16		I	15 - influenza virus vaccine, split virus (incl. purified surface antigen)	0
2001-11-16		I	16 - influenza virus vaccine, whole virus	0
1999-10-19		I	15 - influenza virus vaccine, split virus (incl. purified surface antigen)	0
1999-10-19		I	16 - influenza virus vaccine, whole virus	0
1997-10-24		I	15 - influenza virus vaccine, split virus (incl. purified surface antigen)	0
1997-10-24		I	16 - influenza virus vaccine, whole virus	0
1997-10-15		I	15 - influenza virus vaccine, split virus (incl. purified surface antigen)	0
1997-10-15		I	16 - influenza virus vaccine, whole virus	0
1997-03-06		I	52 - hepatitis A vaccine, adult dosage	2
1996-08-15		I	52 - hepatitis A vaccine, adult dosage	1
1995-02-05		I	53 - typhoid vaccine, parenteral, acetone-killed, dried (U.S. military)	1
1991-12-04		I	37 - yellow fever vaccine	0
1988-12-03		I	9 - tetanus and diphtheria toxoids, adsorbed for adult use	0
1988-12-01		I	2 - poliovirus vaccine, live, oral	0
1983-07-01		I	3 - measles, mumps and rubella virus vaccine	0

APPENDIX L

1985 US Army Request for Proposal for New Anthrax Vaccine

In 1985, the United States Army submitted a "request for proposal" (RFP) to solicit a new anthrax vaccine from the pharmaceutical industry, because "there is no vaccine in current use which will safely and effectively protect military personnel against exposure to this hazardous bacterial agent." The Army discussed the limitations of the current vaccine with its high adverse reaction rate and its questionable efficacy against different strains of anthrax:[9]

DEPARTMENT OF THE ARMY
U.S. ARMY MEDICAL RESEARCH ACQUISITION ACTIVITY
FORT DETRICK, FREDERICK, MD 21701-5014

REPLY TO ATTENTION OF

May 16, 1985

SGRD-RMA-RC

SUBJECT: Request for Proposals (RFP) No. DAMD17-85-R-0078
Date Issued: May 16, 1985
Date Due: July 15, 1985

Gentlemen:

You are invited to submit a proposal in accordance with the requirements of the enclosed RFP No. DAMD17-85-R-0078 for Production of Live Bacillus anthracis Spore Vaccine for Human Use."

Document No. DAMD17-85-R-0078
Page No. 4

SECTION C DESCRIPTION/SPECIFICATIONS/WORK STATEMENT

C.1. Background

C.1.1. There is an operational requirement to develop a safe and effective product which will protect US troops against exposure from virulent strains of Bacillus anthracis. There is no vaccine in current use which will safely and effectively protect military personnel against exposure to this hazardous bacterial agent.

C.1.2. Bacillus anthracis is a Gram positive spore-forming bacterium which in order to be considered fully virulent must not only produce a polyglutamic acid capsule but also produce a tripartite exotoxin. The toxin consists of three distinct proteins; protective antigen, lethal factor and edema factor. None of the protein components alone is biologically active however, protective antigen in combination with edema factor or lethal factor produces localized edema or death respectively in experimental animals.

C.1.3. A licensed vaccine against anthrax, which appears to afford some protection from the disease, is currently available for human use. This vaccine consists of alum-precipitated supernatant material obtained from fermenter cultures of an avirulent strain of Bacillus anthracis. The vaccine is composed primarily of the protective antigen component but does contain trace amounts of the lethal factor and edema factor components. The vaccine is, however, highly reactogenic, requires multiple boosters to maintain immunity and may not be protective against all strains of the anthrax bacillus.

APPENDIX M

Congresswoman Nancy Johnson Press Release

FOR IMMEDIATE RELEASE
SEPTEMBER 12, 1999

CONTACT: DAVID WHITE
Pager: (202) 227-2749

Johnson Battles Defense Department Over Anthrax Vaccine

Fights To End Soldiers' Dishonorable Discharge

WASHINGTON, DC — U.S. Rep. Nancy Johnson urged the Defense Department Friday to end its practice of dishonorably discharging service members for declining to subject themselves to the potentially harmful anthrax vaccine.

In a strongly worded letter sent to Defense Secretary William Cohen, Johnson said "I am writing to express my concern with the Department of Defense's Anthrax Vaccination Program and my vehement opposition to dishonorably discharging service members who leave because they fear the health consequences of the mandatory anthrax vaccine."

In December 1997, the Secretary of Defense announced that all U.S. forces would be inoculated against the potential use of anthrax on the battlefield. To date, no studies have been done to determine the optimum number of doses of the vaccine, nor has the long-term safety of the vaccine been determined, according to the General Accounting Office.

"Every soldier is well aware of the dangers that come with serving in the military," Johnson said. "But I doubt they ever expected their government would force them to take a vaccine that has unknown long-term effects."

As of July 14, 1999, more than 300,000 service members had received at least one dose of the vaccine. After vaccinating 150,000 Gulf War troops, the Defense Department had a unique pool of subjects to study, but due to poor record keeping no large scale research has been conducted.

"Clearly more research is necessary about the vaccine's long-term effects," Johnson said. "I've co-sponsored legislation to impose a moratorium on the vaccine program until further reviews can be done by the National Institutes of Health."

In testimony before the House Subcommittee on National Security, Veterans Affairs and International Relations, Lieutenant Richard Rovet, Health Care Integrator for the Flight Medicine Clinic at Dover Air Force Base, stated that some of the reported symptoms of exposure to vaccine include joint pain, memory impairment, "greyouts," and cardiac problems.

"Forcing our men and women in uniform to choose between this vaccine and their careers fosters distrust and erodes morale," Johnson said. "We shouldn't continue down this path until we have more conclusive evidence."

###

APPENDIX N

Note from H. Ross Perot to Karl Rove

H. R. PEROT
1700 LAKESIDE SQUARE
12377 MERIT DRIVE
DALLAS, TEXAS 75251

March 22, 2001

Mr. Karl Rove
Senior Advisor to the President of the United States
The White House
West Wing, 2nd Floor
1600 Pennsylvania Avenue, NW
Washington, DC 20502

Dear Karl,

Attached (Exhibit A) is the information you requested about the people in the Pentagon whose mission has been to ignore the Gulf War illnesses and label them as stress.

On page 22, the last page, I have included a list of all the acronyms of the different units in the Pentagon.

I suggest that as you read this document, you put the last page beside it so you can easily identify the acronyms.

In addition, I have included (Exhibit B), a booklet prepared by two Air Force Academy graduates on the anthrax vaccine problem, and a packet of news stories (Exhibit C) that reveals a great deal about the company in Michigan, Bioport.

Sincerely,

Ross Perot

RP/bc
Enclosures

APPENDIX O

White House Advisor Karl Rove Memo to DepSecDef Paul Wolfowitz

Karl Rove tasked DoD Undersecretaries Dr. David Chu and Edward Aldridge to review the "political problems" associated with the anthrax vaccine and Gulf War Illness:

THE WHITE HOUSE
WASHINGTON

2001 APR 27 PM 3 54

OSD
WHITE HOUSE SECTION

April 25, 2001

MEMORANDUM FOR PAUL WOLFOWITZ

FROM: KARL ROVE

SUBJECT: GULF WAR SYNDROME AND ANTHRAX

Here is material which has been sent to me by Ross Perot regarding the Gulf War Syndrome, as well as some material on the Anthrax vaccine problem.

He also offered me a packet of materials from the Lydon LaRouche crowd about Richard Armitage, but I turned him down.

I do think we need to examine the issues of both Gulf War Syndrome and the Anthrax vaccine and how they can be dealt with. They are political problems for us.

W00554 01

APPENDIX P

May 5th, 1998, Memo from Fort Detrick US Army Contracting Officer

A DoD functionary confirmed problems with supplemental testing and anthrax vaccine potency. Excerpt: "suspend any further potency testing under the supplemental testing program because the results continue to be all over the board and then must be reported to the FDA":

———————————— Forward Header ————————————

Subject: Suspension of Supplemental Testing

Author: [redacted] at_usamraa_ftdetrck@ftdetrck-ccmail.army.mil > at InternetMail

Date: 5/5/98 7:43 AM

Mike:

Had a voice mail this morning from Tim Wibert with a message that Tony Luttrell wants to suspend any furhter potency testing under the supplemental testing program because the results continue to be all over the board and then must be reported to the FDA. He would like to do some experiments first to resolve the control lot problem and then proceed to complete the testing.

There currently is 826,790 doses ready for shipment at MBPI. I have no idea at this point how long this experimentation would take and how it would affect the schedule for completing the supplemental testing. I will get that from Tim as soon as possible.

Do you want to pursue this strategy or are there some questions you would like to ask of MBPI?

Joe L.

APPENDIX Q
Brigadier General Eddie Cain Emails

Brigadier General Cain email excerpts: DoD "came up flat." DoD's vaccine "was NOT the one tested." Concerns about reporting that "desert storm illnesses were not cause[d] by the anthrax vaccine," when there is "no record of who received the shots." "DoD & the Administration" are in "big time trouble" if they could not address these questions. Finally, admissions that the "DoD is calling all the shots onsight (sic)" at the anthrax vaccine plant:

-----Original Message-----
From: Cain, BG Eddie
Sent: Monday, May 03, 1999 6:05 PM
To: Wade, COL John
Subject: RE: Anthrax Vaccine

Relax John, we should have all anticipated the media picking-up GAO's as oppose to DoD's "soundbites. I still believe we held our own & my people did a superb job getting me ready in such a short period of time. What's done is done; I think the key is to move forward. We keep talking about putting forth an aggressive showing, two key areas we came up flat were the GAO's assertion that #1, the anthrax vaccine licensed was NOT the one tested and # 2, how can DoD say that reported desert storm illnesses were not cause by the anthrax vaccine when we have no record of who received shots. If we can not answer these questions we (DoD & the Administration)are in big time trouble. I will work the BioPort oversight plan, but believe we are digging ourselve a hole that will be too difficult to crawl out off - if we provide anymore oversight, BioPort could technically be called a GOCOM organization. And if you think Congressman Shays was critical of the current relationship between FDA & DoD, wait until he finds out that DoD is calling all the shots onsight.

-----Original Message-----
From: John V. Wade [mailto:WADEJV@acq.osd.mil]
Sent: Monday, May 03, 1999 2:54 PM
To: Cain, BG Eddie; fanelliw@jpobd.osd.mil; gilbreathm@jpobd.osd.mil
Cc: Stanley H Lillie
Subject: Anthrax Vaccine

General Cain,
Just came from meeting with Mr. Oliver, Sue Baily, Pam Berkowsky, Fred Gerber, and reps from LA, OGC, USD(P&R) regarding the Shay's Hearing and follow-on actions. HA worked late into Friday evening, this was the follow-on. The feeling was that while FDA started strong and ended strong, GAO carried the day during the questioning. As a result of this there will likely be a 'single OSD point of contact for Anthrax" assigned. SECDEF will be calling either FDA or HHS with the message "we (DoD/FDA) didn't do to well on this one" requesting that they write to Shays, as will DoD (HA lead). Lots of questions regarding how we were "unprepared to answer/counter GAO's testimony." There will also, most likely be a letter to GAO discussing our displeasure with the "expertise" put on this project.

Here's the web page I mentioned -- doesn't paint a pretty picture:

http://more.abcnews.go.com/sections/us/DailyNews/anthrax990429.html

I understand that there was also an interview in the Hartford Connecticut Chronical (see e-bird last Wed).

Originally Mr. Oliver wanted to "Murder Board" you on AVA tomorrow; later the tasking evolved into a "very tight" point paper by Friday outlining JPO-BD's plan for oversight of BioPort. He also wants a "cradle-to-grave" review of how we got to the current contract.

Please chat this up with Bob, Mike, and Winnie and give me a call in the morning -- we'll need to attack this very aggressively. Also, you know Mr. Oliver's style -- be ready for addtional com's from any imaginable quadrant (suspect he'll be on the phone to LTG's Blanck and Kern).

APPENDIX R

Memo to SecDef Rumsfeld on Minimizing Anthrax Vaccine Use

DoD Undersecretaries Dr. David Chu and Edward Aldridge reviewed the "political problems" associated with the anthrax vaccine and Gulf War illness. The undersecretaries presented recommendations to Defense Secretary Donald Rumsfeld. Highlights included continuing the program only "at a minimum level"; implementing "an acquisition strategy to purchase additional bio-detectors and stockpiles of antibiotics to augment force protection in the absence of an anthrax vaccine"; developing a "coherent institutional process to assess and prioritize biological threats and approve the use of associated countermeasures"; and the development of a "national long-range vaccine that will address the full range of requirements of the DoD, DHHS, and other stakeholders in this plan."

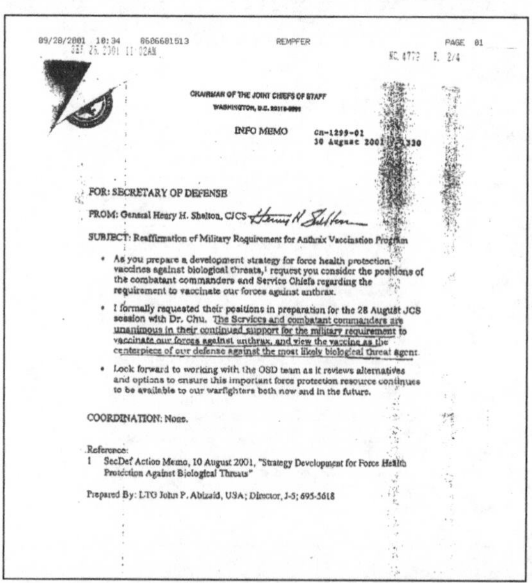

09/28/2001 10:34 8606681513 REMPFER PAGE 01

CHAIRMAN OF THE JOINT CHIEFS OF STAFF
WASHINGTON, D.C.

INFO MEMO CM-1299-01
30 August 2001

FOR: SECRETARY OF DEFENSE

FROM: General Henry H. Shelton, CJCS

SUBJECT: Reaffirmation of Military Requirement for Anthrax Vaccination Program

- As you prepare a development strategy for force health protection vaccines against biological threats,[1] request you consider the positions of the combatant commanders and Service Chiefs regarding the requirement to vaccinate our forces against anthrax.
- I formally requested their positions in preparation for the 28 August JCS session with Dr. Chu. The Services and combatant commanders are unanimous in their continued support for the military requirement to vaccinate our forces against anthrax, and view the vaccine as the centerpiece of our defense against the most likely biological threat agent.
- Look forward to working with the OSD team as it reviews alternatives and options to ensure this important force protection resource continues to be available to our warfighters both now and in the future.

COORDINATION: None.

Reference:
1 SecDef Action Memo, 10 August 2001, "Strategy Development for Force Health Protection Against Biological Threats"

Prepared By: LTG John P. Abizaid, USA; Director, J-5; 695-5618

APPENDIX S

CJCS Memo Dubbing Anthrax Vaccine a "Centerpiece"

09/28/2001 10:34 8606681513 REMPFER PAGE 02
SEP 28, 2001 11:03AM NO 4779 P. 4/4

SECDEF HAS SEEN
AUG 10 2001

ACTION MEMO

August 10, 2001 8:00 AM

[illegible] SECRETARY OF DEFENSE DepSec Action______

[illegible] Mr. E.C. Aldridge, Jr., USD (Acquisition, Technology & Logistics)
Dr. David S. C. Chu, USD (Personnel & Readiness)

[illegible]JECT: Strategy Development for Force Health Protection Against Biological Threats

[illegible]e recommend the following course of action to carry out your plan for anthrax and [illegible]her vaccines required for Force Health Protection:

- USD(AT&L) will review and assess performance of BioPort Corporation and make a determination of future funding to support the strategy outlined below. Report completed by Sept 01.
  The current Anthrax Vaccine Immunization Program will continue at a minimum level (critical personnel and projects only). USD(AT&L) will implement an acquisition strategy to purchase additional bio-detectors and stockpiles of antibiotics to augment force protection in the absence of an anthrax vaccine.
- USD(P&R) will request PA&E to provide the scenario analyses needed for future decisions related to the threat of biological agents. Report to USD(P&R) Dec 01.
- USD(P&R) will request DIA through C3I to provide a comprehensive review of the doctrinal positions of any State or group to use biological weapons as a part of their offensive/defensive strategies. Report to USD(P&R) by Dec 01.
- USD(P&R) will coordinate with appropriate DoD officials to develop a coherent institutional process to assess and prioritize biological threats and approve the use of associated countermeasures. Prepare DoD Directive for coordination by Dec 01.
- USD(AT&L) and USD(P&R) will continue their inter-agency efforts to develop a national long-range vaccine that will address the full range of requirements of the DOD, DHHS, and other stake-holders in this plan. This plan will include an evaluation of the need for a dedicated vaccine facility to serve the national interest. Provide an interim report to SecDef by Dec 01.

[illegible]COMMENDATION: Secretary of Defense approves strategy development by [illegible]tialing this page.

[illegible]ORDINATION: None

[illegible]pared By: Dr. David Chu, USD (Personnel & Readiness), 695-5254

SPL ASSISTANT
SR MA GIAMBAS
MA BUCCI
EXECSEC WHIT[illegible]

APPENDIX T

GAO-02-181T—Changes to the Anthrax Vaccine Manufacturing Process

The GAO testified before the Congress in April 1999, revealing problems with the anthrax vaccine, including unknown long-term safety differences between the tested and "licensed" vaccine, a lack of data for inhalation anthrax, and FDA inspection deficiencies:[10]

United States General Accounting Office

GAO

Testimony
Before the Subcommittee on National Security, Veterans' Affairs, and International Relations, Committee on Government Reform, House of Representatives

For Release on Delivery
Expected at 10:00 a.m., EDT
Tuesday, October 23, 2001

ANTHRAX VACCINE

Changes to the Manufacturing Process

Statement of Nancy Kingsbury, Ph.D., Managing Director, Applied Research and Methods

The anthrax vaccine being given to US military personnel was licensed in 1970. Before the vaccine was licensed, the vaccine and the manufacturing process were changed . . .

First, the manufacturing process changed when MDPH took over. Second, the strain of anthrax that Merck used to grow the original vaccine was changed, and another strain was used to grow the MDPH vaccine. Finally, to increase the yield of the protective antigen (which is believed to be an important part of the vaccine's protective effects), the ingredients used to make vaccine were changed from the original vaccine.

The long-term safety of the vaccine has not yet been studied.

Prior to the time of licensing, no human efficacy testing of the MDPH vaccine was performed. However, a study [Brachman] was done on the efficacy of the original vaccine. This study concluded that the vaccine provided protection to humans against anthrax penetrating the skin. [Cutaneous]

In the 1980s, DoD began testing the efficacy of the licensed vaccine on animals, focusing on its protection against inhalation anthrax. DoD recognizes that correlating the results of animal studies to humans is necessary and told us that it is planning research in this area.

[A] 1991 Army document noted . . . we can conclude that the licensed vaccine is efficacious against cutaneous exposure but that testing still needs to be conducted on inhalation anthrax.

Until 1993, FDA inspectors did not inspect the MDPH facility where the anthrax vaccine was made. According to FDA, access was not granted because its inspectors had not been vaccinated against anthrax. FDA's inspections of the MDPH facility found a number of deficiencies.

In 1998, MDPH closed its plant, which is now being renovated.

APPENDIX U

Minutes and Slides Related to the October 20th, 1995, DoD Meeting

The minutes detailed "the process for modifying the MDPH anthrax vaccine license to . . . expand the indication to include protection against aerosol challenge of spores." In discussing the previous clinical trials, the working group acknowledged, "there was insufficient data to demonstrate protection against inhalation disease":

DEPARTMENT OF THE ARMY
JOINT PROGRAM OFFICE
FOR BIOLOGICAL DEFENSE
5201 Leesburg Pike
Skyline #3, Suite 1200
Falls Church, VA 22041-3203

REPLY TO
ATTENTION OF
SFAE-BD (70-1z)

13 November 1995

MEMORANDUM FOR SEE DISTRIBUTION

SUBJECT: Minutes of the Meeting on Changing the Food and Drug Administration License for the Michigan Department of Public Health (MDPH) Anthrax Vaccine to Meet Military Requirements

A meeting was held on 20 October 1995 to discuss the process for modifying the MDPH anthrax vaccine license to indicate a reduced numbers of injections and to expand the indication to include protection against an aerosol challenge of spores. Attendees (see enclosure 1) met in the conference room of the Joint Program Office for Biological Defense (JPO-BD), located in Suite 1200, Skyline #3, 5201 Leesburg Pike, Falls Church, Virginia 22041.

The meeting was hosted by BG Walter L. Busbee, Joint Program Manager for Biological Defense, and LTC(P) David Danley, Deputy Joint Program Manager for Medical Systems, JPO-BD. BG Busbee opened the meeting with comments about recent events involving the Office of the Secretary of Defense (OSD) and the Joint Staff which could have significant impacts on both the medical and non-medical biological defense programs. He highlighted a key issue concerning the DoD vaccination policy, stating that a tour in some theaters is 13 months, whereas the dosage schedule for anthrax vaccine requires 18 months. This has raised questions among senior military leaders, such as: "How do you protect the force? What sector of the force do you protect? Do you inoculate all 1.5M service members because you don't know exactly who will be deployed?" He charged the audience to objectively, scientifically, and carefully articulate the pros and cons for the different options involving the anthrax vaccination schedule.

COL Arthur Friedlander, Chief, Bacteriology Division, U.S. Army Medical Research Institute for Infectious Diseases (USAMRIID), presented a briefing covering three topics: (1) evidence for a reduction in the number of doses of anthrax vaccine, (2) evidence for vaccine efficacy against an aerosol challenge, and (3) progress toward an in vitro correlate of immunity. He opened the briefing by stating it is impossible to conduct human studies involving inhalation of anthrax spores; therefore, animal models must be developed, and the data related to humans.

With respect to reducing the number of doses in the immunization schedule, COL Friedlander said that the original series of 6 doses was established in the 1950's for an anthrax vaccine similar to but not identical with the MDPH vaccine. Studies of vaccine (not MDPH product) effectiveness in humans working in tanneries showed protection against cutaneous disease, but there was insufficient data to demonstrate protection against inhalational disease. The package insert for MDPH anthrax vaccine does not discriminate between different forms of anthrax, but does describe occupations that may involve anthrax exposure

USAMRIID

PROBLEMS WITH CURRENT MDPH VACCINE

Prolonged immunization schedule

Reactogenicity:

Systemic reactions: 0.7-1.3% ✓

Significant local reactions: 2.4-3.9% (5.7%) ✓

Vaccine components completely undefined in terms of characterization and quantitation of the PA, and other bacterial products and constituents present

Significant lot-to-lot variation in the PA immunogen content

Human trials with similar but not identical vaccine showed protection against cutaneous anthrax but insufficent data to show efficacy against inhalation anthrax

Made from spore-forming strain requiring dedicated production facility

Anthrax Vaccine License Amendment Project Plan

SAIC Project Description

- Provide the Director, Medical Biological Defense Research Program with a project plan to obtain an amendment to the Anthrax Vaccine product license
  - Reduce number of inoculations required
  - Obtain indication for protection against aerosol exposure
- Completed plan due in 30 days (15 Sep 95)

Anthrax Vaccine License Amendment Project Plan

Anthrax Vaccine License Amendment Project Plan

Information Briefing

for

Joint Program Manager

DoD Biological Defense

October 20, 1995

Anthrax Vaccine License Amendment Project Plan

Basis for Current Schedule (Anecdotal -- CDC)

- Based on old (weaker) anthrax vaccine (circa 1950s)
  - Three anthrax cases after first 3 doses (2 at Detrick and 1 at wool mill)
  - Arbitrarily added 3 more doses at 6-month intervals
- MDPH vaccine
  - Higher PA concentration
  - Much more immunogenic
- No schedule change recommended

APPENDIX V

Investigational New Drug (IND) Application for "Inhalation Anthrax"

1996 and 1999 IND applications, Form FDA 1571, filed for the DoD by the Michigan Biologic Products Institute (MBPI) for "inhalation anthrax" indication or approval (in block 7):

DEPARTMENT OF HEALTH AND HUMAN SERVICES
PUBLIC HEALTH SERVICE
FOOD AND DRUG ADMINISTRATION
INVESTIGATIONAL NEW DRUG APPLICATION (IND)
(TITLE 21, CODE OF FEDERAL REGULATIONS (CFR) Part 312)

Form Approved: OMB No. 0910-0014. Expiration Date: November 30, 1995. See OMB Statement on Reverse.

NOTE: No drug may be shipped or clinical investigation begun until an IND for that investigation is in effect (21 CFR 312.40)

1. NAME OF SPONSOR: Michigan Biologic Products Institute (formerly Michigan Department of Public Health)

2. DATE OF SUBMISSION: September 20, 1996

3. ADDRESS (Number, Street, City, State and Zip Code): 3500 N. Martin Luther King Jr. Blvd. Lansing, MI 48906

4. TELEPHONE NUMBER (Include Area Code): (517) 335-9956

5. NAME(S) OF DRUG (Include all available names: Trade, Generic, Chemical, Code): Anthrax Vaccine Adsorbed

6. IND NUMBER (If previously assigned):

7. INDICATION(S) (Covered by this submission)
1. Inhalation Anthrax 2. Change in Route of Administraiton
3. Change in vaccination schedule

8. PHASE(S) OF CLINICAL INVESTIGATION TO BE CONDUCTED: [ ] PHASE 1 [X] PHASE 2 [ ] PHASE 3 [ ] OTHER ______ (Specify)

9. LIST NUMBERS OF ALL INVESTIGATIONAL NEW DRUG APPLICATIONS (21 CFR Part 312), NEW DRUG OR ANTIBIOTIC APPLICATIONS (21 CFR Part 314), DRUG MASTER FILES (21 CFR 314.420), AND PRODUCT LICENSE APPLICATIONS (21 CFR Part 601) REFERRED TO IN THIS APPLICATION.
Establishment License #99
DBS-IND 180
BB-MF 6052
BB-IND 3723

10. IND submissions should be consecutively numbered. The initial IND should be numbered "Serial Number: 000." The next submission (e.g., amendment, report, or correspondence) should be numbered "Serial Number: 001." Subsequent submissions should be numbered consecutively in the order in which they are submitted.

SERIAL NUMBER: 000

11. THIS SUBMISSION CONTAINS THE FOLLOWING: (Check all that apply)
[X] INITIAL INVESTIGATIONAL NEW DRUG APPLICATION (IND) [ ] RESPONSE TO CLINICAL HOLD

PROTOCOL AMENDMENT(S):
[ ] NEW PROTOCOL
[ ] CHANGE IN PROTOCOL
[ ] NEW INVESTIGATOR

INFORMATION AMENDMENT(S):
[ ] CHEMISTRY/MICROBIOLOGY
[ ] PHARMACOLOGY/TOXICOLOGY
[ ] CLINICAL

IND SAFETY REPORT(S):
[ ] INITIAL WRITTEN REPORT
[ ] FOLLOW-UP TO A WRITTEN REPORT

[ ] RESPONSE TO FDA REQUEST FOR INFORMATION [ ] ANNUAL REPORT [ ] GENERAL CORRESPONDENCE

[ ] REQUEST FOR REINSTATEMENT OF IND THAT IS WITHDRAWN, INACTIVATED, TERMINATED OR DISCONTINUED [ ] OTHER ______ (Specify)

CHECK ONLY IF APPLICABLE

JUSTIFICATION STATEMENT MUST BE SUBMITTED WITH APPLICATION FOR ANY CHECKED BELOW. REFER TO THE CITED CFR SECTION FOR FURTHER INFORMATION.
[ ] TREATMENT IND 21 CFR 312.35(b) [ ] TREATMENT PROTOCOL 21 CFR 312.35(a) [ ] CHARGE REQUEST/NOTIFICATION 21 CFR 312.7(d)

FOR FDA USE ONLY

CDR/DBIND/DGD RECEIPT STAMP | DDR RECEIPT STAMP | IND NUMBER ASSIGNED: | DIVISION ASSIGNMENT:

FORM FDA 1571 (12/92) PREVIOUS EDITION IS OBSOLETE. PAGE 1 OF 2

DEPARTMENT OF HEALTH AND HUMAN SERVICES
PUBLIC HEALTH SERVICE
FOOD AND DRUG ADMINISTRATION
INVESTIGATIONAL NEW DRUG APPLICATION (IND)
(TITLE 21, CODE OF FEDERAL REGULATIONS (CFR) PART 312)

Form Approved: OMB No. 0910-0014.
Expiration Date: December 31, 1999
See OMB Statement on Reverse.

NOTE: No drug may be shipped or clinical investigation begun until an IND for that investigation is in effect (21 CFR 312.40).

1. NAME OF SPONSOR
BioPort Corporation

2. DATE OF SUBMISSION

3. ADDRESS *(Number, Street, City, State and Zip Code)*
3500 N. Martin Luther King, Jr. Boulevard
Lansing,MI 48906

4. TELEPHONE NUMBER *(Include Area Code)*
(517) 335-8540

5. NAME(S) OF DRUG *(Include all available names: Trade, Generic, Chemical, Code)*
Anthrax Vaccine Adsorbed

6. IND NUMBER *(If previously assigned)*
BB-IND 6847

7. INDICATION(S) *(Covered by this submission)*
Inhalation Anthrax

8. PHASE(S) OF CLINICAL INVESTIGATION TO BE CONDUCTED:
☐ PHASE 1 ☐ PHASE 2 ☑ PHASE 3 ☐ OTHER ______ *(Specify)*

9. LIST NUMBERS OF ALL INVESTIGATIONAL NEW DRUG APPLICATIONS (21 CFR Part 312), NEW DRUG OR ANTIBIOTIC APPLICATIONS (21 CFR Part 314), DRUG MASTER FILES (21 CFR Part 314.420), AND PRODUCT LICENSE APPLICATIONS (21 CFR Part 601) REFERRED TO IN THIS APPLICATION.
Establishment License #99 BB-IND 3723
DBS-IND 180
BB-MF 6052

ARCHIVAL

10. ***IND submission should be consecutively numbered. The initial IND should be numbered "Serial number: 000." The next submission (e.g., amendment, report, or correspondence) should be numbered "Serial Number: 001." Subsequent submissions should be numbered consecutively in the order in which they are submitted.***

SERIAL NUMBER
009

11. THIS SUBMISSION CONTAINS THE FOLLOWING: *(Check all that apply)*
☐ INITIAL INVESTIGATIONAL NEW DRUG APPLICATION (IND) ☐ RESPONSE TO CLINICAL HOLD

PROTOCOL AMENDMENT(S):
☐ NEW PROTOCOL
☐ CHANGE IN PROTOCOL
☐ NEW INVESTIGATOR

INFORMATION AMENDMENT(S):
☐ CHEMISTRY/MICROBIOLOGY
☐ PHARMACOLOGY/TOXICOLOGY
☐ CLINICAL

IND SAFETY REPORT(S):
☐ INITIAL WRITTEN REPORT
☐ FOLLOW-UP TO A WRITTEN REPORT

☐ RESPONSE TO FDA REQUEST FOR INFORMATION ☐ ANNUAL REPORT ☐ GENERAL CORRESPONDENCE
☐ REQUEST FOR REINSTATEMENT OF IND THAT IS WITHDRAWN, INACTIVATED, TERMINATED OR DISCONTINUED ☑ OTHER Meeting Minutes, 15 December 1998 *(Specify)*

CHECK ONLY IF APPLICABLE

JUSTIFICATION STATEMENT MUST BE SUBMITTED WITH APPLICATION FOR ANY CHECKED BELOW. REFER TO THE CITED CFR SECTION FOR FURTHER INFORMATION.

☐ **TREATMENT IND 21 CFR 312.35(b)** ☐ **TREATMENT PROTOCOL 21 CFR 312.35(a)** ☐ **CHARGE REQUEST/NOTIFICATION 21 CFR312.7(d)**

FOR FDA USE ONLY

CDR/DBIND/DGD RECEIPT STAMP

RECEIVED

CBER/DCC

DDR RECEIPT STAMP

DIVISION ASSIGNMENT:

IND NUMBER ASSIGNED:

FORM FDA 1571 (1/97) PREVIOUS EDITION IS OBSOLETE. PAGE 1 OF 2

APPENDIX W

Federal Bureau of Investigation (FBI) Amerithrax Press Conference

Excerpts from an FBI press release documented the frustrations by a US Army scientist over testing irregularities with the anthrax vaccine—the suspected motive for the 2001 anthrax letter attacks. The FBI alleged the Army scientist's motive to save the troublesome program related to the fact the "anthrax vaccine he was working on was failing" due to potency problems:

Remarks Prepared for Delivery by U.S. Attorney Jeffrey Taylor at Amerithrax Investigation Press Conference

WASHINGTON, D.C.
Wednesday, August 06, 2008

Fifth, as reflected in the court documents, Dr. Ivins had a history of mental health problems and was facing a difficult time professionally in the summer and fall of 2001 because an anthrax vaccine he was working on was failing. The affidavits describe one e-mail to a co-worker in which Dr. Ivins stated that he had "incredible paranoid, delusional thoughts at times," and feared that he might not be able to control his behavior.

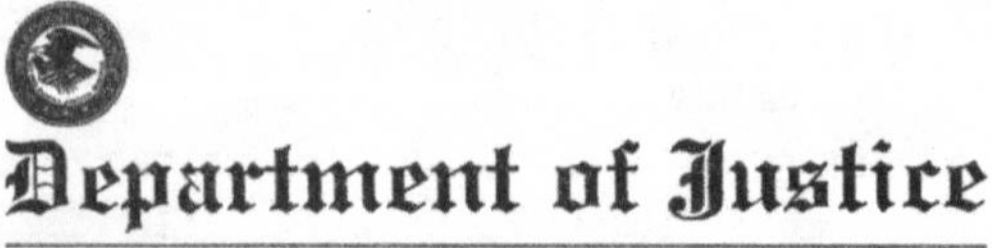
Department of Justice

FOR IMMEDIATE RELEASE
Wednesday, August 6, 2008

MR. TAYLOR: The other question you have, Dr. Ivins is a troubled individual, particularly so at that time. He's very concerned, according to the evidence, that this vaccination program he's been working on may come to an end. He's also very concerned that some have been criticizing and blaming that vaccination program in connection with illnesses suffered by soldiers from, I think, the first Gulf War. So that was going on, according to the evidence, in his mind at that particular time.

With respect to motive, I'll point again to -- with respect to the motive, the troubled nature of Dr. Ivins. And a possible motive is his concern about the end of the vaccination program. And the concerns had been raised, and one theory is that by launching these attacks, he creates a situation, a scenario, where people all of a sudden realize the need to have this vaccine.

FBI Affidavit Analysis Excerpts, page 15

AO106(Rev.5/85) Affidavit for Search Warrant

UNITED STATES DISTRICT COURT
FOR THE DISTRICT OF COLUMBIA

FILED
OCT 31 2007
NANCY MAYER WHITTINGTON, CLERK
U.S. DISTRICT COURT

In the Matter of the Search of

Residence at [redacted]
Frederick, Maryland,
owned by Bruce Edwards Ivins,
DOB [redacted] SSN [redacted]

APPLICATION AND AFFIDAVIT FOR SEARCH WARRANT

CASE NUMBER: 07-524-M-01

Controversy concerning the anthrax vaccine

Beginning shortly after the first Gulf War and through 2001, USAMRIID and Dr. Ivins was the focus of public criticism concerning their introduction of a squalene adjuvant (or additive) to the AVA anthrax vaccine, which was blamed for the Gulf War Syndrome. In 2000 and 2001, as evident by the e-mails above, that same anthrax vaccine was having problems in the production phase at Bioport, a private company in Michigan responsible for manufacturing the vaccine. The Food and Drug Administration (FDA) had suspended further production at Bioport, and the U.S. government, specifically the Department of Defense, was running out of approved lots of the vaccine. The situation placed pressure on select staff members at USAMRIID, including Dr. Ivins, who were part of the Anthrax Potency Integrated Product Team (IPT). The purpose of the IPT was to assist in the resolution of technical issues that was plaguing Bioport's production of approved lots of the vaccines.

In the weeks immediately prior to the attacks, Dr. Ivins became aware that an investigative journalist who worked for NBC News had submitted a Freedom of Information Act (FOIA) requests on USAMRIID seeking detailed information from Dr. Ivins's laboratory notebooks as they related to the AVA vaccine and the use of adjuvants. On August 28, 2001, Dr. Ivins appeared angry about the request providing the following response in an e-mail: "Tell Matsumoto to kiss my ass. We've got better things to do than shine his shoes and pee on command. He's gotten everything from me he will get."

In early 2002, shortly after the anthrax letter attacks, the FDA re-approved the AVA vaccine for human use, production at Bioport resumed, and anthrax research at USAMRIID continued without interruption. As mentioned previously, one of the anthrax letters post marked on September 18, 2001, was addressed to Tom Brokaw, NBC News in New York. Dr. Ivins thereafter received "the highest honor given to Defense Department civilians at a Pentagon ceremony on March 14, 2003" for his work in "getting the anthrax vaccine back into production."

FBI Excerpts from Emails, pages 12–14

June 28, 2000, "Apparently Gore (and maybe even Bush) is considering making the anthrax vaccine for the military voluntary, or even stopping the program. Unfortunately, since the BioPort people aren't scientists, the task of solving their problem has fallen on us. . . . . Believe me, with all the stress of home and work, your email letters to me are valuable beyond what you would ever imagine – and they help me keep my sanity...."

June 29, 2000, "BioPort just tested its final lot of AVA [anthrax vaccine] in a potency test. If it doesn't pass, then there are no more lots to test, and the program will come to a halt. That's bad for everyone concerned, including us. I'm sure that blame will be spread around."

July 6, 2000, "[redacted] I think the **** is about to hit the fan...bigtime. The final lot of AVA, lot 22, isn't passing the potency test, and now there's nothing to back it up. Plus, the control vaccine isn't working. It's just a fine mess. [redacted] are spending probably 95% of our time on this."

August 29, 2000, "[redacted] are 10% of the Bacteriology Division. If we quit, the anthrax program and BioPort would go down the drain. I'm not boasting, [redacted], but the three of us have a combined total of 52 years of research experience with anthrax. You just can't go out and find someone like [redacted] with their knowledge, skill and abilities. Ain't gonna happen."

September 7, 2001, "I was taken off the Special Immunization Program because of what happened last spring, and I've just gotten back on it, getting my anthrax and Yellow fever shots. We are currently finishing up the last of the AVA, and when that is gone, there's nothing to replace it with. I don't know what will happen to the research programs and hot suite work until we get a new lot. There are no approved lots currently available at BioPort. . . . [redacted] has been having us have biweekly meetings on the rPA vaccine progress, and on August 29 I went to the Pentagon – first time there – to go to a meeting in his place on the vaccine. There is a real bag of worms with a new lot of rPA produced by the BDP (a private company) for NCI, who is under contract to USAMRIID. BDP signed a sub-contract with to produce the rPA for a human use vaccine Phase I trial. They were paid and they produced it. Now they are refusing to release it unless the Army pays some incredible sum of money for lawsuit indemnification (about $200,000 per year for the next 50 years). The Army refuses to do that of course, and everything is in Limbo."

APPENDIX X

USAF Reserve Memo Barring Accession after Anthrax Vaccine Refusal

DEPARTMENT OF THE AIR FORCE

AIR FORCE RESERVE COMMAND

MEMORANDUM FOR ALL NAF CCs 12 SEP 2001
ARPC/CC
ALL AFRC RECRUITERS

FROM: AFRC/CC
155 Richard Ray Blvd
Robins AFB, GA 31098-1635

SUBJECT: Accession of Members Who Separated to Avoid Anthrax Vaccination

1. This memorandum is to communicate the policy of the Air Force Reserve regarding accession of members who previously separated to avoid the anthrax vaccine. Pursuant to Air Force Reserve Command Instruction 36-2001, *AFR Recruiting Procedures*, 1 October 1999, paragraph 3.2, accession to the Selected Reserve will be denied to any individual known to have previously separated in order to avoid the anthrax vaccine. I have determined that such accession is contrary to the best interest of the Air Force Reserve. This determination is in accordance with my duty under AFPD 36-20, paragraph 2.3. In the event admission is denied on this basis, it will be documented in accordance with AFRCI 36-2001.

2. Further, any individual involuntarily reassigned to the Non-Affiliated Reserve Section (NARS) as a result of anthrax vaccine avoidance will be denied reassignment to the Selected Reserve. Reassignments for anthrax vaccine avoidance are normally coded more generally as "medical reasons" or "unsatisfactory participation." Therefore, denial of reassignment or accession will be based on actual knowledge of anthrax vaccine avoidance as the underlying reason for the earlier reassignment or separation.

3. When the anthrax vaccine becomes available, applications for accession or reassignment of individuals known to have avoided the anthrax vaccine will be considered on a case-by-case basis, provided they are willing to be vaccinated.

JAMES E. SHERRARD III, Lt Gen, USAF
Commander

cc: HQ AFRC Directors

APPENDIX Y

Under Secretary of Defense Robert L. Wilkie Memo

UNDER SECRETARY OF DEFENSE
4000 DEFENSE PENTAGON
WASHINGTON, D.C. 20301-4000

PERSONNEL AND READINESS

JUL 25 2018

MEMORANDUM FOR SECRETARIES OF THE MILITARY DEPARTMENTS

SUBJECT: Guidance to Military Discharge Review Boards and Boards for Correction of Military / Naval Records Regarding Equity, Injustice, or Clemency Determinations

The Department has evaluated numerous aspects of the Service Discharge Review Boards (DRBs) and Boards for Correction of Military / Naval Records (BCM/NRs) over the last two years. We have redoubled our efforts to ensure veterans are aware of their opportunities to request review of their discharges and other military records. We have initiated several outreach efforts to spread the word and invite feedback from veterans and organizations that assist veterans and active duty members, and issued substantive clarifying guidance on Board consideration of mental health conditions and sexual assault or sexual harassment experiences. And, we have partnered with the Department of Veterans Affairs to develop a web-based tool that provides customized guidance for veterans who want to upgrade their discharges. But our work is not yet done.

Increasing attention is being paid to pardons for criminal convictions and the circumstances under which citizens should be considered for second chances and the restoration of rights forfeited as a result of such convictions. Many states have developed processes for restoring basic civil rights to felons, such as the right to vote, hold office, or sit on a jury, and many states have developed veterans' courts to consider special circumstances associated with military service. States do not have authority, however, to correct military records or discharges.

The Military Departments, operating through DRBs and BCM/NRs, have the authority to upgrade discharges or correct military records to ensure fundamental fairness. DRBs and BCM/NRs have tremendous responsibility and perform their tasks with remarkable professionalism, but further guidance to inform Board decisions on applications based on pardons for criminal convictions is required.

The attached guidance closes this gap and sets clear standards. While not everyone should be pardoned, forgiven, or upgraded, in some cases, fairness dictates that relief should be granted. We trust our Boards to apply this guidance and give appropriate consideration to every application for relief.

Military Department Secretaries will ensure that Board members are familiar with and appropriately trained on this guidance within 90 days. My point of contact is Monica Trucco, Director, Office of Legal Policy, who may be reached at (703) 697-3387 or monica.a.trucco.civ@mail.mil.

Robert L. Wilkie

Attachment:
As stated

cc:
Chairman of the Joint Chiefs of Staff
General Counsel of the Department of Defense
Assistant Secretary of Defense for Legislative Affairs
Assistant Secretary to the Defense for Public Affairs

APPENDIX Z

Sergeant James D. Muhammad's Military Times Corrections Case Article

TROOPS WHO REFUSED THE ANTHRAX VACCINE PAID A VERY HIGH PRICE

By Todd South
tsouth@militarytimes.com

During the first eight years that the Pentagon ran the anthrax vaccination program, hundreds of troops refused the vaccine due to perceived health risks or religious concerns — and many of them paid dearly for that decision.

The penalties ranged widely. Some kept on working, others received nonjudicial punishment, lost rank and pay, saw their careers ended or even faced brig time and dishonorable discharges.

Since then, an unknown fraction of those who were punished have sought to have their records corrected, but only a few have had success. Now, even more than 20 years later, some of those cases remain pending before military record corrections boards.

Numbers are hard to pin down, as service record corrections boards have not comprehensively tracked appeals specifically related to the anthrax vaccine. In many cases, those appeals were denied. But more recently, at least two corrections requests — one in 2019 and the other in 2020 — were granted by the Navy, which awarded two Marine veterans some backpay, rank restoration, discharge upgrades and access to veterans benefits.

Retired Marine Maj. Dale Saran, a former JAG officer, represented one sailor and two Marines who refused the vaccine on Okinawa in 2000.

Saran stayed involved with the issue. He authored the book "United States v. Members of the Armed Forces: The Truth Behind the Department of Defense's Anthrax Vaccine Immunization Program," published in 2020. He also offered legal advice to attorneys with clients facing punishments for refusals.

Saran's clients and many others were often top performers, some early in their careers and others nearing retirement. But once they refused the vaccine, their commands sought to punish them.

"The venom with which they went after people ... I just find that abhorrent," Saran said. Those service members "were charged and threatened, people's lives were destroyed."

HUNDREDS OF APPEALS

Today, military records corrections boards are unable to provide reliable data on the number of appeals involving refusal to get the anthrax shot. When Military Times recently asked about the anthrax cases, the Army, Navy and Air Force were only able to confirm two specific cases.

That's despite language in the 2001 defense spending bill in which Congress told the secretaries for each of the branches to establish a system for "tracking, recording and reporting separations of members of the armed forces" that result from a refusal to participate in the anthrax vaccine immunization program."

Years ago, official Pentagon statements indicated that an estimated 350 servicemembers had refused the vaccine between 1998 to 2000. At least three dozen of them were court-martialed and hundreds left the service to avoid the vaccine, according to Pentagon statements in 2005. Also, at least another 149 troops were forced out of the service for vaccine refusal from 2000 to 2004. Those numbers did not include troops who refused but were allowed to leave the service without punishment or simply allowed to not receive the shot and suffered no repercussions.

Retired Air Force Lt. Col. Thomas Rempfer was a vocal critic of the vaccination program. He left the Connecticut Air National Guard as a captain in 1999 due to his own anthrax vaccine refusal but was able to continue his career in the Air Force Reserve.

Rempfer and fellow Guard pilot Lt. Col. Russ Dingle spearheaded efforts, including congressional testimony and lawsuits, to bring a halt to mandatory vaccinations until several major problems with the program could be fixed.

Dingle died of cancer in 2005, Rempfer continues to push for draft

A voluntary process
U.S. Marines queue up to receive the Moderna coronavirus vaccine April 28 at Camp Hansen on Okinawa, Japan. The services have urged troops to take the shots, but haven't forced them.

legislation — in Dingle's name to honor the late officer's role as the intellectual inspiration of their shared work — that would require the Pentagon to proactively correct the records of service members who were punished in any form for their anthrax vaccine refusal.

To that end, Rempfer has created a website, Hoping4Justice.org, that catalogs key documents and a timeline of events in the anthrax vaccination program and recent success by some service members to have their records corrected nearly two decades later.

"There needs to be a well-advertised effort to let these former service members know there's an opportunity to get their records corrected," Rempfer told Military Times.

Veterans have sought records corrections through their respective boards from the early 2000s through at least the early 2010s, according to court records. Once those were denied, some of those veterans then appealed to the federal civil courts.

In 2004, a court injunction halted the military's mandatory vaccination program, declaring the previous six years of vaccine administration to be illegal.

Yet when individual veterans appealed their cases in federal court, the judges mostly rejected those claims, siding with a ruling by the U.S. Court of Appeals for the Armed Forces that dismissed arguments using the 2004 injunction.

Rempfer has written to each administration since President Barack Obama. He has also created draft legislation for Congress to enact that would push the services to remedy the problem. So far, neither have gained backing.

FROM CAREER AIR FORCE TO SUDDEN DISCHARGE

Senior Airman Jeffrey Bettendorf was loving life in the Air Force. He'd joined six years before and already planned to do a full 30 years. An aerospace ground equipment mechanic, he was stationed at Travis Air Force Base, California, in 1998 when his squadron heard they were going to take a new vaccine against anthrax.

Bettendorf told Military Times that most of his fellow airmen were nonchalant about it, basically seeing it as "just another shot." But he was curious, so he started researching online, where he found government reports, testimony and groups sharing materials about safety issues at the vaccine's manufacturing plant, along with possible linkages to Gulf War Syndrome.

He complied two 3-inch binders full of printed materials and scheduled a meeting with his commander.

The senior airman didn't get far. The commander didn't even look at the binders. The CO told him that if he didn't take the shot he faced a reduction in rank, fines and extra duty until he did.

"Then he sent me to a psychiatrist, saying if I'm going to refuse this vac

IN THE SHADOW OF ANTHRAX

THE VOLUNTARY NATURE OF COVID-19 VACCINATIONS OWES A LOT TO THE MILITARY'S PAST SHORTCOMINGS

By Todd South
tsouth@militarytimes.com

While the military has pursued vaccinating the force against COVID-19 over recent months, taking the shot has remained voluntary.

Even so, the Defense Department has made steady progress in getting the active-duty force vaccinated. Fifty-eight percent of active-duty troops had received at least one dose and 44 percent were fully vaccinated as of May 20, DoD health leaders said.

But not everyone offered the shot has taken it.

In February, reports showed at least a third of troops had declined vaccination. By March, however, acceptance rates were rising, Pentagon officials said.

Each of the services continue to provide incentives, such as fewer liberty restrictions, elimination of movement restrictions before a deployment and maskless work.

But leaders are not allowed to punish or reprimand troops who refuse.

The voluntary COVID-19 vaccine effort stands in stark contrast to the Pentagon's mandatory Anthrax Vaccine Immunization Program, which began in 1998. Despite the questionable safety of the program, all active-duty service members were required to take the series of shots, and troops who refused often faced harsh penalties.

Assessing the threat
Secretary of State Colin Powell holds a vial representing the small amount of anthrax that forced the closure of the U.S. Senate in 2002 during his address to the UN Security Council in February 2003.

How the anthrax vaccination effort was handled casts a long shadow over today's vaccination efforts in the era of coronavirus.

Keeping the COVID-19 vaccine voluntary has been supported by both recent presidential administrations and their defense secretaries, so far. But if President Joe Biden decides that fully vaccinating certain units or the entire force is a vital national security need, the vaccine could go from voluntary to mandatory with the stroke of a pen.

That's because the COVID-19 vaccines are under an emergency use authorization by the U.S. Food and Drug Administration. Only the president can override an EUA and require servicemembers to take a vaccine in that status.

Could that happen now?

A White House spokesman referred questions about military vaccinations to the Defense Department. DoD spokeswoman Lisa Lawrence pointed Military Times to a May 20 press briefing as one of the more recent responses to that question.

At the Pentagon briefing, acting Assistant Secretary of Health Affairs Dr. Terry Adirim said he could not speculate on whether it would be made mandatory, but noted that the department does mandate some licensed vaccines, as all service members know.

A vaccine is not officially licensed, however, while in EUA status.

The anthrax vaccine effort and the EUA program are directly linked.

"The shadow of anthrax comes out of Gulf War Syndrome," said retired Marine Maj. Dale Saran a former JAG officer who represented troops who refused the vaccine.

He called the anthrax vaccination program "institutionally damaging" to the military and said the voluntary COVID vaccination program is the right approach.

"Jamming it down people's throats didn't help anything, either," he said.

ANTHRAX VACCINATION PROGRAM

The Defense Department's Anthrax Vaccine Immunization Program was an effort to inoculate the entire military against the perceived threat of a bioterror attack with a weaponized version of the anthrax spore, a bacteria found in animals that can cause severe, even fatal illness in humans.

But by that time, the anthrax vaccine licensed by the FDA had already been given to some 150,000 U.S. troops during the first Gulf War, according to a 2000 Institute of Medicine publication.

The total-force vaccination effort that came after the war was seen by leadership as a way to save lives and keep units operational should they face

APPENDIX AA

Air Reserve Personnel Center Colonel Promotion Letter

DEPARTMENT OF THE AIR FORCE
HEADQUARTERS AIR RESERVE PERSONNEL CENTER

HQ ARPC/CC
18420 E. Silver Creek Ave., Bldg 390 MS 68
Buckley SFB CO 80011-9502

28 SEP 2021

Colonel Thomas L. Rempfer
5110 East Ramsey Rd
Hereford AZ 85615

Dear Colonel Rempfer,

It is my pleasure to inform you of advancement to colonel on the United States Air Force Reserve Retired List. Congratulations, this achievement attests to your ability and performance in the Air Force Reserve. You can be justifiably proud.

You are authorized to wear the rank of colonel and hold the title of "Colonel." You are entitled to hold a retired ID card that reflects your grade as "Colonel" while the paygrade will reflect "O-5." Advancement on the retired list does not serve as a basis for recomputation of military retired pay or entitle you to back pay.

I wish you continued success in your future.

JENNIE R. JOHNSON, USAF
Brigadier General
Commander

APPENDIX BB

Air Force Notification for Promotion Board and Retroactive Date of Rank

DEPARTMENT OF THE AIR FORCE
HEADQUARTERS AIR RESERVE PERSONNEL CENTER

HQ ARPC/PB
18420 E Silver Creek Ave Bldg 390 MS68
Buckley AFB CO 80011

AUG 0 2 2018

Lt Col Thomas L Rempfer
6058 South Jakemp Trail
Tucson AZ 85747

Dear Lt Col Rempfer

The Air Force Board for Correction of Military Records (AFBCMR) has granted you special selection board (SSB) consideration in lieu of the CY11 Colonel Line and Nonline (PR) Selection Board. You are scheduled to meet the next SSB, currently scheduled for 15 Oct 18, pending Secretary of the Air Force approval.

If selected by the SSB, your date of rank will be 01 Jun 12, as if the original board selected you. Your record will be constructed to appear as it would have for the CY11 board, and will be compared to records of officers who were and were not selected by the original selection board. Results will be released approximately 8-10 weeks after the board adjourns. Contact your Military Personnel Section for results.

You may submit a letter to be included in your record for the SSB. It may only cover events and information that would have been available to the original board on 17 Oct 11, and must include the board identifier V0611A. Letters to the board must be submitted electronically. Your letter must be submitted 10 calendar days before 15 Oct 18, or it will not be accepted. Guidelines for preparing a letter are attached.

If you have any questions, please contact the Policy and Procedures Division (HQ ARPC/PBP) DSN 665-0102 or commercial 210-565-0102, or submit a request through the myPers web site.

Sincerely

MICHAEL P. BURNS, Colonel, USAF
Director, Selection Board Secretariat

2 Attachments:
1. Instructions, Personal Letter to Selection Board
2. OSB

WINGS OF HERITAGE, SHAPING THE FUTURE

APPENDIX CC

Excerpts from the 1994 Civilian Medical Textbook *Vaccines*

The US Army's chief anthrax vaccine researcher at Fort Detrick authored a chapter on the anthrax vaccine. The article was co-authored by Dr. Phillip Brachman and edited by Dr. Stanley Plotkin. Both Brachman and Plotkin were involved with the original vaccine study thirty years earlier in Manchester, NH—location of the first anthrax epidemic. The chapter revealed the shortcomings of the anthrax vaccine later used for the DoD's mandate, including its high reactogenicity or adverse immunological responses and reactions:[11]

> The current vaccine against anthrax is unsatisfactory for several reasons. The vaccine is composed of an undefined crude culture of supernatant adsorbed to aluminum hydroxide. There has been no quantification of the protective antigen content of the vaccine or of any of the other constituents, so the degree of purity is unknown. Standardization is determined by an animal potency test. The undefined nature of the vaccine and the presence of constituents that may be undesirable may account for the level of reactogenicity observed. The vaccine is also less than optimal in that six doses are required over eighteen months, followed by annual boosters. There is also evidence in experimental animals that the vaccine may be less effective against some strains of anthrax. Clearly a vaccine that is completely defined, that is less reactogenic, and that requires one or two doses to produce long-lasting immunity would be highly desirable.

Excerpts from *Vaccines* Textbook, pages 737–738

UNRESOLVED PROBLEMS AND FUTURE DEVELOPMENTS

The incidence of human anthrax in the developed world is very low. The only impetus for the development of an improved human vaccine is the threat to use anthrax as a biological weapon. This horrendous possibility was unfortunately given credence by events in the Gulf War of 1991 and recent revelations concerning the outbreak of anthrax that occurred in 1979 in Sverdlovsk in the former Soviet Union.

The current vaccine against anthrax is unsatisfactory for several reasons. The vaccine is composed of an undefined crude culture supernatant adsorbed to aluminum hydroxide. There has been no quantification of the protective antigen content of the vaccine or of any of the other constituents, so the degree of purity is unknown. Standardization is determined by an animal potency test. The undefined nature of the vaccine and the presence of constituents that may be undesirable may account for the level of reactogenicity observed. The vaccine is also less than optimal in that six doses are required over 18 months, followed by annual boosters. There is also evidence in experimental animals that the vaccine may be less effective against some strains of anthrax.[43, 44, 46] Clearly a vaccine that is completely defined, that is less reactogenic, and that requires one or two doses to produce long-lasting immunity would be highly desirable.

Advances in the molecular pathogenesis of anthrax and further understanding of the structure of protective antigen and its interaction with lethal and edema factors can be expected to lead to significant progress toward

the development of improved vaccines. For example, genetically defined mutations in the cell receptor–binding domain,[56, 57] the protease-sensitive domain,[58] or other parts of the molecule[59] may generate a less toxic protective antigen preparation to be used either alone or as a complex with edema or lethal factor. Similarly, mutations in either edema or lethal toxin may allow nontoxic complexes with protective antigen to be evaluated. In addition, evidence in experimental animals suggests that adjuvants other than aluminum may substantially increase the protective efficacy of protective antigen even after a single dose.[60, 61] Another approach has been to develop live vaccines for human use, because several reports demonstrate that a live vaccine protects experimental animals better than the licensed human protective antigen vaccine does,[42–44, 62] and the precedent exists for using such a vaccine in humans in the former Soviet Union. Live vaccines that are known to protect experimental animals against anthrax include aromatic compound-dependent, toxigenic, unencapsulated strains of *B. anthracis*,[62] *B. subtilis*,[62] and vaccinia,[63] each constructed to contain the cloned protective antigen gene. Although these efforts are in the experimental stage, they may lead to the production of a vaccine that is less reactogenic, requires fewer doses, and provides more efficacious and long-lasting immunity.

APPENDIX DD

Andersen Email to Fauci Challenging "Evolutionary Theory"

From: Fauci, Anthony (NIH/NIAID) [E]
Sent: Sat, 1 Feb 2020 18:43:31 +0000
To: Kristian G. Andersen
Subject: RE: FW: Science: Mining coronavirus genomes for clues to the outbreak's origins

Thanks, Kristian. Talk soon on the call.

From: Kristian G. Andersen (b) (6)>
Sent: Friday, January 31, 2020 10:32 PM
To: Fauci, Anthony (NIH/NIAID) [E] (b) (6)
Cc: Jeremy Farrar (b) (6)>
Subject: Re: FW: Science: Mining coronavirus genomes for clues to the outbreak's origins

Hi Tony,

Thanks for sharing. Yes, I saw this earlier today and both Eddie and myself are actually quoted in it. It's a great article, but the problem is that our phylogenetic analyses aren't able to answer whether the sequences are unusual at individual residues, except if they are completely off. On a phylogenetic tree the virus looks totally normal and the close clustering with bats suggest that bats serve as the reservoir. The unusual features of the virus make up a really small part of the genome (<0.1%) so one has to look really closely at all the sequences to see that some of the features (potentially) look engineered.

We have a good team lined up to look very critically at this, so we should know much more at the end of the weekend. I should mention that after discussions earlier today, Eddie, Bob, Mike, and myself all find the genome inconsistent with expectations from evolutionary theory. But we have to look at this much more closely and there are still further analyses to be done, so those opinions could still change.

Best,
Kristian

APPENDIX EE

1989 Department of Defense Letter to Senator John Glenn

Excerpt from 1989 DoD memo to Senator John Glenn affirmed the Request for Proposal (RFP) premise regarding the safety and efficacy problems with the existing anthrax vaccine:[12]

> Question 14: The 1986 DoD report on the Biological Defense Program states that as a result of the neglect of the program in the 1970's, the U.S. cannot adequately defend itself against "conventional" biological agents such as anthrax. Do you agree with that assessment? If therapies in the form of vaccines and antibiotics are available for treating anthrax, how do you explain DoD's assessment that the U.S. cannot adequately defend its service personnel against anthrax?
>
> Answer: The assessment in the 1986 report is accurate. Current vaccines, particularly the anthrax vaccine, do not readily lend themselves to use in mass troop immunization for a variety of reasons: the requirement in many cases for multiple immunizations to accomplish protective immunity, a higher than desirable rate of reactogenicity, and, in some cases, lack of strong enough efficacy against infection by the aerosol route of exposure. Antibiotics could not be delivered fast enough for mass treatment in the event of a BW attack with anthrax. Such an attack, most likely in the form of an aerosol, would cause pulmonary anthrax, which is difficult to diagnose and has an extremely rapid time course leading to death. Current efforts in vaccine development are directed to addressing the deficiencies in existing vaccines outlined above.

APPENDIX FF

Letter from Colonel George "Bud" Day, USAF, Medal of Honor

G. "Bud" Day

GEORGE E. DAY, P. A.

32 Beal, Parkway SW Fort Walton Beach, Florida 32548-5391

Telephone: (850) 243-1234 - Facsimile: (850) 664-5720

email: LawOffice@vikemisty.gccoxmail.com

June 8, 2011

Major General Hugo Salazar
The Adjutant General of Arizona
5636 East McDowell Road
Phoenix, Arizona 85008-3495

Dear General Salazar:

One of your Arizona residents, and National Guard Officers, Lieutenant Colonel Tom "Buzz" Rempfer, contacted me with disturbing news. He was informed on June 3, 2011 about a unilateral decision by your state's Assistant Adjutant, Brigadier General Michael G. Colangelo, to separate him involuntarily on July 15, 2011. This exemplary officer from the Arizona Air National Guard is one year prior to sanctuary for retirement.

The officer was told he was the subject of an "involuntary tour curtailment," but that he was not being released for cause. The problem is involuntary curtailments by the Air National Guard Instruction (ANGI 36-101) are for "cause" by definition, and can only be utilized after serious adverse personnel action, such as reprimands or Articles 15s. This officer has no negative documentation in his record, and in fact has an exemplary combat record as an MQ-1 Predator UAS Instructor Pilot. This officer has flown F16s, A10s and F117s, and arguably has an experience level unmatched in his outfit.

Making this involuntary separation more perplexing is the fact that his June 3rd notice letter says it is being done based on "force management needs." As an attorney familiar with litigating against the government, and familiar with these processes in the National Guard, this officer would require a board hearing process, a requirement also codified in 10 USC §12683. This does not appear to have occurred, nor is the member being afforded the normal six-month notice and rebuttal process required by regulation for selective non-retentions. Added to this, the member's own unit commander recommended him for retention for two more years in their undermanned unit, all documented via a National Guard Bureau Form 27 (a Federal Retention Evaluation/Recommendation). There is a serious "due process" problem with this action.

Certainly, in this time where unmanned aircraft pilots are in hot demand, and since this officer and instructor pilot is currently one of the most experienced and successful in this combat platform, it seems to defy commonsense that we would allow him to be forced out when no other officers with his reputation and credentials are being let go. This officer is also under a USAF bonus contract, so there are a whole host of significant negative financial reimbursement implications for him and his family, not to mention significant legal liabilities *for the U.S. government if this involuntary termination is not properly constructed.*

I have known Lieutenant Colonel Rempfer for several years from our collective past efforts to assist an Air Force Special Operations NCO with a military corrections case. I know him to be a man of honor who helps people and adheres to Air Force Core Values. My understanding is a similar situation currently exists in the

Arizona Air National Guard – he is under fire by the chain of command for helping several NCO's, and is being accused of not supporting the chain of command. I hope you will agree that this case deserves review and that an involuntary separation of a top notch and courageous officer from the military, with no documented reasons, is not a just precedent. Remember that the Army did this kind of abusive separation to General Billy Mitchell, and have been embarrassed by the decision for decades.

This case also does not appear to be the right way to treat an officer who has been serving continuously for the past five years doing shift work in this valuable mission, while actively engaging the enemy in protecting our troops overseas. It's also no way to treat a distinguished graduate from the Air Force Academy and Naval Postgraduate School who actually does what we teach these men and women – take care of your troops!

As you are probably aware there is an email circulating around which asks, "Where have all of the fighter pilots gone?" It then goes on to talk about the fact that Jimmy Doolittle and Robin Olds, etc., would never make it in the military today and would be passed over or eliminated at Captain or Major now because they were leaders.

I question whether the Arizona Guard wants to go down in history as one of those misguided units that tries to fire a distinguished officer like Buzz Rempfer as if he was just some "at will" employee in a manner completely at odds with established regulations and law. Even Presidents try to do improper things like this. As you will recall, back in the mid-1990s, President Clinton issued an order denying all WW II and Korean War retirees medical care in military hospitals.

As a result of this misguided and illegal action, I filed a lawsuit. In L/Colonel Schism and L/Colonel Reinlie vs. the Attorney General of the United States the Federal Circuit of Appeals in D.C. unanimously ruled against President Clinton. I then co-authored a House and Senate Bill now referred to as Tri-Care for Life, which legislatively restored the free medical care that all of us were promised, both retrospectively and prospectively. I personally walked that bill through the Houses of Congress and the Senate. In the legislative process, Senator John S. McCain III was the Senate sponsor of Tri-Care for Life. The point of this comment is that the system still has checks and balances which must be called into play here.

I request that you review the overall facts of this case, and reverse this flawed decision that has the potential to bring a huge amount of criticism and ill will to the very effective Arizona Guard. I was a POW during Vietnam; the C-97 unit at Phoenix provided considerable help to my wife and four children who were engaged in trying to get decent treatment for us in Hanoi, as well as an accounting. Further, Glendale gave me a welcome party and parade that will never be forgotten by me and the Air Force retirees that now live in metro Phoenix.

Respectfully submitted

Col. Bud Day MOH · AFC

Col. G. "Bud" Day [R] MOH AFC

Cc: Client
Senator John S. McCain III

APPENDIX GG

President Clinton Addressing Previous Military Experimentation

THE WHITE HOUSE

WASHINGTON

In January 1994, after accounts of Cold War-era experiments involving the effects of radiation on humans came to light, I established an independent Advisory Committee on Human Radiation Experiments to investigate these reports. I asked the Committee to determine the truth about this dark chapter in our nation's history.

After taking extensive testimony and conducting numerous public hearings, the Advisory Committee issued its report in October, 1995. The Committee's report included recommendations to make the record of these experiments open to the public, improve ethics in human research today, and right the wrongs of the past inflicted on unknowing citizens. In my remarks when I accepted the report, I promised that it would not be left on the shelf to gather dust. I made a commitment that we would learn fro the lessons that the Committee's report offered and use it as a road map to lead us to better choices in the future.

This document -- my Administration's response to the Advisory Committee's report -- is a milestone in meeting that commitment. We have actively worked to respond to the important recommendations made by the Advisory Committee through a special interagency working group. This group includes representatives from the Executive Office of the President the Departments of Energy, Defense, Health and Human Services, Justice, Veterans Affairs, the National Aeronautics and Space Administration, and the Central Intelligence Agency. The Environmental Protection Agency has also joined the effort. This report reflects the joint progress of these agencies to address the Advisory Committee's recommendations.

My Administration has made significant achievements in opening government and making information more easily available to the citizens to whom it belongs. Agencies have also improved the protections in place for subjects of future human research. Finally, the Federal government i providing redress to those who have suffered from radiation experiments, as recommended by the Advisory Committee.

I emphasize that this document is by no means the end of the journey. Much work remains to be done. I am confident that all of us -- the eminent committee that produced the original report, the Federal officials who worked so hard to support the Committee's efforts and now are implementing its recommendations, and most importantly, the citizens of this great country from whose experiences we have learned so much -- can together help ensure a better world for our children.

My thanks to all of you for a job well done. I pledge my strong support for your continued efforts.

Bill Clinton

APPENDIX HH

Department of Defense Office of Prepublication and Security Review Clearance

DEPARTMENT OF DEFENSE
DEFENSE OFFICE OF PREPUBLICATION AND SECURITY REVIEW
1155 DEFENSE PENTAGON
WASHINGTON, DC 20301-1155

Ref: 22-SB-0155
May 15, 2023

Col (Ret) Thomas Rempfer
trempfer@aol.com

Dear Col (Ret) Rempfer:

This responds to your July 17, 2022, correspondence requesting public release clearance of the manuscript titled, "Unyielding." The manuscript submitted for prepublication security review is **CLEARED** for public release.

This clearance does not include any photograph, picture, exhibit, caption, or other supplemental material not specifically approved by this office, nor does this clearance imply Department of Defense (DoD) endorsement or factual accuracy of the material. The appearance of external hyperlinks does not constitute endorsement by the DoD of the linked websites, or the information, products or services contained therein. The DoD does not exercise any editorial, security, or other control over the information found at these locations.

This office notes that your manuscript may include the names and other personally identifiable information (PII) of former or active duty Service members, DoD employees, and third party individuals, the release of which could be a violation of the privacy rights of these individuals. As the author, you are solely responsible for the release of any PII and its legal implications. If you have not done so already, you may wish to consult these individuals and obtain permission to include their PII in the manuscript.

This office requires that you add the following disclaimers prior to publishing the manuscript: "The views expressed in this publication are those of the author and do not necessarily reflect the official policy or position of the Department of Defense or the U.S. government." and; "The public release clearance of this publication by the Department of Defense does not imply Department of Defense endorsement or factual accuracy of the material."

A copy of the first page of the manuscript with our clearance stamp is enclosed. Please direct any questions regarding this case to Mr. Doug McComb, at 202-774-4825 or douglas.g.mccomb.civ@mail.mil.

Sincerely,

MCCOMB.DOUGLAS.GREGORY.1042611324 Digitally signed by MCCOMB.DOUGLAS.GREGORY.1042611324 Date: 2023.05.15 11:51:42 -04'00'

for George R. Sturgis, Jr.
Chief

Enclosure(s):
As stated

NOTES

Disclaimer

1 Defense Office of Prepublication and Security Review. https://www.esd.whs.mil/DOPSR.

Foreword by Dr. Philip G. Zimbardo

1 Zimbardo P, *The Lucifer Effect: Understanding How Good People Turn Evil* (New York, Random House, 2007). http://www.lucifereffect.com.

Mile 2: The USAF Academy (1983 to 1987)

1 Aldridge E, SecAF, USAF Academy Commencement, USAFA McDermott Library archives, May 27, 1987.

Mile 4: Anthrax Vaccine 101 (1995 to 1998)

1 American Forces Press Service, Jul. 1997. Original link at http://www.defenselink.mil/news /Jul1997/t07311997_t0731coh.html.
2 Schmitt E, "Criticism over blast leads top Air Force General to retire," *New York Times*, Jul. 29, 1997. https://www.nytimes.com/1997/07/29/us/criticism-over-blast-leads-top-air -force-general-to-retire.html.
3 American Forces Press Service, Dec. 1997. Original link at: http://www.defenselink.mil/news /Dec1997/x12181997_x1215mfp.html.
4 "US forces to be inoculated against anthrax," BBC News, Dec. 15, 1997. http://news.bbc .co.uk/2/hi/despatches/39851.stm.
5 Classic instruments of national power use the 'DIME' acronym (diplomatic, informational, military, and economic) and were used with both the anthrax vaccine and the COVID shot indoctrination efforts. https://www.jcs.mil/Portals/36/Documents/Doctrine/jdn_jg/jdn1 _18.pdf.
6 Letter to Congressman Christopher Shays, Apr. 26, 2003.

Mile 5: Tiger Team Alpha (1998 to 1999)

1 Senate Veterans Affairs Committee Staff Report 103–97, Maj. Gen. Ronald Blanck, Commanding General, Walter Reed, p 35, 1994.
2 Senate Veterans Affairs Committee Staff Report 103–97, footnotes 61–63, 1994.
3 Congressional Record, Senator Shelby's Conclusions On The Persian Gulf Syndrome, p S3098, Mar. 17, 1994. http://www.gulflink.osd.mil/czech_french/czfr_refs/n08en014 /s3098.htm.
4 Senate Staff Report 103–97, "Is Military Research Hazardous To Veterans' Health? Lessons Spanning Half A Century," Dec. 8, 1994. https://img1.wsimg.com/blobby/go/4fa7f468 -a250–4088-926e-3c56a998df1f/downloads/1994_12–08_W%20staff%20report.pdf.

5 Air Command and Staff College text reference to R. Kidder's book, *How Good People Make Tough Choices* (New York, Fireside, 1995).
6 Takafuji E, Russell P, "Military Immunizations Past, Present, and Future Prospects," Inf Dis Clinics of N America, 4(1): 143–158, 1990.
7 Connecticut Air National Guard Commander resorted to the word "traitor" early in the controversy to describe his impressions of his military members that refused to participate in the patently illegal anthrax vaccine mandate (Video from FOIA). https://youtu.be/GPJgCc0sMZs.
8 Commander's "Traitor," "Efficacy," "distrustful," "owe answers" video clips, 1998. https://hoping4justice.org/opinions.
9 US Constitution, Article III, Section 3, Clause 1. https://constitution.congress.gov/browse/article-3/section-3/ and Treason clause limitations at https://constitution.congress.gov/browse/essay/artIII-S3-C1-1-1/ALDE_00001225/.
10 Miclon R, "Pilots' decision to quit Guard does not uphold oath," Journal Inquirer, 1999.

Mile 6: Mutual Support (1999)

1 Shaw R, *Fighter Combat, Tactics and Maneuvering* (Annapolis, Naval Institute Press, 1988).
2 SAIC Corporation plan, Sept. 29, 1995.
3 Industrial Capabilities Assessment, Preliminary report by the DoD's JPOBD, Dec. 1997.
4 Evidence Needed to Demonstrate Effectiveness of New Drugs When Human Efficacy Studies Are Not Ethical, 67 FR 37988, May 31, 2002. https://www.fda.gov/emergency-preparedness-and-response/mcm-regulatory-science/animal-rule-summary and https://www.fda.gov/media/88625/download.
5 Font L, Esq., Federal Court hearing, Pittsburgh, PA, Dec. 3, 1999.
6 DoD's Joint Vaccine Acquisition Program (JVAP). https://www.ncbi.nlm.nih.gov/books/NBK98400/ and https://www.ncbi.nlm.nih.gov/books/NBK220960/.
7 *O'Neil v. Secretary of Navy*, 76 F.Supp.2d 641, W.D.PA., Dec. 3, 1999.
8 Salter J, *The Hunters* (Washington, DC, Counterpoint Publishers, 1956).

Mile 7: Information Warfare (1999 to 2000)

1 MDPH letter to CBER responding to FDA inspectional observations made on 12–13 Sept. 1990, Oct. 10, 1990.
2 FDA Form 483 Inspectional Observations, May 4–7, 1993.
3 FDA Form 483 Inspectional Observations, May 31—Jun. 3, 1994.
4 FDA Form 483 Inspectional Observations, Apr. 23—May 5, 1995.
5 FDA Form 483 Inspectional Observations, Nov. 18–27, 1996.
6 Summary of Findings Report, Jan. 14, 1997.
7 NOIR letter from FDA to MBPI, Mar. 11, 1997. Original link at: http://www.fda.gov/cber/infosheets/mich-inf.htm.
8 FDA Form 483 Inspectional Observations, Feb. 4–20, 1998.
9 "Anthrax vaccination program in the Persian Gulf region, officials said it is going well, and troops are getting all the facts," American Forces Press Service, Apr. 1998. Original link at: http://www.defenselink.mil/news/Apr1998/n04201998_9804201.html.
10 Sun-Tzu ping-fa (Sun Tzu *The Art of War*).
11 US Army War College definition of asymmetric attacks. Original link at: http://www.carlisle.army.mil/ssi/pubs/2001/asymetry/asymetry.pdf.

12 Williams TD, "Pilots Quit Guard Unit in Anthrax Argument," *Hartford Courant*, Jan. 15, 1999. https://www.courant.com/1999/01/15/pilots-quit-guard-unit-in-anthrax-argument.
13 Blanck R, "Ignore the Paranoiacs; the Vaccine Is Safe," Army Times Op-Ed (Opinion Editorial), Feb. 22, 1999.
14 American Forces Press Service, Jan. 1999. Original link at: http://www.defenselink.mil/news/Jan1999/t01211999_t121asd_.html.
15 Marine Corps Commandant Gen. Charles C. Krulak speaking to Marines at Camp Pendleton, California, Army Times, p. 3, Mar. 8, 1999.
16 AVIP news reference, Jun. 9, 1999. http://frwebgate.access.gpo.gov/cgi-bin/getdoc.cgi?dbname=106_house_hearings&docid=f:65604.wais.
17 Hartle A, Moral Issues in Military Decision Making, The United States Military Academy.
18 Sun-Tzu, Ping-fa (Sun Tzu, *The Art of War*).
19 Clausewitz C, *Principles of War* (1812).
20 Clausewitz C, *Principles of War* (1812).
21 Clausewitz C, *Principles of War* (1812).
22 Maj Leonard G. Litton, USAF, Air University Principles of War. https://www.airuniversity.af.edu/Portals/10/ASPJ/journals/Chronicles/Litton.pdf
23 Rempfer T, "Anthrax for troops amounts to 'quick fix,'" USA Today, May 9, 2000.
24 DoD anthrax vaccine education campaign funding details, "The Anthrax Vaccine Immunization Program—What have we learned," Committee on Government Reform Hearing, Oct. 3 and 11, 2000. http://frwebgate.access.gpo.gov/cgi-bin/getdoc.cgi?dbname=106_house_hearings&docid=f:73979.pdf.
25 Jasper W, "Vexing Vaccine," *The New American*, Nov. 20, 2000. http://www.thenewamerican.com/tna/2000/11–20-2000/vo16no24_anthrax.htm.
26 GAO-T-NSIAD-99–148, "Medical Readiness: Safety and Efficacy of the Anthrax Vaccine," Apr. 29, 1999. https://www.gao.gov/assets/t-nsiad-99–148.pdf.
27 Bacevich A, "Bad Medicine for Biological Terror," *Orbis*, Spring 2000.
28 "Pentagon background briefing by 'senior defense and Army officials,'" American Forces Press Service, Aug. 5, 1999. Original link at: www.defenselink.mil/Aug1999/x08051999_x0805ant.html.
29 The Armed Forces Officer, US Government Printing Office, Washington, Excerpt: "Either in combat or out, in any situation where a majority of military-trained Americans becomes undutiful, that is sufficient reason for higher authority to resurvey its own judgments, disciplines and line of action," 1950. https://www.google.com/books/edition/The_Armed_Forces_Officer/TM4rAAAAYAAJ?hl=en&gbpv=1&bsq=resurvey.

Mile 8: Unproven Force Protection (1999 to 2000)

1 First Amendment to the Constitution. http://www.house.gov/Constitution/Amend.html.
2 Letter from John Adams to Richard Henry Lee, Nov. 15, 1775. https://founders.archives.gov/documents/Adams/06–03-02–0163.
3 P.L. 105–277, sec. 1603(d), House Report 106–556, fn 1. https://www.govinfo.gov/content/pkg/CRPT-106hrpt556/html/CRPT-106hrpt556.htm.
4 Senate Veterans Affairs Committee Staff Report 103–97, 1994. https://img1.wsimg.com/blobby/go/4fa7f468-a250–4088-926e-3c56a998df1f/downloads/1994_12–08_W%20staff%20report.pdf.

5 Rempfer T, "Anthrax vaccine offers no sure cure in warfare: A fighter pilot questions the effectiveness of the military's immunization program and calls for an independent review." *Baltimore Sun*, Mar. 21, 1999.
6 "The Anthrax Immunization Program," Hearing Before the Subcommittee on National Security, Veterans Affairs, and International Relations of the Committee on Government Reform, House of Representatives, One Hundred Sixth Congress, First Session, Mar. 24, 1999. https://www.google.com/books/edition/The_Anthrax_Immunization_Program/VBZlwi7eZbEC?hl=en.
7 Military Personnel Subcommittee of the House Armed Services Committee hearing, Sept. 30, 1999. http://commdocs.house.gov/committees/security/has273020.000/has273020_0.htm.
8 Hamre J, DepSecDef, American Forces Press Service, Oct. 7, 1999. http://www.defenselink.mil/news/Oct1999/n10071999_9910073.html.
9 GAO-T-NSIAD-99–148, "Medical Readiness: Safety and Efficacy of the Anthrax Vaccine," Apr. 29, 1999. https://www.gao.gov/assets/t-nsiad-99–148.pdf.
10 GAO-02–445, "Anthrax Vaccine: GAO's Survey of Guard and Reserve Pilots and Aircrew," Sept. 20, 2002. https://www.gao.gov/assets/gao-02–445.pdf.
11 Senate Staff Report 103–97, page 35 and note 61–63, Dec. 8, 1994. https://img1.wsimg.com/blobby/go/4fa7f468-a250–4088-926e-3c56a998df1f/downloads/1994_12–08_W staff report.pdf.
12 Senate Staff Report 103–97, note 61–63, Dec. 8, 1994. https://img1.wsimg.com/blobby/go/4fa7f468-a250–4088-926e-3c56a998df1f/downloads/1994_12–08_W staff report.pdf.
13 Dingle-Rempfer testimony. http://frwebgate.access.gpo.gov/cgi-bin/getdoc.cgi?dbname=106_house_hearings&docid=f:57559.pdf.
14 Rempfer testimony. http://frwebgate.access.gpo.gov/cgi-bin/getdoc.cgi?dbname=106_house_hearings&docid=f:57559.wais.
15 C-SPAN Networks, Captain Thomas L. Rempfer video in the C-SPAN Video Library, Testimony on the Anthrax Vaccine Immunization Program, 1999 House Committee, Mar. 24, 1999. https://www.c-span.org/person/?59481/ThomasLRempfer.
16 Department of Defense National Guard Bureau Office of Policy and Liaison, Hearing summary, Mar. 24, 1999.
17 Emergent website, formerly BioPort, and formerly MDPH. https://www.emergentbiosolutions.com/press/news-release-details-emergent-biosolutions-announces-expansion-board-directors-and/ and https://www.emergentbiosolutions.com/press/news-release-details-emergent-biosolutions-board-members-dr-sue-bailey-and-dr-kathryn/.
18 1st Anthrax Vaccine oversight hearing testimony, Mar. 24, 1999, National Security Subcommittee, 106th Congress. http://frwebgate.access.gpo.gov/cgi-bin/getdoc.cgi?dbname=106_house_hearings&docid=f:57559.wais.
19 CBER, "Anthrax Vaccines; Efficacy Testing and Surrogate Markers of Immunity Workshop," Apr. 23, 2002.
20 *War, Morality, and the Military Profession*, 2d ed., Biological Warfare Chapter, Edited by BG Malham M. Wakin, Professor Emeritus, USAF Academy (Boulder, Westview Press, 1986), and Krickus R, "On the Morality of Chemical/Biological War," *Journal of Conflict Resolution*, 9(2), 200–210, Jun. 1965. http://www.jstor.org/stable/173164.

21 Hague Convention, Laws of war: Declaration on the use of projectiles the object of which is the diffusion of asphyxiating or deleterious gases, Jul. 29, 1899. http://avalon.law.yale.edu/19th_century/dec99–02.asp.
22 United Nations, Geneva protocol, 1925. Original link at: http://disarmament.un.org/treatystatus.nsf/44e6eeabc9436b78852568770078d9c0/fd968846cc28913e852568770079dd93.
23 POTUS, National Security Decision Memorandum 35, Nov. 25, 1969. http://www.fas.org/programs/ssp/bio/resource/documents/nsdm-35.pdf.
24 POTUS, National Security Decision Memorandum 44, Feb. 20, 1970. http://www.fas.org/programs/ssp/bio/resource/documents/nsdm-44.pdf.
25 Convention on Bioweapons, Mar. 26, 1975. http://www.state.gov/t/isn/bw/ and http://www.state.gov/www/global/arms/treaties/bwc1.html.
26 Rempfer T, Dingle E, "Op-Ed: Anthrax Vaccine Is A Force Protection Façade," Hartford Courant, Jan. 20, 2000.
27 American Forces Press Service, Dec. 1997. Original link at: http://www.defenselink.mil/news/Dec1997/x12181997_x1215mfp.html
28 Effects-based targeting doctrine. Original link at: http://www.jfcom.mil/about/glossary.htm#E.
29 "Dozens Protest Anthrax Vaccine," Associated Press, Jan. 29, 2000. https://apnews.com/article/b357807ae0d277b77585754cabd0f092.
30 H.R.5546 - National Childhood Vaccine Injury Act of 1986 (NCVIA). https://www.congress.gov/bill/99th-congress/house-bill/5546 and https://www.hrsa.gov/vaccine-compensation/faq#About%20The%20Vicp.
31 Emergent website, formerly BioPort and formerly MDPH. https://investors.emergentbiosolutions.com/board-member/kathryn-zoon.
32 Rempfer T, Testimony, Oct. 12, 1999. https://www.govinfo.gov/content/pkg/CHRG-106hhrg65604/html/CHRG-106hhrg65604.htm.
33 USAF Core values. https://www.airforce.com/mission/vision.
34 "Unproven Force Protection," House Report 106–556, Apr. 3, 2000. https://www.congress.gov/106/crpt/hrpt556/CRPT-106hrpt556.pdf.
35 Public Law 105–277, 105th Congress, Title XVI, sec. 1603(d). https://www.congress.gov/105/plaws/publ277/PLAW-105publ277.pdf.
36 "Unproven Force Protection," House Report 106–556, Apr. 3, 2000. https://www.congress.gov/106/crpt/hrpt556/CRPT-106hrpt556.pdf.
37 American Forces Press Service, Dec. 1997. Original link at: http://www.defenselink.mil/news/Dec1997/x12181997_x1215mfp.html.
38 Shays C, Press Release concerning "Unproven Force Protection," HR 106–556, Mar. 9, 2000.

Mile 9: The Military Times (2000)

1 FDA Form 483, Inspectional Observations, Oct. 10–26, 2000.
2 Hamilton A, Federalist Paper #29, Jan. 9, 1788.
3 SCOTUS, "Perpich v. Department of Defense, 496 US 334 (1990), 496 US 334, No. 89–542, Argued Mar. 27, 1990, Decided Jun. 11, 1990.
4 NGAUS Magazine. Original link at: http://www.ngaus.org/ngmagazine/anthrax399.asp.
5 Hamilton A, Federalist Paper #29, Jan. 9, 1788.

6 Blanck R, "Ignore the Paranoiacs; the Vaccine Is Safe," *Army Times*, Feb. 22, 1999.
7 Saran D, United States v. Members of the Armed Forces: The Truth Behind the Department of Defense's Anthrax Vaccine Immunization Program, Jul. 27, 2020. https://www.amazon.com/United-States-Members-Armed-Forces/dp/1734629304/ref=sr_1_1 and https://childrenshealthdefense.org/defender/dale-saran-military-covid-vaccine-mandate-illegal-defender-podcast/.
8 CT AG press release. Original link at: http://www.cslib.org/attygenl/press/2001/health/dod1.htm.
9 CT AG letter to the FDA. Original link at: http://www.cslib.org/attygenl/press/2001/health/fda.pdf.
10 Williams TD, "Efforts to posthumously promote East Hartford man who opposed military vaccinations," *Journal Inquirer*, Sept. 24, 2008.
11 CT AG letter to the DoD. Original link at: http://www.cslib.org/attygenl/press/2001/health/dod.pdf.
12 Miller K, "Shots In The Dark, What the Pentagon doesn't want you to know about the anthrax vaccine," Army Times, Apr. 9, 2001.
13 PRNewswire by BioPort. Original link at: http://tbutton.prnewswire.com/prn/11690X44058348, SOURCE BioPort Corporation.
14 Maxwell AFB archived article. Original link at: http://www.iwar.org.uk/iwar/resources/usaf/maxwell/students/1996/96–025u.htm.

Mile 10: The OODA Loop (2000 to 2001)

1 DoD press briefing, Jan. 21, 1999.
2 DoD press briefing, Jun. 29, 1999.
3 American Forces Press Service, Aug. 1999. Original link at: http://www.defenselink.mil/news/Aug1999/x08051999_x0805ant.html.
4 Dr. Zoon testimony for the FDA. Original link at: http://www.house.gov/reform/ns/hearings/testimony/statementzoon.htm.
5 Maxwell AFB Air University archives. Original link at: http://www.au.af.mil/au/2025/volume3/chap02/v3c2–1.htm.
6 Maxwell AFB Air University archives. Original link at: http://www.au.af.mil/au/2025/volume3/chap02/v3c2–7.htm#24.
7 Coram R, Boyd: *The Fighter Pilot Who Changed the Art of War* (Boston, Back Bay Books, 2004), cited by Szeligowski R, "Cognifying the OODA Loop: Improved Maritime Decision Making," https://apps.dtic.mil/sti/pdfs/AD1057893.pdf.
8 ORM Analysis of the AVIP. Original link at: http://www.aviationmedicine.com/ORM_analysis_of_the_AVIP.doc.
9 Air University Airpower Chronicles. https://www.airuniversity.af.edu/Portals/10/ASPJ/journals/Chronicles/rempfer.pdf.
10 Hartford Courant Anthrax Vaccine Project. Original link at: http://courant.ctnow.com/projects/anthrax/.
11 Williams, TD, Connecticut Law Tribune, Anthrax Avengers, May 5, 2008. https://www.law.com/ctlawtribune/almID/1202557043478/ and https://www.scoop.co.nz/stories/HL0804/S00145/veteran-battles-pentagons-vaccine.htm.
12 Broad W, Miller J, "Germ Defense Plan in Peril as Its Flaws Are Revealed," *New York Times*, Aug. 7, 1998.

13 Graham B, "Military Chiefs Back Anthrax Inoculations - Initiative Would Affect All of Nation's Forces," *Washington Post*, Oct. 2, 1996.
14 Rempfer T, Dingle R, "Military split over anthrax shots. Point: Suspend, Review Program," *Baltimore Sun*, Jan. 30, 2000.
15 Rempfer T, Sticking Point, *Washington Post*, Outlook section, Jan. 30, 2000. https://www.washingtonpost.com/archive/opinions/2000/01/30/sticking-point/681fd40d-de7d-4f2e-b107–018b07f4d7ae/.
16 Scott J, "Sticking Point," Washington Post, Outlook section, Jan. 30, 2000. https://www.washingtonpost.com/archive/opinions/2000/01/30/sticking-point/3d930aec-da82–41d5-bdf4–20f220191c13/.
17 Colonel John Boyd quote, "He raised his hand and pointed. 'If you go that way you can be somebody. You will have to make compromises and you will have to turn your back on your friends. But you will be a member of the club and you will get promoted and you will get good assignments.' Then Boyd raised his other hand and pointed in another direction. 'Or you can go that way and you can do something- something for your country and for your Air Force and for yourself. If you decide you want to do something, you may not get promoted and you may not get the good assignments and you certainly will not be a favorite of your superiors. But you won't have to compromise yourself. You will be true to your friends and to yourself. And your work might make a difference. To be somebody or to do something.'" https://www.goodreads.com/quotes/120948-tiger-one-day-you-will-come-to-a-fork-in.

Mile 11: Lobbying (1999 to 2001)

1 *Encyclopedia Britannica*, Thomas Paine.
2 Paine T, "The American Crisis," Nos. 2 and 4, Jan. 13, 1777 and Sept. 12, 1777.
3 *Vaccines*, 3rd edition (Philadelphia, Saunders, 1999).
4 Unwin C, et al, "Health of UK Servicemen Who Served in Persian Gulf War," *The Lancet*, 353: 169–78, Jan. 16, 1999. https://www.thelancet.com/journals/lancet/article/PIIS0140–6736(98)11338–7/abstract.
5 "Judge Agrees Anthrax Vaccine Unsafe; Halts Court Martial," Canada News Briefs By The Associated Press, May 5, 2000.
6 Rees A, "Their Dangerous Dose," The Province (Vancouver, Canada) Jun. 25, 2000.
7 "Hundreds of military personnel part of lawsuit over mandatory COVID vaccine policy," National Post, Jun. 26, 2023. https://nationalpost.com/news/canada/hundreds-of-military-part-of-lawsuit-over-mandatory-covid-vaccine/wcm/cfbee875–1110-4258-aab5-c4680d75f364/amp/ and https://nationalpost.com/opinion/military-committee-leads-the-way-on-declaring-vaccine-mandates-unconstitutional.
8 "Interview of Maj. Thomas Rempfer by Heidi Collins on American Morning," CNN, May 29, 2003. http://edition.cnn.com/TRANSCRIPTS/0305/29/ltm.11.html.
9 HASC report 106–22, Sept. 30, 1999. http://commdocs.house.gov/committees/security/has273020.000/has273020_0.htm.
10 Senate Hearing 101–744, "Global Spread of Chemical and Biological Weapons," p 474, 480. https://www.google.com/books/edition/Global_Spread_of_Chemical_and_Biological/tbIRAAAAIAAJ?hl=en&gbpv=1.
11 Telecom with Mr. Lawrence Trahan, DoD IG Hotline Director.

12 Williams TD, "Shays Wants Penalties Dropped in Anthrax-Refusal Cases," *Hartford Courant*, Jul. 13, 2000.

13 House Resolution 2548, "To suspend further implementation of the Department of Defense anthrax vaccination program," pending review by the NIH. https://www.congress.gov/106/bills/hr2548/BILLS-106hr2548ih.pdf.

14 Rempfer T, Dingle R, "ORM and the AVIP." Original link at: http://www.aviationmedicine.com/ORM_analysis_of_the_AVIP.doc.

15 Department of Veterans Affairs, case VAOPGCPREC 4–2002, "HELD: If evidence establishes that an individual suffers from a disabling condition as a result of administration of an anthrax vaccination during inactive duty training, the individual may be considered disabled by an "injury" incurred during such training as the term is used in 38 USC § 101 (24) . . . such an individual may be found to have incurred disability in active military, naval, or air service for purposes of disability compensation under 38 USC § 1110 or 1131," May 14, 2002. https://www.va.gov/ogc/docs/2002/PREC_4–2002.doc.

16 Air Force Chief of Staff memorandum, "Operational Risk Management," Jun. 26, 2002.

17 Excerpts from declassified DoD documents:

Sept. 14, 1990: . . . task from DJS to form a special group to develop proposed PA guidance for the BW Vaccination Program . . . under the auspices of J-5 (Deputy Director for Political Military Affairs—BG Jumper) . . .

Sept. 21, 1990: 'Special Topic' briefed in the TANK to the Operations Deputies by J-4 (RADM Smyth) and J-5 (BG Jumper) . . . Bottom line: decision necessary were no longer "medical" in origin; rather were political, social, and military / operational. Also, no matter what decision made, insufficient vaccines (both AX and BT) to cover all US forces at risk existed.

Sept. 25, 1990: AX Production Charts provided DJS with explanation regarding commencement of production . . .

Oct. 25, 1990: Memo from DDMR (ADM Smyth) on status of AX production . . .

Nov. 2, 1990: TANK informational briefing held with OPSDEPS and Joint Chiefs . . . No change in threat; AX vaccine production maximized . . .

Nov. 9, 1990: Trip to Michigan Department of Public Health Lab (Lansing, MI) by J5 (BG Jumper, COL Fleming) and J4 (ADM Smyth, COL Fry). Purpose: Determine problems and prospects affecting production of BW vaccines. Visited Director of the Lab (Dr. George Anderson) and the Chief of Biologic Products Division (Dr. Robert Myers). . . . Increases in AX vaccine production favorable. . . . here is need for an additional fermenter however. MDPH has suspended production of BT vaccine in favor of AX vaccine.

Nov. 9, 1990: J5 / DDPMA (BG Jumper) formed a working group consisting of DIA, J3, J4, J5 to assure accurate tracking of vaccine production.

Nov. 16, 1990: BG Jumper provided summary of BW threat and general overview of US defensive capabilities (to include vaccines). Briefing showed existing inventories fell short of requirements in the near term.

Nov. 16, 1990: COL Lewis furnished latest information on MDPH fermenter. New fermenter installed and pre-production testing is beginning. Provided to BG Jumper and DJS by DDMR.

Nov. 19, 1990: ASD (HA) memo to SECARMY, "Expansion of Industrial Base for Biological Vaccine Production," . . . on short-term production of AX and BT. . . .

Requested steps be taken on a priority basis to monitor ongoing efforts at MDPH (increased production by 20 Feb 91) . . .

Nov. 19, 1990: Initial information on quantities of antibiotics (doxycycline, ciprofloxacin) . . . furnished by COL Lewis to DDMR and BG Jumper.

Nov. 21, 1990: DA OTSG sent tasking from SECARMY to form Task Force to evaluate ways to increase production of AX and BT vaccines. Implementation Working Group, chaired by BG Blanck, would provide weekly production reports to DASD (MR).

Dec. 3, 1990: J5/BG Jumper outlined course of action needed prior to next TANK session. Need to push toward total integration of all planning efforts associated with BW defensive measures. . . . VCSA has tasked Surgeon General to get plan together (public affairs, psyops, POLMIL, medical, doctrine). Draft memorandum to SECDEF prepared by J5 / COL Fleming requesting SECDEF direct accelerated procurement actions to improve the US biological defensive posture. Memorandum was not finalized.

Dec. 10, 1990: Paper submitted to ADM Smyth and BG Jumper on "Rationale for Antibiotics in Prophylaxis Against Inhalation Anthrax" (Rhesus Monkey Paper). Research effort has been used in considering the use of antibiotics following exposure to AX and before initiation of symptoms. Only one monkey died following treatment with 30 days of Ciprofloxacin antibiotic.

Dec. 14, 1990: Armed Forces Epidemiological Board met to consider the use of antibiotics as an adjunct in countering the threat of inhalation AX. 8 Mar 91—CENTCOM msg (081808ZMar91), . . . indicated vaccination programs for AX and Bot T discontinued due to diminished threat."

18 Air Force Pamphlet 90–902, Original link at http://www.e-publishing.af.mil/pubfiles/af/90/afpam90–902/afpam90–902.pdf, page 43.

19 Encyclopedia Britannica, Categorical syllogisms: "The next more complex form of argument is one with two categorical propositions as premises and one categorical proposition as conclusion."

20 Koritala T, et al, "A narrative review of emergency use authorization versus full FDA approval and its effect on COVID-19 vaccination hesitancy," Infez Med, Sept. 10, 2021. https://www.ncbi.nlm.nih.gov/pmc/articles/PMC8805497/#:~:text=Age%20groups%20and%20EUA&text=It%20is%20likely%20that%20a,cannot%20be%20mandated%20%5B33%5D.

21 Berman J, "'If a student chooses to come to an institution, they agree to abide by the rules': Can colleges force students to get COVID-19 vaccines?," MarketWatch, Excerpt: "For years, the FDA took the position that an EUA product cannot be mandated, this is not a new position, they've held it for years," according to Dorit Rubinstein Reiss, a professor at The University of California Hastings College of the Law, Apr 5, 2021. https://www.marketwatch.com/story/if-a-student-chooses-to-come-to-an-institution-they-agree-to-abide-by-the-rules-can-colleges-be-forced-to-mandate-covid-19-vaccines-11617038227.

22 10 USC §1107, Excerpt, "In the case of the administration of an investigational new drug, or a drug unapproved for its applied use, . . . the requirement that the member provide prior consent to receive the drug in accordance with the prior consent requirement imposed under the Federal Food, Drug, and Cosmetic Act may be waived only by the President." https://uscode.house.gov/view.xhtml?req=federal+food+drug+cosmetic+21+usc+chapter+1&f=treesort&num=332.

Mile 12: The Perot Factor (1999 to 2002)

1 Presidential Special Oversight Board for Department of Defense Investigations of Gulf War Chemical and Biological Incidents, Public Hearing, Day One, Nov. 19, 1998. Original link at: http://www.oversight.ncr.gov/thur.htm.

2 Presidential Special Oversight Board, Dr. Vinh Cam. Original link at: http://www.oversight.ncr.gov/hal_cam.htm.

3 Presidential Special Oversight Board. Original link at: https://www.gulflink.osd.mil/oversight/appdix_a.htm.

4 "Panel Finds Stress a Main Cause of Gulf War Syndrome," Reuters News Service, Dec. 21, 2001.

5 Gulf War and Health: Volume 1, Depleted Uranium, Pyridostigmine Bromide, Sarin, and Vaccines (2000), Institute of Medicine, p.267–287 on anthrax vaccine. https://pubmed.ncbi.nlm.nih.gov/25057724/.

6 Sox H Jr, Statement, Gulf War and Health, Volume 1: Depleted Uranium, Sarin, Pyridostigmine Bromide, and Vaccines, Sept. 7, 2000. http://www4.nationalacademies.org/news.nsf/0a254cd9b53e0bc585256777004e74d3/6d1b07dba5bb852985256ca70072db70?OpenDocument.

7 Email exchange between Brigadier General Eddie Cain and Colonel John Wade, reference 29 APR 99 Congressional testimonies and an ongoing Congressional investigation of the AVIP, May 3, 1999.

8 Cain E, Brigadier General, Email exchange with Col John Wade, reference 29 APR 99 Congressional testimonies and an ongoing Congressional investigation of the AVIP, May 3, 1999.

9 FDA's labeling regulations preclude opinions for implied indications per 21 CFR §201.56, §201.57, §201.56, and §201.57(c)(2), "Indications and Usage. (2) All indications shall be supported by substantial evidence of effectiveness based on adequate and well-controlled studies."

10 10 USC §1107 specifically barred the DoD from inoculating any member without his or her informed consent if the medication was an "investigational new drug" or a "drug unapproved for its applied use."

11 21 USC § 10.85, "an interested person may request an advisory opinion from the Commissioner of Food and Drugs on a matter of general applicability. Original link at: http://www.fda.gov/OHRMS/DOCKETS/98fr/122999c.pdf.

12 Font L. Esq., Federal Court hearing, Pittsburgh, PA., Dec. 3, 1999.

13 Rempfer T, Dingle E, "Intellectual honesty in the ranks sets America's military apart from others," *Sun Herald*, Jun. 13, 2001. Original link at: http://web.sunherald.com/content/biloxi/2001/06/13/opinion/4263451_06132001.htm.

14 House Gov Reform Subcommittee on National Security hearing, "Gulf War Veterans' Illnesses: Health of Coalition Forces," Jan. 24, 2002. https://www.govinfo.gov/content/pkg/CHRG-107hhrg82953/pdf/CHRG-107hhrg82953.pdf.

15 Gilman B, Testimony before the Subcommittee on National Security, Veterans Affairs and International Relations, Jan. 24, 2002. http://frwebgate.access.gpo.gov/cgi-bin/getdoc.cgi?dbname=107_house_hearings&docid=f:82953.wais.

16 Kingsbury N., Testimony before the Subcommittee on National Security, Veterans Affairs and International Relations, Jan. 24, 2002. http://frwebgate.access.gpo.gov/cgi-bin/getdoc.cgi?dbname=107_house_hearings&docid=f:82953.wais.

17 Cincinnati Radiation Litigation, US District Court for the Southern District of Ohio, 874 F. Supp. 796, Excerpt: "The Nuremberg Code is part of the law of humanity. It may be applied in both civil and criminal cases by the federal courts in the United States,". Jan. 11, 1995. https://law.justia.com/cases/federal/district-courts/FSupp/874/796/1478171/.
18 Stennis Center for Public Service, Staff, Charlie Abell. https://history.defense.gov/Portals/70/Documents/oral_history/OH_Trans_ABELLCharlie9–16-2014.pdf?ver=2019–11-04–082102-267.
19 "Why Civilian Control of the Military?" Armed Forces Information Services, May 2, 2001. Original link at: http://dod.gov/news/May2001/n05022001_200105023.html.
20 "Are we treating veterans right?" Senate Hearing 103–647, Nov. 16, 1993. https://ia601306.us.archive.org/26/items/persiangulfwaril00unit/persiangulfwaril00unit.pdf.
21 Online NewsHour Forum: Military Readiness, Sept. 2000.
22 Weiss R., "Demand Growing for Anthrax Vaccine, Fear of Bioterrorism Attack Spurs Requests for Controversial Shot," *Washington Post*, Sept. 29, 2001.Original link at: http://www.washingtonpost.com/wp-dyn/articles/A42952–2001Sep28.html.

Mile 13: Anthrax Letter Attacks (2001)

1 *U.S. ex Rel. Dingle v. BioPort Corp.*, Case No. 5:00-CV-124, Aug. 29, 2002. https://casetext.com/case/us-ex-rel-dingle-v-bioport-corporation.
2 GAO-02–181T, "Anthrax Vaccine: Changes to the Manufacturing Process," Oct. 2001. https://www.gao.gov/products/gao-02–181t.
3 *U.S. ex Rel. Dingle v. BioPort*, 270 F. Supp. 2d 968, Jun. 18, 2003. https://casetext.com/case/us-ex-rel-dingle-v-bioport-corp-wdmich-2003.
4 Email from the office of the Senior Advisor to the President, Sept. 17, 2001.
5 "Scientist's Suicide Linked to Anthrax Inquiry," *New York Times*, Aug. 2, 2008. https://www.nytimes.com/2008/08/02/washington/02anthrax.html and https://www.fbi.gov/history/famous-cases/amerithrax-or-anthrax-investigation and FBI archives. https://www.justice.gov/opa/pr/justice-department-and-fbi-announce-formal-conclusion-investigation-2001-anthrax-attacks and https://www.fbi.gov/history/famous-cases/amerithrax-or-anthrax-investigation.
6 White House Press Briefing, Feb. 25, 2002. Original link at: http://www.whitehouse.gov/news/releases/2002/02/20020225–16.html.
7 "Anthrax in Manchester: Revisiting the Arms Mill Outbreak of 1957," New Hampshire Public Radio, Todd Bookman, Jun. 19, 2017. https://www.nhpr.org/nh-news/2017–06-19/anthrax-in-manchester-revisiting-the-arms-mill-outbreak-of-1957.
8 Plotkin S, et al, "An Epidemic of Inhalation Anthrax, the First in the Twentieth Century," *American Journal of Medicine*, 1960: 992–1001.
9 Details of the anthrax letter attacks. http://www.fas.org/bwc/news/anthraxreport.htm.
10 US Medicine, Sept. 2000.
11 "Sensitive to Troop Worries," Associated Press via *Los Angeles Times*, Apr. 11, 2000.
12 *San Diego Union Tribune*, Feb. 26, 2000. Original link at: http://www.uniontrib.com/news/uniontrib/sat/news/news_1n26mccain.html.
13 DOD News Briefing with Secretary of Defense Donald H. Rumsfeld, Oct. 12, 2001.
14 DoD News Briefing with Defense Secretary Rumsfeld and Gen. Myers, Oct. 18, 2001.
15 DoD News Briefing with Secretary of Defense Rumsfeld, Oct. 25, 2001.
16 DoD News Briefing with Secretary of Defense Rumsfeld, Oct. 28, 2001.

17 DoD News Briefing with Secretary Of Defense Rumsfeld, Oct. 30, 2001.
18 DoD News Briefing with Secretary of Defense Rumsfeld, Feb. 26, 2002.
19 DOD anthrax newsletter, removed from the DoD website in Oct. 1999 following congressional challenge quoting it in a hearing, Jun. 9, 1999.
20 Gugliotta G, "No Decision on Anthrax Vaccine Program," *The Washington Post*, May 20, 2002.
21 *Encyclopedia Britannica*, George Orwell.
22 Orwell G, Nineteen Eighty-Four, part 1, chapter 3 (London, Martin Secker & Warburg, 1949). https://www.george-orwell.org/1984/2.html.
23 Vedantam S, "Lingering Worries Over Vaccine Some Servicemen, Scientists Question Safety, Effectiveness of Anti-Anthrax Shots," *The Washington Post*, Dec. 20, 2001. Original link at: http://www.washingtonpost.com/wp-dyn/articles/A3983–2001Dec19.html.
24 Rosenbaum D, Stolberg, S.G., "As US Offers Anthrax Shots, Safety Debate Begins Again," *New York Times*, Dec. 20, 2001. http://www.nytimes.com/2001/12/20/politics/20ANTH.html?todaysheadlines.
25 DoD News Briefing with Edward C. "Pete" Aldridge, Jr., Mar. 22, 2002.
26 Deputy Secretary of Defense memorandum, "Reintroduction of the Anthrax Vaccine Immunization Program (AVIP)," Jun. 28, 2002.
27 Chief of Staff of the Air Force memorandum, "Air Force Implementation of the Anthrax Vaccine Immunization Program," Oct. 11, 2002. Original link at: http://www.anthrax.mil/media/pdf/AFPlan.pdf.
28 Binns J, Chair for the Department of Veterans Affairs Research Advisory Committee on Gulf War Veterans' Illnesses, Letter to Honorable Leo S. McKay, Dec. 16, 2002. Original link at: http://www.appc1.va.gov/rac-gwvi/docs/Lesson_Learned.doc.
29 Anthrax Vaccine Adsorbed Product Insert, Jan. 31, 2002. Original link at: www.bioport.com/default.asp.
30 Bush G, State of the Union Address, 2002.
31 White House Press Release, Feb. 5, 2002. Original link at: www.whitehouse.gov/news/releases/2002/02/20020205–1.html.
32 Bush G, State of the Union Address, 2003.
33 "DoD's Renewal of Anthrax Vaccine Provokes Criticism," Newsmax, Jul. 4, 2002. https://www.newsmax.com/pre-2008/dod-s-renewal-anthrax-vaccine/2002/07/04/id/667717/.

Mile 14: Citizen Petition 1 (2001)

1 FDA CBER Notice of Intent to Revoke letter for anthrax vaccine. Original link at: http://www.fda.gov/cber/infosheets/mich-inf.htm.
2 John D. Copanos & Sons, Inc. v FDA 854 F.2d 510 (DC Cir. 1988) and American Public Health Association v Veneman 394 F. Supp 1311 (1972), The regulatory threshold for license suspension and the affirmation of the Agency's authority to immediately withdraw a license was established in these two cases when the Secretary "concludes that there is no substantial evidence of efficacy."
3 AP, Neergaard L, "Scientists Advise on Anthrax Vaccine," ABC, Dec. 21, 2001. http://www.wjla.com/showstory.hrb?f=n&s=25584&f1=nat.
4 "Postal, Capitol Hill workers offered vaccine," CNN, Dec. 20, 2001. http://edition.cnn.com/2001/US/12/18/anthrax/.

5 "Postal, Capitol Hill workers offered vaccine," Senator Frist quote, CNN, Dec. 20, 2001. http://www.cnn.com/2001/US/12/18/anthrax/.
6 Connolly C, et al, "Anthrax Exposure Estimates Increased, First Capitol Hill Aides Receive Vaccine Shots," *Washington Post*, Dec. 21, 2001.
7 Navy News Service, Flag Officer Announcements, Sept. 14, 1994. http://www.chinfo.navy.mil/navpalib/news/navnews/nns94/nns94055.txt.
8 Important House Advisory to House Staff, Re Anthrax Vaccinations, message from the Attending Physician, Dec. 19, 2001.
9 Keys R, Taylor G, Letter to FDA Docket 1980n-0208, Mar. 25, 2005.
10 Sen. Jeff Bingaman staffer email, Sent: 4/12/2004 3:54:07 PM EST.
11 SR 278, "Sense of the Senate." https://www.govinfo.gov/content/pkg/CRECB-2003-pt23/html/CRECB-2003-pt23-Pg31936–2.htm.
12 Schumm WR, et al, "Testing new predictive models involving biological warfare attacks on civilians," Medical Veritas, *Journal of Medical Truth*, 1:331–333, Apr. 2005. Original link at: http://www.medicalveritas.com/images/00039.pdf and https://www.researchgate.net/publication/244927409_Testing_new_predictive_models_involving_biological_warfare_attacks_on_civilians.
13 Schumm WR, et al, "Unanswered questions and ethical issues concerning US biodefence research," *J Med Ethics*, 35(10):594–8, Oct. 2009. https://pubmed.ncbi.nlm.nih.gov/19793937/.
14 FDA reply to Dingle/Rempfer Citizen Petition, Aug. 28, 2002. http://www.fda.gov/ohrms/dockets/dailys/02/Sep02/091102/80027a9f.pdf.
15 FDA reply to Dingle/Rempfer Citizen Petition, Aug. 28, 2002. http://www.fda.gov/ohrms/dockets/dailys/02/Sep02/091102/80027a9f.pdf.

Mile 15: Judicial Review (2003 to 2005)

1 SAIC Corporation plan, dated Sept. 29, 1995, and memo from Dr. Anna Johnson-Winegar to Dr. Robert Myers (MDPH), Oct. 5, 1995.
2 Danley D, "Minutes of the Meeting on Changing the Food and Drug Administration License for the Michigan Department of Public Health (MDPH) Anthrax Vaccine to Meet Military Requirements," Nov. 13, 1995.
3 "Industrial Capabilities Assessment, Report for the Production of Anthrax Vaccine," Joint Program Office for Biological Defense, Dec. 1997.
4 10 USC § 1107, Title 10, Subtitle A, Part II, Chapter 55, Sec. 1107. https://www.govinfo.gov/app/details/USCODE-2010-title10/USCODE-2010-title10-subtitleA-partII-chap55-sec1107.
5 *Encyclopedia Britannica*, Judicial review.
6 Jemekiah Barber vs. the US Army, Excerpt: "the issues in this case are beyond the purview of the federal judiciary and that the Court must decline review because the Department of Defense has wide latitude over military personnel decisions," Feb. 2, 2003.
7 Masiello S, "Appellants Brief to the United States Court of Appeals," 10th Circuit, No. 03–1056.
8 Case #1:01CV00941, Declaratory Judgment Act suit, DoD motions, DoD reply, pg. 18, and footnote 5 on pg. 20, Jul. 7, 2003.
9 Ricks T, "The Widening Gap Between the Military and Society," *Atlantic Monthly*, Jul. 1997. www.theatlantic.com/issues/97/milisoc.htm.

10 5 USC § 701(b)(1)(G)
11 Case No. 1:03-CV-254, Western District of Michigan.
12 21 CFR §50.25, Part 50, "Protection of Human subjects, Subpart B, Informed Consent of Human Subjects, Sec. 50.25, Elements of informed consent." https://www.accessdata.fda.gov/scripts/cdrh/cfdocs/cfcfr/cfrsearch.cfm?fr=50.25.
13 "Authorization of Emergency Use of Anthrax Vaccine Adsorbed for Prevention of Inhalation Anthrax by Individuals at Heightened Risk of Exposure Due to Attack With Anthrax," Inaugural EUA for anthrax vaccine, Federal Register, Jan. 27, 2005. https://www.federalregister.gov/documents/2005/02/02/05–2028/authorization-of-emergency-use-of-anthrax-vaccine-adsorbed-for-prevention-of-inhalation-anthrax-by.
14 Public Readiness and Emergency Preparedness (PREP) Act. https://www.congress.gov/109/plaws/publ148/PLAW-109publ148.pdf#page=140.
15 Federal Register, 70 Fed. Reg. 5452, 5455, Feb. 2, 2005 and https://casetext.com/case/doe-v-rumsfeld-4.
16 "DOD Pays Liberty Counsel $1.8 Million for COVID Litigation," Liberty Counsel, Oct. 4, 2023. https://lc.org/newsroom/details/100423-dod-pays-liberty-counsel-dollar18-million-for-covid-litigation and https://www.militarytimes.com/news/your-military/2023/10/09/dod-settles-covid-vaccine-mandate-lawsuits-for-18-million/.

Mile 16: Amerithrax (2001 to 2010)

1 DoD IG DCIS referral to the FBI, May 8, 2002. https://img1.wsimg.com/blobby/go/4fa7f468-a250–4088-926e-3c56a998df1f/downloads/1cr5po1rs_145656.pdf?ver=1707246986777.
2 Senator Daschle office emails between staffers and Maj. Rempfer, 2001. https://img1.wsimg.com/blobby/go/4fa7f468-a250–4088-926e-3c56a998df1f/downloads/1cr5pilop_812721.pdf?ver=1707246986777.
3 "Mr. Fleischer: All indications are that the source of the anthrax is domestic. And I can't give you any more specific information than that. That's part of what the FBI is actively reviewing. And I just can't go beyond that," White House Press Briefing, Feb. 25, 2002. Original link at: http://www.whitehouse.gov/news/releases/2002/02/20020225–16.html.
4 Taylor J, "Transcript of Amerithrax Investigation Press Conference," Excerpts: "summer and fall of 2001 . . . anthrax vaccine . . . was failing." And, "with respect to the motive, the troubled nature of Dr. Ivins. And a possible motive is his concern about the end of the vaccination program." And, "one theory is that by launching these attacks, he creates a situation, a scenario, where people all of a sudden realize the need to have this vaccine," FBI, Aug. 6, 2008. https://www.justice.gov/archive/opa/pr/2008/August/08-opa-697.html and https://www.justice.gov/archives/opa/remarks-prepared-delivery-us-attorney-jeffrey-taylor-amerithrax-investigation-press-conference.
5 FBI Amerithrax report, p. 8. https://www.justice.gov/archive/amerithrax/docs/amx-investigative-summary.pdf and https://img1.wsimg.com/blobby/go/4fa7f468-a250–4088-926e-3c56a998df1f/downloads/1cr5pmtdq_232569.pdf?ver=1707246986777.
6 FBI Amerithrax Report, p 22. https://www.justice.gov/archive/amerithrax/docs/amx-investigative-summary.pdf.
7 Special Interest OIG Hotline Case #84–142. Mar. 8, 2002. Earlier case #'s 79–472 & 473. Jan. 22, and 24, 2001.

8 GAO-02–445. "Anthrax Vaccine. GAO's Survey of Guard And Reserve Pilots and Aircrew," Sept. 2002. https://www.gao.gov/assets/a235626.html.
9 GAO-02–445. "Anthrax Vaccine. GAO's Survey of Guard And Reserve Pilots and Aircrew," Sept. 2002. https://www.gao.gov/assets/a235626.html.
10 GAO-02–181T, "Anthrax Vaccine: Changes to the Manufacturing Process," Oct. 23, 2001. https://www.gao.gov/products/gao-02–181t.
11 GAO-02–181T, "Anthrax Vaccine: Changes to the Manufacturing Process," Oct. 23, 2001. https://www.gao.gov/products/gao-02–181t.
12 GAO-02–445, "Anthrax Vaccine: GAO's Survey of Guard And Reserve Pilots and Aircrew," Sept. 2002. https://www.gao.gov/assets/a235626.html.
13 GAO-03–323T, "Diffuse security threats: Information on US domestic anthrax attacks," Dec. 10, 2002. https://img1.wsimg.com/blobby/go/4fa7f468-a250–4088-926e-3c56a998df1f/downloads/GAO%2003–323T%20Diffuse%20Security%20Threats.pdf?ver=1708361598953 and https://www.gao.gov/assets/gao-02–365.pdf.
14 Morrison P, "Column: Director Scott Z. Burns on why 'The Report' on CIA torture interrogation needed to be a feature film," *Los Angeles Times*, Dec. 11, 2019. https://www.latimes.com/opinion/story/2019–12-11/director-scott-z-burns-the-report-cia-torture-interrogation-feature-film.
15 "Scientist Is Paid Millions by US in Anthrax Suit," *New York Times*. https://www.nytimes.com/2008/06/28/washington/28hatfill.html.
16 Sword and Pen interview by Military Veterans in Journalism, timestamp 11:15, Aug. 2, 2022. https://podcasters.spotify.com/pod/show/swordandpen/episodes/Kyle-Rempfer-Military-Times-e1m0dlv.
17 Albert Einstein, Quoted in Des MacHale (London, Wisdom, 2002).
18 *Encyclopedia Britannica*, Albert Einstein.

Mile 17: Learning to Overcome (2006 to 2015)

1 Ackerman R, "Game-Changing Environment Vexes Planners," Signal Online Exclusive, Jan. 26, 2011. https://www.afcea.org/signal-media/west-2011-online-show-daily-game-changing-environment-vexes-planners.
2 Original figures posted on FBO.gov. https://globalbiodefense.com/2016/12/09/cdc-awards-1-billion-biothrax-deliveries-strategic-national-stockpile/ and https://www.emergentbiosolutions.com/story/emergent-biosolutions-lands-911-million-contract-anthrax-vaccine.
3 Rempfer T, "Anthrax Vaccine as a Component of the Strategic National Stockpile: A Dilemma for Homeland Security," DTIC.mil report number ADA514307, Naval Postgraduate School, Dec. 1, 2009. https://apps.dtic.mil/sti/citations/ADA514307 and https://www.hsdl.org/c/abstract/?docid=30641.
4 Rempfer T, "The Anthrax Vaccine: A Dilemma for Homeland Security," HSAJ 5, Article 3, May 2009. https://www.hsaj.org/articles/102.
5 NPS CHDS, "Outstanding Thesis Award for cohort 0803/0804 goes to Lt. Col. Thomas Rempfer for his in-depth study of the events leading to current policies that control the production and distribution of the Anthrax vaccine." https://calhoun.nps.edu/handle/10945/48423.
6 "Silo thinking in vaccine stockpiling persists," Center for Homeland Defense and Security, Apr. 13, 2010, updated May 28, 2015. https://www.chds.us/c/silo-thinking-in-vaccine-stockpiling-persists.

7 The 2018 National Biodefense Strategy. https://trumpwhitehouse.archives.gov/wp-content/uploads/2018/09/National-Biodefense-Strategy.pdf and https://www.dhs.gov/archive/coronavirus/presidents-biodefense-strategy and by 2022 the administration proposed a new National Biodefense Strategy projected to cost $88 billion over five years. https://www.whitehouse.gov/wp-content/uploads/2022/10/National-Biodefense-Strategy-and-Implementation-Plan-Final.pdf and https://www.whitehouse.gov/briefing-room/presidential-actions/2022/10/18/national-security-memorandum-on-countering-biological-threats-enhancing-pandemic-preparedness-and-achieving-global-health-security/.

8 White House Archives, President Donald J. Trump is Strengthening America's Biodefense, Sept. 18, 2018. https://trumpwhitehouse.archives.gov/briefings-statements/president-donald-j-trump-strengthening-americas-biodefense/ and https://www.presidency.ucsb.edu/documents/fact-sheet-president-donald-j-trump-strengthening-americas-biodefense.

9 Whistle-Blowing: An Unyielding Duty, Jul. 23, 2020. https://www.chds.us/ed/alumni-short-talks-whistle-blowing-an-unyielding-duty/.

Mile 18: How to Make You Whole (2011 to 2022)

1 Thomas L. Rempfer FOIA case ruling. http://www.state.ct.us/foi/2001fd/20010110/FIC2000–303.htm.

2 Russell E. Dingle FOIA case ruling. http://www.state.ct.us/foi/2001fd/20010110/FIC2000–304.htm.

3 Williams TD, "Dodd Questions Handling Of Anthrax Complaint," *Hartford Courant*, Mar. 09, 2001.

4 "An Act Prohibiting the Administration of Experimental Drugs and Vaccines to Members of the Connecticut Militia," HB06079, Jan. 17, 2001. Public Safety Committee Testimony, Feb. 27, 2001. https://www.cga.ct.gov/2001/psdata/chr/2001PS-00227-R001100-CHR.htm.

5 Williams TD, "Shays Wants Penalties Dropped in Anthrax-Refusal Cases," Hartford Courant, Jul. 13, 2000. https://www.courant.com/news/connecticut/hc-xpm-2000–07-13–0007130478-story.html.

6 Grossman E, "Judge advances anthrax vaccine refusal case, Pentagon must reconsider exonerating two military pilots discharged after resisting inoculations prior to FDA approval," Govexec, Mar. 24, 2008. https://www.govexec.com/defense/2008/03/judge-advances-anthrax-vaccine-refusal-case/26550/.

7 South T, "Troops who refused anthrax vaccine paid a high price," *Military Times*, Jun. 17, 2021. https://www.militarytimes.com/news/pentagon-congress/2021/06/17/troops-who-refused-anthrax-vaccine-paid-a-high-price/.

8 South T, "The shadow of anthrax: The voluntary COVID-19 vaccination effort owes much to past failures," *Army Times*, Jun. 17, 2021. https://www.armytimes.com/news/your-army/2021/06/17/the-shadow-of-anthrax-the-voluntary-covid-19-vaccination-effort-owes-much-to-past-failures/.

9 Muhammad J, "Marine punished for refusing anthrax vaccine says 'justice prevails' in his case," *Marine Times*, Jul. 11, 2021. https://www.militarytimes.com/opinion/2021/07/11/marine-punished-for-refusing-anthrax-vaccine-says-justice-prevails-in-his-case/.

10 Lt Col Russ Dingle Memorial Anthrax Vaccine Justice Act (MAVJA) Download, updated later as the Memorial Anthrax and COVID Vaccine Justice Act (MACJA). https://img1.wsimg.com/blobby/go/4fa7f468-a250–4088-926e-3c56a998df1f/downloads/MACJA%20RECORD%20CORRECTION%20BILL%20DRAFT.pdf?ver=1674168293789.

11 Lt Col Russ Dingle Memorial Anthrax Vaccine Justice Act Record Correction Bill Presentation. https://img1.wsimg.com/blobby/go/4fa7f468-a250–4088-926e-3c56a998df1f/downloads/MAVJA%20Record%20Correction%20Bill%20Presentation.pdf?ver=1624388846967.
12 USAF Inspector General Report of Investigation, p. 25.
13 Department of Defense Report of Investigation, pgs. 12–29.
14 10 USC §628(d)(2) & §14502. https://www.govinfo.gov/content/pkg/USCODE-2011-title10/html/USCODE-2011-title10-subtitleA-partII-chap36-subchapIII-sec628.htm and 10 USC 14502(e)(2) and https://www.govinfo.gov/content/pkg/USCODE-2011-title10/html/USCODE-2011-title10-subtitleE-partIII-chap1407-sec14502.htm.

Mile 19: PART TWO—Pandemic (2019 to 2023)

1 Pandolfo C, "The government paid hundreds of media companies to advertise the COVID-19 vaccines while those same outlets provided positive coverage of the vaccines," *The Blaze*, Mar. 3, 2022 https://www.theblaze.com/news/review-the-federal-government-paid-media-companies-to-advertise-for-the-vaccines and https://lc.org/newsroom/details/030722-biden-administration-paid-media-dollar1-billion-for-covid-shot-propaganda-1 and https://www.congress.gov/bill/117th-congress/house-bill/1319/text and https://wecandothis.hhs.gov/resource/we-can-do-campaign-background.
2 Great Barrington Declaration authors: Dr. Martin Kulldorff, Harvard University; Dr. Sunetra Gupta, Oxford University; Dr. Jay Bhattacharya, Stanford University Medical School, Oct. 4, 2020. https://gbdeclaration.org/.
3 Michael Levitt, The Nobel Prize in Chemistry, 2013. https://www.nobelprize.org/prizes/chemistry/2013/levitt/facts/.
4 Qiu L, "Theory About US-Funded Bioweapons Labs in Ukraine Is Unfounded," *New York Times* Fact Check, Mar. 11, 2022. https://www.nytimes.com/2022/03/11/us/politics/us-bioweapons-ukraine-misinformation.html.
5 Price N, "The Kremlin's Allegations of Chemical and Biological Weapons Laboratories in Ukraine," US State Department Press Statement, Mar. 9, 2022. https://www.state.gov/the-kremlins-allegations-of-chemical-and-biological-weapons-laboratories-in-ukraine/.
6 Mercola J, "Bioweapons Expert Speaks Out About US Biolabs in Ukraine," Apr. 9, 2022. https://articles.mercola.com/sites/articles/archive/2022/04/09/biolabs-in-ukraine.aspx and https://www.youtube.com/watch?v=s56zAchvk3w and https://www.heraldopenaccess.us/openaccess/coronavirus-is-a-biological-warfare-weapon.
7 "DoD Fact Sheet on WMD Threat Reduction Efforts with Ukraine, Russia and Other Former Soviet Union Countries," Jun. 9, 2022. https://www.defense.gov/News/Releases/Release/Article/3057517/fact-sheet-on-wmd-threat-reduction-efforts-with-ukraine-russia-and-other-former/.
8 "Defense Department Records Reveal US Funding of Anthrax Laboratory Activities in Ukraine," *Judicial Watch*, Nov. 10, 2022. https://www.judicialwatch.org/dod-records-anthrax-lab/ and https://www.judicialwatch.org/videos/new-defense-department-records-reveal-u-s-funding-of-anthrax-laboratory-activities-in-ukraine/.
9 Basu Z, "Ex-CDC director says he believes coronavirus originated in Wuhan lab," *Axios*, Mar. 26, 2021. https://www.axios.com/wuhan-lab-coronavirus-cdc-director-c599cf7b-9e30–4314-909b-a9bdac28ead6.html.

10 Piper G, "'Fauci knows' he funded gain-of-function research, 'misled Congress,' former CDC director says," *Just the News*, Sept. 16, 2022. https://justthenews.com/government/federal-agencies/fauci-knows-he-funded-gain-function-research-misled-congress-former-cdc and https://www.msn.com/en-us/news/politics/ex-cdc-director-says-unredacted-fauci-gain-of-function-email-reveals-aggressive-attempt-to-change-narrative/ar-AA1dZPfw.
11 National Archives, President Dwight D. Eisenhower's Farewell Address, warnings against the establishment of a "military-industrial complex." Jan. 17, 1961. https://www.archives.gov/milestone-documents/president-dwight-d-eisenhowers-farewell-address.
12 Gøtzsche PC, "Origin of COVID-19: The biggest cover up in medical history," Institute for Scientific Freedom, Copenhagen, Oct. 6, 2023. https://www.scientificfreedom.dk/wp-content/uploads/2023/10/Gotzsche-Origin-of-COVID-19-The-biggest-cover-up-in-medical-history.pdf.
13 McCarthyism and The "Red Scare," Notes from the President Dwight D. Eisenhower Presidential Library, "The American Heritage Dictionary gives the definition of McCarthyism as: 1. The political practice of publicizing accusations of disloyalty or subversion with insufficient regard to evidence; and 2. The use of methods of investigation and accusation regarded as unfair, in order to suppress opposition." https://www.eisenhowerlibrary.gov/research/online-documents/mccarthyism-red-scare and Prelude to McCarthyism: The Making of a Blacklist, Vol. 38, No. 3, 2006. https://www.archives.gov/publications/prologue/2006/fall/agloso.html.
14 CDC, The Threat of an Anthrax Attack, with no explanation about the singular anthrax attack in US history relating to a motive to resuscitate the failing anthrax vaccine program by a US Army scientist. https://www.cdc.gov/anthrax/bioterrorism/threat.html.
15 Senate Hearing 101–744, "1989 Global Spread of Chemical and Biological Weapons Hearings before the Committee on Governmental Affairs," 1989. https://www.google.com/books/edition/Global_Spread_of_Chemical_and_Biological/tbIRAAAAIAAJ?hl=en&gbpv=1.
16 *Vaccines*, ed. Plotkin and Mortimer, Chapter 26 (United Kingdom, W.B. Saunders Company, 1994).
17 Caldera L, Secretary of the Army. Memorandum of Decision, Sept. 3, 1998.
18 *Vaccines*, ed. Plotkin and Mortimer, p. 737 (United Kingdom, W.B. Saunders Company, 1994).
19 "Industrial Capabilities Assessment, Summary Report for the Production of Anthrax Vaccine," Preliminary report prepared by the Joint Program Office for Biological Defense (JPOBD), Falls Church, VA, Dec. 1997.
20 Gans J, "CDC relaxes COVID restrictions for international travelers," *The Hill*, Apr. 27, 2023. https://thehill.com/policy/healthcare/3976562-cdc-relaxes-covid-restrictions-for-international-travelers/.
21 Claypool R, "Written and verbal testimony before the House Government Reform Committee," Jul. 21, 1999.
22 Zimbardo P, *The Lucifer Effect: Understanding How Good People Turn Evil* (New York, Random House, 2007) Foreword.
23 Zimbardo P, *The Lucifer Effect: Understanding How Good People Turn Evil* (New York, Random House, 2007) 443.
24 Maslow, AH, *The Psychology of Science: A Reconnaissance* (United Kingdom, Harper & Row, 1966) p. x, https://www.google.com/books/edition/The_Psychology_of_Science

/qitgAAAAMAAJ?hl=en&gbpv=1&bsq=hammer and Kaplan A, *The Conduct of Inquiry: Methodology for Behavioral Science* (San Francisco: Chandler Publishing Co. 1964), 28, https://books.google.com/books?id=OYe6fsXSP3IC&newbks=0&lpg=PA28&pg=PA28#v=onepage&q=hammer&f=false.

25 Breckenridge J, Zimbardo P, The strategy of terrorism and the psychology of mass-mediated fear, In B. Bongar (Ed.), *Psychology of Terrorism,* (New York, Oxford University Press, 2007) 122.

26 Breckenridge J, Zimbardo P, The strategy of terrorism and the psychology of mass-mediated fear, In B. Bongar (Ed.), *Psychology of Terrorism,* (New York, Oxford University Press, 2007) 122.

27 Public Health Emergency Declarations. https://www.phe.gov/Preparedness/legal/Pages/phedeclaration.aspx and https://www.whitehouse.gov/briefing-room/presidential-actions/2022/02/18/notice-on-the-continuation-of-the-national-emergency-concerning-the-coronavirus-disease-2019-covid-19-pandemic-2/ and https://www.federalregister.gov/documents/2020/03/18/2020–05794/declaring-a-national-emergency-concerning-the-novel-coronavirus-disease-covid-19-outbreak and 50 USC §1622 and https://uscode.house.gov/view.xhtml?req=granuleid:USC-prelim-title50-section1622&num=0&edition=prelim and https://aspr.hhs.gov/legal/PHE/Pages/covid19–13Oct2022.aspx and https://aspr.hhs.gov/legal/PHE/Pages/covid19–11Jan23.aspx.

28 Zimbardo P, A situationist perspective on the psychology of evil: Understanding how good people are transformed into perpetrators, In A. Miller (Ed.), *The Social, Psychology of Good and Evil,* (New York, Guilford Press, 2005), 21–498.

29 Zimbardo P, *The Lucifer Effect: Understanding How Good People Turn Evil* (New York, Random House, 2007) and Lecture on the psychology of fear management and terrorism, Naval Postgraduate School Center, Jun. 24, 2009.

30 Zimbardo P, *The Lucifer Effect: Understanding How Good People Turn Evil* (New York, Random House, 2007), 480–82.

31 Zimbardo P, *The Lucifer Effect: Understanding How Good People Turn Evil* (New York, Random House, 2007), 211.

32 Milgram S, *Obedience to Authority: an Experimental View* (London, Tavistock, 1974), p. 123.

33 Mulroney T, et al, "N1-methylpseudouridylation of mRNA causes +1 ribosomal frameshifting," Nature, 2023. https://www.nature.com/articles/s41586–023-06800–3 and https://www.cdc.gov/coronavirus/2019-ncov/variants/ and https://www.bbc.com/news/health-67625180 and https://www.telegraph.co.uk/news/2023/12/06/mrna-jabs-modena-pfizer-quarter-unintended-response/.

34 "'Covid-19 vaccine administration must stop'—Dr Aseem Malhotra's MUST READ paper on mRNA vaccines," BizNews, Sept. 26, 2022. https://www.biznews.com/health/2022/09/26/mrna-vaccines-malhotra and Dr. Aseem Malhotra who promoted COVID-19 vaccine on TV calls for its immediate suspension. https://youtu.be/MtE0I5FqHPs.

35 "Tucker Carlson interviews Dr Aseem Malhotra on the corruption of medicine by Big Pharma," Fox News video. https://www.youtube.com/watch?v=w3MPnBpfrRk.

Mile 20: Fauci Effect (1984 to 2022)

1 AMA medical education experts discuss the Fauci effect, Dec. 15, 2020. https://www.ama-assn.org/medical-students/preparing-medical-school/ama-medical-education-experts-discuss-fauci-effect.

2 Browne E, "Fauci Was 'Untruthful' to Congress About Wuhan Lab Research, New Documents Appear To Show," *Newsweek*, Sept. 9, 2021. https://www.newsweek.com/fauci-untruthful-congress-wuhan-lab-research-documents-show-gain-function-1627351 and https://www.congress.gov/117/meeting/house/114270/documents/HHRG-117-GO24–20211201-SD004.pdf.

3 Spencer C, "Sen. Rand Paul referred Anthony Fauci to the Justice Department for allegedly lying to Congress," *The Hill*, Jul. 26, 2021. https://thehill.com/changing-america/well-being/prevention-cures/564803-rand-paul-sends-official-criminal-referral-on/.

4 Paul R, *The Great COVID Cover-Up* (New York, Skyhorse/Regnery, 2023). https://www.regnery.com/9781684515134/deception/.

5 US State and Interior Department FOIAs by US Right to Know. https://usrtk.org/wp-content/uploads/2022/03/State-Department-FOIA-An-Analysis-of-Circumstantial-Evidence-for-Wuhan-Labs-as-the-Source-of-the-Coronavirus.pdf and https://2017–2021.state.gov/fact-sheet-activity-at-the-wuhan-institute-of-virology/index.html and https://usrtk.org/covid-19-origins/scientists-proposed-making-viruses-with-unique-features-of-sars-cov-2-in-wuhan/.

6 "WHO's Scientific Advisory Group for the Origins of Novel Pathogens (SAGO) Preliminary Report," WHO, Jun. 9, 2022. https://cdn.who.int/media/docs/default-source/scientific-advisory-group-on-the-origins-of-novel-pathogens/sago-report-09062022.pdf.

7 "FBI Investigated Fauci Agency!," Judicial Watch archive, Jul. 15, 2022. https://www.judicialwatch.org/fbi-investigated-fauci-agency/.

8 Dr. George Fareed on TPC with Tommy Carrigan, Discussing "Glacial pace" of the truth, Jun. 2, 2023. https://rumble.com/v2rwkfc-dr.-george-fareed-dept.-hhs-vaccine-corruption.html.

9 Zimmer C, et al, "Fight Over Covid's Origins Renews Debate on Risks of Lab Work," *New York Times*, Jun. 20, 2021. https://www.nytimes.com/2021/06/20/science/covid-lab-leak-wuhan.html.

10 Knudsen H, "Rand Paul Grills Fauci on Gain-of-Function: You Changed the Definition to 'Cover Your Ass,'" Breitbart, Nov. 4, 2021. https://www.breitbart.com/politics/2021/11/04/rand-paul-grills-fauci-on-gain-of-function-you-changed-the-definition-to-cover-your-ass/.

11 NIH on Enhanced Potential Pandemic Pathogens. https://www.nih.gov/news-events/research-involving-potential-pandemic-pathogens.

12 NIH grant search for Ralph S. Baric. https://reporter.nih.gov/search/B4pLYUlqdkCEaIGuOHlhWg/projects?agencies=NIAID.

13 Baric R, et al, No See'm technology explained: "No See'm sites can be used to insert foreign genes into viral, eukaryotic, or microbial genomes or vectors, simultaneously removing all evidence of the restriction sites that were used in the recombinant DNA manipulation," A59, Journal of Virology, 76(21):11065–78, Nov. 2002. https://www.ncbi.nlm.nih.gov/pmc/articles/PMC136593/.

14 Menachery V, Yount B, Debbink K, Agnihothram S, Gralinski L, Plante J, Graham R, Scobey T, Ge X, Donaldson E, Randell S, Lanzavecchia A, Marasco W, Shi Z, Baric R, "A SARS-like cluster of circulating bat coronaviruses shows potential for human emergence," Nat Med, (12):1508–13, Dec. 21, 2015. https://www.ncbi.nlm.nih.gov/pmc/articles/PMC4797993/.

15 Brewster J, "Here's What Dr. Fauci Has Said About Covid's Origins And The Lab Leak Theory," *Forbes*, Jun. 16, 2021. https://www.forbes.com/sites/jackbrewster/2021/06/16/heres-what-dr-fauci-has-said-about-covids-origins-and-the-lab-leak-theory/.

16 Calisher C, Carroll D, Colwell R, Corley R, Daszak P, Drosten C, et al, "Statement in support of the scientists, public health professionals, and medical professionals of China combatting COVID-19," *The Lancet*, Volume 395, issue 10226, e42-e43, Feb. 19, 2020. https://www.thelancet.com/journals/lancet/article/PIIS0140–6736(20)30418–9/fulltext.

17 Calisher C, Carroll D, Colwell R, Corley R, Daszak P, Drosten C, et al, "Science, not speculation, is essential to determine how SARS-CoV-2 reached humans," *The Lancet*, Volume 398, issue 10296, P209–211, Jul. 5, 2021. https://www.thelancet.com/journals/lancet/article/PIIS0140–6736(21)01419–7/fulltext.

18 Van Helden J, et al, "An appeal for an objective, open, and transparent scientific debate about the origin of SARS-CoV-2," *The Lancet*, Volume 398, issue 10309, P1402–1404, Sept. 17, 2021 https://www.thelancet.com/journals/lancet/article/PIIS0140–6736(21)02019–5/fulltext.

19 Andersen K, Twitter account, "account that no longer exists," https://twitter.com/K_G_Andersen/status/1377826266850553857.

20 "Congressional Select Subcommittee's Investigation into the Proximal Origin of SARS-CoV-2," Congress, Mar. 5, 2023. https://oversight.house.gov/wp-content/uploads/2023/03/2023.03.05-SSCP-Memo-Re.-New-Evidence.Proximal-Origin.pdf and https://oversight.house.gov/release/wenstrup-releases-alarming-new-report-on-proximal-origin-authors-nih-suppression-of-the-covid-19-lab-leak-hypothesis/ and https://oversight.house.gov/release/wenstrup-hhss-year-long-campaign-of-stonewalling-congressional-oversight-ends-today/.

21 Andersen, K, "New NIAID-funded center established," Aug. 27, 2020. https://andersen-lab.com/new-niaid-funded-center-established/.

22 Andersen K, et al, "The proximal origin of SARS-CoV-2," Nat Med, 450–452, 2020. https://www.nature.com/articles/S41591–020-0820–9.

23 "COVERUP: Fauci, China, Harvard Suppressed Lab Leak," Breaking Points with Krystal and Saagar, Sept. 20, 2022. https://www.youtube.com/watch?v=JwX3U1fZDWQ and https://childrenshealthdefense.org/defender/covid-origins-cover-up-cola/.

24 Markson S, "US Department of Health official who conspired with Anthony Fauci to downplay COVID lab-leak theory reveals 'agonising' over his actions," Sky News, Nov. 27, 2023. https://www.skynews.com.au/australia-news/us-department-of-health-official-who-conspired-with-anthony-fauci-to-downplay-covid-lableak-theory-reveals-agonising-over-his-actions/news-story/f568f544d4b5eb05fb26dc3e198f50ae and https://www.skynews.com.au/world-news/us-intelligence-official-linked-to-who-was-critical-in-downplaying-covid-lab-leak-theory-during-joe-bidens-90day-probe-into-virus-origins/news-story/70cec8fe1513491a421d45b12b45a8e7 and https://www.skynews.com.au/world-news/world-exclusive-biden-probe-censored-expert-claims-that-covid-was-likely-genetically-engineered-in-a-laboratory/news-story/54b70e4d95974d528d8f754d1323232d.

25 Eban K, "In Major Shift, NIH Admits Funding Risky Virus Research in Wuhan," *Vanity Fair*, Oct. 22, 2021. https://www.vanityfair.com/news/2021/10/nih-admits-funding-risky-virus-research-in-wuhan.

26 Thacker P, "The covid-19 lab leak hypothesis: did the media fall victim to a misinformation campaign?," BMJ, 374:n1656, Jul. 8, 2021. https://www.bmj.com/content/374/bmj.n1656.

27 Thacker P, "Covid-19: Lancet investigation into origin of pandemic shuts down over bias risk," BMJ, 375:n2635, Nov. 2, 2021. https://www.bmj.com/content/375/bmj.n2414 and https://www.bmj.com/content/375/bmj.n2635.
28 Burdick S, "Fauci Not 'Telling It Like It Is,' Chair of COVID-19 Commission Tells RFK Jr.," Origins of COVID-19 With Jeffrey Sachs, Aug. 25, 2022. https://childrenshealth defense.org/defender/jeffrey-sachs-fauci-covid-rfk-jr-podcast/.
29 "Fauci: No scientific evidence the coronavirus was made in a Chinese lab," *National Geographic*, May 4, 2020. https://www.nationalgeographic.com/science/article/anthony-fauci-no-scientific-evidence-the-coronavirus-was-made-in-a-chinese-lab-cvd.
30 Sachs J, "Finding the Origins of COVID-19 and Preventing Future Pandemics," Columbia University, Jun. 22, 2021. https://www.jeffsachs.org/newspaper-articles/cp24mtcpswg yty5st4pm29mwh6dt2d.
31 Sachs J, et al, "The Lancet Commission on lessons for the future from the COVID-19 pandemic," The Lancet Commission, Vol 400, Issue 10359, P1224–1280, Oct. 8, 2022. https://www.thelancet.com/journals/lancet/article/PIIS0140–6736(22)01585–9/fulltext.
32 Winters N, "COVID 'Most Likely' Escaped Wuhan Lab, Possibly 'Genetically Engineered, British MPs Told," *The National Pulse*, Dec. 15, 2021. https://thenationalpulse.com/news/mps-briefed-on-covid-lab-leak-origins/.
33 Moderna, Inc. and BioNTech disclosures to the US Securities and Exchange Commission, "mRNA is considered a gene therapy product by the FDA," Nov. 9, 2018. https://www.sec.gov/Archives/edgar/data/1682852/000119312518323562/d577473ds1.htm and Sept. 9, 2019. https://www.sec.gov/Archives/edgar/data/1776985/000119312519241112/d635330df1.htm.
34 Ambati Balamurali K, et al, "MSH3 Homology and Potential Recombination Link to SARS-CoV-2 Furin Cleavage Site," Frontiers in Virology, Vol. 2, Feb. 21, 2022. https://www.frontiersin.org/article/10.3389/fviro.2022.834808.
35 "Multiple news reports suggesting natural virus origins," https://www.nytimes.com/interactive/2022/02/26/science/covid-virus-wuhan-origins.html and https://www.cnn.com/2022/02/26/health/coronavirus-origins-studies/index.html and https://www.theguardian.com/world/2022/feb/26/coronavirus-wuhan-market-chinese-lab-studies.
36 Mack E, "Covid-19 Patients Zero In Wuhan Identified, Boosting Lab Leak Theory," *Forbes*, Jun. 15, 2023. https://www.forbes.com/sites/ericmack/2023/06/15/covid-19-patients-zero-in-wuhan-identified-boosting-lab-leak-theory/.
37 Bowser B, "Officials are taking another look at the controversial anthrax vaccine," PBS News Hour, Dec. 4, 2001. https://www.pbs.org/newshour/show/the-anthrax-vaccine.
38 "Can BioShield effectively procure medical countermeasures that safeguard the nation?," Congress, House Committee on Homeland Security, Subcommittee on Emerging Threats, Cybersecurity and Science and Technology, HR 110–23, p. 48–50, Apr. 18, 2007. https://www.govinfo.gov/content/pkg/CHRG-110hhrg43559/html/CHRG-110hhrg43559.htm.
39 Eisenhower D, Executive Order 10631, Code of Conduct for members of the Armed Forces of the United States, 20 FR 6057, 3 CFR, 1954-1958, p. 266, Aug. 17, 1955. https://www.archives.gov/federal-register/codification/executive-order/10631.html.
40 Dr. Anthony Fauci interview excerpt published by the House Select Subcommittee on the Coronavirus Pandemic via their Twitter account, Nov. 30, 2023. https://twitter.com/COVIDSelect/status/1730198243202502925.

41 "Rand Paul: This is the biggest coverup in the history of science," FOX Business, Sept. 20, 2022. https://video.foxbusiness.com/v/6312572522112#sp=show-clips.

42 In 1943 George Orwell became the literary editor of Tribune, writing a column called As I Please, where he fine-tuned the ideas that would eventually become 1984. http://galileo.phys.virginia.edu/classes/inv_inn.usm/orwell3.html.

43 MacKinnon D, "What must never be asked about COVID-19 and vaccines—nor ever revealed," *The Hill*, Opinion, Nov. 13, 2021. https://thehill.com/opinion/white-house/581001-what-must-never-be-asked-about-covid-and-vaccines-nor-ever-revealed.

44 Valverde M, "Social media post falsely claims a federal law would require vaccination against COVID-19," PolitiFact check, Dec. 11, 2020. https://www.politifact.com/factchecks/2020/dec/11/facebook-posts/post-falsely-claims-federal-law-would-require-vacc/.

45 Hellmann J, "Fauci says he does not see US mandating COVID-19 vaccination for general public," *The Hill*, Aug. 18, 2020. https://thehill.com/policy/healthcare/512542-fauci-says-he-does-not-see-us-mandating-covid-19-vaccination-for-general?rl=1.

46 The Trusted News Initiative (TNI) partners. https://www.bbc.com/mediacentre/2020/trusted-news-initiative-vaccine-disinformation and https://www.bbc.co.uk/mediacentre/latestnews/2020/coronavirus-trusted-news.

47 Maddow R, "Rachel Maddow spreading vaccine misinformation," MSNBC. https://www.youtube.com/watch?v=kMrTG06qi40.

48 Pandolfo C, "Federal government paid hundreds of media companies to advertise the COVID vaccines while those outlets provided positive coverage of the vaccines," *The Blaze*, Mar. 03, 2022. https://www.theblaze.com/news/review-the-federal-government-paid-media-companies-to-advertise-for-the-vaccines and https://www.theblaze.com/news/review-the-federal-government-paid-media-companies-to-advertise-for-the-vaccines.

49 Winters N, "Pfizer is Funding Facebook's Fact-Checking Partner," *The National Pulse*, Feb. 25, 2022. https://thenationalpulse.com/2022/02/25/pfizer-funds-facebook-fact-checking-partner-combatting-covid-19-misinformation/.

50 OpenTheBooks.com is a project of American Transparency—a 501(c)3 nonprofit, nonpartisan charitable organization, Investigation. https://www.openthebooks.com/substack-investigation-faucis-royalties-and-the-350-million-royalty-payment-stream-hidden-by-nih/.

51 Demasi M, "From FDA to MHRA: are drug regulators for hire?," *BMJ*, 377:o1538, Jun. 29, 2022. https://www.bmj.com/content/377/bmj.o1538.

52 Dougherty M, "Anthony Fauci: I am the science," National Review, Nov. 29, 2022. https://www.nationalreview.com/2021/11/anthony-fauci-i-am-the-science/.

53 Richard L, "Last year, Fauci said 'you cannot force someone' to get COVID-19 vaccine," Yahoo News, Aug. 13, 2021. https://cc.bingj.com/last-fauci-said-cannot-force-212300017.html?q=Yahoo+news.+https%3a%2f%2fwww.yahoo.com%2fnow%2flast-fauci-said-cannot-force-212300017.html&d=4799424450026262&mkt=en-US&setlang=en-US&w=ZPzzBJ4UfPqGhqs326c8l8eKNDhLD-Yi.

54 Herper M, et al, "FDA advisory panel recommends Pfizer Covid-19 vaccine be authorized for children," Stat news, Oct. 26, 2021. https://www.statnews.com/2021/10/26/fda-vrbpac-pfizer-vaccine-covid-kids/.

55 United States Senate, Sen. Hrg. 101–744, Re Secretary of Defense Robert B. Barker letter to former US Sen. John Glenn, Feb. 10, 1989. https://www.google.com/books/edition/Global_Spread_of_Chemical_and_Biological/tbIRAAAAIAAJ?hl=en&gbpv=1.

56 Vaccines, pp. 773–996 (United Kingdom, W.B. Saunders Company, 1994).
57 Vaccines, pp. 629–636 (United Kingdom, W.B. Saunders Company, 1998).
58 "Countering anthrax: Vaccines and immunoglobulins," Clinical Infectious Diseases, 46(1), pp. 129–134, Jan. 1, 2008.
59 Nevradakis M, "Key Bioweapons Official Publicly Accuses Fauci of 'Denial and Deception' on COVID Origins," *The Defender*, Jan. 5, 2024. https://childrenshealthdefense.org/defender/bioweapons-official-fauci-denial-deception-covid-origins/.
60 "Military Documents About Gain-of-Function Contradict Fauci Testimony Under Oath," Project Veritas, Jan. 10, 2022. https://www.projectveritas.com/news/military-documents-about-gain-of-function-contradict-fauci-testimony-under/.
61 "How DARPA seeded the ground for a rapid COVID-19 cure," Yahoo Finance video, Dec. 17, 2020. https://finance.yahoo.com/video/darpa-seeded-ground-rapid-covid-214655580.html.
62 President Obama White House archives, Policy Guidance for Departmental Development of Review Mechanisms for Potential Pandemic Pathogen [PPP], Jan. 9, 2017. https://obamawhitehouse.archives.gov/sites/default/files/microsites/ostp/p3co-finalguidance statement.pdf and http://www.phe.gov/s3/dualuse/Documents/gain-of-function.pdf.
63 "SARS-COV-2 Origins Investigation . . . ," Officer's Report to the Inspector General of the DoD, Project Veritas, Aug. 13, 2021. https://assets.ctfassets.net/syq3snmxclc9/2mVob3c1aDd8CNvVnyei6n/95af7dbfd2958d4c2b8494048b4889b5/JAG_Docs_pt1_Og_WATERMARK_OVER_Redacted.pdf.
64 Sen. Ron Johnson letter to SecDef Austin, Feb. 1, 2022. https://www.ronjohnson.senate.gov/services/files/FB6DDD42–4755-4FDC-BEE9–50E402911E02.
65 Crapo M, "Let's Do Away With Federal Overreach—Vaccine Mandates A Good Place To Start," US Senator Mike Crapo, May 16, 2022. https://www.crapo.senate.gov/news/in-the-news/weekly-column-lets-do-away-with-federal-overreach_vaccine-mandates-a-good-place-to-start.
66 Huff A, *Truth about Wuhan* (New York, Skyhorse, 2022). https://www.skyhorsepublishing.com/9781510773882/the-truth-about-wuhan/.
67 Renz Law, Dr. Andrew Huff affidavit, Sept. 12, 2022. https://tomrenz.substack.com/api/v1/file/42e49d09–33f1–4548-a7f2–7f87037f1d00.pdf.
68 Huff A, COVID-19 Origins, Oct. 21, 2022. https://rumble.com/v1p3053-jim-hoft-interviews-dr.-andrew-huff-on-covid-19-origins.html.
69 Calvert J, "What really went on inside the Wuhan lab weeks before Covid erupted," *The Times*, Jun. 11, 2023. https://www.theaustralian.com.au/world/the-times/what-really-went-on-inside-the-wuhan-lab-weeks-before-covid-erupted/news-story/96a38b1d31eed7e7b01c6407b481d3f5.
70 "Analyzing the potential for future bat coronavirus emergence . . . ," NIH Report, Contact PI/Project Leader: Daszak P, Awardee Organization: EcoHealth Alliance, Inc., Sept. 21, 2022. https://reporter.nih.gov/search/0jAp779zVkaN-DEsKnKa5A/project-details/10522470.
71 "Judicial Watch Obtains Records Showing NIAID under Dr. Fauci Gave Wuhan Lab $826k for Bat Coronavirus Research From 2014 to 2019," Judicial Watch, Jun. 4, 2021. https://www.judicialwatch.org/fauci-wuhan-826k/ and https://www.judicialwatch.org/covid-19-vaccine-campaign/.

72 Fauci A, "Louisiana, Missouri Release Full Fauci Deposition Transcript," Nov. 23rd, 2022. https://agjefflandry.com/Article/13094.
73 National Institute of Health Mission statement. https://www.nih.gov/about-nih/what-we-do/mission-goals.
74 "Transcript: All In with Chris Hayes, 5/17/21, Guests: Anthony Fauci," MSNBC, May 17, 2021. https://www.msnbc.com/transcripts/transcript-all-chris-hayes-5–17-21-n1267740.
75 Connolly G, "'No hospitalisations and no deaths': All three US vaccines 'highly efficacious,'" *Independent*, Feb. 28, 2021. https://www.independent.co.uk/news/world/americas/us-politics/vaccine-covid-fauci-deaths-b1808878.html.
76 Wilson G, "'Scandalous': Pfizer Exec Tells EU Lawmaker COVID Jab Was Never Tested To Show It Blocked Transmission," *The Daily Wire*, Oct. 12, 2022. https://www.dailywire.com/news/scandalous-pfizer-exec-tells-eu-lawmaker-covid-jab-was-never-tested-to-show-it-blocked-transmission.
77 Choi J, "Fauci: Vaccinated people become 'dead ends' for the coronavirus," *The Hill*, May 16, 2022. https://thehill.com/homenews/sunday-talk-shows/553773-fauci-vaccinated-people-become-dead-ends-for-the-coronavirus/.
78 "Masks off? Fauci confirms 'extremely low' risk of transmission, infection for vaccinated," MSNBC video, May 17, 2021. https://www.msnbc.com/all-in/watch/dr-fauci-confirms-extremely-low-risk-of-transmission-and-infection-for-vaccinated-112213061906.
79 Redshaw M, "Fauci Tests Positive for COVID After 4 Doses, Gets Grilled at Senate Hearing on Response to Pandemic," *The Defender*, Jun. 16, 2022. https://childrenshealthdefense.org/defender/fauci-positive-covid-senate-hearing-pandemic/.

Mile 21: Citizen Petition 2 (2021)

1 Robert Francis Kennedy. https://www.jfklibrary.org/learn/about-jfk/the-kennedy-family/robert-f-kennedy.
2 Kennedy, Robert F, *American Values: Lessons I Learned from My Family* (United States, HarperCollins, 2018). https://www.google.com/books/edition/American_Values/xHDUDAAAQBAJ?hl=en.
3 Washington Post poll, see question #10, Nov. 16, 2021. https://www.washingtonpost.com/context/nov-7–10-2021-washington-post-abc-news-poll/160508a9-cea2–4433-92c8–6dc51838721e/.
4 Center for countering digital hate; anti-vax watch. https://www.prnewswire.com/news-releases/disinformation-dozen-two-thirds-of-online-anti-vaccine-content-originates-from-top-12-anti-vax-leaders-301255060.html.
5 Eisenstein C, "The Term 'Vaccine Hesitancy' Is Patronizing and Presumptuous. Here's Why," *The Defender*, Nov. 15, 2021. https://childrenshealthdefense.org/defender/vaccine-hesitancy-medical-paradigms/.
6 Alliance for Human Research Protection (AHRP), upholding humanitarian values and ethical standards enshrined in the Hippocratic Oath and the Nuremberg Code: "Voluntary informed consent of the human subject is absolutely essential." https://ahrp.org/board/vera-sharav/.
7 Citizen Petition from Scientific Advisory Board on behalf of Children's Health Defense, Document ID FDA-2021-P-0460–0001. https://www.regulations.gov/document/FDA-2021-P-0460–0001.
8 21 CFR §10.30, Citizen Petition. https://www.law.cornell.edu/cfr/text/21/10.30.

9 21 USC §360bbb-3. https://www.govinfo.gov/app/details/USCODE-2011-title21/USCODE-2011-title21-chap9-subchapV-partE-sec360bbb-3.

10 21 C.F.R. §201.200(e)(1), "Labeling Claims for Drugs in Drug Efficacy Study and Disclosure of drug efficacy study evaluations in labeling and advertising." https://www.ecfr.gov/current/title-21/chapter-I/subchapter-C/part-201/subpart-F#201.200.

11 70 Federal Register, "Authorization of Emergency Use of Anthrax Vaccine Adsorbed for Prevention of Inhalation Anthrax by Individuals at Heightened Risk of Exposure Due to Attack With Anthrax," pp. 5452–5455, Feb. 2, 2005. https://www.federalregister.gov/documents/2005/02/02/05–2028/authorization-of-emergency-use-of-anthrax-vaccine-adsorbed-for-prevention-of-inhalation-anthrax-by.

12 Public Law 108–136, "National Defense Authorization Act for Fiscal Year 2004," Nov. 24, 2003. https://www.govinfo.gov/content/pkg/PLAW-108publ136/pdf/PLAW-108publ136.pdf and https://www.congress.gov/108/plaws/publ136/PLAW-108publ136.htm.

13 10 US Code § 1107a. "Emergency use products." https://www.govinfo.gov/app/details/USCODE-2010-title10/USCODE-2010-title10-subtitleA-partII-chap55-sec1107a/context and https://uscode.house.gov/view.xhtml?hl=false&edition=prelim&req=granuleid%3AUSC-prelim-title21-section360bbb-3&num=0&saved=%7CZ3JhbnVsZWlkOlVTQy1wcmVsaW0tdGl0bGUyMS1zZWN0aW9uMzYwYmJiLTNh%7C%7C%7C0%7Cfalse%7Cprelim.

14 CDC VAERS data. https://www.cdc.gov/vaccinesafety/ensuringsafety/monitoring/vaers/access-VAERS-data.html.

15 Independent VAERS analysis using CDC data as of Dec. 2023. https://openvaers.com/covid-data and https://www.medalerts.org/vaersdb/findfield.php?TABLE=ON&GROUP1=CAT&EVENTS=ON&VAX=COVID19 and https://vaersanalysis.info/.

16 VAERS performance report by Harvard Pilgrim Health Care and Lazarus R, DHHS grant report: "Adverse events from drugs and vaccines are common, but underreported," and "CDC contacts were no longer available and the CDC consultants responsible for receiving data were no longer responsive." https://digital.ahrq.gov/sites/default/files/docs/publication/r18hs017045-lazarus-final-report-2011.pdf.

17 Lazarus R, "Electronic Support for Public Health - Vaccine Adverse Event Reporting System," Harvard Pilgrim Health Care, Inc., Grant No. R18 HS017045), p. 6, 2010). https://digital.ahrq.gov/sites/default/files/docs/publication/r18hs017045-lazarus-final-report-2011.pdf.

18 World Health Organization (WHO) collaborating center for international drug monitoring, Uppsala, Sweden, Search covid-19 vaccine. http://www.vigiaccess.org/.

19 Fraiman J, et al, "Serious Adverse Events of Special Interest Following mRNA Vaccination in Randomized Trials," Jun. 2022. https://ssrn.com/abstract=4125239 and https://www.sciencedirect.com/science/article/pii/S0264410X22010283.

20 Mathieu E, et al, "Coronavirus Pandemic (COVID-19)," Published online at OurWorldInData.org, 2024. https://ourworldindata.org/covid-vaccinations.

21 Kory P, "Huge Number of Vax Deaths & It's Getting Worse," Jan. 8, 2022. https://rumble.com/vs7c9i-huge-number-of-vax-deaths-and-its-getting-worse-dr.-pierre-kory.html.

22 Gundry S, "mRNA COVID Vaccines Dramatically Increase Endothelial Inflammatory Markers and ACS Risk as Measured by the PULS Cardiac Test: a Warning," *Circulation*, volume 144, issue sup 1, Nov. 16, 2021. https://www.ahajournals.org/doi/10.1161/circ.144.suppl_1.10712.

23 Rose J, et al, "Determinants of COVID-19 vaccine-induced myocarditis," Therapeutic Advances in Drug Safety, 2024. https://journals.sagepub.com/doi/10.1177/2042098 6241226566#.

24 Oster M, et al, "Myocarditis Cases Reported After mRNA-Based COVID-19 Vaccination in the US From Dec. 2020 to Aug. 2021," *JAMA*, 327(4):331–340, Jan. 25, 2022. https: //pubmed.ncbi.nlm.nih.gov/35076665/.

25 Witberg G, et al, "Myocarditis after Covid-19 Vaccination in a Large Health Care Organization," *New England Journal of Medicine*, Dec. 2021. https://pubmed.ncbi.nlm.nih .gov/34614329/.

26 "Confidential Pfizer Documents Reveal 'Evidence' Suggesting 'Increased Risk of Myocarditis' Following COVID-19 Vaccinations," Project Veritas, Mar. 16, 2023. https://www .projectveritas.com/news/breaking-confidential-pfizer-documents-reveal-pharmaceutical -giant-had/.

27 Schreckenberg R, et al, "Cardiac side effects of RNA-based SARS-CoV-2 vaccines: Hidden cardiotoxic effects of mRNA-1273 and BNT162b2 on ventricular myocyte function and structure," Br J Pharmacol, Oct. 12, 2023. https://pubmed.ncbi.nlm.nih.gov/37828636/.

28 Schwab C, et al, "Autopsy-based histopathological characterization of myocarditis after anti-SARS-CoV-2-vaccination," Clin Res Cardiol, Nov. 27, 2022. https://link.springer .com/article/10.1007/s00392–022-02129–5 and Hulscher, N, et al, "A Systemic Review of Autopsy Findings in Deaths after COVID-19 Vaccination," Findings excerpt: "A total of 240 deaths (73.9%) were independently adjudicated as directly due to or significantly contributed to by COVID-19 vaccination," Jul. 6, 2023. https://zenodo.org/record /8120771 and Hulscher N, et al, "Autopsy findings in cases of fatal COVID-19 vaccine-induced myocarditis," ESC Heart Fail, Jan. 14, 2024. https://pubmed.ncbi.nlm.nih .gov/38221509/.

29 Blaylock R, "COVID UPDATE: What is the truth?," Surg Neurol Int., Apr. 2022. https: //www.ncbi.nlm.nih.gov/pmc/articles/PMC9062939/.

30 Montgomery J, Ryan M, Engler R, et al, "Myocarditis Following Immunization With mRNA COVID-19 Vaccines in Members of the US Military," JAMA Cardiology, 6(10):1202–1206, Jun. 29, 2021. https://jamanetwork.com/journals/jamacardiology /fullarticle/2781601 and https://www.military.com/daily-news/2021/06/30/dod-confirms-rare-heart-inflammation-cases-linked-covid-19-vaccines.html.

31 Cadegiani F, "Catecholamines are the key trigger of mRNA SARS-CoV-2 and mRNA COVID-19 vaccine-induced myocarditis and sudden deaths," Feb. 24, 2022. https: //www.researchgate.net/publication/358834540_Catecholamines_are_the_key_trigger _of_mRNA_SARS-CoV-2_and_mRNA_COVID-19_vaccine-induced_myocarditis _and_sudden_deaths_a_compelling_hypothesis_supported_by_epidemiological _anatomopathological_molecular

32 Cadegiani F, "Catecholamines Are the Key Trigger of COVID-19 mRNA Vaccine-Induced Myocarditis: A Compelling Hypothesis Supported by Epidemiological, Anatomopathological, Molecular, and Physiological Findings," *Cureus*, 11;14(8):e27883, Aug. 2022. https://pubmed.ncbi.nlm.nih.gov/35971401/ and https://pubs.rsna.org/doi /10.1148/radiol.230743.

33 Almamlouk R, et al, "COVID-19-Associated cardiac pathology at the postmortem evaluation: a collaborative systematic review," Clin Microbiol Infect, 28(8):1066–1075, Aug. 2022. https://www.ncbi.nlm.nih.gov/pmc/articles/PMC8941843/.

34 Praneel J, et al, "Increased heart rate response to exercise following Pfizer COVID-19 vaccination with no change in cardiac output or stroke volume," *Journal of Applied Physiology*, 133:4, 985–985, 2022. https://journals.physiology.org/doi/full/10.1152/japplphysiol.00281.2022.
35 "Survey shows fans vary in how they think Kyrie Irving, Aaron Rodgers, Novak Djokovic handled COVID-19 vaccination status," ESPN, Jan. 28, 2022. https://www.espn.com/espn/story/_/id/33168664/survey-shows-fans-vary-how-think-kyrie-irving-aaron-rodgers-novak-djokovic-handled-covid-19-vaccination-status and https://www.foxnews.com/sports/novak-djokovic-kyrie-irving-aaron-rodgers-modern-day-heirs-muhammad-ali-clay-travis and https://slate.com/news-and-politics/2021/06/unvaccinated-cole-beasley-blasts-nfl-covid-protocols.html.
36 Seneff, S., et al, "Worse Than the Disease? Reviewing Some Possible Unintended Consequences of the mRNA Vaccines Against COVID-19," *International Journal of Vaccine Theory, Practice, and Research*, 2(1), 38–79. 2021. https://ijvtpr.com/index.php/IJVTPR/article/view/23.
37 Dowd E, *"Cause Unknown," The Epidemic of Sudden Deaths in 2021 & 2022* (New York, Skyhorse, 2022). https://www.theyliedpeopledied.com/.
38 Dowd E, "Humanity Projects," Vaccine Damage, Excess Mortality, Disabilities, etc., Phinance Technologies, Mar. 2023. https://phinancetechnologies.com/HumanityProjects/Projects.htm.
39 Marks D, "Former BlackRock Advisor Tells RFK, Jr.: 'FDA Is in on the Cover-Up.'" *The Defender*, Mar. 18, 2022. https://childrenshealthdefense.org/defender/chd-tv-rfk-jr-defender-blackrock-edward-dowd-fda-cover-up/ and https://totalityofevidence.com/edward-dowd/.
40 Dowd E, "Pfizer Fraud and Wall Street," Mar.24, 2022. https://podcasts.apple.com/us/podcast/pfizer-fraud-and-wall-street-with-ed-dowd/id1552000243?i=1000554431124.
41 "Group Life COVID-19 Mortality Survey - Updated through September 2021," Society of Actuaries Group Life COVID-19 Mortality Survey, See table 5.7 for accelerations in mortality coincident with increased rates of COVID vaccinations for ages 25–64 in Q3 2021. https://www.soa.org/resources/experience-studies/2022/group-life-covid-19-mortality/.
42 Steyn M, "The Steyn Line," Beginning at timestamp 7:50 to 22:55 for U.K. Health Security Organization data analysis and call for a Royal Commission on COVID vaccine policy, Apr. 21, 2023. https://youtu.be/a8kdH2Xgf-k and U.K. Office for National Statistics, Deaths in England and Wales. https://www.ons.gov.uk/peoplepopulationandcommunity/birthsdeathsandmarriages/deaths/bulletins/deathsregisteredweeklyinenglandandwalesprovisional/weekending30december2022#main-points.
43 United Kingdom Statistics Authority for Parliament, Deaths by vaccination status, England. https://www.ons.gov.uk/peoplepopulationandcommunity/birthsdeathsandmarriages/deaths/datasets/deathsbyvaccinationstatusengland and analysis at https://childrenshealthdefense.org/defender/mortality-rates-rise-covid-vaccines-cola/.
44 Provincial Respiratory Surveillance Report. https://www.gov.mb.ca/health/publichealth/surveillance/covid-19/2022/week_30/index.html and https://correlation-canada.org/wp-content/uploads/2023/09/2023–09-17-Correlation-Covid-vaccine-mortality-Southern-Hemisphere-cor.pdf.

45 CDC website. https://www.cdc.gov/coronavirus/2019-ncov/vaccines/effectiveness/why-measure-effectiveness/breakthrough-cases.html.

46 CDC website. https://www.cdc.gov/vaccines/covid-19/health-departments/breakthrough-cases.html.

47 Crout R, et al, "History of the FDA," FDA website, Nov. 12, 1997. https://www.fda.gov/media/116721/download and https://www.ecfr.gov/current/title-21/chapter-I/subchapter-C/part-201/subpart-F.

48 Barbaro M, "An interview with Dr. Anthony Fauci," *New York Times*, Nov. 12, 2021. https://www.nytimes.com/2021/11/12/podcasts/the-daily/anthony-fauci-vaccine-mandates-booster-shots.html.

49 Goldberg Y, et al, "Protection and waning of natural and hybrid COVID-19 immunity," MedRxiv, Dec. 4, 2021. https://doi.org/10.1101/2021.12.04.21267114.

50 CDC definitions. https://www.cdc.gov/vaccines/vac-gen/imz-basics.htm.

51 Department of Health & Human Services and BARDA Portal at MedicalCountermeasures.gov. https://www.medicalcountermeasures.gov/newsroom/2020/pfizer-1/ and "What are Medical Countermeasures?," FDA. https://www.fda.gov/emergency-preparedness-and-response/about-mcmi/what-are-medical-countermeasures.

52 Pfizer-BioNTech COVID-19 Vaccine EUA Letter of Authorization reissued 02–25-2021. https://www.fda.gov/media/144412/download.

53 "FDA Emergency Use Authorization," FDA. https://www.fda.gov/emergency-preparedness-and-response/mcm-legal-regulatory-and-policy-framework/emergency-use-authorization and https://www.fda.gov/medical-devices/coronavirus-disease-2019-covid-19-emergency-use-authorizations-medical-devices/faqs-emergency-use-authorizations-euas-medical-devices-during-covid-19-pandemic.

54 Johnson R, et al, US Senate letter to FDA Commissioner Hahn, Aug. 18, 2020. https://www.hsgac.senate.gov/imo/media/doc/2020–08-18%20RHJ%20Letter%20to%20FDA%20on%20HCQ%20+%20CQ.pdf.

55 Efimenko I, et al, "Treatment with Ivermectin Is Associated with Decreased Mortality in COVID-19 Patients," *International Journal of Infectious Diseases*, Volume 116, Supplement, Page S40, 2022. https://www.sciencedirect.com/science/article/pii/S1201971221009887

56 Parvez S, et al, "Insights from a computational analysis of the SARS-CoV-2 Omicron variant: Host-pathogen interaction, pathogenicity and possible therapeutics," Jan. 20, 2022. https://doi.org/10.48550/arXiv.2201.08176.

57 "Court revives doctors' lawsuit saying FDA overstepped its authority with anti-ivermectin campaign," AP, Sept. 1, 2023. https://apnews.com/article/coronavirus-ivermectin-fda-doctors-lawsuit-bbc8d4fc726c08940ae4b0dad70170e0.

58 Mary Talley Bowden v. US Dept of Health and Human Services, et al., Case No. 3:22-cv-184, March 21, 2024, https://www.documentcloud.org/documents/24495557-fda-stipulation-of-dismissal.

59 Ahmed S, et al, "A five-day course of ivermectin for the treatment of COVID-19 may reduce the duration of illness," *Int J Infect Dis*, Feb. 2021. https://pubmed.ncbi.nlm.nih.gov/33278625/ and https://pubmed.ncbi.nlm.nih.gov/33038449/ and https://pubmed.ncbi.nlm.nih.gov/33341233/ and https://pubmed.ncbi.nlm.nih.gov/33864232/.

60 "Peter McCullough, MD testifies to Texas Senate HHS Committee," Mar. 10, 2021. https://www.youtube.com/watch?v=QAHi3lX3oGM and https://www.youtube.com/watch?v=QAHi3lX3oGM and Modern Medicine's Great Controversy explaining the

"grand deception," Aug. 20, 2023. https://www.youtube.com/watch?v=rG38_53SEbU and Parry PI, et al, 'Spikeopathy': COVID-19 Spike Protein Is Pathogenic, from Both Virus and Vaccine mRNA, Biomedicines, Aug. 17, 2023. https://www.ncbi.nlm.nih.gov/pmc/articles/PMC10452662/.

61 Mead M, et al, "COVID-19 mRNA Vaccines: Lessons Learned from the Registrational Trials and Global Vaccination Campaign," Cureus, Jan. 2024. https://pubmed.ncbi.nlm.nih.gov/38274635/ and https://www.cureus.com/articles/203052-covid-19-mrna-vaccines-lessons-learned-from-the-registrational-trials-and-global-vaccination-campaign#!/ and https://img1.wsimg.com/blobby/go/4fa7f468-a250–4088-926e-3c56a998df1f/downloads/Mead%20et%20all%20Cureus%20Moratorium.pdf?ver=1708371037707.

62 Iskander J, et al, "Monitoring the safety of annual and pandemic influenza vaccines: lessons from the US experience," Swine influenza vaccine Guillain–Barré syndrome (GBS) complications, Expert Review of Vaccines, 7:1, 75–82, Jan. 9, 2014. https://www.tandfonline.com/doi/abs/10.1586/14760584.7.1.75.

63 "Dramatic Testimony in Military Shot Mandate Case," Green Beret Flight Surgeon Lt Col Peter Chambers, Liberty Counsel, Mar. 11, 2022. https://lc.org/newsroom/details/031122-dramatic-testimony-on-military-shot-mandate-case-1.

64 Sigoloff S, "After Hours with Dr. Sigoloff," https://podcasts.apple.com/us/podcast/after-hours-with-dr-sigoloff/id1601073627 and "Army surgeon says military ignored warnings," Nov. 5, 2021. https://americanmilitarynews.com/2021/11/army-surgeon-says-military-ignored-her-warnings-over-covid-19-vaccine-injuries/.

65 Army LtCol and Dr. Theresa Long testimony, Sept. 17, 2022. https://rumble.com/v1lbl6l-army-ltcol-theresa-long-md-full-testimony.html.

66 Ogunjimi OB, et al, "Guillain-Barré Syndrome Induced by Vaccination Against COVID-19: A Systematic Review and Meta-Analysis," Cureus, Apr. 14, 2023. https://pubmed.ncbi.nlm.nih.gov/37193456/.

67 Neustadt R, et al, *The Swine Flu Affair: Decision-Making on a Slippery Disease* (Washington, DC, National Academies Press, 1978). https://pubmed.ncbi.nlm.nih.gov/25032342/.

68 Sencer D, et al, "Reflections on the 1976 Swine Flu Vaccination Program," Emerging Infectious Diseases, 12(1), 29–33, 2006. https://wwwnc.cdc.gov/eid/article/12/1/05–1007_article.

69 Rep. Panetta requested waiver of informed consent. https://www.documentcloud.org/documents/20521870-panetta_dod-covid-vaccine-waiver.

70 10 USC §1107a. https://www.govinfo.gov/app/details/USCODE-2010-title10/USCODE-2010-title10-subtitleA-partII-chap55-sec1107a/summary.

71 21 USC §360bbb-3: Authorization for medical products for use in emergencies. https://www.govinfo.gov/content/pkg/USCODE-2010-title21/pdf/USCODE-2010-title21-chap9-subchapV-partE-sec360bbb-3.pdf.

72 "FDA EUA of Medical Products and Related Authorities, Guidance for Industry and Stakeholders," FDA, Jan. 2017. https://www.fda.gov/regulatory-information/search-fda-guidance-documents/emergency-use-authorization-medical-products-and-related-authorities.

73 21 CFR §50.25, Informed Consent of Human Subjects. https://www.accessdata.fda.gov/scripts/cdrh/cfdocs/cfcfr/cfrsearch.cfm?fr=50.25.

74 45 CFR §46, Protection of Human Subjects, https://www.ecfr.gov/current/title-45/subtitle-A/subchapter-A/part-46.

75 Inaugural EUA for anthrax vaccine, Federal Register, Jan. 27, 2005. https://www.federal register.gov/documents/2005/02/02/05–2028/authorization-of-emergency-use-of-anthrax -vaccine-adsorbed-for-prevention-of-inhalation-anthrax-by.

76 Public Law 114–255, 21st Century Cures Act, Sec 3024, Informed Consent Waiver, Dec. 13, 2016. INVESThttps://www.govinfo.gov/content/pkg/PLAW-114publ255/pdf/PLAW -114publ255.pdf.

77 Inaugural EUA for anthrax vaccine, Federal Register, Jan. 27, 2005. https://www.federal register.gov/documents/2005/02/02/05–2028/authorization-of-emergency-use-of-anthrax -vaccine-adsorbed-for-prevention-of-inhalation-anthrax-by.

78 DoJ Acting Assistant Attorney General, Office of Legal Counsel, Jul. 6, 2021. https://www .justice.gov/olc/file/1415446/download.

79 Bloomberg Law Court Opinions, *Jacobson v. Massachusetts*, 197 U.S. 11, 25 S. Ct. 358, 49 L. Ed. 643, Feb. 20, 1905. https://www.bloomberglaw.com/public/desktop/document /JacobsonvMassachusetts197US1125SCt35849LEd6431905CourtOpinion/.

80 Bridges J, et al, vs. Houston Methodist Hospital, case dismissal, Jun. 21, 2021. https: //www.documentcloud.org/documents/20860669-houston-methodist-lawsuit-order-of -dismissal.

81 Doctrine of Preemption. https://www.law.cornell.edu/wex/preemption.

82 US Constitution, Art. VI., § 2, "This Constitution, and the Laws of the United States which shall be made in Pursuance thereof; and all Treaties made, or which shall be made, under the Authority of the United States, shall be the supreme Law of the Land; and the Judges in every State shall be bound thereby . . . " https://www.archives.gov/founding-docs /constitution-transcript.

83 Manchester J, "Biden: Coronavirus vaccine should not be mandatory," *The Hill*, Dec. 4, 2020. https://thehill.com/homenews/campaign/528834-biden-coronavirus-vaccine-should -not-be-mandatory.

84 Federal Food, Drug, and Cosmetic Act, Authorization for medical products use in emergencies, 21 USC §360bbb-3. https://uscode.house.gov/view.xhtml?hl=false&edition=p relim&req=granuleid%3AUSC-prelim-title21-section360bbb-3&num=0&saved=%7CZ3J hbnVsZWlkOlVTQy1wcmVsaW0tdGl0bGUyMS1zZWN0aW9uMzYwYmJiLT Nh%7C%7C%7C0%7Cfalse%7Cprelim.

85 FDA mission statement. https://www.fda.gov/about-fda/what-we-do#mission.

86 FDA website. https://www.fda.gov/about-fda/fda-history-exhibits/80-years-federal-food -drug-and-cosmetic-act.

87 FDA website. https://www.fda.gov/consumers/consumer-updates/kefauver-harris -amendments-revolutionized-drug-development and https://www.fda.gov/about-fda /histories-product-regulation/promoting-safe-effective-drugs-100-years.

88 Rempfer T, Colonel, USAF, retired, Interview and Op-Ed discussing the disturbing trend of decreased drug authorization standards, in support of AFLDS' White Coat Summit organized by medical freedom advocate Dr. Simone Gold, Jul. 27, 2024. https://www .whitecoatsummit.com/videos/2023 and https://americasfrontlinedoctors.org/videos/post /the-eua-anthrax-and-covid-vaccines-by-colonel-tom-rempfer and https://coloradofreepress .com/adulteration-of-vaccine-safety-and-efficacy/.

89 Biologics Review Preamble, 37 FR 16679. https://archives.federalregister.gov/issue_slice /1972/8/18/16678–16682.pdf#page=2.

90 Kennedy RF Jr., *The Real Anthony Fauci: Bill Gates, Big Pharma, and the Global War on Democracy and Public Health* (New York, Skyhorse, 2021). https://www.skyhorsepublishing.com/9781510766808/the-real-anthony-fauci/.

91 Kennedy RF Jr., *The Wuhan Cover-up And the Terrifying Bioweapons Arms Race* (New York, Skyhorse, 2023). https://www.skyhorsepublishing.com/9781510773981/the-wuhan-cover-up/.

92 Keith T, "RFK Jr. says he's running for president as an independent," NPR, Oct. 9, 2023. https://www.npr.org/2023/10/09/1204681962/rfk-jr-independent and https://www.kennedy24.com/.

Mile 22: Natural Immunity? (2021)

1 Reese C, "Germ theory vs terrain theory," Nutritionist Resource, Sept. 27, 2022. https://www.nutritionist-resource.org.uk/memberarticles/germ-theory-vs-terrain-theory-in-relation-to-the-coronavirus.

2 Novavax vaccine authorization news. https://www.forbes.com/sites/carlieporterfield/2021/12/17/who-grants-novavax-covid-vaccine-emergency-use-approval-as-omicron-fears-deepen and https://www.cnbc.com/2022/07/13/fda-authorizes-novavax-covid-vaccine-for-adults-.html.

3 Di Chiara C, et al, "Long-term Immune Response to SARS-CoV-2 Infection Among Children and Adults After Mild Infection," *JAMA*, Jul. 1, 2022. https://pubmed.ncbi.nlm.nih.gov/35816313/.

4 14 CFR §61.53. https://www.ecfr.gov/current/title-14/chapter-I/subchapter-D/part-61/subpart-A/section-61.53.

5 FAA Guide for Aviation Medical Examiners (GAME), Do Not Issue—Do Not Fly Guidance to FAA AME's (Aviation Medical Examiners), Dec. 28, 2022. https://www.faa.gov/about/office_org/headquarters_offices/avs/offices/aam/ame/guide/pharm/dni_dnf/ and https://www.faa.gov/about/office_org/headquarters_offices/avs/offices/aam/ame/guide/media/AME_GUIDE.pdf.

6 FAA GAME for AME's incorrectly implies all COVID vaccines are FDA approved, pg. 407, Dec. 28, 2022. https://www.faa.gov/about/office_org/headquarters_offices/avs/offices/aam/ame/guide/media/AME_GUIDE.pdf.

7 Kojima N, et al, "Protective immunity after recovery from SARS-CoV-2 infection," *The Lancet, Infectious Diseases*, Volume 22, issue 1, P12–14, Jan. 1, 2022. https://www.thelancet.com/journals/laninf/article/PIIS1473–3099(21)00676–9/fulltext and https://doi.org/10.1016/S1473–3099(21)00676–9.

8 CDC's V-safe After Vaccination Health Checker. https://www.cdc.gov/coronavirus/2019-ncov/vaccines/safety/vsafe.html.

9 CDC's raw v-safe data obtained by the Informed Consent Action Network (ICAN). https://www.icandecide.org/v-safe-data/.

10 ICAN website. https://www.icandecide.org/ican_press/ican-eviscerates-cdc-in-formal-exchange-regarding-natural-immunity/.

11 Alexander P, "160 Plus Research Studies Affirm Naturally Acquired Immunity to Covid-19," Brownstone Institute, Oct. 17, 2021. https://brownstone.org/articles/research-studies-affirm-naturally-acquired-immunity/ and https://brownstone.org/articles/what-is-medical-freedom-exactly/.

12 Flynn P, "Natural Immunity—140 Studies of Validation," The Wellness Way, Dec. 20, 2021. https://www.thewellnessway.com/natural-immunity-140-studies-of-validation/.

13 Association of American Physicians and Surgeons. https://aapsonline.org/.

14 CDC website. https://www.cdc.gov/coronavirus/2019-ncov/cases-updates/burden.html and https://www.worldometers.info/coronavirus/country/us/.

15 CDC reported COVID-19 Vaccinations in the US. https://covid.cdc.gov/covid-data-tracker/#vaccinations_vacc-total-admin-rate-total.

16 Makary M, "Natural immunity to covid is powerful, policymakers seem afraid to say so," *Washington Post* Opinion, Sept. 15, 2021. https://www.washingtonpost.com/outlook/2021/09/15/natural-immunity-vaccine-mandate/.

17 Servellita V, et al, "Predominance of antibody-resistant SARS-CoV-2 variants in vaccine breakthrough cases from the San Francisco Bay Area, California," Nat Microbiol 7, 277–288, Jan. 10, 2022. https://www.medrxiv.org/content/10.1101/2021.08.19.21262139v1.

18 WHO Statement on COVID-19 vaccines. Jan. 11, 2022. https://www.who.int/news/item/11–01-2022-interim-statement-on-covid-19-vaccines-in-the-context-of-the-circulation-of-the-omicron-sars-cov-2-variant-from-the-who-technical-advisory-group-on-covid-19-vaccine-composition.

19 Gøtzsche P, et al, "Serious harms of the COVID-19 vaccines: a systematic review," CSH, BMJ, Yale COVID-19 SARS-CoV-2 preprints from medRxiv, Mar. 22, 2023. https://www.medrxiv.org/content/10.1101/2022.12.06.22283145v2.

20 Eythorsson E, et al, "Rate of SARS-CoV-2 Reinfection During an Omicron Wave in Iceland," *JAMA* Network Open Research Letter Infectious Diseases, Aug. 3, 2022. https://jamanetwork.com/journals/jamanetworkopen/article-abstract/2794886.

21 Yamamoto K, "Adverse effects of COVID-19 vaccines," Virology J., Jun. 5, 2022. https://www.ncbi.nlm.nih.gov/pmc/articles/PMC9167431/.

22 Liu L, et al, "Striking antibody evasion manifested by the Omicron variant of SARS-CoV-2," *Nature*, 602, 676–681, 2022. https://www.nature.com/articles/s41586–021-04388–0 and https://www.cuimc.columbia.edu/news/new-study-adds-more-evidence-omicron-immune-evasion and https://doi.org/10.1016/S2666–7568(21)00276–2.

23 Seneff S, et al, "Innate immune suppression by SARS-CoV-2 mRNA vaccinations: The role of G-quadruplexes, exosomes, and MicroRNAs," *Food and Chemical Toxicology*, Volume 164, Jun. 2022. https://pubmed.ncbi.nlm.nih.gov/35436552/.

24 Gao FX, et al, "Extended SARS-CoV-2 RBD booster vaccination induces humoral and cellular immune tolerance in mice," *Science*, 22;25(12):105479, Dec. 22, 2022. https://pubmed.ncbi.nlm.nih.gov/36338436/.

25 Wadman M, "Having SARS-CoV-2 once confers much greater immunity than a vaccine—but vaccination remains vital," *Science*, Vol 373, Issue 6559. https://www.science.org/content/article/having-sars-cov-2-once-confers-much-greater-immunity-vaccine-vaccination-remains-vital.

26 Gazit S, et al, "Comparing SARS-CoV-2 natural immunity to vaccine-induced immunity: reinfections versus breakthrough infections," *Clinical Infectious Diseases*, Apr. 5, 2022. https://academic.oup.com/cid/article/75/1/e545/6563799.

27 Altarawneh H, et al, "Effects of Previous Infection and Vaccination on Symptomatic Omicron Infections," *New England Journal of Medicine*, Jun. 28, 2022. https://www.nejm.org/doi/full/10.1056/NEJMoa2203965.

28 "Town Hall With President Joe Biden," CNN, Jul. 21, 2021. https://transcripts.cnn.com/show/se/date/2021–07-21/segment/01.

29 White House transcript, Jul. 22, 2021. https://www.whitehouse.gov/briefing-room/speeches-remarks/2021/07/22/remarks-by-president-biden-in-a-cnn-town-hall-with-don-lemon/.

30 "Joe Biden Announces Six-Point Plan to Fight Pandemic amid Delta Surge and Stalled Vaccination Rate," Interview of Sanjay Gupta with Anthony Fauci on Anderson Cooper 360 Degrees, Sept. 9, 2021. https://transcripts.cnn.com/show/acd/date/2021–09-09/segment/01.

31 "Dr. Fauci Best Vaccination Is Infection," Dr. Fauci on C-SPAN, Oct. 11, 2004. https://www.c-span.org/video/?c5009217/user-clip-dr-fauci-vaccination-infection.

32 White House transcript, Sept. 9, 2021. https://www.whitehouse.gov/briefing-room/speeches-remarks/2021/09/09/remarks-by-president-biden-on-fighting-the-covid-19-pandemic-3/.

33 Günter K, "COVID-19: stigmatising the unvaccinated is not justified," *The Lancet*, Volume 398, Issue 10314, 1871, Nov. 20, 2021. https://www.thelancet.com/journals/lancet/article/PIIS0140–6736(21)02243–1/fulltext.

34 Shrestha NK, et al, "Necessity of COVID-19 vaccination in previously infected individuals," *Clinical Infectious Diseases*, Jul. 1, 2022. https://academic.oup.com/cid/article/75/1/e662/6507165.

35 Mathioudakis A, et al, "Self-Reported Real-World Safety and Reactogenicity of COVID-19 Vaccines: A Vaccine Recipient Survey," Life (Basel), 11(3):249, Mar. 17, 2021. https://pubmed.ncbi.nlm.nih.gov/33803014/.

36 Shrestha NK, et al, "Risk of Coronavirus Disease 2019 (COVID-19) among Those Up-to-Date and Not Up-to-Date on COVID-19 Vaccination," Plos One, Nov. 8, 2023. https://journals.plos.org/plosone/article?id=10.1371/journal.pone.0293449.

37 Shrestha NK, et al, "Effectiveness of the Coronavirus Disease 2019 (COVID-19) Bivalent Vaccine," Clinical *Infectious Diseases*, Jun. 2023. https://academic.oup.com/ofid/article/10/6/ofad209/7131292.

38 Skydsgaard N, et al, "Iceland to lift all COVID-19 restrictions," Reuters, Feb. 23, 2022. https://www.reuters.com/business/healthcare-pharmaceuticals/iceland-lift-all-covid-19-restrictions-friday-media-reports-2022–02-23/ and https://www.nbcnews.com/health/health-news/natural-immunity-protective-covid-vaccine-severe-illness-rcna71027 and https://www.thelancet.com/journals/lancet/article/PIIS0140–6736(22)02465–5/fulltext.

39 Danish Health Authority, Vaccination recommendation for covid-19. https://www.sst.dk/en/english/corona-eng/vaccination-against-covid-19.

Mile 23: Federal Mandates (2021 to 2023)

1 "Update on Implementation of COVID-19 Vaccination Requirement for Federal Employees," The White House, Dec. 9, 2021. https://www.whitehouse.gov/briefing-room/presidential-actions/2021/09/09/executive-order-on-ensuring-adequate-covid-safety-protocols-for-federal-contractors/, and https://www.whitehouse.gov/covidplan/.

2 "Statement on the Status of the OSHA COVID-19 Vaccination and Testing ETS," Jan. 25, 2022. https://www.osha.gov/coronavirus/ets2.

3 "J&J COVID vaccine factory forced to trash even more doses," AP, Aug. 11, 2022. https://apnews.com/article/covid-science-health-congress-2fbc5e039df0e5b1d581927ea11adc5b

and https://www.reuters.com/business/healthcare-pharmaceuticals/us-fda-finds-control-lapses-moderna-manufacturing-plant-2023–12-15/.

4 Emergent BioSolutions COVID vaccine series by Pulitzer Prize winning author Christopher Hamby, *New York Times*. https://www.nytimes.com/topic/company/emergent-biosolutions-inc and https://www.nytimes.com/2021/03/06/us/emergent-biosolutions-anthrax-coronavirus.html and https://www.nytimes.com/topic/company/emergent-biosolutions-inc.

5 Tebor C, "FDA restricts use of Johnson & Johnson COVID vaccine due to blood clot risk," *USA Today*, May 5, 2022. https://www.usatoday.com/story/news/health/2022/05/05/fda-covid-vaccine-johnson-johnson-restriction-blood-clot-risk/9667674002/ and https://www.usatoday.com/story/news/health/2021/04/13/covid-us-recommends-pause-j-j-vaccine-after-reports-blood-clots/7200817002/.

6 "Janssen (Johnson & Johnson) COVID-19 Vaccine," CDC, May 7, 2023. https://archive.cdc.gov/#/details?url=https://www.cdc.gov/vaccines/covid-19/info-by-product/janssen/index.html and https://oversightdemocrats.house.gov/news/press-releases/chairs-clyburn-maloney-reveal-that-quality-failures-by-emergent-biosolutions.

7 42 USC 247d-6d: Targeted liability protections for pandemic and epidemic products and security countermeasures. https://uscode.house.gov/view.xhtml?req=granuleid:USC-2010-title42-section247d-6d&num=0&edition=2010.

8 "No FDA-Approved COVID-19 Shot Is Available," Liberty Counsel, Jan. 31, 2022. https://lc.org/newsroom/details/013122-no-fdaapproved-covid19-shot-is-available-1.

9 Pfizer-BioNTech Regulatory Information and Letter of Authorization (Reissued), Apr. 23, 2023. https://www.fda.gov/media/150386/download.

10 "Remarks by President Biden on Fighting the COVID-19 Pandemic," The White House, Sept. 9, 2021. https://www.whitehouse.gov/briefing-room/speeches-remarks/2021/09/09/remarks-by-president-biden-on-fighting-the-covid-19-pandemic-3/.

11 "Remarks by President Biden After Meeting with Members of the COVID-19 Response Team," The White House, Dec. 16, 2021. https://www.whitehouse.gov/briefing-room/speeches-remarks/2021/12/16/remarks-by-president-biden-after-meeting-with-members-of-the-covid-19-response-team/ and https://www.whitehouse.gov/briefing-room/press-briefings/2021/12/17/press-briefing-by-white-house-covid-19-response-team-and-public-health-officials-74/.

12 President Biden's State of the Union Address, Mar. 1, 2022. https://www.whitehouse.gov/state-of-the-union-2022/.

13 Remarks by President Biden on Fighting the COVID-19 Pandemic, Sept. 9, 2021. https://www.whitehouse.gov/briefing-room/speeches-remarks/2021/09/09/remarks-by-president-biden-on-fighting-the-covid-19-pandemic-3/.

14 Declaration of Independence, Jul. 4, 1776. https://www.archives.gov/founding-docs/declaration-transcript.

15 Bill of Rights, First Amendment, Dec. 15, 1791. https://www.archives.gov/founding-docs/bill-of-rights-transcript#toc-amendment-i and https://www.archives.gov/founding-docs/bill-of-rights and https://www.archives.gov/founding-docs/bill-of-rights-transcript#toc-the-preamble-to-the-bill-of-rights.

16 Constitution, Fifth and Fourteenth Amendments. https://www.archives.gov/founding-docs/bill-of-rights-transcript#toc-amendment-v and https://www.archives.gov/milestone-documents/14th-amendment.

17 5th Circuit Court of Appeals Stay. https://www.ca5.uscourts.gov/opinions/pub/21/21–60845-CV0.pdf.

18 5th Circuit Court of Appeals Stay, pg. 12, citing 54 Fed. Reg. 23,042, 23,045, May 30, 1989. https://www.ca5.uscourts.gov/opinions/pub/21/21–60845-CV0.pdf.

19 "Supreme Court Justices Hear Case on Vaccine Mandate," C-SPAN, Jan. 7, 2022. https://www.c-span.org/video/?516919–1/justices-hear-case-vaccine-mandate-health-care-workers.

20 SCOTUS, *National Federation of Independent Business, et al, v. Dept of Labor, OSHA, et al,* Case No. 21A244. https://www.supremecourt.gov/oral_arguments/argument_transcripts/2021/21a244_kifl.pdf.

21 Chen J, et al, "Omicron Variant (B.1.1.529): Infectivity, Vaccine Breakthrough, and Antibody Resistance," J. Chem. Inf. Model, 62, 2, 412–422, Jan. 6, 2022. https://pubs.acs.org/doi/10.1021/acs.jcim.1c01451.

22 Willett B, et al, "The hyper-transmissible SARS-CoV-2 Omicron variant exhibits significant antigenic change, vaccine escape and a switch in cell entry mechanism," medRxiv, Jan. 03, 2022. https://www.semanticscholar.org/paper/The-hyper-transmissible-SARS-CoV-2-Omicron-variant-Willett-Grove/87e3e2c7749c1ce1333612a1c57b4af0c96a12ea.

23 SCOTUS ruling on OSHA mandate, Jan. 13, 2022. https://www.supremecourt.gov/opinions/21pdf/21a244_hgci.pdf.

24 Reed A, "Covid Jab Mandate Foes Cite 'Major Questions' to Keep Case Alive," *Bloomberg Law*, Nov. 18, 2022. https://news.bloomberglaw.com/health-law-and-business/covid-jab-mandate-foes-cite-major-questions-to-keep-case-alive.

25 Interim final rule; withdrawal of COVID–19 Vaccination and Testing; Emergency Temporary Standard, Docket No. OSHA-2020–0007, Federal Register. https://public-inspection.federalregister.gov/2022–01532.pdf.

26 Brauner L, "Federal Covid-19 Updates: OSHA Withdraws Vaccination-or-Testing Requirement After Supreme Court Defeat, CMS Vaccine Mandate Stands, And Federal Contractor Mandate Halted," Mondaq, Schnader Harrison Segal & Lewis LLP, Mar. 3, 2022. https://www.mondaq.com/unitedstates/health—safety/1167964/federal-covid-19-updates-osha-withdraws-vaccination-or-testing-requirement-after-supreme-court-defeat-cms-vaccine-mandate-stands-and-federal-contractor-mandate-halted.

27 OSHA withdrawal of COVID-19 Vaccination and Testing; Emergency Temporary Standard, Jan. 26, 2022. https://www.federalregister.gov/documents/2022/01/26/2022–01532/covid-19-vaccination-and-testing-emergency-temporary-standard.

28 EO 14043, Executive Order for Federal Employees, Sept. 9, 2021. https://www.whitehouse.gov/briefing-room/presidential-actions/2021/09/09/executive-order-on-requiring-coronavirus-disease-2019-vaccination-for-federal-employees/.

29 Feds for Medical Freedom, et al, Plaintiffs v. Joseph R. Biden, Jr, et al, Defendants. https://www.documentcloud.org/documents/21183444-judge-jeffrey-brown-injunction-against-fed-employee-vaccine-mandate.

30 Myers M, "DoD is pausing civilian COVID vaccine mandate after court ruling," *Military Times*, Jan. 27, 2022. https://www.militarytimes.com/news/pentagon-congress/2022/01/27/dod-is-pausing-civilian-covid-vaccine-mandate-after-court-ruling/.

31 Sneed T, "Appeals court refuses to reinstate federal employee vaccine mandate while it reviews case," CNN, Feb. 9, 2022. https://www.cnn.com/2022/02/09/politics/appeals-court-federal-employee-vaccine-mandate/index.html.

32 Shepardson D, et al, "US court reinstates Biden federal employee COVID vaccine mandate," Reuters, Apr. 7, 2022. https://www.reuters.com/world/us/us-court-reinstates-biden-federal-employee-covid-vaccine-mandate-2022–04-07/.

33 Durkee A, "Biden Vaccine Mandate For Federal Employees Blocked Again As Appeals Court Dissolves Earlier Ruling," *Forbes*, Jun. 27, 2022. https://www.forbes.com/sites/alisondurkee/2022/06/27/biden-vaccine-mandate-for-federal-employees-blocked-again-as-appeals-court-dissolves-earlier-ruling/ and https://aboutblaw.com/3FE and https://www.ca5.uscourts.gov/opinions/pub/22/22–40043-CV3.pdf.

34 Smith Z, "Federal Vaccine Mandate On Hold? White House Reportedly Delays Enforcement," *Forbes*, Apr. 8, 2022. https://www.forbes.com/sites/zacharysmith/2022/04/08/federal-vaccine-mandate-on-hold-white-house-reportedly-delays-enforcement/.

35 Morris K, "States file lawsuit against Biden admin to halt vaccine mandate for federal contractors; The lawsuit alleges the vaccine mandate on federal workers is unconstitutional," Fox News, Oct. 29, 2021. https://www.foxnews.com/politics/10-states-file-lawsuit-against-biden-admin-to-halt-vaccine-mandate-for-fed-contractors.

36 EO 14042, Executive Order for Federal Contractors, Sept. 9, 2021. https://www.whitehouse.gov/briefing-room/presidential-actions/2021/09/09/executive-order-on-ensuring-adequate-covid-safety-protocols-for-federal-contractors/.

37 Georgia v. Biden, Southern District of Georgia., No. 21-cv-00163. https://aboutblaw.com/0LV and https://news.bloomberglaw.com/health-law-and-business/biden-vaccine-mandate-for-federal-contractors-blocked-nationwide and https://news.bloomberglaw.com/daily-labor-report/contractors-in-uncharted-water-with-narrowed-vaccine-mandate.

38 *Kentucky v. Biden, Georgia v. Biden, Louisiana v. Biden, Missouri v. Biden, Florida v. Nelson* (NASA Administrator), *Brnovich v. Biden*, all enjoined the federal contractor mandate, Aug. 5, 2022. https://www.jdsupra.com/legalnews/whatever-happened-to-that-federal-4699914/ and https://www.jdsupra.com/legalnews/federal-contractor-vaccine-mandate-6802648/.

39 *US Court of Appeals for the Sixth Circuit, KY v. Joseph R. Biden*, Jan. 5, 2022. https://www.opn.ca6.uscourts.gov/opinions.pdf/22a0002p-06.pdf and Courthouse News Service, Sixth Circuit upholds block on vaccine mandate for federal contractors in three states, Jan. 12, 2023. https://www.courthousenews.com/sixth-circuit-upholds-block-on-vaccine-mandate-for-federal-contractors-in-three-states/.

40 Fifth Circuit Court of Appeals, Excerpt: "Congress has not authorized the issuance of this mandate, whether the President may nonetheless exercise this power. We hold that he may not. Accordingly, we AFFIRM the district court's grant of an injunction," Dec. 19, 2022. https://www.ca5.uscourts.gov/opinions/pub/22/22–30019-CV0.pdf.

41 Safer Federal Workforce Task Force update on Federal Contractor mandate, Sept. 2022. https://www.saferfederalworkforce.gov/contractors/.

42 *Doe v. Rumsfeld* fee ruling, Aug. 21, 2007.

43 Acting Assistant Attorney General, Office of Legal Counsel, Jul. 6, 2021. https://www.justice.gov/olc/file/1415446/download and https://www.justice.gov/sites/default/files/opinions/attachments/2021/07/26/2021–07-06-mand-vax.pdf.

44 "FDA Approval of the Pfizer-BioNTech COVID-19 Vaccine: Frequently Asked Questions," Congressional Research Service, R46913, Sept. 29, 2021. https://crsreports.congress.gov/product/pdf/R/R46913.

45 "Simple Search Results for: Comirnaty," FDA Purple Book, confirming no biosimilar and no interchangeable data for Comirnaty, excerpt: "Biosimilar(s) No biosimilar data at this time,

Interchangeable(s) No interchangeable data at this time, Reference Product(s) Proprietary Name Comirnaty." https://purplebooksearch.fda.gov/results?query=COVID-19%20Vaccine,%20mRNA&title=Comirnaty.

46 FDA re-issued EUA for Pfizer-BioNTech. https://www.fda.gov/media/150386/download.

47 "DoD anthrax vaccine immunization program," Senate Armed Services Committee, Hearing #106–886, pg. 107, Apr. 13, 2000. https://www.google.com/books/edition/Department_of_Defense_Anthrax_Vaccine_Im/AX-RU_AHMp8C?hl=en&gbpv=1&bsq=per%20se.

48 Secretary of Defense COVID-19 Mandate, Aug. 24, 2021. https://media.defense.gov/2021/Aug/25/2002838826/-1/-1/0/MEMORANDUM-FOR-MANDATORY-CORONAVIRUS-DISEASE-2019-VACCINATION-OF-DEPARTMENT-OF-DEFENSE-SERVICE-MEMBERS.PDF and https://www.defense.gov/News/News-Stories/Article/Article/2746111/secretary-of-defense-mandates-covid-19-vaccinations-for-service-members/ and Aug. 9, 2021. https://www.usafa.edu/app/uploads/MESSAGE-TO-THE-FORCE-MEMO-VACCINE.pdf.

49 *Coker, et al., v. Austin*, Case No. 3:21-cv-01211-AW, May 20, 2022. https://storage.courtlistener.com/recap/gov.uscourts.flnd.409961/gov.uscourts.flnd.409961.88.1.pdf.

50 Pfizer-BioNTech website affirmed re-issued EUAs, excerpt: "Pfizer-BioNTech COVID-19 Vaccine (2023–2024 Formula) has not been approved or licensed by FDA, but has been authorized for emergency use by FDA, under an EUA." https://www.pfizer.com/products/product-detail/pfizer-biontech-covid-19-vaccine.

51 "FDA Labeling Requirements—Misbranding," FDA. https://www.fda.gov/medical-devices/general-device-labeling-requirements/labeling-requirements-misbranding.

52 Defense Health Agency Implementation of Department of Defense (DoD) Coronavirus Disease 2019 (COVID-19) Vaccination Program Implementation, Jun. 16, 2022. https://health.mil/Reference-Center/Policies/2022/06/16/DHA-IPM-20–004.

53 Winkie D, "Oklahoma Guard goes rogue, rejects COVID vaccine mandate after sudden change of command," *Army Times*, Nov. 12, 2021. https://www.armytimes.com/news/your-army/2021/11/12/oklahoma-guard-goes-rogue-rejects-covid-vaccine-mandate-after-sudden-change-of-command/.

54 Tritten T, et al, "Pentagon Says It Can Overrule Oklahoma Guard on Vaccinations," Military.com, Nov. 15, 2021. https://www.military.com/daily-news/2021/11/15/pentagon-says-it-can-overrule-oklahoma-guard-vaccinations.html.

55 Hensch M, "Adjutant General: Vaccine Mandate a 'Threat' to Readiness," NGAUS, Aug. 9, 2022. https://www.ngaus.org/newsroom/adjutant-general-vaccine-mandate-threat-readiness.

56 Texas Governor Abbott Signs COVID Vaccine Freedom Bill, Nov. 10, 2023. https://gov.texas.gov/news/post/governor-abbott-signs-covid-vaccine-freedom-bill-at-governors-mansion and Texas Attorney General Ken Paxton Sues Pfizer for Misrepresenting COVID-19 Vaccine Efficacy and Conspiring to Censor Public Discourse, Nov. 30, 2023. https://www.texasattorneygeneral.gov/news/releases/attorney-general-ken-paxton-sues-pfizer-misrepresenting-covid-19-vaccine-efficacy-and-conspiring and https://www.texasattorneygeneral.gov/sites/default/files/images/press/Pfizer%20Vaccine%20Petition%20Filed.pdf.

57 Florida Department of Health Alert on mRNA COVID-19 Vaccine Safety, Feb. 15, 2023. https://www.floridahealth.gov/newsroom/2023/02/20230215-updated-health-alert.pr.html.

58 State Surgeon General Dr. Joseph A. Ladapo Issues New mRNA COVID-19 Vaccine Guidance, Oct. 7, 2022. https://www.floridahealth.gov/newsroom/2022/10/20220512-guidance-mrna-covid19-vaccine.pr.html.
59 FL Gov DeSantis and Surgeon General Ladapo letter to FDA Commissioner Califf and CDC Director Walensky, May 10, 2023. https://www.floridahealth.gov/_documents/newsroom/press-releases/2023/05/20230510-florida-department-of-health-letter-to-fda-and-cdc.pdf.
60 Governor Ron DeSantis and Joseph A. Ladapo, MD, PhD, issue State Surgeon General Guidance for COVID-19 Boosters, Sept. 13, 2023. https://floridahealthcovid19.gov/wp-content/uploads/2023/09/20230913-booster-guidance-final.pdf.
61 Florida Dept of Health, Florida State Surgeon General Calls for Halt in the Use of COVID-19 mRNA Vaccines, Jan. 3, 2024. https://www.floridahealth.gov/newsroom/2024/01/20240103-halt-use-covid19-mrna-vaccines.pr.html.
62 Chiu S, et al, "Changes of ECG parameters after BNT162b2 vaccine in the senior high school students," Eur J Pediatr, Jan. 5, 2023. https://www.ncbi.nlm.nih.gov/pmc/articles/PMC9813456/.
63 Knudsen B, et al, "COVID-19 vaccine induced myocarditis in young males: A systematic review," Eur J Clin Invest, Dec. 8, 2022. https://onlinelibrary.wiley.com/doi/epdf/10.1111/eci.13947 and Oster ME, et al, Myocarditis Cases Reported After mRNA-Based COVID-19 Vaccination in the US, Dec. 2020 to Aug. 2021, JAMA, 327(4):331–340, Jan. 25, 2022. https://pubmed.ncbi.nlm.nih.gov/35076665/.
64 Guetzkow, J, "CDC Finally Released Its VAERS Safety Monitoring Analyses for COVID Vaccines via FOIA, And now it's clear why they tried to hide them," Daily Clout, Jan. 5, 2023. https://dailyclout.io/cdc-finally-released-its-vaers-safety-monitoring-analyses-for-covid-vaccines-via-foia/ and CDC & FDA Identify Preliminary COVID-19 Vaccine Safety Signal for Persons Aged 65 Years and Older, Jan. 13, 2023. https://www.cdc.gov/coronavirus/2019-ncov/vaccines/safety/bivalent-boosters.html.
65 Berild J, et al, "Analysis of Thromboembolic and Thrombocytopenic Events After the AZD1222, BNT162b2, and MRNA-1273 COVID-19 Vaccines in 3 Nordic Countries," JAMA Netw Open, 2022. https://jamanetwork.com/journals/jamanetworkopen/fullarticle/2793348.
66 Hulscher N, et al, "Delayed Fatal Pulmonary Hemorrhage Following COVID-19 Vaccination: Case Report, Batch Analysis, And Proposed Autopsy Checklist," Preprints, Feb. 20, 2024. https://www.preprints.org/manuscript/202402.1096/v1.
67 Wong K, "'It's Done Just to Break': Troops Speak Out Against COVID Vaccine Mandate," Breitbart, Oct. 18, 2022. https://www.breitbart.com/politics/2022/10/18/its-done-just-to break-troops-speak-out-against-covid-vaccine-mandate/ and "Former Troops Punished over Biden's Vaccine Mandate Sue for Billions in Lost Wages," Nov. 20, 2023. https://www.breitbart.com/politics/2023/11/20/former-troops-punished-over-bidens-vaccine-mandate-sue-for-billions-in-lost-wages/.
68 Michels L, et al, "Ignoring the Law with Vaccine Mandates," America Out Loud, Attorney John "Lou" Michels and Colonel Tom "Buzz" Rempfer interview by the Truth For Health Foundation, Nov. 26, 2022. https://www.americaoutloud.com/ignoring-the-law-with-vaccine-mandates/ and https://creativedestructionmedia.com/video/2023/01/14/io-episode-155-dod-vax-mandate-can-members-sue-military/.

69 *Wilson et al v.* Austin III, et al, and *Clements et al v. Austin, III, et al.* https://dockets.justia.com/docket/texas/txedce/4:2022cv00438/214840 and https://dockets.justia.com/docket/south-carolina/scdce/2:2022cv02069/272903 and https://militarybackpay.com/ and Bassen v. USA, No. 23–211C, *Botello v. USA*, No. 23–174C, and *Harkins v. USA*, No. 23–1238C, U.S. Court of Federal Claims.

70 Dyer O, "US judge halts compulsory anthrax vaccination for soldiers," BMJ, 329(7474):1062, Nov. 6, 2004. https://www.ncbi.nlm.nih.gov/pmc/articles/PMC526141/.

71 Army Regulation 40–562, BUMEDINST 6230.15B, AFI 48–110_IP, CG COMDTINST M6230.4G, Medical Services, Immunizations and Chemoprophylaxis for the Prevention of Infectious Diseases. https://armypubs.army.mil/epubs/DR_pubs/DR_a/pdf/web/r40_562.pdf.

72 Members of the Armed Forces for Liberty (MAFL), Wilson, et al v. Austin III, et al, No. 4: 22-cv-438 (E.D. TX), in opposition to the DoD vaccine mandates. https://mafl.knack.com/mafl#home/ and https://dockets.justia.com/docket/texas/txedce/4:2022cv00438/214840.

73 United States Court of Appeals for the Fifth Circuit, Feb. 17, 20222. https://www.documentcloud.org/documents/21261598–5coa-united-vaccine-mandate-per-curiam?responsive=1&title=1 and Fifth Circuit en banc ruling, 13 to 4, citing "rule of whim" in denying review, Aug. 18, 2022. https://law.justia.com/cases/federal/appellate-courts/ca5/21–11159/21–11159-2022–08-18.html.

74 Hals T, et al, "U.S. judge upholds United Airlines' COVID-19 vaccine mandate for employees," Reuters, Nov. 8, 2021. https://www.reuters.com/business/aerospace-defense/us-judge-upholds-united-airlines-covid-19-vaccine-mandate-2021–11-08/.

75 Case 21–1074, Sambrano, et al, v. United Airlines, Inc. https://www.govinfo.gov/app/details/USCOURTS-txnd-4_21-cv-01074/context and https://twitter.com/SenTedCruz/status/1471254546278227972.

76 SCOTUS, Groff v. Dejoy, Jun. 29, 2023. https://www.supremecourt.gov/opinions/22pdf/22–174_k536.pdf.

77 Vliet L, "The Ill Fated DOD Anthrax Vaccine Mandate & What It Means For the EUA COVID Experimental Gene Therapy Shot Mandate," Truth for Health Foundation interview with Colonel Thomas L. Rempfer, May 23, 2022. https://www.truthforhealth.org/2022/05/the-ill-fated-dod-anthrax-vaccine-mandate-what-it-means-for-the-eua-covid-experimental-gene-therapy-shot-mandate/ and https://rumble.com/v2uesgi-flag-day-press-conference.html.

78 "The Department of Defense Anthrax Vaccine Immunization Program: Unproven Force Protection," House Government Reform Committee, Apr. 3, 2000. https://www.congress.gov/congressional-report/106th-congress/house-report/556/1.

79 Dr. Mattias Desmet. https://biblio.ugent.be/person/801001743835 and https://www.chelseagreen.com/writer/mattias-desmet/.

80 Shrestha NK, et al, "Necessity of COVID-19 vaccination in previously infected individuals," medRxiv, Jun. 1, 2021. https://doi.org/10.1101/2021.06.01.21258176 and Mathioudakis AG, et al, Self-Reported Real-World Safety and Reactogenicity of COVID-19 Vaccines: A Vaccine Recipient Survey, Life (Basel), Mar. 17, 2021. https://pubmed.ncbi.nlm.nih.gov/33803014/.

81 White House Press Briefing by Press Secretary Jen Psaki, Jul. 27, 2021. https://www.whitehouse.gov/briefing-room/press-briefings/2021/07/27/press-briefing-by-press-secretary-jen-psaki-july-27–2021/.

82 Cohen D, et al, "Biden on '60 Minutes': 'The pandemic is over," Politico, Sept. 18, 2022. https://www.politico.com/news/2022/09/18/joe-biden-pandemic-60-minutes-00057423.
83 50 USC §1622, National emergencies, Termination Methods. https://uscode.house.gov/view.xhtml?req=granuleid:USC-prelim-title50-section1622&num=0&edition=prelim.
84 "Moving Beyond COVID-19 Vaccination Requirements for Federal Workers," The White House, Executive Order 14099, May 9, 2023. https://www.whitehouse.gov/briefing-room/presidential-actions/2023/05/09/executive-order-on-moving-beyond-covid-19-vaccination-requirements-for-federal-workers/ and https://www.whitehouse.gov/briefing-room/statements-releases/2023/05/01/the-biden-administration-will-end-covid-19-vaccination-requirements-for-federal-employees-contractors-international-travelers-head-start-educators-and-cms-certified-facilities/ and https://public-inspection.federalregister.gov/2023–11449.pdf.
85 Baletti B, "Breaking: Family of 24-Year-Old Who Died From COVID Vaccine Sues DOD in 'Groundbreaking Case,'" *The Defender*, May 31, 2023. https://childrenshealthdefense.org/defender/george-watts-jr-pfizer-covid-vaccine-injury/.
86 "FDA Authorizes Changes to Simplify Use of Bivalent mRNA COVID-19 Vaccines," FDA News Release, Apr. 18, 2023. https://www.fda.gov/news-events/press-announcements/coronavirus-covid-19-update-fda-authorizes-changes-simplify-use-bivalent-mrna-covid-19-vaccines and https://www.fda.gov/vaccines-blood-biologics/coronavirus-covid-19-cber-regulated-biologics/janssen-covid-19-vaccine.
87 Gann E, *Fate Is the Hunter: A Pilot's Memoir* (New York, Simon & Schuster, 1986).

Mile 24: www.Hoping4Justice.org

1 Myers S, "Armed Services opt to discharge those who refuse vaccine," *New York Times*, Mar. 11, 1999.
2 Hoping 4 Justice tab dedicated to COVID vaccine patent illegality. https://hoping4justice.org/covid.
3 Manual for Courts-Martial (MCM), United States (2019 Edition). https://jsc.defense.gov/Portals/99/Documents/2019%20MCM%20(Final)%20(20190108).pdf?ver=2019–01-11–115724-610.
4 Vliet E, "Department of Defense Disasters—From Anthrax to COVID-19," America Out Loud Whistleblower Report by the Truth for Health Foundation, Oct. 18, 2022. https://www.americaoutloud.com/department-of-defense-disasters-from-anthrax-to-covid-19/.
5 Senate Bill 1605, National Defense Authorization Act for Fiscal Year 2022. https://www.congress.gov/bill/117th-congress/senate-bill/1605/text?r=44&s=1.
6 H.R.291, Vaccine Discharge Parity Act, Note: This bill ensures that members of the Armed Forces who were granted a general discharge under honorable conditions are eligible for Department of Veterans Affairs (DVA) educational assistance under the Montgomery GI Bill-Active Duty and Post-9/11 GI Bill programs. https://www.congress.gov/bill/118th-congress/house-bill/291/cosponsors.
7 Orwell G, "As I Please," Tribune, Feb. 4, 1944. http://galileo.phys.virginia.edu/classes/inv_inn.usm/orwell3.html.
8 *Encyclopedia Britannica*, George Orwell's 1984.
9 The DoD IG agreed that testimony was "less than accurate" and "lacked the necessary element of 'straightforwardness,' and so was inconsistent with guidelines for honesty as set forth by the Joint Ethics Regulations (JER);" Jaeger, R.W., Wisconsin State Journal,

Jun. 19, 1999; DOD IG Complaint #74–998, filed Jan. 14, 2000; DoD IG transcript, Interview with Connecticut Air National Guard Commander, Jan. 18, 2001; Attrition testimony to Congress at https://www.govinfo.gov/content/pkg/CHRG-106hhrg63501/html/CHRG-106hhrg63501.htm and DoD Joint Ethics Regulation (JER), see section 84.47, Ethical values, https://www.govinfo.gov/content/pkg/FR-1994–03-21/html/94–5975.htm.

10 10 USC §1178(a) modification by the 2001 NDAA. https://uscode.house.gov/view.xhtml?req=granuleid:USC-prelim-title10-section1178&num=0&edition=prelim.

11 Chu D, Letter from Under Secretary of Defense for Personnel and Readiness to Rep. Christopher Shays, Chair of the Subcomm. on National Security. Sept. 26, 2003.

12 DoD public comments. http://www.signonsandiego.com/news/military/20031001–9999_1mi1anthrax.html.

13 Department of Energy Executive Summary responding to the Advisory Committee on Human Radiation Experiments Report, Oct. 3, 1995. http://tis.eh.doe.gov/ohre/roadmap/whitehouse/exesum.html.

14 Bush G, "President's Remarks on Signing the Veterans' Compensation Amendments of 1991 and the Agent Orange Act of 1991," Feb. 6, 1999. http://bushlibrary.tamu.edu/papers/1991/91020600.html.

15 Fierce Health. https://www.fiercehealthcare.com/hospitals/how-many-employees-have-hospitals-lost-to-vaccine-mandates-numbers-so-far.

16 Reusch J, et al, "Inability to work following COVID-19 vaccination among healthcare workers—an important aspect for future booster vaccinations," Public Health, vol 222, 186–195, Sept. 2023. https://www.sciencedirect.com/science/article/abs/pii/S0033350623002470.

17 US Navy unvaccinated as of Dec. 29, 2021. https://www.navy.mil/US-Navy-COVID-19-Updates/.

18 Horton A, "Marine Corps compliance with vaccine mandate on course to be military's worst," *Washington Post*, Nov. 21, 2021. https://www.washingtonpost.com/national-security/2021/11/21/vacine-mandate-marine-corps/.

19 Athey P, "14% of Marine reservists unvaccinated as deadline passes," *Marine Times*, Dec. 30, 2021. https://www.marinecorpstimes.com/news/your-marine-corps/2021/12/30/14-percent-of-marine-reserves-unvaccinated-as-deadline-passes/.

20 Cohen R, "Unvaccinated active duty airmen will soon be barred from moving to new assignments," *Air Force Times*. Nov. 21, 2021. https://www.airforcetimes.com/news/your-air-force/2021/11/24/unvaccinated-active-duty-airmen-will-soon-be-barred-from-moving-to-new-assignments/ and https://www.defense.gov/Spotlights/Coronavirus-DoD-Response/.

21 Hensch M, "Army Guardsmen Risk Service Before COVID-19 Deadline," NGAUS Newsroom, *Washington Report*, Jun. 28, 2022. https://www.ngaus.org/newsroom/army-guardsmen-risk-service-covid-19-deadline.

22 US Army unvaccinated. https://www.army.mil/article/252821/active_army_achieves_98_percent_vaccination_rate_with_less_than_one_percent_refusal_rate.

23 Beynon S, "Army Cuts Off More Than 60K Unvaccinated Guard and Reserve Soldiers from Pay and Benefits," Military.com, Jul. 6, 2022. https://www.military.com/daily-news/2022/07/06/army-cuts-off-more-60k-unvaccinated-guard-and-reserve-soldiers-pay-and-benefits.html.

24 US Air Force unvaccinated. https://www.af.mil/News/Article-Display/Article/2831845/daf-covid-19-statistics-dec-7–2021/.

25 Novelly T, "Limited Approval of Religious Exemptions for COVID-19 Vaccine," Military.com, Nov. 30, 2022. https://www.military.com/daily-news/2022/11/30/new-court-ruling-slams-air-forces-limited-approval-of-religious-exemptions-covid-19-vaccine.html and https://www.opn.ca6.uscourts.gov/opinions.pdf/22a0255p-06.pdf.

26 DoD letter to Rep Mike Rogers on mandate statistics, Feb. 27, 2023. https://banks.house.gov/uploadedfiles/osd_vaccine_discharge_letter.pdf.

27 Liebermann O, "Only 43 of more than 8,000 discharged from US military for refusing Covid vaccine have rejoined," CNN, Oct. 2, 2023. https://www.cnn.com/2023/10/02/politics/us-military-covid-vaccine/index.html.

28 Bardosh K, et al, "COVID-19 vaccine boosters for young adults: a risk benefit assessment and ethical analysis of mandate policies," *Journal of Medical Ethics*, Dec. 2022. https://jme.bmj.com/content/early/2022/12/05/jme-2022–108449.

29 Shane L, et al, "Defense Secretary Lloyd Austin has coronavirus, symptoms 'mild,'" Military Times, Jan. 2, 2022 and Aug. 15, 2022. https://www.militarytimes.com/news/pentagon-congress/2022/01/02/defense-secretary-lloyd-austin-has-coronavirus-symptoms-mild/ and https://www.militarytimes.com/news/pentagon-congress/2022/08/15/austin-quarantining-with-second-bout-of-covid-19/.

30 Eythorsson E, et al, "Rate of SARS-CoV-2 Reinfection During an Omicron Wave in Iceland," JAMA Research Letter, Infectious Diseases, Aug. 2022. https://jamanetwork.com/journals/jamanetworkopen/article-abstract/2794886.

31 Brown W, "Marine Colonel Asks Why the Vaxx," COVID 19 vaccine questions for the Commandant of the Marine Corps, Dec. 30, 2021. https://rumble.com/embed/vp2d9t/.

32 Wong K, "'I've Been Grounded': Six Marine Pilots Speak Out Against COVID Vaccine Mandate," Breitbart, Apr. 11, 2022. https://www.breitbart.com/politics/2022/04/11/ive-been-grounded-marine-pilots-speak-out-covid-vaccine-mandate/.

33 Rahman K, "Over 200 Service Members Sign Letter Against Military Leaders," *Newsweek*, Jan. 2, 2024. https://www.newsweek.com/service-members-military-leaders-covid-vaccine-mandate-1857027 and Declaration of Military Accountability, Jan. 1, 2024. https://militaryaccountability.net/.

34 Green R, *Defending the Constitution Behind Enemy Lines* (New York, Skyhorse, 2023). https://www.skyhorsepublishing.com/9781510778078/defending-the-constitution-behind-enemy-lines/.

35 Klein B, "Biden continues to test positive for Covid-19 Wednesday following rebound diagnosis," CNN, Aug. 3, 2022. https://www.cnn.com/2022/08/03/politics/biden-still-positive-covid-19/index.html.

Mile 25: Whistleblower 1–2-3

1 Washington Headquarters Services published DD-149. https://www.esd.whs.mil/Portals/54/Documents/DD/forms/dd/dd0149.pdf.

2 Phelps J, "In battle over COVID-19 mandate, Army general served with criminal complaint by whistleblower," *American Family News*, Dec. 7, 2023. https://afn.net/national-security/2023/12/07/in-battle-over-covid-19-mandate-army-general-served-with-criminal-complaint-by-whistleblower/

3 Russell D, *The Real RFK Jr.: Trials of a Truth Warrior* (New York, Skyhorse, 2023). https://www.skyhorsepublishing.com/9781510776098/the-real-rfk-jr/.
4 MacKinnon J, "CNN medical analyst who said the unvaccinated should not be able to travel changes her tune, emphasizes benefits of natural immunity," *The Blaze*, Dec. 22, 2022. https://www.theblaze.com/news/cnn-medical-analyst-who-said-the-unvaccinated-should-not-be-able-to-travel-changes-her-tune.
5 Shrestha N, et al, "Effectiveness of the Coronavirus Disease 2019 (COVID-19) Bivalent Vaccine," Cleveland Clinic, Dec. 17, 2022. https://www.medrxiv.org/content/10.1101/2022.12.17.22283625v1.full.pdf and https://academic.oup.com/ofid/article/10/6/ofad209/7131292.
6 Demasi M, "FDA oversight of clinical trials is "grossly inadequate," say experts," BMJ Investigation, Nov. 16, 2022. https://www.bmj.com/content/379/bmj.o2628.
7 Schmeling M, et al, "Batch-dependent safety of the BNT162b2 mRNA COVID-19 vaccine," Eur J Clin Invest, Mar. 30, 2023. https://pubmed.ncbi.nlm.nih.gov/36997290/.
8 595 U. S. ____ (2021) 1, Gorsuch, J., dissenting, SCOTUS, No. 21A145, Dr. A, et al, Applicants v. Kathy Hochul, Governor of NY, et al, on application for injunctive relief, Dec. 13, 2021. https://www.supremecourt.gov/opinions/21pdf/21a145_gfbi.pdf.
9 595 U. S. ____ (2021) 1, Gorsuch, J., dissenting, SCOTUS, No. 21A145, *Dr. A, et al, Applicants v. Kathy Hochul, Governor of NY, et al*, on application for injunctive relief, Dec. 13, 2021. https://www.supremecourt.gov/opinions/21pdf/21a145_gfbi.pdf.
10 NY Supreme Court, Police Benevolent Association vs. the City of New York, et al, Sept. 23, 2022. https://iapps.courts.state.ny.us/nyscef/ViewDocument?docIndex=jJE8DRs3dmN80YfmamRllA==.
11 NY Supreme Court, *Garvey, et al, vs. the City of New York*, et al, Oct. 24, 2022. https://iapps.courts.state.ny.us/fbem/DocumentDisplayServlet?documentId=JK5E3gx5XV1/ku37jnWR_PLUS_w==&system=prod and https://childrenshealthdefense.org/wp-content/uploads/september-2023-new-york-teachers-decision-order.pdf.
12 "BREAKING: CHD Defeats NY State Healthcare Workers COVID Mandate!," Children's Health Defense, Jan. 13, 2023. https://childrenshealthdefense.org/press-release/breaking-chd-defeats-ny-state-healthcare-workers-covid-mandate/ and https://childrenshealthdefense.org/wp-content/uploads/2023–1-13-doc-86-decision-and-order.pdf and https://www.jdsupra.com/legalnews/new-york-department-of-health-1602868/ and https://www.health.ny.gov/press/releases/2023/2023–05-24_statement.htm.
13 598 U. S. ____ (2023), Statement of Justice Gorsuch, SCOTUS, AZ, et al, v. Mayorkas, DHS, et al, Writ of Certiorari, No. 22–592, Decided May 18, 2023. https://www.supremecourt.gov/opinions/22pdf/22–592_5hd5.pdf.
14 Supreme Court of the United States opinion in *Lloyd J. Austin, III, Secretary of Defense, et al, v. U. S. NAVY SEALS*, 1–26, et al, Mar. 25, 2022. https://www.supremecourt.gov/opinions/21pdf/21a477_1bo2.pdf.
15 Cohen R, "Federal judge temporarily halts Air Force's COVID-19 vaccine mandate," Re Religious Freedom Restoration Act (RFRA) cases, *Air Force Times*, Jul. 15, 2022. https://www.airforcetimes.com/news/your-air-force/2022/07/15/federal-judge-temporarily-halts-air-forces-covid-19-vaccine-mandate/ and https://lc.org/newsroom/details/081922-us-marines-win-class-protection-from-shot-mandate-1.
16 CDC COVID-19 Tracker, Vaccinations in the United States, updated May 11, 2023 https://covid.cdc.gov/covid-data-tracker/ and https://www.reuters.com/business/healthcare

-pharmaceuticals/about-4-mln-americans-got-their-updated-covid-vaccines-sept-2023–10-04/.

17 Biden J, President of the U.S., et al., v. Feds for Medical Freedom, Petition for a Writ of Certiorari, Justice Jackson, Excerpt: “In my view, the party seeking vacatur has not established equitable entitlement to that remedy,” Dec. 11, 2023. https://www.supremecourt.gov/orders/courtorders/121123zor_e29g.pdf.

18 “Infodemic,” WHO, Excerpt: “An infodemic is too much . . . false or misleading information.” https://www.who.int/health-topics/infodemic.

Mile 26: HEED Your Training

1 USAF Flight Safety Magazine, and video produced by the Hanscom AFB Safety Office. https://youtu.be/NdX_tqdMj3g.

2 USAF Flight Safety Magazine, ORM is not a “Ka-ching,” Jun. 2004. chrome-extension://efaidnbmnnnibpcajpcglclefindmkaj/viewer.html?pdfurl=https%3A%2F%2Fwww.safety.af.mil%2FPortals%2F71%2Fdocuments%2FMagazines%2FFSM%2Ffsmjun04.pdf%3Fver%3D2017–03-28–100957-797&clen=5743677&chunk=true.

3 Tanaka A, et al, Raw data for “Unnatural evolutionary processes of SARS-CoV-2 variants and possibility of deliberate natural selection,” Zenodo, Aug. 15, 2023. https://doi.org/10.5281/zenodo.8254894 and https://zenodo.org/record/8254894.

4 2022 National Biodefense Strategy and Implementation Plan. https://www.whitehouse.gov/wp-content/uploads/2022/10/National-Biodefense-Strategy-and-Implementation-Plan-Final.pdf.

5 FDA’s Proposed Framework for Addressing Future COVID-19 Vaccine Strain Composition. https://www.fda.gov/media/157466/download.

6 “Bill Signed: H.J.Res. 7,” Excerpt: “the President signed into law: H.J. Res. 7, which terminates the national emergency related to the COVID-19 pandemic,” The White House, Apr. 10, 2023. https://www.whitehouse.gov/briefing-room/legislation/2023/04/10/bill-signed-h-j-res-7/ and https://aspr.hhs.gov/legal/PHE/Pages/covid19–11Jan23.aspx.

7 DHHS Fact Sheet: HHS Intent to Amend the Declaration Under the PREP Act for Medical Countermeasures Against COVID-19, Apr. 14, 2023. https://www.hhs.gov/about/news/2023/04/14/factsheet-hhs-announces-amend-declaration-prep-act-medical-countermeasures-against-covid19.html.

8 Mandowara K, et al, “U.S. committee recommends COVID shot for CDC’s free vaccine program,” Reuters, Oct. 19, 2022. https://www.reuters.com/world/us/us-committee-recommends-covid-shot-cdcs-free-vaccine-program-2022–10-19/.

9 “Child and Adolescent Immunization Schedule, Recommendations for Ages 18 Years or Younger,” CDC, 2023. https://www.cdc.gov/vaccines/schedules/hcp/imz/child-adolescent-compliant.html#note-covid-19 and https://www.cdc.gov/vaccines/schedules/easy-to-read/child-easyread.html.

10 “Rate Of Americans Who Have Received The Bivalent Vaccine Isn’t What Experts Expected,” MSN, Oct. 3, 2022. https://www.msn.com/en-us/health/other/the-rate-of-americans-who-have-received-the-bivalent-vaccine-isn-t-what-experts-expected/ar-AA12yaNn.

11 US Surgeon General Dr. Vivek Murthy request for research, data, and personal experiences related to health misinformation. https://www.hhs.gov/surgeongeneral/health-misinformation-rfi/index.html.

12 House Judiciary Committee and the Select Subcommittee on the Weaponization of the Federal Government, The Weaponization of CISA: How a 'Cybersecurity' Agency Colluded with Big Tech and 'Disinformation' Partners to Censor Americans; interim staff report details how the Cybersecurity and Infrastructure Security Agency (CISA), originally intended to protect pipelines and other critical infrastructure from cyberattacks, expanded its mission to surveil and censor Americans' speech on social media; the report outlined collusion between CISA, Big Tech, and government-funded third parties to conduct censorship by proxy and cover up CISA's unconstitutional activities, Jun. 26, 2023. https://judiciary.house.gov/sites/evo-subsites/republicans-judiciary.house.gov/files/evo-media-document/cisa-staff-report6–26-23.pdf and https://judiciary.house.gov/media/press-releases/new-report-reveals-cisa-tried-cover-censorship-practices and https://int.nyt.com/data/documenttools/injunction-in-missouri-et-al-v/7ba314723d052bc4/full.pdf and https://www.ca5.uscourts.gov/opinions/pub/23/23–30445-CV0.pdf and Supreme Court https://www.supremecourt.gov/opinions/23pdf/23a243_7l48.pdf.

13 US District Court Western District of Louisianna, Missouri et al v. Biden, No. 3:22-cv-1213, Judge Terry A. Doughty, Jul. 10, 2023. https://storage.courtlistener.com/recap/gov.uscourts.lawd.189520/gov.uscourts.lawd.189520.301.0.pdf.

14 United States v. Alvarez, 567 U.S. 709, Orwell-Oceania metaphor, Jun. 28, 2012. https://supreme.justia.com/cases/federal/us/567/709/.

15 H.R.5546 - National Childhood Vaccine Injury Act of 1986 (NCVIA). https://www.congress.gov/bill/99th-congress/house-bill/5546 and https://www.hrsa.gov/vaccine-compensation/faq#About%20The%20Vicp.

16 Barbara Loe Fisher, President, National Vaccine Information Center (NVIC). https://www.nvic.org/about/staff-volunteers/barbara-loe-fisher.

17 Bruesewitz v. Wyeth LLC, 562 U.S. 223, Feb. 22, 2011. https://www.lexisnexis.com/community/casebrief/p/casebrief-bruesewitz-v-wyeth-llc.

18 Federal Register, Determination of Public Health Emergency or Credible Risk of Future Public Health Emergency and Liability Immunity due to the "credible risk that the spread of Bacillus anthracis . . . and the resulting disease or conditions may in the future constitute a public health emergency, Dec. 23, 2023. https://www.federalregister.gov/d/2022–28010/p-7.

19 Zacks Equity Research, "Emergent (EBS) Gets $235.8M DoD Contract for Anthrax Vaccine," Yahoo Finance, Jan. 12, 2024. https://finance.yahoo.com/news/emergent-ebs-gets-235–8m-192000831.html.

20 Charbonneau D, "'Natural Immunity Works and Pfizer Knows It,' FDA Document Dump Reveals," *The Defender*, Apr. 6, 2022. https://childrenshealthdefense.org/defender/natural-immunity-pfizer-fda-document-dump/ and https://phmpt.org/pfizers-documents/.

21 Dr. Fauci on C-SPAN, Oct. 11, 2004. https://www.c-span.org/video/?c5009217/user-clip-dr-fauci-vaccination-infection.

22 Chen D, et al, "Role of spike in the pathogenic and antigenic behavior of SARS-CoV-2 BA.1 Omicron," BioRxiv, Oct. 14, 2022. https://www.biorxiv.org/content/10.1101/2022.10.13.512134v1 and https://www.foxnews.com/us/boston-university-lethal-covid-strain-lab.

23 Global Covid Summit, International Alliance of Physicians and Medical Scientists Declaration. https://globalcovidsummit.org/news/declaration-iv-restore-scientific-integrity and https://doctorsandscientistsdeclaration.org/.

24 NIH.gov Announcements, Sept. 13, 2021. https://dailymed.nlm.nih.gov/dailymed/dailymed-announcements-details.cfm?date=2021–09-13.

25 Kates J, et al, "How Much Could COVID-19 Vaccines Cost the U.S. After Commercialization?," KFF, Mar. 10, 2023. https://www.kff.org/coronavirus-covid-19/issue-brief/how-much-could-covid-19-vaccines-cost-the-u-s-after-commercialization/.

26 Federal Register, Eleventh Amendment to Declaration Under the Public Readiness and Emergency Preparedness Act for Medical Countermeasures Against COVID-19, May 12, 2023. https://www.federalregister.gov/documents/2023/05/12/2023–10216/eleventh-amendment-to-declaration-under-the-public-readiness-and-emergency-preparedness-act-for.

27 FDA website, FDA Takes Action on Updated mRNA COVID-19 Vaccines, Sept 11, 2023. https://www.fda.gov/news-events/press-announcements/fda-takes-action-updated-mrna-covid-19-vaccines-better-protect-against-currently-circulating.

28 Adverse Events Reported after COVID-19 Vaccination. https://www.cdc.gov/coronavirus/2019-ncov/vaccines/safety/adverse-events.html.

29 Countermeasures Injury Compensation Program (CICP) Data. https://www.hrsa.gov/cicp and https://www.hrsa.gov/cicp/cicp-data and https://reason.com/2023/11/21/lawsuit-covid-vaccine-injury-claims-diverted-to-unconstitutional-kangaroo-court/.

30 Krumholz H, et al, "Post-Vaccination Syndrome: A Descriptive Analysis of Reported Symptoms and Patient Experiences After Covid-19 Immunization," medRxiv, Nov. 9, 2023. https://www.medrxiv.org/content/10.1101/2023.11.09.23298266v1 and Safavi F, et al, Neuropathic symptoms with SARS-CoV-2 vaccination, medRxiv, May 17, 2022. https://www.medrxiv.org/content/10.1101/2022.05.16.22274439v1.

31 Polykretis P, et al, "Autoimmune inflammatory reactions triggered by the COVID-19 genetic vaccines in terminally differentiated tissues Autoimmune inflammatory reactions triggered by the COVID-19 genetic vaccines in terminally differentiated tissues," Autoimmunity, Dec. 2023. https://pubmed.ncbi.nlm.nih.gov/37710966/ and Nune A, et al, "New-Onset Rheumatic Immune-Mediated Inflammatory Diseases Following SARS-CoV-2 Vaccinations until May 2023: A Systematic Review," Oct. 8, 2023. https://www.mdpi.com/2076–393X/11/10/1571.

32 CHD Bus, Capt Kelli Donley tells her story, https://rumble.com/v3fqlgp-anthrax-vaccine-destroyed-my-life-chd-bus-stories.html.

33 Benjamin M, "Father of dead soldier claims Army coverup," UPI, Aug. 7, 2003. https://www.upi.com/Odd_News/2003/08/07/Father-of-dead-soldier-claims-Army-coverup/37551060294351/.

34 Harvey L, "Anthrax vaccine a deadly defense," *Northwest Indiana Times*, Jun. 22, 2003. http://www.nwitimes.com/articles/2003/06/22/news/top_news/8096364ebe0cdc9d86256d4c007d7c23.txt and "Coroner rules vaccinations contributed to reservist's death," Times Online, Jun. 12, 2003. http://www.thetimesonline.com/articles/2003/06/12/news/region_and_state/5d5ad8063c5d300b86256d42007bf6db.prt.

35 Benjamin M, "Pentagon Slow In Vaccine Death," UPI, Aug. 18, 2003. http://www.upi.com/print.cfm?StoryID=20030818–060641-9420r and https://www.upi.com/Odd_News/2003/08/18/Doctor-Pentagon-slow-in-vaccine-death/UPI-94201061246588/.

36 "'Still a little unclear what she died from,'" *Daily Press*, Dec. 06, 2005. https://www.dailypress.com/news/dp-anth-day3-lacydec03-story.html.

37 "The Anthrax Vaccine Immunization Program—What Have We Learned," Committee on Government Reform, Oct. 3, 2000. https://www.govinfo.gov/content/pkg/CHRG-106hhrg73979/html/CHRG-106hhrg73979.htm.

38 Evenson A, et al, "BioPort to track reactions to vaccine—FDA concerned that certain batches may have adverse effects," Nov. 10, 2000. www.lsjxtra.com/news/bioport10.html.

39 Rempfer T, Anthrax Opinion Disserves Troops, *Army* and *Marines Corps Times*, Opinion Letters, Aug. 2, 1999.

40 Rempfer T and Dingle R, Don't hold troops accountable for others' mistakes on anthrax vaccine, *Army Times*, Nov. 4, 2002.

41 Thorp J, et al, "COVID-19 Vaccines: The Impact on Pregnancy Outcomes and Menstrual Function," Dec. 30, 2022. https://www.preprints.org/manuscript/202209.0430/v2 and Peer-reviewed version at "https://www.jpands.org/vol28no1/thorp.pdf.

42 Wesselink A, et al, Prospective Cohort Study of COVID-19 Vaccination, SARS-CoV-2 Infection, and Fertility," Am J Epidemiol, 191(8):1383–1395, Jul. 23, 2022. https://pubmed.ncbi.nlm.nih.gov/35051292/.

43 Bujard M, et al, "Fertility declines near the end of the COVID-19 pandemic: Evidence of the 2022 birth declines in Germany and Sweden," Sonstige Publikationen, Jun. 2022. https://www.bib.bund.de/Publikation/2022/Fertility-declines-near-the-end-of-the-COVID-19-pandemic-Evidence-of-the-2022-birth-declines-in-Germany-and-Sweden.html.

44 Föhse F, et al, "The BNT162b2 mRNA vaccine against SARS-CoV-2 reprograms both adaptive and innate immune responses," medRxiv, May 3, 2021. https://doi.org/10.1101/2021.05.03.21256520 and https://www.biorxiv.org/content/10.1101/2022.03.16.484616v2.full.

45 Halma M, et al, "The Novelty of mRNA Viral Vaccines and Potential Harms: A Scoping Review," MDPI, 6(2):220–235, Apr. 17, 2023. https://doi.org/10.3390/j6020017.

46 Eens S, et al, "B-cell lymphoblastic lymphoma following intravenous BNT162b2 mRNA booster in a BALB/c mouse: A case report," Front Oncol, May 2023. https://pubmed.ncbi.nlm.nih.gov/37197431/.

47 Kakovan M, et al, "Stroke Associated with COVID-19 Vaccines," J Stroke Cerebrovasc Dis, 31(6):106440, Jun. 2022. https://pubmed.ncbi.nlm.nih.gov/35339857/ and https://www.cdc.gov/coronavirus/2019-ncov/vaccines/safety/bivalent-boosters.html.

48 Aldén M, et al, "Intracellular Reverse Transcription of Pfizer BioNTech COVID-19 mRNA Vaccine BNT162b2 In Vitro in Human Liver Cell Line," Current Issues in Molecular Biology, 44, no. 3: 1115–1126, 2022. https://doi.org/10.3390/cimb44030073.

49 Willyard C, "Scientists debate evidence for a micro-clot hypothesis," Nature, Aug. 24, 2022. https://www.nature.com/articles/d41586–022-02286–7.

50 Yale Medicine, Amyloidosis. https://www.yalemedicine.org/conditions/amyloidosis.

51 Safavi F, et al, "Neuropathic symptoms with SARS-CoV-2 vaccination," NIH, May 17, 2022. https://pubmed.ncbi.nlm.nih.gov/35611338/.

52 Rong Z, et al, "SARS-CoV-2 Spike Protein Accumulation in the Skull-Meninges Brain Axis: Potential Implications for Long-Term Neurological Complications in post-COVID-19," BioRxiv, Apr. 5, 2023. https://www.biorxiv.org/content/10.1101/2023.04.04.535604v1.full.pdf.

53 McKernan K, et al, "Sequencing of Bivalent Moderna and Pfizer mRNA Vaccines Reveals Nanogram to Microgram Quantities of Expression Vector dsDNA Per Dose," OSF Preprints, Apr. 10, 2023. https://osf.io/b9t7m/ and Speicher D, et al, DNA fragments

detected in monovalent and bivalent Pfizer/BioNTech and Moderna modRNA COVID-19 vaccines from Ontario, Canada: Exploratory dose response relationship with serious adverse events, OSF Preprint, Oct. 18, 2023. https://osf.io/mjc97/.

54 Uversky V, et al, "IgG4 Antibodies Induced by Repeated Vaccination May Generate Immune Tolerance to the SARS-CoV-2 Spike Protein," Vaccines, 11(5):991, May 17, 2023. https://www.mdpi.com/2076–393X/11/5/991.

55 Mörz M, "Case Report: Encephalitis and Myocarditis after BNT162b2 mRNA Vaccination against COVID-19," Excerpt: "encephalitis and encephalomyelitis have been reported in connection with the gene-based COVID-19 vaccines, with many being considered causally related to vaccination. However, this is the first report to demonstrate the presence of the spike protein within the encephalitic lesions and to attribute it to vaccination rather than infection. These findings corroborate a causative role of the gene-based COVID-19 vaccines, and this diagnostic approach is relevant to potentially vaccine-induced damage to other organs as well," MDPI, Oct. 1, 2022. https://www.mdpi.com/2076–393X/10/10/1651.

56 Garrido I, et al, "Autoimmune hepatitis after COVID-19 vaccine—more than a coincidence," J Autoimmun, Dec. 2021. https://pubmed.ncbi.nlm.nih.gov/34717185/.

57 Watanabe S, et al, "SARS-CoV-2 vaccine and increased myocarditis mortality risk: A population based comparative study in Japan," Excerpt: "SARS-CoV-2 vaccination was associated with higher risk of myocarditis death, not only in young adults but also in all age groups including the elderly. Considering healthy vaccinee effect, the risk may be 4 times or higher than the apparent risk of myocarditis death. Underreporting should also be considered. Based on this study, risk of myocarditis following SARS-CoV-2 vaccination may be more serious than that reported previously," Oct.13, 2022. https://www.medrxiv.org/content/10.1101/2022.10.13.22281036v1.

58 Sriwastava S, et al, "COVID-19 Vaccination and Neurological Manifestations," *Brain Science*, 12(3):407, Mar. 18, 2022. https://pubmed.ncbi.nlm.nih.gov/35326363/.

59 Li J, et al, "Risk assessment of retinal vascular occlusion after COVID-19 vaccination," *Nature Portfolio Journal Vaccines*, 8, 64, May 2, 2023. https://doi.org/10.1038/s41541–023-00661–7.

60 Kyriakopoulos A, et al, "Mitogen Activated Protein Kinase (MAPK) Activation, p53, and Autophagy Inhibition Characterize the Severe Acute Respiratory Syndrome Coronavirus 2 (SARS-CoV-2) Spike Protein Induced Neurotoxicity," Excerpt: "Overall results suggest that neurodegeneration is in part due to the intensity and duration of spike protein exposure . . . the neurologically damaging effects can be cumulatively spike-protein dependent, whether exposure is by natural infection or, more substantially, by repeated mRNA vaccination." Cureus, Dec. 9, 2022. https://www.ncbi.nlm.nih.gov/pmc/articles/PMC9733976/.

61 Park Y, et al, "Correlation between COVID-19 vaccination and inflammatory musculoskeletal disorders," medRxiv, Nov. 14, 2023. https://www.medrxiv.org/content/10.1101/2023.11.14.23298544v1.

62 Qui Y, et al, "Covid-19 vaccination can induce multiple sclerosis via cross-reactive CD4+ T cells recognizing SARS-CoV-2 spike protein and myelin peptides," Multiple Sclerosis Journal, 28(3 Supplement):776, 2022. https://pesquisa.bvsalud.org/global-literature-on-novel-coronavirus-2019-ncov/resource/pt/covidwho-2138820 and https://pubmed.ncbi.nlm.nih.gov/37515255/.

63 Willyard C, "Scientists debate evidence for a micro-clot hypothesis," *Nature*, Aug. 24, 2022. https://www.nature.com/articles/d41586–022-02286–7.

64 Joint explanatory material accompanying the National Defense Authorization Act for Fiscal Year 2023. Sec. 525 includes "Rescission of COVID-19 vaccination mandate." Excerpt: "The agreement includes a provision that would require the Secretary of Defense to rescind the mandate that members of the Armed Forces be vaccinated against COVID-19 . . . We note that the Department of Defense has mechanisms to correct a servicemember's military record for discharge due to failure to receive the COVID-19 vaccine. In addition, the military departments have the ability to consider applications for reinstatement of servicemembers who were previously separated for refusing the vaccine. We would support efforts by the Secretary to ensure that the military departments have a consistent process in place to consider such requests for correction of military records and reinstatement." https://rules.house.gov/sites/democrats.rules.house.gov/files/BILLS-117HR7776EAS-RCP117–70-JES.pdf.

65 Scott R, et al, Senator's letter to the SecDef, Dec. 16, 2022. https://www.lee.senate.gov/services/files/F5D0B207-CA97–41F9-A55E-5C39779EC675.

66 Rempfer T, et al, "Justice demands correction of military records for vaccine refusers, *Military Times* Op-Ed, Dec. 28, 2022. https://www.militarytimes.com/opinion/commentary/2022/12/28/justice-demands-correction-of-military-records-for-vaccine-refusers/.

67 Bishop R, et al, "Congress Agrees to Stop Unlawful COVID Vaccines, But Damage to our Troops Must Be Rectified," America Out Loud, Dec. 9, 2022. https://www.americaoutloud.com/congress-agrees-to-stop-unlawful-covid-vaccines-but-damage-to-our-troops-must-be-rectified/ and Dress B, "2024 NDAA draft defense bill orders Pentagon to review reinstatement of troops fired for COVID-19 refusal," *The Hill*, Dec. 7, 2023. https://thehill.com/policy/defense/4347700-defense-bill-reinstatement-troops-covid-19/.

68 "Biden Signs National Defense Authorization Act Into Law," DoD News, Dec. 23, 2022. https://www.defense.gov/News/News-Stories/Article/Article/3252968/biden-signs-national-defense-authorization-act-into-law/.

69 DOD Mandate Rescission Memo, Jan. 10, 2023. https://www.defense.gov/News/Releases/Release/Article/3264323/dod-rescinds-covid-19-vaccination-mandate/ and https://media.defense.gov/2023/Jan/10/2003143118/-1/-1/1/SECRETARY-OF-DEFENSE-MEMO-ON-RESCISSION-OF-CORONAVIRUS-DISEASE-2019-VACCINATION-REQUIREMENTS-FOR-MEMBERS-OF-THE-ARMED-FORCES.PDF.

70 Archie A, "The CDC will no longer issue COVID-19 vaccination cards," NPR, Oct. 5, 2023. https://www.npr.org/2023/10/05/1203924997/covid-19-vaccination-cards-cdc.

71 FY 2024 NDAA, Conference Report, Accompanying HR 2670, Dec. 2023. https://armedservices.house.gov/sites/republicans.armedservices.house.gov/files/FY24%20NDAA%20Conference%20Report%20-%20%20FINAL.pdf.

72 Muhammad J, et al, "Congress Using the NDAA Legislates End to COVID Vaccine Mandates," America Out Loud interview by Truth For Health Team, Dec. 11, 2022. https://www.americaoutloud.com/congress-using-the-ndaa-legislates-end-to-covid-vaccine-mandates/ and https://dailycaller.com/2023/01/18/vaccine-mandate-repairs-reenlist-lost-trust/.

73 Samuels B, "Pence: Discharged military members should be reinstated," *The Hill*, Jan. 12, 2023. https://thehill.com/homenews/administration/3810810-pence-discharged-military-members-should-be-reinstated-get-back-pay-over-vaccine-mandates/.

74 U.S. District Court Northern District of TX, Public Health and Medical Professionals for Transparency vs FDA, Civil Action No. 4:22-cv-915-P, see fn #4, Apr. 12, 2023. https://www.sirillp.com/wp-content/uploads/2023/05/countered-b9ef852b371b2b0a9d3ae928e9849a48.pdf and https://aaronsiri.substack.com/p/how-did-our-vaccine-oversight-system.

75 Woods L, 1986 USAF Academy graduate. https://creativedestructionmedia.com/news/health-freedom/2023/04/14/col-tom-rempfer-usaf-ret-encourages-participation-in-chicago-event-are-the-skies-safe/ and https://creativedestructionmedia.com/video/2023/04/22/livestream-1230pm-est-the-globalists-in-plain-sight-the-history-of-the-pandemic-playbook-part-1/ and https://creativedestructionmedia.com/video/2023/04/30/livestream-1230pm-est-the-globalists-in-plain-sight-anthrax-history-part-ii/.

76 Risch H, "Lessons From Anthrax Were Not Learned, Why Should Lessons From COVID Be Learned?," America Out Loud interview with Colonel Thomas Rempfer, Jun. 3, 2023. https://www.americaoutloud.com/lessons-from-anthrax-were-not-learned-why-should-lessons-from-covid-be-learned/ and https://www.americaoutloud.com/bidens-gargantuan-fraud-on-covid-mandates/ and https://www.americaoutloud.com/a-pandemic-of-intentional-malfeasance-that-cost-millions-of-lives/ and https://www.truthforhealth.org/2023/09/fda-repeats-covid-fraud/.

Finish Line—The Debrief

1 Bloom's cognitive domains or taxonomy originally was represented by six different cognitive domain levels: (1) knowledge, (2) comprehension, (3) application, (4) analysis, (5) synthesis, and (6) evaluation. https://www.britannica.com/topic/Blooms-taxonomy.

2 Rancourt D, et al, "COVID-19 vaccine-associated mortality in the Southern Hemisphere," Sept. 17, 2023. https://denisrancourt.substack.com/p/covid-19-vaccine-associated-mortality.

3 Dr. Harvey Risch, Professor Emeritus, Yale Institute for Global Health. https://ysph.yale.edu/profile/harvey-risch/.

4 Dr. George Fareed, Espoused the need to engage in a healthy "dialectic" with respect to COVID vaccines and mandates, Co-author of Overcoming the COVID Darkness, along with Dr. Brian Tyson. https://overcomingcoviddarkness.com/about/.

5 Dr. Elizabeth Lee Vliet. https://vivelifecenter.com/about/founder-chief-medical-officer/ and https://www.truthforhealth.org/author/leevlietmd/.

6 NIH-NIAID, Retirement of Anthony S. Fauci, M.D., Aug. 22, 2022. https://www.niaid.nih.gov/news-events/statement-anthony-s-fauci-md.

7 Klein B, "Fauci to leave federal government in December after decades as nation's top infectious disease expert," CNN, Aug. 23, 2022. https://www.cnn.com/2022/08/22/politics/anthony-fauci-government/index.html.

8 "'This Week' Transcript 4–10-22: White House National Security Advisors Jake Sullivan, Kyiv Mayor Vitali Klitschko & Dr. Anthony Fauci," ABC News, Apr. 10, 2022. https://abcnews.go.com/Politics/week-transcript-10–22-white-house-national-security/story?id=83984068.

9 "Dr. Birx Says She Knew of Natural COVID-19 Reinfections as Early as December 2020," C-SPAN, House Oversight and Reform Subcommittee on Select Coronavirus Crisis, White

House COVID-19 Coordinator Dr. Deborah Birx testimony, Jun. 23, 2022. https://www.c-span.org/video/?c5021092/dr-birx-knew-natural-covid-19-reinfections-early-december-2020.

10 Atlas S, "America's COVID Response Was Based on Lies," Newsweek, Opinion by Dr. Scott W. Atlas, Senior Fellow, Stanford's Hoover Institution, Mar. 6, 2023. https://www.newsweek.com/america-covid-response-was-based-lies-opinion-1785177.

11 Mazziotta J, "CDC Walks Back Claim that Vaccinated People Do Not Carry COVID: 'The Evidence Isn't Clear,'" *People*, Apr. 2, 2021. https://people.com/health/vaccinated-people-do-not-appear-carry-spread-covid-19/.

12 Gumbrecht J, et al, "CDC director tests positive for Covid-19," CNN, Oct. 22, 2022. https://www.cnn.com/2022/10/22/health/cdc-director-positive-covid/index.html and CDC walks back claim that vaccinated people can't carry COVID-19. https://youtu.be/aMmF9rdAXLc.

13 Walensky R, "CDC Director: Science Isn't Settled," Video of CDC Director Washington University, St. Louis, excerpt: "I have frequently said, Science is the foundation of everything we do. That is entirely true. But I think the public heard that as Science is foolproof, science is black and white. We get the answer and then we make the decision based on the answer. But the truth is science is gray. And science is not always immediate. Sometimes it takes months and years to actually find out the answer, and you have to make decisions before you have that answer. We might be faulted for not making exactly the right decision, in the moment. I'm okay with that. But I don't want to be faulted for not making a decision. Because not making a decision in and of itself is a decision," Mar. 3, 2022. https://rumble.com/vwh4f3-cdc-director-science-isnt-settled-its-black-and-white.html and timestamp 29:00 at https://livestream.com/accounts/7945443/events/10161457/videos/229680766.

14 Aitken P, "CDC Director Walensky avoids question on vaccine mandate for essential workers," Fox News, CDC Director Walensky quotes, Oct. 24, 2021. https://www.foxnews.com/health/cdc-director-walensky-supports-vaccine-mandate-for-essential-workers and https://www.nytimes.com/2021/07/31/us/virus-unvaccinated-americans.html.

15 Stobbe M, "CDC's Walensky resigns," AP, May 5, 2023. https://apnews.com/article/rochelle-walensky-resigns-cdc-f0175f772389e6466d6b449a5ce7b25c.

16 Massetti G, et al, "Summary of Guidance for Minimizing the Impact of COVID-19 on Individual Persons, Communities, and Health Care Systems — United States, August 2022," CDC, MMWR, Aug. 19, 2022. https://www.cdc.gov/mmwr/volumes/71/wr/mm7133e1.htm?s_cid=mm7133e1_w and https://www.cdc.gov/coronavirus/2019-ncov/science/science-briefs/vaccine-induced-immunity.html.

17 "PBS News Hour, Dr. Fauci on why the US is 'out of the pandemic phase.'" PBS transcript, Apr. 26, 2022. https://www.pbs.org/newshour/show/dr-fauci-on-why-the-u-s-is-out-of-the-pandemic-phase-2.

18 Pfizer, "Omicron-adapted COVID-19 vaccines, subject to authorization [EUA] from the US Food and Drug Administration (FDA)," Jun. 29, 2022. https://www.pfizer.com/news/press-release/press-release-detail/pfizer-and-biontech-announce-new-agreement-us-government.

19 Meadows M, "Promoting Safe & Effective Drugs for 100 Years," FDA, Feb. 2006. https://www.fda.gov/about-fda/histories-product-regulation/promoting-safe-effective-drugs-100-years and https://www.fda.gov/consumers/consumer-updates/kefauver-harris-amendments-revolutionized-drug-development.

20 FDA Authorizes Bivalent (mixing omicron and ancestral mRNA) COVID-19 Vaccines Boosters, Aug. 31, 2022. https://www.fda.gov/news-events/press-announcements/coronavirus-covid-19-update-fda-authorizes-moderna-pfizer-biontech-bivalent-covid-19-vaccines-use.

21 Golding H, et al, "What Is the Predictive Value of Animal Models for Vaccine Efficacy in Humans?," Cold Spring Harb Perspect Biol, Apr. 2, 2018. https://pubmed.ncbi.nlm.nih.gov/28348035/.

22 Biologics Review Preamble, 37 FR 16679. https://archives.federalregister.gov/issue_slice/1972/8/18/16678–16682.pdf#page=2.

23 America's Frontline Doctor's Whitecoat Summit, The EUA: Anthrax & Covid Vaccines by Colonel Tom Rempfer, Jul. 25, 2023. https://www.whitecoatsummit.com/videos/2023 and https://americasfrontlinedoctors.org/videos/post/the-eua-anthrax-and-covid-vaccines-by-colonel-tom-rempfer.

24 Bard J, "Why the military can use emergency powers to treat service members with trial COVID-19 drugs," *The Conversation*, May 11, 2020. https://theconversation.com/why-the-military-can-use-emergency-powers-to-treat-service-members-with-trial-covid-19-drugs-135876.

25 Rep Thomas Massie on Military Vaccine Requirements, Jul. 27, 2022. https://www.youtube.com/watch?v=zHtRpzcm6TI.

26 House Report 118–272, Nov. 23, 2023. https://www.congress.gov/118/crpt/hrpt272/CRPT-118hrpt272.pdf.

27 Sen. Ron Johnson, Increase in Medical Diagnoses Among Military Personnel? https://www.ronjohnson.senate.gov/2022/2/sen-johnson-to-secretary-austin-has-dod-seen-an-increase-in-medical-diagnoses-among-military-personnel.

28 10-Minute Summary of Senator Ron Johnson's 3-Hour Washington D.C. COVID-19 Roundtable, Dec. 10, 2022. https://rumble.com/v200qsq-10-minute-summary-of-u.s.-senator-ron-johnsons-3-hour-washington-d.c.-covid.html.

29 Fox Business, 8:05 timestamp, Aug. 11, 2023. https://rumble.com/v36j01b-senator-ron-johnson-on-mornings-with-maria-8.11.23.html.

30 Sen. Marshall Bills Protect Service Academy Members Who Refuse COVID Vaccination, Jun. 2, 2022. https://www.marshall.senate.gov/newsroom/press-releases/sen-marshall-bills-protect-service-academy-members-who-refuse-covid-vaccination/ and https://www.washingtontimes.com/news/2023/dec/6/gop-senators-file-bill-reinstate-pilots-ousted-vac/.

31 King R, "Florida Supreme Court gives DeSantis go-ahead on COVID-19 vaccine grand jury," *Washington Examiner*, Dec. 22, 2022. https://www.washingtonexaminer.com/policy/healthcare/florida-supreme-court-approves-desantis-vaccine-jury.

32 Gulf War Exposures Hearing before the Subcomm. on Health of the Comm. On Veterans' Affairs, US House of Rep., 110th Congress, 1st session, US GPO, Washington, DC, Jul. 26, 2007. https://www.govinfo.gov/content/pkg/CHRG-110hhrg37476/html/CHRG-110hhrg37476.htm.

33 "World Today, Iraq's anthrax source traced back to Britain," Aug. 10, 2005. https://www.abc.net.au/worldtoday/content/2005/s1434633.htm.

34 FBI Amerithrax report, p. 8. https://www.justice.gov/archive/amerithrax/docs/amx-investigative-summary.pdf and https://vault.fbi.gov/Amerithrax and https://www.justice.gov/archive/index-amerithrax.html.

35 Bruttel V, et al, "Endonuclease fingerprint indicates synthetic origin of SARS-CoV-2," BioRxiv, 2022. https://www.biorxiv.org/content/10.1101/2022.10.18.512756v1.

36 FBI, Anthrax Investigation, Closing a Chapter, Aug. 6, 2008. https://archives.fbi.gov/archives/news/stories/2008/august/amerithrax080608a.

37 An Analysis of the Origins of the COVID-19 Pandemic Interim Report, Senate Committee on Health Education, Labor and Pensions, Minority Oversight Staff, Oct. 27, 2022. https://www.help.senate.gov/imo/media/doc/report_an_analysis_of_the_origins_of_covid-19_102722.pdf.

38 Gordon M, et al, "Lab Leak Most Likely Origin of Covid-19 Pandemic, Energy Department Now Says, U.S. agency's revised assessment is based on new intelligence," WSJ, Feb. 26, 2023. https://www.wsj.com/articles/covid-origin-china-lab-leak-807b7b0a.

39 Director of National Intelligence, National Intelligence Council, Updated Assessment on COVID-19 Origins. https://www.intelligence.gov/assets/documents/702%20Documents/declassified/Declassified-Assessment-on-COVID-19-Origins.pdf and https://www.c-span.org/video/?527424–1/house-hearing-origins-covid-19 and https://abcnews.go.com/US/video/preponderance-evidence-covid-19-originated-lab-leak-report-98667632 and https://www.foxnews.com/politics/these-biden-admin-agencies-have-admitted-covid-lab-leak-plausible and https://www.heritage.org/public-health/commentary/biden-administration-fails-share-intel-covid-19-origins-independent.

40 Sen. Marshall, COVID Origins, Apr. 17, 2023. https://www.marshall.senate.gov/newsroom/press-releases/sen-marshall-releases-bombshell-covid-19-origins-report/ and https://www.marshall.senate.gov/wp-content/uploads/MWG-EXECUTIVE-SUMMARY-4.17-Final-Version.pdf.

41 CDC Director Robert Redfield testimony, March 8, 2023. https://oversight.house.gov/wp-content/uploads/2023/03/2023.03.08-Statement-of-Dr.-Robert-R-Redfield88.pdf and https://www.c-span.org/video/?526520–1/house-hearing-origins-covid-19.

42 Sabes A, "FBI director says pandemic 'most likely' originated from Chinese lab, Department of Energy also concluded pandemic likely originated from lab leak," Fox News, Mar. 1, 2023. https://www.foxnews.com/politics/fbi-director-says-covid-pandemic-most-likely-originated-chinese-lab.

43 "Preparing For the Future By Learning From the Past: Examining COVID Policy Decisions," House Committee on Oversight and Accountability, Dr. Jay Bhattacharya, M.D., Ph.D., Stanford University; Dr. Martin Kulldorff, Ph.D., Harvard University; Dr. Marty Makary, M.D., M.P.H., Johns Hopkins University, Dr. Makary quote referenced the government being "the greatest perpetrator of misinformation," Feb. 28, 2023. https://oversight.house.gov/roundtable/preparing-for-the-future-by-learning-from-the-past-examining-covid-policy-decisions/.

44 United States Court of Appeals for the Third Circuit, Case No. 22–2970, *Children's Health Defense, et al, v. Rutgers, et al*, On Appeal from the United States District Court for the District of New Jersey (D.C. No. 3–21-cv-15333), excerpt: "there is no fundamental right to refuse vaccination," Feb. 15, 2024. https://law.justia.com/cases/federal/appellate-courts/ca3/22–2970/22–2970-2024–02-15.html.

45 Hancock J, et al, "Declaration of Independence," Jul. 4, 1776. https://www.archives.gov/founding-docs/declaration-transcript.

46 MacKinnon J, "After advocating for COVID-19 vaccination for over a year, Ben Shapiro says he was deceived: 'We were lied to by everyone'" The Blaze, Oct. 25, 2022. https:

//www.theblaze.com/news/after-advocating-for-covid-19-vaccination-for-over-a-year-double-vaxxed-ben-shapiro-admits-he-was-deceived-we-were-lied-to-by-everyone.

47 Childrens Health Defense Military Chapter interview with Guardians of Warriors and Champions of Health, Brad Miller and David Beckerman, Nov. 29, 2023. https://rumble.com/v3xsvj7-gow-ep3-colonel-retired-tom-buzz-rempfer.html.

48 Tucker on Twitter, "Ep. 6, Bobby Kennedy is winning," aviation analogy timestamp at 5:45, Jun. 22, 2023. https://twitter.com/TuckerCarlson/status/1672014260480901120?t=h_u2S9mZo93IgdTv5wskxQ&s=09.

49 "CDC Director (Walensky) Calls for Agency's Overhaul," *Barrons*, Aug. 17, 2022. https://www.barrons.com/articles/cdc-director-calls-for-agencys-overhaul-51660754646.

Epilogue

1 "US Armed Forces—A Great Place to Start,"1988 DoD recruitment advertisement. https://www.youtube.com/watch?v=YKSKzPLHtO4.

2 "Direct Order," Directed by Scott Miller, Narrated by Michael Douglas and Linda Hamilton. https://www.imdb.com/title/tt1422806/ and https://rumble.com/v28wj24-direct-order-anthrax-vaccine-doccumentary.html and https://www.youtube.com/watch?v=wDDMsvErsQw.

Appendix

1 Senate Veterans Affairs Committee Staff Report 103–97, Maj. Gen. Ronald Blanck, Commanding General, Walter Reed Army Hospital, to Committee staff, 414 Russell Senate Office Bldg., Washington, DC, p 35, 1994.

2 Senate Veterans Affairs Committee Staff Report 103–97, footnotes 61–63, 1994.

3 Congressional Record, Senator Shelby's Conclusions On The Persian Gulf Syndrome, p S3098, Mar. 17, 1994. http://www.gulflink.osd.mil/czech_french/czfr_refs/n08en014/s3098.htm.

4 Senate Staff Report 103–97, "Is Military Research Hazardous To Veterans' Health? Lessons Spanning Half A Century," Dec. 8, 1994. https://img1.wsimg.com/blobby/go/4fa7f468-a250–4088-926e-3c56a998df1f/downloads/1994_12–08_W%20staff%20report.pdf.

5 Food and Drug Administration Proposed Rule, 37 FR 51001, Dec. 13, 1985.

6 FDA Notice of Intent to Revoke manufacturer's license originally available at http://www.fda.gov/cber/infosheets/mich-inf.htm and https://img1.wsimg.com/blobby/go/4fa7f468-a250–4088-926e-3c56a998df1f/downloads/1cr5obl0u_869077.pdf?ver=1708371037439 and https://img1.wsimg.com/blobby/go/4fa7f468-a250–4088-926e-3c56a998df1f/downloads/1cr5odu7g_688835.pdf?ver=1708371037439.

7 FDA Form 483 Inspectional Observations, Nov. 15–23, 1999. https://img1.wsimg.com/blobby/go/4fa7f468-a250–4088-926e-3c56a998df1f/downloads/1cr5oifuk_539962.pdf?ver=1708371037439.

8 Caldera L, Secretary of the Army, Memorandum of Decision, Sept. 3, 1998.

9 Request for Proposals (RFP) No. DAMD 17–85-R-0078, US Army Medical Research Acquisition Activity, Fort Detrick, MD, May 16, 1985. https://img1.wsimg.com/blobby/go/4fa7f468-a250–4088-926e-3c56a998df1f/downloads/1cr5o7gje_801494.pdf?ver=1708371037434.

10 GAO-T-NSIAD-99–148, "Medical Readiness: Safety and Efficacy of the Anthrax Vaccine," Apr. 29, 1999. https://www.gao.gov/assets/t-nsiad-99–148.pdf.

11 *Vaccines* (United Kingdom, W.B. Saunders Company, 1994), 737.

12 Senate Hearing 101–744, Letter from former Assistant Secretary of Defense Robert B. Barker to former US Senator John Glenn, Chair of the Senate Governmental Affairs Committee, p. 474, 480, Aug. 24, 1989. https://www.google.com/books/edition/Global_Spread_of_Chemical_and_Biological/tbIRAAAAIAAJ?hl=en&gbpv=1 and https://books.google.com/books/content?id=tbIRAAAAIAAJ&pg=PA480&img=1&zoom=3&hl=en&bul=1&sig=ACfU3U1O1vk7IuYlym0esRLW9uCus5yXkg&ci=49%2C308%2C779%2C689&edge=0.

INDEX

ABOUT THE AUTHOR

Colonel Tom "Buzz" Rempfer, USAF retired, ended his military career instructing and evaluating MQ-1 Predator and MQ-9 Reaper missions. Past flying assignments included duty in Asia, the Middle East, and Europe as an F-16 flight lead, F-117 instructor pilot, C-130 aircraft commander, and A-10 forward air controller, with a total flight time of over 5,500 hours. He earned a master's degree through the Naval Postgraduate School's Center for Homeland Defense and Security and received the program's Outstanding Thesis Award. He also served on the USAF Chief of Staff's Cyberspace Task Force. Colonel Rempfer was a distinguished graduate from pilot training and the USAF Academy, where he was an all-American boxer. Colonel Rempfer served in a retired capacity on the Military Advisory Council for Representative Ann Kirkpatrick (D-AZ-2). Tom's civilian career as a commercial pilot included flying the B-787, B-777, B-767, B-757, B-737, A-300, DC-10, MD-80, and ATR-42 aircraft. In his civilian capacity, he volunteered with his airline pilot union as a peer pilot, and served by assisting with aeromedical and disability matters. Buzz continues to crisscross the oceans, never veering from his course, telling his and Russ Dingle's unyielding story to a captive audience of gracious pilots.